"This impressive volume brings together many of the world's leading thinkers and practitioners on global water policy. The chapters survey the water and governance challenges in a number of important countries and regions. The contributors grapple with the complexity of how problems related to water potentially create security concerns for the United States and what, if anything, the U.S. government can do to help others and thereby help itself."
—*Joshua Busby, Associate Professor of Public Affairs at the University of Texas at Austin and author of* Water and U.S. National Security *(Council on Foreign Relations 2017)*

"Climate change has driven home what we should have known for a long time: water is the root of much of the world's instability and conflict. And there is no one better than David Reed to apply his enormous experience across a score of regions. His careful book drives home a second truth we should have known: water is the problem but mismanagement is the curse. Accordingly, he provides a framework for a critical shift in policy, from providing water to managing watersheds, almost always involving more than one country."
—*Greg Treverton, former chair of the U.S. National Intelligence Council*

Water, Security and U.S. Foreign Policy

The prosperity and national security of the United States depend directly on the prosperity and stability of both partner and competing countries around the world. Today, U.S. interests are under rising pressure from water scarcity, extreme weather events and water-driven ecological change in key geographies of strategic interest to the U.S. Those water-driven stresses are undermining economic productivity, weakening governance systems and fraying social cohesion in scores of countries and, in the process, undermining the vitality of rural livelihoods, fostering local and ethnic conflicts, driving broad migratory movements and contributing to the growth of insurgencies and terrorist networks.

While the U.S. intelligence community has steadily expanded natural resource concerns in their global threat analyses, our overseas development assistance remains locked into provision of water and hygienic services rather than responding to the full sweep of global water challenges including governance and policy failures, growing conflicts over water and the need for promoting sustainable transboundary water arrangements in partner countries. A fundamental departure from the past is urgently needed.

Based on 18 case studies, *Water, Security and U.S. Foreign Policy* provides an analytical framework to help policy makers, scholars and researchers studying the intersection of U.S. foreign policy with the environment and sustainability issues, interpret the impacts of water-driven social disruptions on the stability of partner governments and U.S. interests abroad. The book also delivers specific recommendations to reorient U.S. development and diplomatic engagements that can forestall and prevent social disruptions and ensuing threats to U.S. prosperity and national security.

David Reed is Senior Policy Advisor for WWF-US.

Water, Security and U.S. Foreign Policy

Edited by David Reed

NEW YORK AND LONDON

First published 2017
by Routledge
711 Third Avenue, New York, NY 10017

and by Routledge
2 Park Square, Milton Park, Abingdon, Oxon OX14 4RN

Routledge is an imprint of the Taylor & Francis Group, an informa business

© 2017 Taylor & Francis

The right of David Reed to be identified as the author of the editorial material, and of the authors for their individual chapters, has been asserted in accordance with sections 77 and 78 of the Copyright, Designs and Patents Act 1988.

All rights reserved. No part of this book may be reprinted or reproduced or utilised in any form or by any electronic, mechanical, or other means, now known or hereafter invented, including photocopying and recording, or in any information storage or retrieval system, without permission in writing from the publishers.

Trademark notice: Product or corporate names may be trademarks or registered trademarks, and are used only for identification and explanation without intent to infringe.

Library of Congress Cataloging-in-Publication Data
A catalog record for this book has been requested

ISBN: 978-1-138-05149-2 (hbk)
ISBN: 978-1-138-05151-5 (pbk)
ISBN: 978-1-315-16827-2 (ebk)

Typeset in Times New Roman
by Apex CoVantage, LLC

Printed and bound in the United States of America by
Edwards Brothers Malloy on sustainably sourced paper

Contents

Foreword: General James L. Jones x
Acknowledgments xii
List of Acronyms xv

PART I
In Search of a Mission 1

1 In Search of a Mission 3
   DAVID REED

PART II
Water and Social Disruptions 35

   Framing Note: The Social Dimensions of Water 35
   DAVID REED

2 Development and Diplomacy: Water, the SDGs,
   and U.S. Foreign Policy 39
   ERIKA WEINTHAL, FARAH F. HEGAZI, AND LESHA B. M. WITMER

3 Climate Variability, Water and Security in El Salvador 56
   HERMAN ROSA

4 Panama: Water Security and Social Conflicts within the
   Context of Climate Change 76
   ARIEL CUSCHNIR

5 Mexico's Pursuit of Water Security 91
   ROMÁN GÓMEZ GONZÁLEZ COSÍO

6 Who Stole the Water: Water, Security, and U.S. Foreign
   Policy in Guatemala   110
   EDUARDO STEIN WITH LILIAN MARQUEZ

7 Water Stress, Instability, and Violent Extremism in Nigeria   128
   MARCUS KING

8 Water Resources, Climate Change, and the
   Destabilization of Modern Mesopotamia   149
   PETER GLEICK

9 Iran's Impending Water Crisis   168
   DAVID MICHEL

10 Dammed If You Do and Damned If You Don't:
   Afghanistan's Water Woes   189
   GLEN HEARNS

11 Winter Is Coming: U.S. Strategic Interests and the
   Water-Energy-Agriculture Conundrum in Central Asia   206
   RICHARD KYLE PAISLEY

12 The Perils of Denial: Challenges for
   a Water-Secure Pakistan   222
   ALI HASNAIN SAYED, CHELSEA N. SPANGLER, AND MUHAMMAD
   FAIZAN USMAN

13 Water Scarcity and Regional Security in India   237
   CECILIA TORTAJADA, UDISHA SAKLANI, AND ASIT K. BISWAS

14 Water-Energy Nexus in the Himalayas   253
   KEITH SCHNEIDER

15 A Perfect Storm in the Greater Mekong Subregion:
   Climate Change Impacts on Food, Water, and Energy   272
   ARJUN THAPAN

16 Building Resilience for Peace: Water, Security, and
   Strategic Interests in Mindanao, Philippines   288
   ROGER-MARK DE SOUZA

PART III
Financing Water Infrastructure   307

Framing Note: Persistent Challenges   307
E. PATRICK COADY

17 Helping Weak Water Utilities Climb the Financial Ladder   311
ALDO BAIETTI

**18 Financing Water and Sewer Infrastructure in the
Developing World**   326
WILLIAM STREETER

19 A New Chapter in Developing Water Infrastructure   344
MARC JEULAND

PART IV
New Challenges, New Directions   365

20 Paths of Influence   367
DAVID REED

**21 Recommendations for Water, Security and
U.S. Foreign Policy**   396
DAVID REED

List of Contributors   401
Index   406

Foreword

David Reed and the World Wildlife Fund have turned an urgently needed spotlight on the future-defining issue of global water security. The report provides a powerful rationale for why addressing global water security must be central to the U.S. national security strategy and the shaping of our foreign policy.

During most of my military service, U.S. national security was defined by the long, twilight struggle against communism and the Soviet military threat. Security was expressed in the calculus of comparative troop strength, weapons count, and nuclear throw weight. Today's threats are exponentially more diverse and more complex than in earlier times. What's clear is that modernity demands a far richer conception of national and international security—one less reliant on reaction and far more focused on anticipation and prevention—one that centers on disarming the root causes and major multipliers of conflict and instability.

Viewed from that lens, what comes into sharp relief is that the premier strategic threat to global security is not a particular country, ideology, or weapon. Rather, it is failing to satisfy human needs and wants, especially for the basics including food, energy, water, and human dignity.

Today, water scarcity is dangerously weakening vulnerable societies, creating fragile states. Lack of access to resources, in particular water, whether caused by mismanagement or inequitable distribution, is a major driver of poverty, conflict, and extremism. As I write, 1.8 billion human beings have no access to clean water. The bulk of this population clings to life in the most unstable, violence-torn regions on Earth, where terrorist and insurgent enterprises are working to capture operating space and followers by leveraging human want and desperation to achieve a perverted form of legitimacy.

Water scarcity's role in creating fragile states and sharpening interstate conflict, in catalyzing disruptive mass migration, and in predisposing populations for exploitation and conflict is not some abstract, distant danger. It is happening now. Without question, water stress exacerbated by extreme weather events has contributed significantly to conflicts and instability across the Middle East and in Africa while fueling social conflict in the Northern Triangle in Central America, Iran, Pakistan, India, and other countries around the world. Resource scarcity and imbalance feed insurgency and help fuel groups such as

Boko Haram, the Islamic State, the Taliban, and Abu Sayyaf. Conditions in the geographies where these malefactors operate will have enormous impact on the reach of global peace and development, as well as our own prosperity and security, in the decades to come.

The trends could not be more challenging. By the end of the century, the global population will reach 11 billion people. The bulk of this growth will take place in the world's most arid and inherently unstable regions. In only three decades, 40% of the world's population will live in a stressed water basin. Meanwhile, within that time, humankind must produce 50% more food and double energy production—activities requiring massive water resources—while contending with disruptive ecological changes, making the mission more difficult. This is why James Clapper, former U.S. Director of National Intelligence, reports that major deficits in freshwater availability will (and I emphasize "will"—not "may") contribute to instability in nations across the globe.

Yet, with American leadership, no task is too great. In this century as it was in the last, shaping a world of peace and prosperity rather than of desperation and conflict will require American leadership at its best. No other country can come close to what we can do. We must set our sights on making water security and wise resource management a cornerstone of global conflict prevention and re-stabilization. That commitment must be core to everything we do—our diplomacy, policies, practices, and innovations—and thereby must promote wise stewardship of the natural systems required to sustain human well-being. The recommendations offered by David Reed must inform a sensible national strategy that addresses the rapidly emerging threats to our national interests—a battle plan for a development revolution to help foster peace and stability. The approaches and architecture he suggests will help the United States utilize our development and diplomatic tools to prevent and mitigate future conflicts, and win a better future. With America in the lead, we can, we must, and, I believe, we will rise to the monumental challenges and opportunities before us.

General James L. Jones

Acknowledgments

As I close preparation of this publication, the United States is poised to experience major political change. The exact direction of our domestic political realignment and the longer-term impacts on allies and competitors alike around the world are uncertain and unpredictable. As new U.S. government priorities and policies take hold, however, there is absolute certainty that the ecological changes and accompanying social disruptions explored in the following chapters will continue to leave a deepening mark on the prosperity and stability of countries around the world and, consequently, on our own prosperity and security. While some leaders from the public realm and the private sector have chosen to sidestep, even deny, the continued march of ecological and associated social change, many others have embraced the challenges of uncertainty: they accept the risks of acting without having absolute understanding of what their actions may lead to but have decided to act in an informed, responsible manner to avoid greater future harm to the human enterprise.

I must first express my boundless pride and gratitude to WWF-US for being at the forefront of those organizations willing to take risks and make sustained investments to protect the ecological integrity of our planet and its diverse human communities. The embrace of an uncompromised ecological vision and a culture of taking risks begins with the leadership of the organization, with Carter Roberts, its president and chief executive officer, and Marcia Marsh, the chief operating officer. To them, I owe an immeasurable debt of gratitude for their confidence in me and their willingness to explore the dynamics of this emerging theme of environment and security. Tom Dillon, senior vice president for forests and fresh water, David McCauley, senior vice president for policy and government affairs, and Karin Krchnak, director of the freshwater program, marshaled the resources of their respective programs to make sure that the full commitment and experience of the organization were directed in support of this effort. Their commitment allowed me to form a Water and Security Working Group that met weekly for more than a year and a half to study emerging issues, debate conflicting perspectives, review drafts, formulate recommendations, and build partnerships with external organizations and experts. Members of the Working Group, to whom I am very indebted for their good spirit, rigorous minds, critical thinking, and unflagging encouragement, include

Ryan Bartlett, Sarah Davidson, Sarah Freeman, Lauren Kovach, Brent Nordstrom, America Pintabutr, Todd Shelton, Kate Simma, and Allegra Wrocklage.

Foremost, however, I must extend my heartfelt appreciation to Chelsea Spangler, who, day in and day out, drove this endeavor to its successful conclusion. I thank her for coordinating work with the full stable of authors scattered around the world; developing relations with many external partners, such as the Woodrow Wilson Center, Duke University, and George Washington University; and managing the finances of the project. Chelsea has also been an invaluable contributing writer, editor, and researcher at virtually all stages of the project. Deadlines and standards would never have been met without her steadfast dedication.

In the earliest stages of this project, a number of renowned experts led discussions of the Water and Security Working Group, posing challenges and setting research targets to help mature our views. Those experts include Ken Conca, Rich Engel, Marc Jeuland, Marcus King, Jack May, Aaron Salzberg, Rod Schoonover, and Erika Weinthal, many of whom became contributing authors to this book.

The research contributions of Chris McGahey were instrumental in framing the background analysis presented in Chapter 1, "In Search of a Mission." Chris interviewed dozens of experts inside USAID, the U.S. State Department, and other offices of government to ferret out information otherwise lost to folders and files now long since discarded. My thanks in equal measure for his help in developing specific recommendations.

To each of the authors of the 18 contributions, please receive my appreciation for expanding my own limited knowledge about these rapidly emerging water challenges that are shaping societies around the planet. I owe particular thanks to Marcus King for his help in developing the paths of influence graphic applied to each of the geographic studies.

Over the course of the year, three experts have provided invaluable guidance in shaping the approach and views presented in this publication: John Raidt challenged the language of analysis and recommendations in ways that invited broader engagement from military, private sector, and policy leaders across the full political spectrum. Aaron Salzberg shared decades of his own experience and that of other water experts from diplomatic and development communities, thereby challenging facile conclusions and self-serving points of view about U.S. government practices. Cheryl Rosenblum, approaching the water issue in the context of the new challenges facing the security community, opened a rich and challenging exchange with members of the Military Advisory Board and the full range of experts in the Center for Naval Analyses. I extend my many thanks to each of them for pushing and provoking me to explore new paths of inquiry and to reach out to other important leaders with divergent points of view.

It has been an honor and pleasure to work again with Patrick Coady, former U.S. executive director to the World Bank. In addition to writing the framing note for the three chapters dealing with water infrastructure finance, Pat worked closely with those financial analysts to sharpen their analysis and recommendations.

I also extend my thanks to Jonathan Haskett and Irving Mintzer from the Johns Hopkins University School for Advanced International Studies (SAIS) for coordinating and supporting the work of three graduate students—Katie Pogue, Yiran Lu, and Maren Wenzel—who provided important research contributions during the earliest months of the project. To Josh Busby, my thanks for his support and critical comments along the way. To Andrea Kutter, my heartfelt gratitude for her critical guidance and support over the past year and a half.

Despite the guidance and advice from this remarkable range of thought leaders, I, in the last measure, bear responsibility for the views presented herein.

David Reed
April 2017
Washington, DC

Acronyms

ACP	Panama Canal Authority (in English)
ADB	Asian Development Bank
AIIB	Asian Infrastructure Investment Bank
ANDS	Afghan National Development Strategy
ANERA	American Near East Refugee Aid
APEC	Asia-Pacific Economic Cooperation
AQIM	Al Qaeda in the Islamic Maghreb
ARMM	Autonomous Region in Muslim Mindanao
ASEAN	Association of Southeast Asian Nations
BOT	build, operate, and transfer
CA	Central Asia
CAB	Comprehensive Agreement on the Bangsamoro
CAPEX	capital expenditures
CARSI	Central American Regional Security Initiative
CAT	catastrophic risk
CCRIF	Caribbean Catastrophe Risk Insurance Facility
CDP	city development plan
CGIAR	Consultative Group on International Agricultural Research
CIAT	Centro Internacional de Agricultura Tropical
CNA	Center for Naval Analyses
COMEST	World Commission on the Ethics of Scientific Knowledge and Technology
CONAGUA	National Water Council (Panama), also National Water Commission (Comisión Nacional del Agua, Mexico)
CPA	Coalition Provisional Authority
DNI	Director of National Intelligence
DSCR	debt service coverage ratio
EIA	environmental impact assessment
ENCCP	National Strategy for Climate Change (in English)
EPIRA	Electric Power Industry Reform Act
ERC	Energy Regulatory Commission
FACTS	Foreign Assistance Coordination and Tracking System
FAO	Food and Agriculture Organization of the United Nations

FARC	Revolutionary Armed Forces of Colombia
FCCC	United Nations Framework Convention on Climate Change
FDI	foreign direct investment
FMSO	Foreign Military Studies Office
FX	foreign exchange
GAP	Great Anatolia Project (Turkey)
GDP	gross domestic product
GERD	Grand Ethiopian Renaissance Dam
GNI	gross national income
GMS	Greater Mekong subregion
GON	Government of Nigeria
GWP	Global Water Partnership
HDI	Human Development Index
IADB	Inter-American Development Bank
IBNET	International Benchmarking Network for Water and Sanitation Utilities
IBWC	International Boundary and Water Commission
ICG	International Crisis Group
ICWC	Interstate Commission for Water Coordination
IDAAN	Instituto de Acueductos y Alcantarillados Nacionales
IFAS	International Fund for Saving the Aral Sea
IFC	International Finance Corporation
IFI	international financial institution
IITA	International Institute of Tropical Agriculture (Nigeria)
IJC	International Joint Commission
IPCC	Intergovernmental Panel on Climate Change
ISC	Inter-State Council (India)
ISIS	Islamic State of Iraq and Syria
ISWD Act	Inter-State Water Disputes Act (India)
IUCN	International Union for Conservation of Nature and Natural Resources
IWRM	integrated water resources management
JCPOA	Joint Comprehensive Plan of Action
JMP	Joint Monitoring Programme
JNNURM	Jawaharlal Nehru National Urban Renewal Mission
KPWF	Kenya Pooled Water Fund
KRG	Kurdish Regional Government
KWDT	Krishna Water Disputes Tribunal (India)
LEAD	Leadership for Environment and Development
LGU	local government unit
LMCCD	Lancang-Mekong River Community of Common Destiny
LMI	Lower Mekong Initiative
MAF	million acre feet
MDB	multilateral development bank
MDGs	Millennium Development Goals
MEND	Movement for the Emancipation of the Niger Delta

MILF	Moro Islamic Liberation Front
MoUD	Ministry of Urban Development (India)
MRC	Mekong River Commission
MRV	monitoring, reporting, and verification
MSP	multi-stakeholder platform
NAFTA	North American Free Trade Agreement
NATO	North Atlantic Treaty Organization
NDA	Niger Delta Avengers
ND-GAIN	Notre Dame Global Adaptation Index
NDPVF	Niger Delta People's Volunteer Force
NDV	Niger Delta Vigilante
NHPC	National Hydropower Corporation
NIC	U.S. National Intelligence Council
NPV	negative net present value
OBA	output-based aid
OBC	Other Backward Class (India)
OCHA	UN Office for the Coordination of Humanitarian Affairs
ODA	official development aid
OECD	Organisation for Economic Co-operation and Development
PCRWR	Pakistan Council for Research in Water Resources
PNSH	National Water Security Plan (English translation)
PPPs	public-private partnerships
QDR	U.S. Quadrennial Defense Review
RBO	river basin organization
RKVY	Rashtriya Krishi Vikas Yojana
SAIS	Johns Hopkins University School for Advanced International Studies
SDGs	Sustainable Development Goals
SEA	strategic environmental assessment
SPV	special purpose vehicle
TA	technical assistance
TARWR	total actual renewable water resources
TMC	thousand million cubic meters
TSR	Targeted Subsidies Reform Act (Iran)
TVA	Tennessee Valley Authority
ULB	urban local body
UMPP	Ultra Mega Power Plants
UNDP	United Nations Development Program
UN DPA	United Nations Department of Political Affairs
UNECE	United Nations Economic Commission for Europe
UNEP	UN Environment Programme
UNRCCA	United Nations Regional Centre for Preventive Diplomacy for Central Asia
UNSGAB	United Nations Secretary-General's Advisory Board on Water and Sanitation
USACE	U.S. Army Corps of Engineers

USAID	U.S. Agency for International Development
USG	U.S. Government
USWP	U.S. Water Partnership
WACC	weighted average cost of capital
WASH	water, sanitation, and hygiene
WASREB	Water Services Regulatory Board (Kenya)
WCD	World Commission on Dams
WfWA	Water for the World Act
WRM	water resource management
WSPF	Water and Sanitation Pooled Fund
WSS	water supply and sanitation
WSSD	World Summit on Sustainable Development
WWF	World Wildlife Fund

Part I
In Search of a Mission

1 In Search of a Mission

David Reed

Continental America's water abundance has provided the natural-resource wealth to build and sustain the world's largest economy and to fuel a centuries-long rise in our country's living standards. For more than two centuries, engineers, hydrologists, agribusinesses, family farmers, politicians and government agents have pushed, pulled, stored and moved water across the continent to satisfy an insatiable demand for fresh water. Over the course of those two-plus centuries, we have become masters of hard-infrastructure construction, leaders in irrigation technology innovation and major suppliers to commodity markets that stretch around the world. We have accrued an unmatched set of water management assets: sophisticated and flexible regulatory instruments; water management institutions at local, regional, and national levels; consultation protocols; standards and technical capacities for new remote sensing; and data management and modeling technologies. Our national experience has been replete with risk-taking, extraordinary success, costly failure, institution-building, social displacement, violence, sabotage, treaty signing and peacemaking.

While richly endowed in general, U.S. water wealth is distributed highly unevenly across our land, with the 100th meridian and Continental Divide cleaving hard lines between regions that enjoy a natural abundance of water and those geographies marked by scarcity and weather extremes. Regardless of those sharp contrasts in natural endowment, no Congress and no president, with the exception of President Jimmy Carter in 1977, has dared deny the financial, legal and technical resources needed by states and local communities to ensure delivery of clean water in a stable, timely manner.[1] The construction of the world's largest water-moving infrastructure and the steady growth of our water management systems were the result not only of deals cut in smoke-filled rooms of politicians and titans of American industry—those steel and concrete behemoths were the result of a long-standing recognition that America's prosperity and economic well-being depended on the delivery of safe water to America's farmers, industrialists, miners and households without impediment.

As the United States now stands as the world's sole superpower, it would seem axiomatic that our centuries-long consensus on the centrality of water to America's prosperity would extend naturally to the international sphere. Our domestic economy is inextricably interwoven into all corners of the world. Commodity supply lines of U.S. importers, traders and processors reach into small farmer communities and engage agribusinesses on all continents, providing American families with a steady provision of every imaginable source of nourishment. In the world's largest economy, our prosperity depends directly on the prosperity and stability of our partners and competitors around the world.

In equal measure, U.S. traders and processors provide millions of farmers around the world with stable markets and economic opportunities. Water is the lifeline of those small-hold farmers and agribusinesses alike. The loss of those agricultural jobs, collapse of rural livelihoods in scores of developing countries and deprivation of safe water for the hundreds of millions of recent immigrants to the teeming cities of Southeast Asia and Africa can disrupt those global supply lines on which our economy depends. Moreover, as the Syrian civil war painfully reminds us, drought and weather extremes can unleash a chain of migration and social displacement that undermines governments without regard for political or ideological disposition. Lack of water at the local level can lead to conflicts that deepen ethnic and cultural cleavages, increase state fragility and raise tensions among neighboring countries along shared watercourses. The interconnectedness of the global economy and the attendant reliance on the stability of political allies and competitors alike weds our national security in ways previously unknown to the provision and sustainable management of fresh water in distant stretches of the planet.

It is precisely through the rising impacts of climate change that the links between sustainable water management and our national security multiply. For example, the stress and dislocations brought by climate change on societies with population bulges of unemployed youth in rural and urban areas are often sufficient to test the viability of local governments as they struggle to provide basic services to their constituencies. When dispossessed youth are captured by the ideological appeals of prophetic insurgencies, that explosive mixture can combust into social upheaval, swell the ranks of the insurgent movements and erode the rule of law at national and regional levels. As that increasingly familiar scenario unfolds, it is no longer possible to address water scarcity and extreme weather events through water partnerships, technical assistance, flood control programs and water infrastructure construction. The full range of U.S. development, diplomatic and kinetic measures are needed to attenuate the threats and to help societies build the economic foundations on which future stability and prosperity depend. In short order, new conditions and dynamics such as these will oblige our government to reconsider what constitutes a sound U.S. foreign policy as our planet's climate continues to change.

Paradoxically, our government's history in addressing rising global water challenges has been, at best, a checkered affair. U.S. water policy abroad has been inconsistent, incoherent and ineffective to the point that, despite our goodwill, we are not able to provide technical, material and policy support that will enable our partners to respond to protracted drought and extreme weather events whose destructive impacts are reinforced by deficient policies and weak institutions. In the following pages, I explore a 25-year kaleidoscopic history of confused and competing U.S. approaches that are implemented through our bilateral development assistance, channeled through multilateral institutions and translated through diplomatic and military engagements in geographies deemed strategically important. Perhaps the most remarkable feature of this exploration is how competing domestic policies and leading political figures have both handcuffed and provided leadership in shaping U.S. water engagements around the world since 1992. As a foretaste of this complex history, I begin this exploration with a minor incident that unfolded shortly after the U.S. invasion of Iraq in 2003.

A Nigh-Forgotten Incident

The chroniclers of the buildup, kinetic intervention and aftermath of the U.S. invasion of Iraq have laid bare the chronology, justifications and generally catastrophic outcomes of that 2003 military adventure. Sadly, the full cost of the misadventure has yet to be tallied as the country and the region remain locked in strife and human suffering. U.S. security interests still rest in the balance. One incident in the U.S. reconstruction effort has largely remained out of public view. That nigh-forgotten incident deserves a bit more attention for the simple reason that it sheds light on the central concern of this book: namely, the role of the U.S. government in responding to the ever-growing challenges to U.S. security associated with access to and the management of water around the world.

A mere four months after the U.S. assault began in March 2003, the U.S. Army Corps of Engineers (USACE) was advising the newly formed Iraqi Ministry of Water Resources to shape its strategy for rebuilding the country's water infrastructure and management regime.[2] Though shocked at the parlous state of primary infrastructure, the corps' representative claimed that "Iraq can become the contemporary 'California of the Middle East.'"[3] This is an extraordinary statement not only because of its fantasticality but also because it reveals a great deal about how operative assumptions derived from our domestic experience managing water are translated into our engagements with international partners. On a first level, it is hard to imagine that an agency centrally responsible for helping manage our own country's waters could hold the conviction that the hydrological regime in California is sustainable and warrants replication in an even drier part of the world with millennia of water management experience. True, the recent five-year drought besetting California had not yet focused our national attention, but the repeated history of drought in the state and the region over the past century certainly should have been well-known to anyone vested with the responsibilities bearing on Iraq's water future. Second, what understanding of regional dynamics could have led leaders of the post-invasion, reconstruction regime to ignore the fact that Iraq is held hostage to the water allocation decisions of Turkey and Syria, both upstream neighbors whose dams and reservoirs on the Tigris and Euphrates rivers control Iraq's access to those waters? A simple unforeseen provocation with those neighbors could further reduce flows to Iraq, which had already experienced the negative consequence of its neighbors' expanding water infrastructure and consumption hundreds of miles upstream.

On a third level, it is even more difficult to understand how the U.S. advisors could promote the "privatization of water facilities"[4] in a country ravaged by civil unrest, where private enterprise had long been throttled under a state-driven economy and where creating a market-driven economy would require years of institutional and regulatory development. Then again, since privatizing the whole of Iraq's economy was a goal set by the United States, it would seem natural that water management should also be brought under private-sector management. Finally, while the 2006–11 drought that precipitated the Syrian civil war lay several years in the future, the impacts of climate change on the driest region on the planet were predicted years earlier, with ample warnings about the attendant strains on the region's water supplies. Given that such scientific warnings found little resonance in the U.S. administration at that time, reconstruction plans did not allow such distractions to be woven into the reconstruction of a country halfway around the world.

One clear lesson that this nigh-forgotten incident provides is that it is not possible to separate the policies and programs of U.S. overseas engagements from the centuries-long domestic experiences of managing water within our own borders. As I will explore below, government agencies and entire sectors of our economy were built on the premise of unlimited water availability and massive water infrastructure construction to solve our nation's water challenges. Environmental constraints affecting one region of the country could be countered with unmatched engineering prowess and the capability to move unprecedented volumes of water across the plains, valleys and mountains. So why not in distant lands? With that shaky assumption, our market economy could be counted on to maximize water development and efficiency, with results sufficient to satisfy ever-growing domestic demand from mining, industry, agriculture, municipalities and household consumption. So why not apply that economic assumption to other quarters of the world?

So as not to repeat the oversights reflected in the "California dreaming" incident, this chapter will carry the reader through three context-setting discussions. First, I will reference four convergent trends associated with water use and management that pose new challenges to every community, every company and virtually every government on the planet. Second, I will explore three seminal lessons from our own domestic experience that spotlight the core of the American approach to managing our country's water resources that we have at times ill-advisedly imposed on development and policy abroad. Those experiences allow us to reflect on the legal foundations, expansive engineering ambitions and competing management approaches we have taken to governing our nation's water. Finally, my focal point will be to examine in greater detail the three pathways through which the U.S. has managed our water engagements in countries around the world. Those pathways include our bilateral development assistance; U.S. policies and practices transmitted through and with the multilateral development community; and the defense, development and diplomatic approach in regions deemed strategically important to our national security.

Four Convergent Trends

The world of water as we have known it is gone: Evaporated. Drained. Salinized. Desalinated. Allocated. Captured. Stored. Packaged. Bottled. Depleted. Contaminated. Shared. Appropriated. Diverted. Fracked. Heated. Cooled. Lifted. Injected.

The relative security in the volume, quality and accessibility of the world's freshwater stocks that we once enjoyed has changed dramatically as water scarcity and variability have become the "new normal" in virtually all regions of the world. Underlying that shift from relative supply stability to increasing scarcity and variability lie four global trends that converge with a vengeance, albeit highly unevenly, on communities around the planet.

In many ways, this new normal should not be surprising given that three of the convergent trends have accompanied evolution of the human enterprise for centuries, if not millennia. What have changed are the scale and attendant impacts of these trends that now intersect with unprecedented effect. The first is the simple fact that the steadily burgeoning human population has increased pressure on water stocks. The more human beings on the planet, the greater the demand for water. From a global population of 3 billion in the year 1960, the planet now hosts 7.3 billion humans, with another 2 billion expected by 2050.[5] Demographic trends will continue to generate increased demand for water for decades to come.

The second trend is the ever-expanding size, complexity and interconnectedness of the global economy. While the human enterprise's total output was valued at $1.365 trillion in the year 1960, today global output reaches $73.9 trillion in total value and is expected to triple by 2050.[6,7] Despite efforts to diminish the natural-resource intensity of production, the ever-growing enormity and sophistication of agricultural, extractive, industrial and cybernetic production have placed unparalleled demands on a relatively fixed volume of fresh water. Moreover, as the global economy has expanded along with growing demands on available water, so too have contaminants and pollution, which have diminished the quality and availability of fresh water for the economic juggernaut. As the global economy continues its steady expansion, so too will diverse economic pressures increase demand on water resources.

The third trend is the perpetuation of a centuries-old operative assumption that fresh water is limitless. This assumption has been translated into governance regimes built on a range of laws, regulations, treaties and institutions erected decades—often centuries—ago that, with a few notable exceptions, remain in full effect today.[8] Water management regimes across the planet continue to operate on pre-scarcity, pre–climate variability assumptions. Even the most complex, sophisticated water transportation system in the world that moves the mountain snowmelt in the Rockies to California and across the Southwest remains predicated largely on inaccurate data and assumptions that today render the system's viability uncertain. Perhaps the greatest resistance to changing water policy in countries around the world resides in the stubborn unwillingness or incapability to embrace market-based pricing that accounts for the full cost of providing and managing water. Businesses, governments and households have enjoyed a virtually untouched subsidy reproduced by the myth of water's boundlessness. The repeated failure to adjust policies and institutions to manage water in the face of scarcity, weather extremes and climate variability carries potential disastrous consequences the world over.

The fourth convergent trend, climate change, is comparatively new. Yet, despite its recent influence, climate change exponentially intensifies the negative consequences of the other three factors. As though the three aforementioned convergent trends weren't enough to subvert the sound functioning of economic and social systems, climate change has injected a degree of uncertainty and variability on a global scale that lies beyond the realm of human experience. A widely accepted axiom of climate change is that water is the medium through which the human community will experience the impacts of climate change. What sound science holds as a certainty is that the impacts will be more extreme and unpredictable with each passing year. As the climate changes with greater intensity, the entirety of human systems will experience more frequent disarticulation and disruption, with unpredictable consequences for the human enterprise. Moreover, all evidence shows that the greatest burden of those impacts will fall most immediately and directly on the vulnerable communities around the world least able to defend themselves.

Due caution is warranted not to overstate the challenges and proclaim that humankind is at cataclysm's doorstep. Indeed, we have pursued these policies for decades, if not centuries and still the human enterprise remains remarkably intact and resilient. Yet today we are witness to the steady growth around the world of chronic water crises with the potential to evolve acute threats to economic and social stability worldwide. That spirit is appropriately captured by former Environmental Protection Agency Administrator Bill Reilly, who stated:

> Water is life. We ignore that simple truth at our peril. It may well take more widespread droughts, more massive spring floods, more shocking news of water contamination or other dramatic findings to galvanize action. The game now is to be armed with analyses and ideas so when the political moment is ripe, the country's political leaders know what they can and they should do.[9]

Woven together, these four trends constitute the global context that invariably influences the motives, strategies, priorities and consequences of the U.S. government's overseas engagements related to the provision and management of water. An equally important force shaping the U.S. government's overseas water engagements is the decades of domestic experience, turmoil and experimentation in managing our own country's water challenges. In the following section I will explore three domestic experiences that are woven into the strategies and approaches our government has employed in shaping our international water engagements for the past 25 years. The first domestic experience regards the legal foundations of U.S. transboundary relations that emerged from treaty arrangements between the United States and Mexico, followed by agreements with Canada. The second experience is the dam-building period between 1930 and 1970. That experience began with a domestic "nation building" exercise entrusted to the Tennessee Valley Authority (TVA) in 1933, followed by construction of the world's five largest dams and thousands of lesser infrastructure projects across our country, from California and the Southwest to the Northern Plains and the Southeast. The third experience is the decades-old push and pull between the federal and state governments as they have sought to maintain control over planning and managing the nation's water resources.

The Domestic Foundations

Legal Underpinnings

The first international water treaty involving the U.S. government was signed with Mexico in 1906 in response to years of contentious border incidents in which users on both sides of the Rio Grande and Colorado River lodged claims against each other. The 1906 Boundary Waters Convention vested the U.S. State Department with responsibility for ensuring compliance with the treaty. But the constantly changing contours of the rivers, coupled with steadily expanding infrastructure in the river basins flowing into Mexico, obliged the countries to sign a new treaty in 1944. The treaty authorized the newly formed International Boundary and Water Commission (IBWC) to monitor and regularly update arrangements between the two countries to reflect the constantly changing character of problems along the waterways.[10] A similar agreement, the Boundary Waters Treaty of 1909, was signed between the United States and Great Britain, then acting on behalf of Canada, a dominion of the British Crown, calling for the creation of the International Joint Commission (IJC) to manage conflicts along our northern border. The treaty originated in contentious interactions in two cases relating to the apportionment of waters between the two countries: the Milk and St. Mary rivers in Montana and the Niagara River and the level of Lake Erie.[11]

Those two treaties and their corresponding management arrangements have shaped U.S. overseas water engagements in two ways. First, the institutional arrangements embedded in the treaties have provided the government with the flexibility to address the ever-changing watercourse boundaries between the three neighboring countries. Second, both treaties were initially constructed on the Harmon Doctrine of absolute sovereignty. That doctrine gave the upstream water user, in this case the United States, the absolute right to use water within its borders any way it chooses, regardless of the impact on Mexico, the downstream user.[12] Needing to keep peace and amicable relations with our neighbors quickly obliged the United States government to repudiate that 1890s ruling by Judge Harmon and, as a consequence, recognize Mexico's legitimate claims for damages caused by U.S. actions. From that time forth, the U.S. has actively promoted the basic tenet that the needs and interests of all riverine countries must be addressed in managing transboundary watersheds and has supported that tenet as a basic principle in current international agreements on international watercourses. It warrants mention that only China, Turkey and Burundi, all three upstream nations, invoke the principles associated with the Harmon Doctrine in guiding their relations with neighboring countries.[13]

The Epoch of U.S. Dam Building

The second major experience that has shaped U.S. involvement in international water affairs is grounded in the dam-building period that supercharged American economic development between 1930 and 1970, when more than 30,000 dams, large and small, were constructed largely through the U.S. Bureau of Reclamation and the Army Corps of Engineers.[14] That unparalleled construction frenzy provided the abundant energy for American industry that eventually allowed the U.S. military to prevail in World War II. The precipitating event for the ensuing dam-building epoch came in the form of the Mississippi Flood of 1927, the most destructive river flood in U.S. history, which inundated 27,000 square miles, displaced 700,000 people and fueled the northward Great Migration of hundreds of thousands of African-Americans who lost homes in Louisiana, Mississippi and Arkansas. That event prompted President Herbert Hoover to charge the U.S. Army Corps of Engineers with the task of "taming the Mississippi," despite the corps' earlier assurance that its painstakingly constructed Mississippi levee system would contain the raging river.

Two competing water infrastructure development approaches were tested during that 40-year period. The first was the Tennessee Valley Authority (TVA), premised on a comprehensive, participatory regional development model for the South that was viewed in the 1930s as the country's most economically backward region.[15] Born out of the Great Depression, the TVA became a "regional development agency" and, at the same time, the U.S. government's successful experience in internal "nation building."[16] Over a 15-year period, the TVA evolved into a flood-control system, a generator of hydroelectric power through some 30 dams, an irrigation system, a network of navigable waterways, a producer of fertilizer, a public education system, a program to combat soil erosion and deforestation and a malaria eradication program.[17] Of equal importance, the TVA embraced from the outset the principle that it should be a demonstration model of grassroots democracy and, as such, repudiate the authoritarian regimes emerging in the 1930s in Germany and the Soviet Union, both on accelerated industrialization paths.[18]

By the end of World War II, the U.S. government, joined by the Rockefeller Foundation, heralded the technical accomplishments and the participatory democracy of the TVA as a development model to be replicated across the developing world and China as a viable alternative to the Soviet authoritarian model.[19] The successful outcomes of the TVA prompted President Franklin D. Roosevelt in 1937 to call for the replication of the TVA experience in no fewer than seven other major watersheds in the United States, including the Columbia and Missouri basins. Immediately, private utilities, having opposed the TVA as a "socialist" experiment, rose to protest further federal government expansion in water governance affairs. States, fearing "encroachment on their water rights and powers," made common cause with privately held water and power utilities and succeeded in scuttling TVA's expansion.[20] Although domestic political opposition foreclosed subsequent applications inside the United States, the TVA approach influenced the design and strategies of modernization programs across the developing world for several decades.[21]

The second model, more frequently employed throughout that 40-year period, was driven foremost by construction of large hydrologic infrastructure that supported expansion of irrigation systems for U.S. agroindustry. In the long term, that model was financed by the sale of hydroelectric energy to rapidly growing cities and industries in the American West and Southwest. The Hoover Dam, operational in 1936, "inaugurated a seminal turning point in water history—the age of giant, multipurpose dams" that allowed the United States to acquire "greater command of its water resources than any other nation."[22] The Hoover Dam became the blueprint for scores of large dams, including the "five largest structures on Earth, all dams, under construction in the western United States" in the 1930s.[23]

The pending collapse of agricultural production in California's Central Valley during the 1930s from unrelenting groundwater pumping provoked a subsequent round of massive government-sponsored water infrastructure investments in that state.[24] Having drawn down the aquifers the length of the valley through unlimited water pumping, farmers demanded support from the federal government to transfer water from the resource-rich northern part of the state through huge canals and two large dams into what would soon become the country's most productive agricultural region. The collapse of Midwestern agriculture following the onset of the Dust Bowl in the mid-1930s injected additional justification into Roosevelt's support for the landscape-changing program. The Californian New Deal projects were further bolstered by supplemental water transfer projects in the 1960s that moved water through aqueducts and a battery of pumping stations across the Sierra Nevada to fertile lands in the Central Valley.[25]

The seminal feature of this unparalleled infrastructure-building period—in which some 30,000 dams were constructed—was the central role that the federal government, largely through the Bureau of Reclamation and the Army Corps of Engineers, played in planning, organizing and financing each of the large-scale projects. By the same token, in both the TVA and the Southwest, private construction, agribusiness and energy companies exerted unrelenting pressure to mobilize political support at the federal and state levels and to shape the design of the infrastructure. Both the design and the siting of the massive projects determined who were to be the long-term beneficiaries, what companies would receive the federal subsidies and who would shoulder the costs of the government-funded projects. Perhaps the most singular lesson drawn from this epoch was confidence in America's unlimited ability to build massive infrastructure projects to transfer water, tame the power of rivers and spawn unprecedented levels of human welfare. There was, in fact, no project too large, no volume of water too huge to be put under the control of steel, concrete and new technology to serve human needs. That conviction has shaped U.S. overseas water engagements to the present time, regardless of the fact that infrastructure design is often predicated on unreliable data and deficient cost-benefit assumptions, technology and design do not address current hydrological and climatic conditions and the four mentioned trends, notably climate change, pose fundamental questions as to the viability of current infrastructure approaches.

Competing Jurisdictional Interests

The third experience relevant to our overseas water engagements is our government's checkered history and vacillating approach in creating efficient, effective management of our country's water. During the New Deal, President Roosevelt created the National Resources Planning Board to strengthen coordination and efficient resource use prior to and during the war years. One clear example of the board's impact was the planning and implementation of the TVA. By 1943, that very success prompted private companies and state governments to pressure Congress to dismantle the board "for fear of excessive centralization of power," "which was seen as an overture to state socialism."[26,27] Indeed, the board was disbanded. However, in 1950, the need to harmonize the conflicting water management systems at local, state, watershed and federal levels prompted President Harry Truman to issue Executive Order 10095, creating the President's Water Resources Policy Commission. President Truman's order skirted congressional will through the carefully drawn scope and purpose of the commission, which was mandated to study and make recommendations regarding the "development, utilization and conservation of water resources."[28]

A decade and a half later, Congress passed the Water Resources Planning Act of 1965, establishing the Federal Water Resources Council. The council was authorized to "establish principles, standards and procedures for federal water resources planning."[29] The council, a cabinet-level agency, organized regional planning and management commissions that, in turn, conducted comprehensive data collection and analysis on water supply and demand, from which river basin commissions developed and implemented appropriate management systems. As the council was beginning its work in 1965, the U.S. Bureau of the Budget, the predecessor of the current Office of Management and Budget, advised Congress that authorization language proposed for the Lower Colorado River Basin Development Fund and associated dams raised basic questions about future water policies in the United States. Unable to resolve broader water policy issues relating to water development in the Colorado River basin, Congress created in 1968 the National Water Commission, composed of citizens who did not serve in the federal government and had no direct interest in or commitment to any federal agency or program. The commission's mandate was to "review present and anticipated national water resource problems" and to "consider economic and social consequences of water resource development, including, for example, the impact of water resource development on regional economic growth, on institutional arrangements and on esthetic values affecting the quality of life of the American people."[30]

In 1973, the commission released a nearly 600-page study that highlighted ways of strengthening collaboration among agencies and levels of government and emphasized the need to replace publicly funded water projects with the "users pay" and "beneficiary pays" principle, in essence eliminating the huge public subsidies to agribusinesses and individual farmers.[31] The recommendations are far-reaching and have considerable resonance for the challenges the country faces today. However, the recommendations, if implemented, threatened the institutional authority of many agencies actively engaged in building water infrastructure and managing domestic waters. Rather than risk losing budgets and authority, implicated government agencies took no action, allowing the recommendations to languish and sorely needed reforms to go unattended. Little stakeholder resistance arose when President Ronald Reagan, following the advice of Secretary of the Interior James Watt, dismantled the commission in 1981 as part of his drive to reduce the reach of the federal government. As with the disappearance of the federal planning body, so fared the six river basin commissions on the Ohio, Missouri, upper Mississippi rivers, the Great Lakes and the Pacific Northwest, along with their recommendations for improved water management in those major American watersheds.

The turf-related tensions and ambiguous authority among the various levels of government in managing our domestic water resources continue today. Each level of government brings a unique set of perspectives, priorities and contributions, and of course, outcomes. With strong federal involvement, broader cooperation, planning and rationalization of water use tend to prevail. Under local and state jurisdictions, local control and tailored management regimes best reflect the needs of local users. The question yet to be answered is how long it will be before the four convergent trends intensify their impacts such that water scarcity, weather extremes and climate variability oblige citizens and government to align planning and management systems at the federal, state and watershed levels. Will intensified conflicts at local and subnational levels provoke stronger interventions and new management regimes? Will deepening water scarcity and weather extremes encourage federal and state planners to balance our hard infrastructure solutions with "soft-path approaches" to address those new challenges?[32] And will lessons drawn from our domestic responses to the new challenges come too late to help far-more-vulnerable developing countries adapt to the impacts of climate change that are already destabilizing their societies? Or will U.S. innovation and testing spawn new management regimes and approaches that can help guide other countries and communities as they wrestle with scarcity and weather extremes? Nor can we exclude the strong possibility that experimentation and innovation in other countries will offer solutions for our own communities. In the meantime, we have no national water strategy to set priorities, harmonize competing governance regimes and coordinate the multiplicity of programs and actions unfolding at local, state and federal levels.

The Political Cauldron

These and other experiences manifest themselves in the country's institutional approach to dealing with water-related planning, scarcity and conflict. But none of those experiences prepared the country for the political turmoil and paralysis that began in 1994 and continue today, severely constraining U.S. government engagement in global water affairs.

It is important to signal that for 15 years prior to 1994, the United States had established a strong leadership role in shaping global water politics. In fact, for some 16 years, from 1978 to 1994, the United States was providing the technical, financial and policy leadership through which we earned widespread support from developing countries and development agencies alike. Our international standing emanated from the U.S. Agency for International Development's (USAID) Water and Sanitation for Health Program, carried out with scores of partner countries in the developing world. In turn, that program reflected U.S. commitment to supporting emerging international water policies such as those forged at the UN Water Conference held in Mar del Plata, Argentina, in 1977. With the decade of the 1980s having been declared the UN International Decade for Drinking Water Supply and Sanitation, the U.S. government's programs enjoyed attention and support from the international community.

Although domestic political pressures were pushing President George H. W. Bush to weaken federal environmental regulations, he did join 108 other heads of state at the UN Conference on the Environment and Development in 1992 in Rio de Janeiro.[33] That historic event represented the culmination of rising global anxiety and protest movements, challenging the failure of governments and international development agencies to anticipate and mitigate the environmental costs associated with economic growth. As regards international water policy, water leaders convened the Dublin Conference in Ireland six months prior to Rio to hammer out four operational principles, which remain the bedrock of international water management today. The Dublin Principles recognized water as a finite and vulnerable resource, called for participatory water management arrangements, underscored the fundamental role of women in water provision and management and also affirmed that water should be recognized and managed as an economic good.[34]

Just as the international community was translating the outcomes of Dublin and Rio through a global strategic plan called Agenda 21, the newly elected President Bill Clinton came under siege at home from political opponents. President Clinton released his first National Security Strategy in 1994, in which he laid the policy foundations for the "economization" of U.S. foreign policy. Over the course of his two terms, President Clinton pursued policies predicated on the tenet that U.S. global influence was built on the strength of our economy. This prompted him to expand trade relations with developed and developing countries alike. Moreover, through subsequent National Security Strategies, President Clinton embraced and further expanded on the centrality of sustainable development, the conceptual foundation of the international consensus on environment and development, as a basic pillar of U.S. overseas engagements.[35]

U.S. Development Assistance Under Siege

The Republican party's response to that proposition and President Clinton's expansionist policies more generally were unveiled in 1994 when Newt Gingrich, then Speaker of the House, released the Contract with America. That new vision of the United States and its engagements overseas led to, among other major policy shifts, a full-throated reordering of U.S. development assistance around the world.

From his position as chair of the Senate Foreign Relations Committee between 1995 and 2001, Senator Jesse Helms, the senior Republican senator from North Carolina, became the spearhead for promoting the Contract with America in international affairs. The core of Helms' proposal in challenging the Clinton agenda was his vow, made in the early days of the 104th Congress, to abolish three agencies: USAID, the Arms Control and Disarmament Agency and the U.S. Information Agency.[36] His intent was to merge the three agencies into the State Department, thereby making those agencies more direct instruments of U.S. foreign policy. Moreover, Senator Helms proposed ending any direct U.S. governmental role in alleviating poverty in poor countries, seeking instead to establish an International Development Foundation that would issue block grants to private agencies that, in turn, would manage aid programs.[37]

The battle between Senator Helms and Senator John Kerry, then ranking minority member of the Senate Foreign Relations Committee, was enjoined and, in the process, paralyzed approvals for ambassadorial nominees, arms control treaties and international development initiatives. After a year of wrangling, a compromise reached in 1995 avoided total dismemberment of the three agencies but required a $1.7 billion funding reduction for the three agencies and the State Department that would become effective by 2000. As a result of that agreement, USAID reduced its active country programs from 120 to 70 countries, effectively closing 50 country programs, and slashed USAID's workforce of 11,500 employees to less than 8,000. USAID remained an independent agency, but its budgetary autonomy was drastically altered when, in 2006, it became answerable for the first time to the Secretary of State rather than to the Office of Management and Budget.[38] During this period of conflicting pressures, USAID was driven to focus on yearly, performance-based funding cycles as opposed to long-term strategic plans. The development agency also shifted to quantifying and defending its disbursements, as the State Department was obliged by Congress to codify all of its expenditures and achievements through eventual definition of hundreds of indicators that do not necessarily reflect realities on the ground.[39]

The drive to quantify the impacts of U.S. expenditures was justifiable given the executive branch's long-standing failure to provide consistent information on the benefits of U.S. development assistance. By the same token, however, the push to quantify benefits had the perverse impact of steadily reducing U.S. government investments in more-complex water management activities that are essential for developing countries to meet the rising demands from rapidly growing, urbanizing economies and populations. It is significantly easier to quantify how U.S. taxpayer dollars resulted in more families or schools having clean water than it is to quantify the benefits resulting from improved water regulations, strengthened institutions or upstream watershed management programs. Under the new reporting pressures, USAID's investment portfolio focused increasingly on water sanitation and hygiene projects, which delivered clean water to users while not necessarily addressing the underlying institutional, infrastructure and policy changes to ensure delivery of water over the long term.

By the time George W. Bush became president in 2001, three movements increased pressure on Congress to continue its drive to quantify development assistance expenditures in the water and other sectors. First, from the outset of his administration and in response to a significant domestic political constituency, President Bush actively encouraged more involvement of faith-based organizations in delivering U.S. governmental assistance to developing countries.[40] The enthusiastic response and sustained lobbying activities from faith-based organizations, backed by prominent political figures including Senator Sam Brownback, further sharpened congressional and USAID focus on water delivery and toilet construction rather than sustainable water operation and management. All the while, the increased government funding flowed through the books of these faith-based and other civil society organizations, creating vested interests in perpetuating the emerging priorities.

The second pressure arose from an important engagement by the Bush administration in UN processes—development and endorsement of the Millennium Development Goals (MDGs), a global initiative launched at the beginning of the millennium to coordinate and accelerate efforts to support development progress. Paradoxically, the MDGs' emphasis on achieving quantifiable outcomes, including "halving the global population without sustainable access to safe drinking water and basic sanitation," reinforced USAID's focus on counting tap and toilet construction as the linchpin of U.S. water development assistance, further shifting emphasis away from water management.[41]

A third impetus for reinforcing the government's commitment to water provision arose from the office of Senator Bill Frist, who drew from his extensive experience and stature in the medical field. As Senate majority leader, Senator Frist exhorted his colleagues to give proper attention to the human and security costs of ignoring the plight of millions without access to clean water, stating:

> Water basins do not follow national borders, and conflict over them will escalate as safe water becomes even scarcer. These conflicts may come to threaten our national security. Modest, pragmatic clean water projects that yield real measurable benefits will make things better.[42]

While invoking the opportunity to address the multifaceted dimensions of promoting sustainable water management, his operational focus reverted to providing water access in keeping with established priorities of delivering water services through U.S. development assistance.

His exhortations coincided with the administration's continued drive to improve the State Department's accounting metrics of development expenditures. This impetus was picked up and carried to fruition by then Secretary of State Condoleezza Rice and became more formally known as the "F-process." This initiative, formalized in 2006, provided state agencies, USAID and other development assistance agencies with long-overdue, reliable statistics that could account for and demonstrate how U.S. tax dollars were being translated into real improvements for millions of assistance recipients around the world. The new reporting system, called the Foreign Assistance Coordination and Tracking System (FACTS), allowed the State Department to counter congressional detractors who argued that foreign assistance does not contribute to U.S. national interests.

The F-process, however, subjected development activities to yearly performance-based funding cycles while stimulating expansion of mandated congressional "earmarks" for objectives supported by lobbying constituencies. This change proved to be quite negative because USAID country missions could no longer respond positively to the priorities and strategies established at the local level by host governments, such as improving upstream water management, strengthening local and regional water authorities or resolving water-access conflicts among communities. Country missions were locked into the processes and priorities set by Congress. In short, the changes sent the message that recipient countries simply had to follow U.S. government priorities or development assistance would be curtailed.

This 10-year period marked a critical turning point, at which time U.S. development assistance shifted from creating conducive environments for long-term development trajectories to focusing on service delivery. The paradox of this fundamental shift resides in the fact that dozens of other development agencies, organizations and companies could and were delivering comparable water sanitation and hygienic services. However, no other government or development agency had the technical and scientific competence, capacity-building know-how, institutional reach, diplomatic stature and financial resources to help dozens of countries address longer-term water management challenges. While the United States could have provided the global software and cohesive strategy to help carry developing countries through the sharpening water challenges shaping their development paths, this convergence of political pressures reduced the United States' contribution to being a highly competent installer of water facilities in water-stressed geographies. Moreover, the option of being a service provider diverted U.S. resources away from pursuing strategic U.S. interests in promoting economic development and addressing humanitarian challenges while also promoting peace and security.

During the George W. Bush administration, the battle for control over international water policy swung between Congress and the Department of State amid the larger budgetary and policy fights. In its efforts to influence freshwater policy, the State Department and USAID released a new strategy, "Addressing Water Challenges in the Developing World: A Framework for Action," as required by the Water for the Poor Act of 2008.[43,44] That framework was an "expression of the overall U.S. Government approach to the world's water challenges developed to lay out the guiding principles for U.S. foreign assistance in the water sector, embracing the government's broad and interrelated portfolio of water expertise and approaches." However, it did "not define the full scope of needed interventions to ensure that water resources are available to meet the entire range of human development needs," nor provide "a broader interagency strategy, reaching beyond a framework focused on the State Department and USAID."[45] The framework collapsed on release, leaving the development agency without a clear blueprint to respond to the rising international concerns about freshwater availability.

The next significant reorientation in the government's engagement on global water affairs took place between 2010 and 2012 under Secretary of State Hillary Clinton and significantly elevated the profile of water in U.S. foreign policy operations:

1 In a 2010 speech on World Water Day, Secretary Clinton defined "five streams of action that make up our approach to water action," recognizing that the water crisis is a health crisis, a farming crisis, an economic crisis, a climate crisis and increasingly a political crisis. She further emphasized water's role in promoting U.S. overseas interests and declared "clean water is a matter of human security. It's also a matter of national security"[46]
2 The State Department/USAID signed a memorandum of understanding (MOU) with the World Bank emphasizing the need for balance between providing access to needy communities and sustainable water management.[47]
3 The Secretary requested a "National Intelligence Council estimate on national security implications of water security up to the year 2040."[48]

4   The State Department established the U.S. Water Partnership (USWP), which purported to strengthen cooperation between U.S. companies and investors with the U.S. government, civil society organizations, technical experts and academia.

Of those initiatives, identifying water as a national security interest and requesting the intelligence community's assessment may have proved to be the most influential. The intelligence community's analysis highlighted the geographies and mechanisms through which water problems—including scarcity, floods and water quality—can directly affect U.S. national interests. For reasons unknown, the MOU with the World Bank generated no detectable response from USAID and no enduring outcomes. Now converted into a nongovernmental organization (NGO), the USWP continues to promote broader engagements linking the private sector to public functions and initiatives of academia and civil society.

With the departure of Secretary Clinton in 2012, the salience of freshwater issues in State Department policies and programs diminished while the department shifted focus to mitigating and adapting to climate change, under Secretary John Kerry. Attempting to keep abreast of the State Department's strengthened focus on water and security, in 2011 USAID released its *2013–2018 Water and Development Strategy*.[49] In its initial framing, the strategy invoked the State Department's higher-level policy statements, while in operational terms USAID's priorities focused on providing water, sanitation and hygienic services and improving water for agriculture to support the U.S. government's food security initiative, Feed the Future. Broader transboundary water management, policy reform, water pricing and finance issues are referenced in passing but do not constitute significant operational priorities.

Beyond Secretary Clinton's laser focus on the centrality of water to U.S. security interests, the Obama administration took little direct action to strengthen overseas water engagements. Even so, the president's 2014 Executive Order on Climate-Resilient International Development set an operational and conceptual framework requiring that all U.S. international development work integrate climate resilience into all bilateral and multilateral "strategies, planning, programs, projects, investments, overseas facilities and related funding decisions."[50] As water is the principal vehicle through which climate change affects human communities, the executive order requires development assistance programs to pay increased attention to the ways extreme weather and climate variability, including water, will affect communities, governments and companies. It is important to note that the international dimension is complemented by two companion executive orders: Preparing the United States for the Impacts of Climate Change and Incorporating Natural Infrastructure[51] and Ecosystem Services in Federal Decision-Making.[52]

That history notwithstanding, the opportunities for the U.S. government to embrace a more comprehensive approach to addressing the rapidly rising water crises reside in the requirements of the Senator Paul Simon Water for the World Act (WfWA) signed by President Barack Obama in 2014. That act expresses the sense of Congress that

1. water and sanitation are critically important resources that impact many aspects of human life and;
2. the United States should be a global leader in helping provide sustainable access to clean water and sanitation for the world's most vulnerable populations.

Most notably, the WfWA also

> directs the President, not later than October 1, 2017, through the Secretary of State, the USAID Administrator and heads of other federal departments and agencies, to submit a single government-wide Global Water Strategy to Congress that describes how the United States intends to: (i) increase access to safe water, sanitation, and hygiene in high priority countries; (ii) improve management of water resources and watersheds in such countries; and (iii) work to prevent and resolve intra- and trans-boundary conflicts over water resources in such countries.[53]

This vision offers a clear opportunity for a major break from the past. However, if a new and more balanced path is to be pursued, three issues must be resolved. The act's mandate covers only "high priority countries" that lack access to safe water, sanitation and hygiene, meaning that only countries for which U.S. water service provision is prioritized, not the full range of countries facing immediate and longer-term water management challenges, are to be considered. Second, more than a dozen agencies and departments involved in constructing the new strategy must reach consensus, a necessary condition that has proved elusive over past decades. And third, a new policy and accompanying institutional reform must overcome bureaucratic resistance and inconsistency that have crippled U.S. water work for 80 years. Recent stirrings in USAID, including a request for proposals to implement a new, five-year Sustainable Water Partnership with funding of $65 million, may indicate increased interest in preventing conflict and enhancing resilience in countries particularly vulnerable to increased competition at local and watershed levels.[54]

U.S. Policy Influence on Multilateral Development Banks

As the foundations of the government's bilateral assistance policies were undergoing major changes in the early 1990s, the contours of U.S. water policy expressed through multilateral development institutions were simultaneously taking form, notably through engagement with the World Bank Group. If domestic politics and the Contract with America shaped the U.S. government's approach to development assistance, the recurrent and deepening debt crises besetting developing countries in the 1980s and 90s set the context for a dramatic shift in international economic policy driven by the U.S. Treasury Department.

During the 1970s and 80s, steady support from the World Bank Group and other multilateral institutions provided a steady flow of financing to state-driven economies of scores of young countries still emerging from the shadow of colonialism. Without robust, domestic entrepreneurial classes, those struggling governments relied heavily on centralized planning and state-controlled enterprises to engender domestic economic activity and to build links to the international economy. After several decades, the limits of that state-driven economic model were clear: economic inefficiencies, recurrent budget deficits and protectionist policies eventually sapped the vitality from those countries gripped by deepening indebtedness to public and private lenders alike. By the end of the 1990s, growing debt overhang, recurrent defaults on lending obligations and rising political discontent with stagnant economies presaged a draconian shift in prevailing macroeconomic policy across the developing world.

That change was driven by the policies of Prime Minister Margaret Thatcher and President Ronald Reagan, whose domestic economic policies of reducing governmental intervention in the economy were matched in the international sphere by a single-minded thrust to replace state-driven with market-based economies across the developing world. Under the mantle of the Washington Consensus, the multilateral financial institutions, with the World Bank Group at the forefront, imposed a series of conditionalities on borrowing countries to ensure that governments rapidly divested economic power to domestic and international investors.[55] Through structural adjustment loans, scores upon scores of borrowing countries opened their borders to foreign investment, cut expenditures for social programs and dismantled regulatory authority of government agencies, all considered requisites for reestablishing fiscal balance.[56] The 1990s witnessed the selling-off of state-owned enterprises, ranging from airlines and hotel chains to public utilities and mining companies alike.

Beginning with the first structural adjustment loan to Turkey in 1989, a full decade of work in the 1990s was needed to restore basic macroeconomic balances in highly indebted countries. The hundreds of billions of dollars in adjustment lending provided ample incentive to shift scores of developing countries irrevocably away from centrally planned, state-managed economies as market dynamics took hold. By the same token, for more than a decade World Bank adjustment lending failed to address the environmental costs of these fundamental economic changes.[57] Opening countries' borders to foreign investment in extractive industries was often accompanied by weakening or dismantling of government regulations and enforcement capacity.[58] Not infrequently, foreign investors had a strong hand in writing new mining laws, forest codes and fishing regulations. Moreover, as economic structures changed, farmers and urban workers were pushed out of jobs, often returning to rural communities where survival needs led to increased pressure on forests, fishing grounds and rural livelihoods. While structural adjustment lending improved standard economic indicators in scores of developing countries, the legacy of ignoring the environmental impacts of outward-oriented economic growth remains present today, often registered in weak regulatory codes and ineffective enforcement capacity.

By the end of the 1990s, with market economy principles shaping economic life in a growing number of developing countries, the World Bank Group sharpened its targeting of public utilities, notably of publicly owned energy and water utilities. The shift in focus was not accidental. In most developing countries, public water utilities were often viewed as a source of government revenue to finance a wide range of development activities. Water utilities claimed the notorious distinction of gross wastage and inefficiencies, poor maintenance, recurrent budget deficits and failure to extend networks to rapidly growing urban centers. Moreover, governments did not have the capital resources to meet the burgeoning water and sanitation demands of urban dwellers, who had little recourse but to turn to informal service providers that often operated without regulatory controls.

As early as 1993, the World Bank's water policy, "Improving Water Resources Management," stated that public financial support would come at the price of obtaining concessions from borrowing governments "aimed at improving the efficiency of publicly owned enterprises and reducing bulging government deficits in order to reduce the public deficit," among the nonnegotiable principles of the Washington Consensus.[59] Guided by that strategy, virtually all World Bank water lending operations for the water sector thereafter were conditional on some form of devolution of management or assets to private operators and cost-recovery principles.[60]

As private water companies started taking control, public protest and controversy soon followed. By the mid-1990s, flawed privatization experiences in Bolivia, the Philippines and Argentina, among others, provoked public unrest, some shutting down local governments as poor communities protested water rate increases not matched by improved service. The protest tide intensified, acquiring the proportions of a global anti-privatization protest movement. By the turn of the millennium, international lending agencies and private investors, as well as national and municipal governments, were forced to step back from the privatization-at-all-costs drive.

In 2003, a decade after embracing its initial strategy, the World Bank released a revised strategy, "Water Resources Sector Strategy: Strategic Directions for World Bank Engagement," that downplayed the ideological underpinnings previously shaped by U.S. policy. In that less ideological context, the World Bank embraced an inclusive approach to innovative hybrid financial arrangements in which governments and commercial lenders distributed responsibility and risk in keeping with the conditions and opportunities of each locale. Moreover, the World Bank recognized that water subsidies can be socially desirable and economically efficient when transparency and accountability ensure avoidance of moral hazard and rent-seeking behavior. Those revised guidelines opened the doors to a wide range of innovative financial arrangements and experimentation.

The collision of the U.S.-driven, neoliberal ideology of the Reagan-Thatcher period with the needs and realities of struggling communities around the world gave way to a protracted period of uncertainty and testing. While innovative pilot programs involving international development agencies and private investors have given rise to a number of successful prototypes, the international community remains far from having established a replicable, effective partnership arrangement to finance and manage water access for communities around the world.

At the same time that the privatization issue rose to trouble international financial institutions, a parallel global water controversy embroiled multilateral development banks and national governments as well. This time the issue focused on the role of large dams in the development process across the developing world. The controversy arose from a multitude of localities in India, Brazil, Indonesia and elsewhere as scores of national governments borrowed from international financial institutions in hopes of providing longer-term responses to rising public demand for water and energy. While the ideological positions of those favoring or opposing large dams hardened over the course of the 1990s, initial protests arose from two major sources. First, hundreds of thousands of inhabitants were displaced as the huge infrastructure projects took form. For reference, the U.S. experience with the TVA some 50 years earlier also required relocating some 125,000 people from the valleys and hills to be flooded by the massive system of dams on the Tennessee and other rivers. The second grievance, though one with a marked difference between the TVA and experiences in other localities, was the lack of participation and transparency in the planning and construction of the hydroelectric plants in developing countries where national governments and lenders often acted with impunity. Protests erupted, often forcing shutdowns at the construction sites.

The flash point of the global debate was the 2002 release of the World Commission on Dams (WCD) report, the product of a five-year research and policy development process that involved national governments, the World Bank, academics and civil society organizations from countries around the world. The majority of participants, including representatives from developing-country governments, academia and NGOs, fully endorsed the commission's recommendations. The World Bank, counting on support from the United States, recognized the validity of concerns that had sparked the controversy but did not embrace the full slate of the proposed 26 guidelines. Instead, the World Bank committed to working with "the government and developer on applying the relevant guidelines in a practical, efficient and timely manner."[61] The World Bank responded further by saying that a more effective response would be to ensure stricter compliance with its environmental and social safeguards and procedures in its own lending operations and thereby address many of the complainants' original concerns. As we explore in subsequent chapters, the World Bank's ensuing stricter adherence to those safeguard protections has led more than a few developing countries to forgo World Bank loans in favor of lenders who are willing to overlook such environmental and social protections.

U.S. Geostrategic Interests

There are a limited number of geographies where the full range of U.S. foreign policy assets—defense, diplomatic and development tools—are integrated to counter declared U.S. security threats. Generally, the engagement of U.S. military assets frequently receives greatest public attention because soldiers' lives and national treasures are under direct threat. Despite the comparatively diminished public visibility, both the diplomatic and the development dimensions of the U.S. foreign policy apparatus are fully engaged in responding to the identified threats, including responding to pressing water-provision and management problems.

The region of greatest concern is the Middle East, where U.S. interests are deep and diverse and where water scarcity has created explosive conditions that have destabilized political regimes. By some measures, the region has been running out of water since the 1970s.[62] Data on water is elusive, in part because governments often consider such information to be a national security matter. However, a 2013 analysis of NASA satellite images demonstrated the extent of the unsustainable drawdown: Across the Fertile Crescent, an amount of water equivalent to the Dead Sea was lost between 2003 and 2009.[63] Water scarcity experienced today by inhabitants, in the tens of millions, in the Middle East is not entirely due to inevitable natural causes. Poor water management has played a major role too, creating an explosive situation that will only be exacerbated by climate change. Unsustainable extraction practices, particularly from underground aquifers and government policies that provide cheap, subsidized water have concealed the crisis for decades.

Director of National Intelligence James Clapper underscored the role that water scarcity played in increasing regional volatility in the *2015 Worldwide Threat Assessment*, an annual summary of global threats to U.S. interests. He stated that "risks to freshwater supplies—due to shortages, poor quality, floods and climate change—are growing," particularly in North Africa and the Middle East, "where lack of adequate water might be a destabilizing factor." The report points out that "terrorist organizations might also increasingly seek to control or degrade water infrastructure to gain revenue or influence populations."[64] In equal measure, the 2014 Quadrennial Defense Review points to climate change—which is expected to have major impacts on water availability—as a threat multiplier, aggravating other stressors.[65]

Iraq

As referenced in the opening paragraphs of this chapter, the initial post-invasion optimism of 2003 was matched by a congressional appropriation of $4.3 billion for water and public works as a key element in U.S. reconstruction efforts.[66] Within a year's time, that buoyant view of a post–Saddam Hussein world gave way to the Coalition Provisional Authority (CPA), which was beset by confusion, lack of strategic leadership and a 50% reduction in infrastructure funding in order to respond to security threats from a poorly defined but rapidly expanding insurgency. In those early days, James Stephenson, Iraq director for USAID from 2004 to 2005, urged Ambassador Paul Bremer to focus on providing programs to improve the quality of life and win the confidence of local communities through employment opportunities and provision of basic survival needs, including food, water, energy and health care. Instead, the rudderless CPA awarded infrastructure projects to U.S.-based multinational corporations—including Halliburton, Bechtel and Parsons—despite the certainty that the projects would be "slow to develop, generate little employment and be largely invisible to the average Iraqi."[67] Moreover, the initial impulse to "privatize" Iraqi utilities and future infrastructure construction foundered because of fears of corruption and potential links to the Baathist party.[68]

A more fundamental reason turned the U.S. government away from water and general "humanitarian projects" in the early years of reconstruction: When the anticipated "massive surge" of refugees and humanitarian needs failed to materialize, invasion planners turned their attention to restoring electricity and oil production. As Paul Wolfowitz promised, once Iraq's oil industry was restored and generating over $30 billion per year, Iraq's natural-resource wealth, not American taxpayer money, would finance the reconstruction effort.[69] The Bush administration's belief that Iraq's oil industry, once restored, could finance the reconstruction of the country proved disastrously wrong despite payments of more than $10 billion to Halliburton, KBR, Parsons and other multinational corporations to get the oil facilities back online. Not only would widespread security threats and the parlous state of Iraq's oil infrastructure quickly sour that plan, but foreign contractors were unable to provide a steady flow of water needed to extract the petroleum. The first $2.4 billion appropriation by Congress was followed by a second $18.4 billion and additional commitments thereafter to forestall the country's economic collapse.[70] That drawdown on America's treasure did not cease for years. Prioritizing the restoration of oil production, while a necessary long-term goal, diverted attention from the daily needs of the Iraqi populace engaged in daily struggle for survival, the deprivations of which further fueled the growing insurgency.

One assessment of the USACE's contributions to rebuilding the water sector claims that reconstruction efforts provided clean water to more than 4.7 million people in Iraq.[71] However, by 2011, the lofty goals of the post-invasion days remained distant promises as U.S. investments had little impact on providing increased water access to the Iraqi population or improving water quality. Former U.S. Commander of Central Command General Anthony Zinni, commenting in 2013 on the current water challenges in Iraq, stated, "The situation in Iraq is deteriorating much faster than expected," with prospects for a looming crisis that will involve "an ugly mix of human suffering, governmental instability, population movement, and a rise in extremist violence."[72]

Despite the withdrawal of U.S. combat forces from Iraq in 2011, water-related issues pulled the U.S. military back into the country in 2014 when the specter of the Islamic State's (IS) consolidating control over the country's largest dams loomed ominously. Iraq avoided a major humanitarian crisis in April 2014 when Islamic insurgents first seized the Fallujah Dam and then closed the dam's 10 gates in an effort to flood, then dislodge, Iraqi pro-government troops seeking to retake the city of Fallujah. The willingness of the IS to use water as a weapon of war was reminiscent of Saddam Hussein's "punitive hydro-engineering" when he drained the 1,000 square kilometers of the Huwaiza marshes and destroyed 70 marsh villages in the early 1990s.[73] Ultimately, only the introduction of Kurdish Pesh-merga forces to bolster the Iraqi army, coupled with steady U.S. air strikes, deprived the insurgents of that prized target in Fallujah.[74]

Some four months later, Sunni militants seized the Mosul and Haditha dams, the country's two largest, but this time refrained from using water for its military advantage. For one, had the IS destroyed the dams, that violent act would have stripped the self-proclaimed caliphate of its veneer of being a viable alternative to then Prime Minister Maliki's government and it would have denied hundreds of thousands of Iraqis water to irrigate fields and satisfy basic needs. In addition, the insurgents had learned that efficient water management and steady provision could be a major source of revenue for the emergent state through fees imposed on domestic consumption and irrigation. Today, geographic control and battle lines continue to shift, leaving reconstructing the country's shattered water infrastructure a task for the future.

A more general uncertainty prevails about Iraq's ability to address its future water needs. The first unknown is the political aspirations of the Kurdish Regional Government (KRG). Situated in the northwestern part of the country, Iraqi Kurdistan has abundant groundwater resources, with high recharge capacity, that provide much of the water flowing into the Tigris, on which southern regions of the country depend. With two important dams built on tributaries of the Tigris—the Dokan Dam on the Little Zab River and the Derbendi Khan Dam on the Diyala River—the KRG has launched feasibility studies on no fewer than three other dams designed to support agriculture and development programs in the semiautonomous region.[75] Should aspirations shift from remaining as an autonomous region to becoming an independent state, the future of regional water access and management would shift from a tripartite negotiation among Turkey, Syria and Iraq to a negotiation among four parties, at which the KRG would take a seat at the table.[76]

The second unknown involves the future impacts of climate change. The country has already suffered from antiquated, inefficient irrigation systems, a lack of coherent water management policies and reduced flows caused by Turkish and Syrian dams on the Tigris and Euphrates. Declining irrigation waters have already accelerated farm abandonment such that by 2013, domestic production could satisfy only 30% of Iraq's food needs.[77] With the United Nations rating Iraq as one of the Arab region's countries most vulnerable to climate change, these recent indicators may portend serious economic and social dislocations in coming years.

The Jordan River Basin

In pursuit of geostrategic objectives, the United States has not only been a mere respondent to water-driven crises in the Middle East; it has also played an active role in sharpening regional water tensions. This dynamic has played out over past decades in the Jordan River basin. The Jordan River is small, with only about 5% of the flow of the Euphrates and a length of only 250 kilometers. Of the five claimants—Israel, Syria, Lebanon, Jordan and Palestine—Israel has successfully garnered control of most of the Jordan basin's water through war and threat of war.[78]

Disputes among the claimants trace back to Israel's earliest days, when, for example, President Dwight Eisenhower launched the Johnston Mission in 1955 in hopes of forging an accord for the multinational development of the Jordan River basin. Israel and Jordan worked within the allocation framework proposed by the Unified (Johnston) Plan, with Israel constructing the National Water Carrier by 1964 and Jordan completing the King Abdullah Canal by 1966. Those agreements remain in effect some 50 years after signature. However, the Arab League opposed the Unified Plan on grounds that it legitimized Israel's water claims. The disagreements precipitated a series of localized conflicts between Israel, Syria and neighboring Arab states that, over the course of several years, culminated in the 1967 Six-Day War.[79] Often called the first modern water war, it allowed Israel to seize the Golan Heights and the West Bank, resulting in Israel's quadrupling its territory and doubling its supply of fresh water.[80]

One outcome of the 1967 war is that Palestinian communities no longer have access to the lower Jordan River. Traditionally, West Bank occupants relied on rainfall and shallow wells for domestic water use, agriculture and grazing. Israeli control has placed strict limitations on Palestinian use of water while allowing the Mountain Aquifer to be tapped for drinking water for Israel and to support the development of intensive agriculture in Jewish settlements.[81] Israel and the West Bank settlements take about 89% of the water extracted from the aquifer, leaving the Palestinian population with the rest, according to a 2013 United Nations country mission report.[82]

Herein lies a basic contradiction in current U.S. policy. On one hand, U.S. policy supports Israel's current approach to securing its borders, which includes condoning practices that deny Palestinians access to water resources on which their future development opportunities depend. Since the 1995 Oslo II Agreement, the Israeli posture has created a water management regime that significantly skews water access opportunities to Israeli settlers and other citizens, relative to Palestinians, by a ratio of five to one.[83] That approach also ensures Israel access to 89% of the West Bank's aquifer, leaving the remaining 11% for the Palestinians. Over the past several decades, Israeli policy has included demolition of a wide range of household and communal initiatives seeking to provide basic water services to water-stressed families.[84] Further, destruction of Palestinian water infrastructure has been a recurring part of Israeli responses to cross-border aggressions by Palestinians.[85]

On the other hand, the "U.S. Government is the leading provider of bilateral development assistance to the Palestinians," including "for water resources and infrastructure," according to USAID.[86] By way of recent example, the United States announced in spring 2015 that "USAID has upgraded water distribution networks and installed more than 866 kilometers of water pipelines to provide access to clean water to more than 1 million people."[87] That support is channeled directly to the Palestinian Authority, often to improve water management capacity, as well as to civil society organizations, such as American Near East Refugee Aid (ANERA), with long-standing humanitarian and relief programs for Palestinians.[88]

By taking this security-focused approach, U.S. policy has addressed episodically, at best, critical factors shaping access to water that are now contributing to conflict and further destabilization of the region. Within countries, politically driven allocation of water aggravates inequality and perpetuates injustice. Regionally, powerful states—and now nonstate actors such as IS—are able to appropriate water, peaceably or through military means, at the expense of other countries and peoples. Yet, unequivocally, the specters of deepening water scarcity and conflict, particularly as climate change impacts deepen their hold, portend a region unable to satisfy basic human needs, while military conflicts may expand.

If our fundamental interest in the region is stability, the United States has little hope of achieving that goal unless it places providing fresh water to all citizens of the region as a top, long-term priority. In light of the role that protracted drought played in precipitating Syria's civil war, U.S. policy in the region and beyond must recognize that water insecurity can be a volatile accelerant of social instability. Nor can U.S. policy and active engagements ignore the multiple uses of water as a weapon of war and a source of collective punishment. In short, ensuring access to water to communities across the region is neither a tertiary concern nor an optional response: water access is a primary requisite for stabilizing societies and an opportunity for constructive engagement with all parties across the region. Achieving U.S. stabilization goals requires active engagement to strengthen access, participatory governance and a longer-term, supply management strategy, even before conflicts have fully subsided across the region.

In Summary

This brief overview, admittedly partial and incomplete, underscores several key facts. First, through its multifaceted overseas engagements, the United States has demonstrated a constant lack of understanding of the nature of the problems facing our partners around the world. For several decades, the United States has been an important contributor to global efforts to address the persistent challenges of water scarcity and lack of access over past decades. The government's bilateral efforts to provide fresh water and sanitation services have improved the lives of millions of families living in water-scarce regions of the world. The United States has been fully aligned with and remains a major contributor to the United Nations' water Millennium Development Goal and is working to adapt to the recently agreed-upon Sustainable Development Goals. The United States has provided widely distributed technical, engineering and data collection assistance to strengthen water management regimes through bilateral and multilateral programs. By the same token, U.S. development assistance focuses on a tightly circumscribed set of activities that do not respond flexibly to the approaches and stated priorities of national governments. Nor are resources committed through predictable, multiyear programs that will ensure redress of long-standing problems. Moreover, strictly constructed appropriations prevent the United States from addressing the more fundamental challenges of ensuring long-term access to water through modern policies, efficient regulations and strong institutions able to enforce prevailing laws and conflict mitigation programs. To date, little if any support has been channeled into soft-infrastructure programs as necessary complements to hard-infrastructure construction, in which the United States is an established leader. While U.S. government development assistance strategies and programs frequently invoke notions of sustainable water management, conflict mitigation and capacity building, on-the-ground operations consistently revert to water provision and sanitation-service delivery. Strategic documents underscore the importance of creating enabling policy and institutional and market conditions, while actual funding prioritizes specific infrastructure projects to provide fresh water for targeted populations.

Second, the United States, working through the multilateral financial institutions, was the determinant actor in reshaping global macroeconomic policy that led to improving the economic performance of scores of countries as they became more deeply integrated into the global economy. Despite efforts of the World Bank's U.S. executive director, the structural reforms programs hosted by the leading multilateral lending institution ignored environmental impacts that today continue to weaken developing countries' economic performance and burden their citizens with mounting environmental debt. Initial efforts to privatize municipal water utilities, an integral part of the global macroeconomic structuring process, collided head on with the needs of water-stressed communities in the developing world that were unable to shoulder the financial burden of privately managed water systems. That initial setback notwithstanding, the United States has been supportive of more recent efforts to encourage hybrid financing arrangements linking public sector finance with private investment when tailored to local needs and conditions. U.S. financial support channeled through the World Bank Group and other multilateral institutions remains central to helping developing countries mature domestic capital markets to finance infrastructure and improve the overall performance of local utilities so that they become more attractive to private investors. U.S. support for the more inclusive, longer-term water management approaches promoted by the World Bank Group and other multilateral institutions is divorced from the narrower service provision focus of U.S. bilateral assistance.

Third, U.S. efforts to address water scarcity and variability in geographies of strategic importance have generated highly checkered results. On one hand, the government has employed military force to deny terrorist networks water resources that insurgents have used to impose collective punishment on communities resisting IS rule and the self-proclaimed Islamic caliphate has used to extract taxes to finance its Islamist regime. On the other hand, the United States has supported a hydro-hegemon—Israel—through its military, diplomatic and development assets while also providing water access to the Palestinian communities deprived of access to their water rights. Reactive, inconsistent responses without attention to identified human and national security risks seem poorly adapted to the many geographies facing water scarcity and weather extremes.

In short, U.S. government engagement in water provision and management over the past decades has been abundant, fraught with inconsistencies and highly susceptible to domestic politics. At times, U.S. engagements have drawn heavily on domestic experience, notably through the engineering and technical dimensions; at other times, U.S. engagements have been driven by short-term responses to crises and immediate threats to U.S. security interests. At times, the United States has pursued multilateral cooperation through the international development agencies and the United Nations; at other times, the United States has pursued a go-it-alone approach resistant to the lessons and policy advances of other countries, both north and south. Moreover, it is hard to find any organic link or mandated coordination that weaves together the three arms of U.S. overseas water engagements, resulting in conflicting policies and practices among the various elements of its defense, diplomacy and development establishments. While offices of the U.S. diplomatic corps highlight the threats of climate change and the military does battle on the ground in societies already experiencing the dislocations of climate impacts, U.S. development assistance programs lack consistency in responding to deepening environmental changes.

This background sets the context for the subsequent sections of this publication, in which we will present fresh analysis of the emerging challenges and at times threats, to U.S. global interests that are tied to the rising pressures of water scarcity and extreme weather events.

Notes

1 Marc Reisner, *Cadillac Desert* (New York: Penguin Books, 1993), pp. 306–323.
2 Thomas Ricks, *Fiasco: The American Military Adventure in Iraq, 2003 to 2005* (New York: Penguin Group, 2007), p. 145.
3 Quoted in Frederick Lorenz and Edward J. Erickson, *Strategic Water: Iraq and Security Planning in the Euphrates-Tigris Basin* (Quantico, VA: Marine Corps University Press, 2013), p. 117.
4 Lorenz and Erickson, *Strategic Water*, p. 118.
5 "World Population," U.S. Census Bureau, last modified July 2015, www.census.gov/population/international/data/worldpop/table_population.php.

6 "World Economic Outlook: Too Slow for Too Long," International Monetary Fund, April 2016, accessed July 2016, www.imf.org/external/pubs/ft/weo/2016/01/pdf/tblparta.pdf.
7 "The World in 2050: Will the Shift in Global Economic Power Continue?" *PWC*, February 2016, accessed July 2016, www.pwc.com/gx/en/issues/the-economy/assets/world-in-2050-february-2015.pdf.
8 Two countries that have recently changed their constitutions to address water governance are South Africa and Brazil. See Ken Conca, *Governing Water* (Cambridge: The MIT Press, 2005), pp. 257–371.
9 William K. Reilly, in Juliet Christian-Smith and Peter H. Gleick, *A Twenty-First Century U.S. Water Policy* (New York: Oxford University Press, 2012), p. ix.
10 Allie Umoff, *An Analysis of the 1944 U.S.-Mexico Water Treaty: Its Part, Present, and Future* (Davis: University of California, 2008).
11 John Harrison, "Boundary Waters Treaty," *Columbia River History Project*, Northwest Power & Conservation Council (Portland, October 31, 2002), p. 1, nwcouncil.org/history/BoundaryWatersTreaty
12 Ibid., p. 1.
13 Emily Green, *Whose Water Is It, Anyway? California Water Rights, Explained*, KCET, April 7, 2015, www.kcet.org/redefine/whose-water-is-it-anyway-california-water-rights-explained.
14 Steven Solomon, *Water: The Epic Struggle for Wealth, Power, and Civilization* (New York: Harper, 2010), www.downsizinggovernment.org/interior/cutting-bureau-reclamation#_ednref47.
15 Bruce Schulman, *From Cotton Belt to Sunbelt* (Durham: Duke University Press, 1994), p. 3.
16 David Ekbladh, *The Great American Mission: Modernization and the Construction of an American World Order* (Princeton: Princeton University Press, 2010).
17 Ekbladh, *The Great American Mission* p. 49.
18 Ekbladh, *The Great American Mission*, p. 60; David Ekbladh, "Mr. TVA": Grass-Roots Development, David Lilienthal, and the Rise and Fall of the Tennessee Valley Authority as a Symbol for U.S. Overseas Development, 1933-1973. In *Diplomatic History*, December 2002.
19 Ekbladh, *The Great American Mission*, p. 71. The longer-term impact of the TVA is reflected in Tanzanian President Julius Nyerere's villagization plans following independence from Great Britain in the 1960s. See James C. Scott, *Seeing Like a State: How Certain Schemes to Improve the Human Condition Have Failed* (New Haven: Yale University Press, 1999), pp. 223–261.
20 "Water Resources Planning Act Passed by Congress." In *CQ Almanac*, https://library.cqpress.com/cqalmanac/document.php?id=cqal65-1257794.
21 Scott, *Seeing Like a State*, p. 224.
22 Solomon, *Water*, p. 338.
23 Solomon, *Water*, p. 338.
24 Reisner, *Cadillac Desert*, p. 9.
25 Reisner, *Cadillac Desert*, pp. 336–342.
26 "Water Resources Planning Act Passed by Congress."
27 David Nye, *Electrifying America: Social Meanings of a New Technology, 1880–1940* (Cambridge, MA: MIT Press, 1990), p. 298.
28 Executive Order 10095, Harry S. Truman, http://trumanlibrary.org/executiveorders/index.php?pid=74.
29 Theodore Schad, "The National Water Commission and the Water Resources Council." *American Water Works Association*, Vol. 63, No.2 (February 1971), p. 109; Paper submitted to National Water Commission Annual Conference, Washington, DC, June 1970, p. 1.

30 Public Law 90-515, September 26, 1968; Sec.3.(a). www.uscode.house.gov/statutes/pl/90/515.
31 Betsy A. Cody and Nicole T. Carter, *35 Years of Water Policy: The 1973 National Water Commission and Present Challenges* (Washington, DC: Congressional Research Service, 2009), pp. 9–10.
32 See Christian-Smith and Gleick, *A Twenty-First Century U.S. Water Policy*, pp. xvii–xxi, for a discussion on the "soft path for water."
33 "UN Conference on Environment and Development," United Nations, revised May 1997, accessed July 2016, www.un.org/geninfo/bp/enviro.html.
34 "The Dublin Statement on Water and Sustainable Development," www.wmo.int/pages/prog/hwrp/documents/english/icwedece.html
35 "Lawmakers Look for Deep Cuts in Foreign Affairs Spending." In *CQ Almanac 1995*, 51st ed., 10-3-10-9. Washington, DC: Congressional Quarterly, 1996, www.library.cqpress.com/cqalmanac/document.php?id=cqal95-1099581.
36 John Norris, "The Clashes of the 1990s," www.devex.com/news/the-clashes-of-the-1990s-83341; *Washington Times*, "Helms Proposal Calls for USAID to be Shut Down," Washington, DC, January 12, 2001.
37 "Lawmakers Look for Deep Cuts in Foreign Affairs Spending."
38 Government Accounting Office, *Foreign Aid Reform: Comprehensive Strategy, Interagency Coordination, and Operational Improvements Would Bolster Current Efforts*, 2009, www.gao.gov/products/GAO-09-192.
39 Ryan Weddle, "What's Next for the 'F' Process?" 2009, www.devex.com/news/what-s-next-for-the-f-process-59510.
40 Diana B. Henriques and Andrew W. Lehrenmay, "Religious Groups Reap Federal Aid for Pet Projects," *New York Times,* May 13, 2007.
41 United Nations Foundation, *What We Do: The Millennium Development Goals*, New York, 2012, www.unfoundation.org/what-we-do/issues/mdgs.html
42 Congressional Record: Volume 150, Part 19. November 20, 2004, p. 25301.
43 U.S. Department of State, Bureau of Oceans, Environment, and Science (2008). Senator Paul Simon Water for the Poor Act Report to Congress. Annex A. June 2008.
44 "Foreign Aid Reform: Comprehensive Strategy, Interagency Coordination, and Operational Improvements Would Bolster Current Efforts," April 23, 2009, www.gao.gov/products/GAO-09-192
45 Erik Peterson and Rachel Posner, *Global Water Futures: A Roadmap for Future U.S. Policy* (Washington, DC: CSIS Press, 2008), p. 15.
46 Hillary Rodham Clinton, Remarks in Honor of World Water Day, Washington, DC, March 22, 2013, www.state.gove/secretary/20092013clinton/rm/2010/03/138737.htm
47 Council on Foreign Relations, "Fact Sheet: US Government and World Bank Memorandum of Understanding on Water Cooperation," New York, March 22, 2011. www.cfr.org/world/fact-sheet-us-government-world-bank-memorandum-understanding-water-cooperation/p24475
48 National Intelligence Community, "Global Water Security," Washington, DC, February 2, 2012, www.dni.gov/files/documents/Special%20Report_ICA%20Global%20Water%20Security.pdf
49 "USAID Water and Development Strategy, 2013–2018," www.usaid.gov/sites/default/files/documents/1865/USAID_Water_Strategy_3.pdf
50 The White House, "2014 Executive Order—Climate-Resilient International Development," p. 3, www.whitehouse.gov/the-press-office/2014/09/23/executive-order-climate-resilient-international-development.
51 The White House, "Preparing the United States for the Impacts of Climate Change," 2013, www.whitehouse.gov/the-press-office/2013/11/01/executive-order-preparing-united-states-impacts-climate-change.
52 The White House, "Incorporating Natural Infrastructure and Ecosystem Services in Federal Decision-Making," 2015, www.whitehouse.gov/blog/2015/10/07/incorporating-natural-infrastructure-and-ecosystem-services-federal-decision-making.

53 Summary: H.R. 2901–113th Congress (2013–2014), www.congress.gov.
54 USAID, "Sustainable Water Partnership Agency for International Development," 2016, www.grants.gov/web/grants/view-opportunity.html?oppId=283058.
55 John Williamson, "What Washington Means by Policy Reform." In *Latin American Adjustment: How Much Has Happened?* Washington, DC: Peterson Institute for International Economics, 1990. John Williamson, "The Strange History of the Washington Consensus." *Journal of Post- Keynesian Economics*, Vol. 27, No. 2 (2004–2005), pp. 195–205, piie.com/publications/papers/williamson0904-2.pdf.
56 Ernest Stern, "Evaluation and Lessons of Adjustment Lending." In *Restructuring Economies in Distress*, ed. Vinod Thomas, Ajay Chhibber, Mansoor Dailami, and Jaime de Melo (New York: Oxford University Press, 1991), pp. 4–5.
57 David Reed, *Structural Adjustment, the Environment, and Sustainable Development* (London: Earthscan, 1996); David Reed, *Structural Adjustment and the Environment* (London: Earthscan, 1992).
58 David Reed, *Economic Change, Governance & Natural Resource Wealth* (London: Earthscan, 2001); David Reed, *Towards a Just South Africa: The Political Economy of Natural Resource Wealth* (Pretoria: CSIR, 2003).
59 Adriana Damianova, *The Role of Public and Private Sectors in Absorbing the Risks Associated With Climate Change in Water Utilities and Water Infrastructure* (Washington, DC, 2016, unpublished report prepared for WWF), p. 15.
60 World Bank, 1993, Improving Water Resources Management.
61 "The World Bank Position on the Report of the World Commission on Dams," The World Bank, accessed July 2016, http://siteresources.worldbank.org/INTWRD/903857-1112344791813/20424179/TheWBPositionontheReportoftheWCD.pdf, p. 1.
62 J. A. Allan, *The Middle East Water Question: Hydropolitics and the global economy* (London and New York: I. B. Tauris Publishers, 2002).
63 The Waters of Babylon are Running Dry," *The Economist,* March 2013, www.economist.com/news/middle-east-and-africa/21573158-waters-babylon-are-running-dry-less-fertile-crescent.
64 James Clapper, director of National Intelligence, *2015 Worldwide Threat Assessment of the US Intelligence Community* (Office of the Director of National Intelligence, February 26, 2015), www.dni.gov/files/documents/Unclassified_2015_ATA_SFR_-_SASC_FINAL.pdf, p. 15.
65 Department of Defense, *Quadrennial Defense Review 2014*, www.defense.gov/pubs/2014_Quadrennial_Defense_Review.pdf.
66 Lorenz and Erickson, *Strategic Water*, pp. 117–119.
67 James Stephenson, *Losing the Golden Hour* (Dulles, VA: Potomac Books, 2007), p. 30.
68 James N. R. Walser, *The U.S. Army Corps of Engineers in Stability Operations* (West Point, NY: United States Military Academy, 2008), p. 44.
69 T. Christian Miller, *Blood Money* (New York: Little, Brown and Company, 2007), p. 43.
70 L. Paul Bremer, *My Year in Iraq* (New York: Simon & Schuster, 2006), pp. 111–112. T. Christian Miller, *Blood Money*, pp. 72–108.
71 Walser, *The U.S. Army Corps of Engineers in Stability Operations*, p. 48.
72 General Anthony Zinni, foreword to *Strategic Water*, p. v.
73 Michael Wood, "Saddam Drains the Life of the Marsh Arabs," *The Independent*, www.independent.co.uk/news/world/saddam-drains-the-life-of-the-marsh-arabs-the-arabs-of-southern-iraq-cannot-endure-their-villages-1463823.html
74 Erin Cunningham, "For Islamic State, Water Is a Weapon," *Washington Post*, October 8, 2014. www.washingtonpost.com/world/middle_east/islamic-state-jihadists-are-using-water-as-a-weapon-in-iraq/2014/10/06/aead6792-79ec-4c7c-8f2f-fd7b95765d09_story.html.
75 Lorenz and Erickson, *Strategic Water*, p. 145.

76 Lorenz and Erickson, *Strategic Water*, pp. 154–156.
77 Hassan Janabi, "Climate Change Impact on Iraqi Water and Agriculture Sectors," *Iraqi Economists Network*, March 2013. www.iraqieconomists.net/en/2013/04/05/climate-change-impact-on-iraqi-water-and-agriculture-sectors/.
78 Brahma Chellaney, *Water: Asia's New Battleground* (Washington, DC: Georgetown University Press, 2011), pp. 251–254; Brahma Chellaney, *Water, Peace, and War: Confronting the Global Water Crisis* (Lanham, MD: Rowman & Littlefield, 2013), pp. 46–55.
79 Muhammed Amin al-Husayni, "The Johnston Mission: Arab Higher Committee for Palestine Rejects Johnston Plan," *Jewish Virtual Library*, August 18, 1955, www.jewishvirtuallibrary.org/jsource/History/ahcrej.html.
80 Ariel Sharon with David Chanoff, *Warrior: An Autobiography* (New York: Simon and Schuster, 1989), pp. 166–167.
81 Mark Zeitoun, *Power and Water in the Middle East* (New York: I.B Tauris, 2012), pp. 44–62.
82 United Nations General Assembly, Human Rights Council, "Human Rights Situation in Palestine and Other Occupied Arab Territories," 24th sess., Agenda item 7, September 11, 2013.
83 Jihan Abdalla, "Israel Denies Palestinians Equal Water Access," *AL Monitor*, April 8, 2013, www.al-monitor.com/pulse/originals/2013/04/westbank-water-restrictions-israel.html, M. Zeitoun, *Power and Water in the Middle East*, p. 70.
84 E. Koek, "Turning Off the Palestinian Taps: Israel's Predominance Over the Allocation and Management of Palestinian Water Resources." In *Water Scarcity, Security and Democracy*, ed. Francesca de Chatel, Gail Holst-Warhaft, and Tammo Steenhuis (Ithaca, NY: Cornell University, 2014).
85 M. Zeitoun, *Power and Water in the Middle East*, pp. 87–98.
86 "Water and Infrastructure," USAID, last updated November 18, 2014, www.usaid.gov/west-bank-and-gaza/water-and-infrastructure.
87 Ibid.
88 "PRESS RELEASE: USAID Awards ANERA Major Infrastructure Development Program in West Bank and Gaza," *ANERA*, July 8, 2013, www.anera.org/resources/press-releases/press-release-usaid-awards-anera-major-infrastructure-development-program-in-west-bank-and-gaza/.

Part II
Water and Social Disruptions

Framing Note: The Social Dimensions of Water

David Reed

Water is foundational. It is the source of all life. Quite simply, access to water determines who lives and who doesn't, who will prosper and who won't. Because it is foundational, water permeates every aspect of society and shapes the many social functions we have created to maintain our health and well-being. We refer to the multitude of ways through which water flows into and shapes our lives as the *social dimensions* of water.

Water is also the principal medium through which our rapidly changing climate will alter livelihoods, standards of living, and relations with neighbors and neighboring countries. As our planet experiences major ecological change, a society's ability to ensure delivery of safe water will either allow that social order to maintain its integrity and cohesion or will reveal and deepen social fractures. Successful management of stressed water resources can strengthen systems of governance and create new economic opportunities. In contrast, deficient water governance can intensify conflicts among water-deprived communities and sharpen social cleavages. Management of water will expose both opportunities and vulnerabilities that can directly and indirectly influence U.S. interests in a country or region. These are the issues that we explore in this section.

In the following chapters, many experts from Central America, Africa, South Asia, Mesopotamia and the Greater Mekong explore the social dimensions of water. These experts examine the ways that water imposes its fluid will on communities, companies and countries without bias or predisposition. To begin, the analysts identify how, in the course of a country's historical development, water-related issues have shaped production systems, institutions and relations among social groups in a broad range of countries of interest to the United States. With that historical reference as a starting point, the analysts highlight significant ecological changes that each country or region is undergoing as a result of governance and management challenges, rising competition and conflict among groups in their societies and climate change. Recurring ecological changes across the many countries we have examined include, not surprisingly, more frequent and intense droughts and floods, stronger and recurring extreme weather events, lowering of water tables and outright collapse of specific ecosystems, among others.

Each chapter then analyzes how those ecological changes are transmitted into the human systems that their respective countries rely on for their prosperity and security. At times, changing ecological conditions disrupt the livelihoods of millions of rural farming families, make providing water to expanding urban areas more difficult and expensive and can destroy major infrastructure for generating and transmitting energy, transporting goods and people and carrying water to fields and farms. Ultimately, those impacts on the productive capacity and institutions of each society reach down to the community and household levels, where individuals and families make decisions about how best to protect their interests and standards of living. Responses include migrating to cities or neighboring countries, confronting governments or powerful economic actors that seem to be depriving people of water access, and joining criminal networks or insurgencies that promise relief from growing burdens.

In a final step, the analysts bring us to understand how this complex set of changes and interactions can affect U.S. interests in those countries or regions. In some countries, U.S. interests are directly affected through growing migration pressures, weakening of close allies' economic productivity and social stability, or eroding of U.S. military and geostrategic engagements. In other regions, U.S. interests are less direct, conveyed through prospects of interstate conflict and rising influence of other major powers.

We open this part of the book with a chapter on the Sustainable Development Goals, notably SDG 6, which envisions how the international community can mobilize and coordinate its resources to ensure that all communities will have access to safe water and sanitation services necessary for productive lives. A key feature of this chapter is the recognition of how the international community's approach to water security and sanitation has changed in recent years, becoming more inclusive and cutting across other development goals.

Thereafter, we delve into the challenges facing four countries in Mesoamerica—El Salvador, Panama, Mexico, and Guatemala—to explore how they are coping with the five years of El Niño–related drought and extreme weather events that have devastated landscapes and rural livelihoods. The stakes for the United States are immediate in that northward migration, coupled with rising urban lawlessness and expansion of drug cartels, can reverberate quickly across borders and into our society.

The second geography we explore is Nigeria in the "Arc of Instability," where water stress is a rapidly rising problem and where U.S. interests are at play as the United States works with governments across the region to contain and suppress insurgencies, including Boko Haram.

From Africa, we shift to the Middle East, a region rife with conflicts in which water remains a central factor, as it has for millennia past, but is now reaching a new frontier with the added impacts of climate change. U.S. military engagements and subsequent state recovery in Mesopotamia and Afghanistan, accompanied now by U.S. interests in the stability of Iran, cannot succeed unless sustained water provision and accountable governance can be ensured, especially in the face of warming trends that make provision increasingly difficult. The neighboring five countries of Central Asia—Kazakhstan, Kyrgyzstan, Tajikistan, Turkmenistan and Uzbekistan—represent a region of contention between the Russian Federation and the People's Republic of China, where U.S. interests reside in promoting a regional water agreement following the collapse of prior water-energy agreements put in place by the Soviet Union.

As we move eastward to South Asia, the water and diplomatic challenges in Pakistan and India rise to heightened levels of intensity. Both countries have been severely damaged by recurrent floods and expanding drought that have affected hundreds of millions of poor farmers across the subcontinent. In addition, recent territorial tensions have spilled over to the Indus Basin Accord, which, while having stabilized water relations over past decades, now threatens to unravel in the face of cross-border incursions. India remains a vital U.S. ally in stabilizing the region in the face of rapidly expanding Chinese influence.

The final geography we consider is the Asia-Pacific region. The Greater Mekong subregion is undergoing unprecedented infrastructure construction of hydroelectric plants, roads and rail lines and port facilities. The region is caught between the two economic powerhouses of India and China, and it holds the key to a future U.S. presence in the Asia-Pacific region. Transboundary water issues are rising as the social and environmental costs and benefits of rapid infrastructure expansion become clearer. The Philippines figures high in the U.S. government's strategic pivot to the Asia-Pacific region despite rising uncertainty about future U.S. diplomatic and military relations. Extreme weather events and drought sharpen water conflicts in Mindanao and fuel insurgent groups affiliated with the Islamic State, adding another element of instability to U.S. relations in the region.

At the end of each country or regional study, we have included a corresponding graphic, *Paths of Influence*, that interprets how ecological changes affecting the geography are transmitted to the country's social systems and influence choices that communities and families make. Any graphic representation is necessarily reductive. That said, we also believe that these graphic interpretations highlight the most salient social impacts and outcomes that ultimately translate into impacts on U.S. security interests.

2 Development and Diplomacy
Water, the SDGs, and U.S. Foreign Policy

Erika Weinthal, Farah F. Hegazi, and Lesha B. M. Witmer

In September 2015, UN member states agreed to a new global agenda: "Transforming our World: The 2030 Agenda for Sustainable Development."[1] Through a more comprehensive approach to sustainable development, this new agenda moved beyond the Millennium Development Goals (MDGs), which provided the groundwork for eradicating extreme poverty and fostering sustainable development, to introduce the five *P*'s: people, planet, prosperity, peace, and partnership.[2] Composed of 17 goals and 169 targets (see Table 2.1), the Sustainable Development Goals (SDGs) offer a bold and transformative framework for linking environment, development, and security.

From 2015 until 2030, the SDGs will shape development policy across UN member states, as governments take steps and measure progress toward meeting the specific SDGs; official development assistance and private investment (including philanthropy) will be directed toward supporting the SDGs. SDG 6 is specifically devoted to water and sanitation, not only calling for universal and equitable access to water and sanitation, but also incorporating water quality, encouraging international cooperation, and protecting water-related ecosystems. For the United States, this water goal will influence foreign policy, as it is relevant for multiple countries of strategic interest including Afghanistan, Pakistan, Ethiopia, Iraq, and Syria. An assessment of global water security by the Office of the Director of National Intelligence (DNI) found that if global water problems—including access to water and sanitation—are not managed successfully, water stresses can "contribute to or aggravate existing problems such as poverty, social tensions, environmental degradation, ineffectual leadership, and weak political institutions."[3] From a security and stability perspective, water interventions are also integrated into U.S. counterinsurgency programming in countries such as Afghanistan to help stabilize population movement, increase food security, abet public health, contribute to energy access, and build confidence in government institutions.[4]

What set the SDGs apart from the MDGs are their universality and interdependence, highlighting how inextricably linked peace, security, environmental integrity, and development are. Through the lens of the SDGs, with a particular focus on water, this chapter highlights their relevance for the crosscutting

Table 2.1 UN Sustainable Development Goals

UN Sustainable Development Goals
1 No poverty
2 Zero hunger
3 Good health and well-being
4 Quality education
5 Gender equality
6 Clean water and sanitation
7 Affordable and clean energy
8 Decent work and economic growth
9 Industry, innovation, and infrastructure
10 Reduced inequalities
11 Sustainable cities and communities
12 Responsible consumption and production
13 Climate action
14 Life below water
15 Life on land
16 Peace, justice, and strong institutions
17 Partnership for the goals

Source: www.un.org

themes of this volume: water and conflict, financing water management systems, and climate change impacts on social stability.

That the SDGs were designed to be universal, interdependent, and comprehensive is their strength; yet it also adds complexity, as implementation will require unprecedented global cooperation and partnerships, creative financing, climate change leadership, and a commitment to human rights and equity by both states and other major stakeholders. The United States has an opportunity to demonstrate leadership in addressing these challenges.

In order to map the role of the SDGs in addressing global water sustainability challenges for U.S. foreign policy, the rest of the chapter proceeds as follows. The next section reviews the evolution of SDG 6 from the previous MDG for water and sanitation, taking into account design mechanisms focused on universality, interdependence, and partnerships. Second, the chapter outlines the main challenges moving forward with implementing SDG 6 and water-related targets. Third, the chapter turns to a discussion of areas where the United States can take steps globally through international cooperation and through partnerships to facilitate the SDGs' implementation. The chapter concludes with a discussion of why it is important for the United States to implement the SDGs.

Moving From the MDGs to SDGs for Water

The Millennium Declaration and the MDGs were adopted by (then) 189 member states at the 2000 UN Millennium Summit with the purpose of improving the lives of the world's population living in poverty through reducing child mortality, improving maternal health, combating major diseases, increasing gender equality, providing universal primary education, and improving environmental sustainability.[5] Although Agenda 2030 references the MDGs, the SDGs embody a fundamentally different agenda: They are designed to be universal, interdependent, and mutually reinforcing.[6] Combined, the SDGs are more expansive, spanning development, cooperation, security, and environment. They cover topics as diverse and complex as ending poverty; improving access to water, sanitation, and energy; combating climate change; reducing inequality; and fostering peaceful and inclusive societies. That they are distinct from the prior MDGs and incorporate the "Rio" agenda implies that governments, including the United States, will need to reconsider strategies and policies at the global and national levels for implementation.

The water and sanitation targets under the MDGs largely emphasized expanding access to improved services, with an emphasis on taps and toilets. Specifically, MDG 7 called for halving the proportion of people worldwide without adequate access to safe drinking water and sanitation by 2015.[7] On March 6, 2012, the United Nations announced that the global community had met the first MDG target of access to improved drinking water in advance of the 2015 deadline, albeit noting that the world community was still far from meeting the target for access to improved sanitation.[8] Practice, however, showed otherwise—the MDGs did not measure the amount and quality of water flowing from taps. Zawahri and colleagues, in particular, found that the Joint Monitoring Programme's (JMP) assessment methodologies, which focused on counting "taps," overstated coverage rates by not focusing on issues of access, affordability, and quality. Thus, even when countries in the Middle East and North Africa were seen to have high rates of "improved" access to drinking water in both rural and urban areas, this did not account for whether they had access to safe drinking water as well as a continuous and affordable supply.[9]

Despite applauding this important milestone for having met the MDG target for drinking water, U.S. policy-makers have recognized the need to address the ways in which the MDGs have fallen short to set new goals to meet the needs of people still living without access to safe water and to improve aid effectiveness;[10] there are still at least 663 million people who continue to rely upon unimproved water sources, with tremendous inequalities across the globe, and 1.8 billion people who use "drinking water sources contaminated with faecal matter."[11] AquaFed, an umbrella organization of private utilities, noted that because the MDG targets emphasized "improved water sources," it was unlikely that the MDG target as originally envisioned was reached, as the organization remained uncertain about the number of people who "have access to water that is really safe."[12]

Furthermore, whereas the MDGs focused on legibility and measurement, the mantra "what you cannot measure, you cannot implement" will no longer be and cannot be the guiding principle; rather, countries will need to develop new methodologies to assess a broader set of indicators that account for qualitative dimensions, such as quality, equity, and accountability. Solving the distribution problem is not enough, and as such, the qualitative dimension of access to water and sanitation will have a large impact on the design and implementation of drinking water programs. Targets 6.1, 6.2, and 6.3 focus on quality along with universal, equitable, and adequate access to safe and affordable drinking water for all by 2030 to ensure that "no one will be left behind." Protecting drinking water sources is a crucial difference between the sets of goals and is especially embodied in Target 6.5.

The SDGs emerged from a consultative and inclusive process using open working groups and stakeholder engagement.[13] In contrast to the MDGs, which were expert driven, the SDGs were designed to be "bottom up," taking into consideration input and criticism from a wide range of stakeholders. Open Working Group 3—specific to water and sanitation—opened the process of developing the dedicated water goal to a variety of stakeholders, including the International Chamber of Commerce, WaterAid, World Wildlife Fund, Women for Water Partnership, and Local Governments for Sustainability, among others.[14]

Goal 17 and Agenda 2030 further demonstrate the inclusive approach of the SDGs, giving very clear guidance on stakeholder involvement in both further deliberations and implementation. To enhance the commitment to implement SDG 6 and water-related targets based on inclusion, transparency, and collaboration, UN Secretary-General Ban Ki-moon and the president of the World Bank Group, Dr. Jim Yong Kim, convened in early 2016 a High Level Panel on Water of 11 sitting heads of state and government, which, in September 2016, issued an action plan to achieve the water SDG.[15]

A human rights approach is also integral to the SDGs, and the human rights to water and sanitation are explicitly mentioned in Agenda 2030. It is widely recognized that women in developing countries often bear the brunt of the limited realization of a human right to water. Because women and girls are primarily responsible for water collection, SDGs 5 and 6 combined will assist to achieve gender equality and the empowerment of all women and girls.[16] A case study from Ghana highlights the importance of gender equality (Target 5.1) and its connections to improving access to water supply (Target 6.1) and girls' participation in education (Targets 4.1, 4.2, 4.5): simply, reducing the amount of time required to walk to a water source by 15 minutes could enhance girls' school attendance by 8%–12%.[17] But this is only part of the equation—not involving women in water management is a serious shortcoming that significantly impacts women's participation in the labor market in the water sector.[18] Studies from cases by Women for Water and the UN University show that access to water (and sanitation) empowers women economically and socially.[19] The combined implementation of SDGs 5 and 6 can lead to implementing the Dublin Statement's Principle 3.[20]

In contrast to the MDGs, not only is water a dedicated goal under the SDGs, but also SDG 6 and other water-related targets in the SDGs—16 directly, and 31 including indirect effects—address the entire hydrological cycle in an integrated manner and cover everything from drinking water availability and sanitation to quality and broader water-related ecosystem health. From a water-management perspective, the implications are sweeping, requiring policy-makers to look at water not only as a local issue, with management taking place at the subnational level, but also as an issue on which national governments should take an integrative approach across levels. Precisely because SDG 6 brings together the environment and development agendas in a comprehensive manner, it too necessitates placing water management within a global context. For example, to meet SDG 6 and other SDGs, policy-makers will need to grapple with multiple causal pathways such as the effects of droughts on migration or how poor water management combined with climate change will affect the hydrological cycle at regional and global scales. In trying to reach the goal on food security, virtual water issues will need addressing.[21] In doing so, policy-makers will need to speak to why cooperation between states matters and the implications of climate change impacts on water availability and quality.

The SDGs' implementation is designed to affect how water is managed at all scales, requiring global and transboundary coordination. Implementing the water SDG will require countries to reexamine current governance structures and stakeholder involvement mechanisms at global, regional, and national levels. Overall, SDG 6 will require transformative solutions that will have implications for development cooperation programs and finance mechanisms, as governments will need to focus on both capacity development and governance mechanisms.

Challenges Moving Forward

Cross-Sectoral Linkages Across Goals

The main challenge in implementing the SDGs so that they support development and facilitate cooperation rather than conflict is contending with the inherent tension in their design. On the one hand, they appear to be dedicated goals when considering their operationalization; yet on the other, they were designed to take a more comprehensive and integrated approach to sustainability and development. Simply put, the goals are intended to be mutually reinforcing. Unlike the MDGs that encouraged "silo" policies, the SDGs require a systemic approach (i.e., integrated policy development) across and within goals.[22]

When working on the water SDG, states will be unable to achieve all the targets if they look at them in isolation. UN Water noted that

> [a] paradigm shift is needed from the focus of the [MDGs] on drinking water and sanitation to recognize that Goal 6 targets are wider in scope, highly interdependent and will require considerably more effort to achieve. To realize "availability and sustainable management of water and sanitation for all," it is essential to manage competing demands for water resources and to exploit synergies between water uses, reuse and recycling, and ecosystem protection and ambient water quality.[23]

The most notable of the Goal 6 targets is to operationalize and include integrated water resources management (IWRM)[24] and transboundary cooperation, offering a clear target on connecting the needs of different nations, sectors, and stakeholders.[25,26]

While SDG 6 focuses primarily on water (and sanitation), it also has links across the other 16 goals, including on poverty reduction, equality, and governance [1, 10, 16]; agriculture [2]; health [3]; education [4]; gender [5]; energy [7]; the economy and infrastructure [8–12]; climate change and resilience [13]; and the environment [14, 15], such that in order to achieve many of the targets, states will also need to meet the targets under SDG 6.[27] This means that organizations charged with implementing water targets will need to have a broader set of expertise and the ability to work across issue areas that may encompass not only water, sanitation, and hygiene (WASH), but also infrastructure, diplomacy, and financing.

Although the SDGs were designed to be interdependent, it remains unclear whether all countries will have the capacity to evaluate water in relationship to the other goals; some countries may need to prioritize some goals over others, depending on their institutional and financial resources. As USAID works in partnership with governments, it will then need to pay attention to conflicts that may arise between governmental agencies and organizations tasked with implementation in integrating these goals/targets, so as not to undermine country programming goals.

An integrated partnership approach will, however, be a challenge for governments, ministries, and UN agencies because governance structures will need to be revamped to enhance policy coherence and implementation coordination: the current way of compartmentalizing and funding water-related issues in silos and projects is unlikely to support effective implementation of national plans to reach the SDG targets.[28] The operationalization of the integrated partnerships approach will require extensive coordination between ministries, organizations, and the private sector. Such complex partnerships will be challenging to manage, especially where there is no prior history of cooperation or coordination; however, they are the only way forward.

Integrating a human rights–based approach is seen as fundamental and underscores not only the importance of the universality of access to water and sanitation, but also the importance of water and sanitation to women's empowerment.[29] Yet in most cases the connection between SDGs 5 and 6 has not been emphasized enough. Women's local and traditional knowledge as water managers should be accounted for, and policies should be rephrased to involve women as actors in water management rather than to depict them as victims and beneficiaries. Implementing Principle 3 of the Dublin Statement, in particular, will be possible only by integrating SDGs 5 and 6 in programming. Access to water is key for economic development. It is only by focusing on more than domestic water use, and addressing agricultural use, use by small and medium-sized enterprises, and the role of women in all water use that gender equality will be achieved.

Because the SDGs are integrative and crosscutting, there are several areas where conflicts might arise and trade-offs will need to be adjudicated.[30] For example, meeting Targets 6.3 (water quality), 6.4 (sustainable water use), and 6.6 (ecosystems) could create challenges for achieving the energy, work, industry, and urban sustainability SDGs (7, 8, 9, and 11, respectively). Meeting those goals will involve technical and infrastructure solutions designed to realize multiple goals without jeopardizing the achievement of others.

Fragile States, Conflict, and the SDGs

What was notable about the MDGs and their success rates was that of the 34 countries furthest from reaching the MDGs, 22 were in or emerging from conflict.[31] Governments (including the U.S.) and other organizations designing interventions for the water SDG will have to pay attention to the variation in both state institutions/capacity and whether countries are conflict-affected or emerging from conflict, since achieving the SDGs will be a greater challenge in post-conflict settings.[32] Because the MDGs were tailored to development with a focus on poverty alleviation, their implementation was difficult in settings following conflict, owing to the breakdown of infrastructure and the need to rebuild institutions.

The SDGs will need to grapple with the challenges that fragile states are likely to face in meeting them, especially for water and sanitation because nearly 1.2 billion people live in conflict-affected and fragile states, and about 70% of fragile states have seen conflict since 1989.[33,34] Technical solutions will not be enough due to the lack of governance structures. If progress is to be made on the SDGs in fragile states, greater attention will need to be paid to the linkages between governance and fragility, including governance and public (basic) service provision. More so, as lessons from examining the viability of building effective, sustainable water management in post-conflict settings have shown, governments and organizations will need to consider how to sequence interventions, which may be challenging given that the SDGs are designed to be comprehensive rather than a pick-and-choose arrangement.[35]

What is often taken for granted is that once a country has introduced improved water and sanitation, then it has met its target, as was the case with the MDGs. Yet, what the conflicts in the Middle East show is that even when countries were at the forefront of providing water and sanitation, political conflict can set some countries years back and result in development in reverse.[36] Whereas the JMP concluded in 2008 and again in 2010 that, with few exceptions, the Middle East and North Africa were on track toward meeting the targets,[37,38] Syria, for example, is in many ways starting over with the SDGs. Prior to the start of the civil war, access in Syria to improved water and sanitation stood at 90.1% and 95.7%, respectively.[39] Today, the media have reported that "access to safe water" has decreased by 50% and that almost "half of the total production capacity" has been "lost or damaged."[40,41] In addition, and highlighting the interconnectedness of the SDGs, the targeting of power plants during the conflict has resulted in reduced water supply as, for example, pumping stations come offline.

Because large numbers of the population may be displaced and in refugee camps that lie outside of traditional state institutions, providing access to water and sanitation may look vastly different than in a peaceful state. Rebuilding states while introducing the SDGs will likely create additional challenges for donors and state institutions in considering forms of financing since it may be harder to attract foreign investment and engage in public-private partnerships.

Providing assistance in the water sector has been essential to rebuilding peace and security in conflict-affected countries such as Afghanistan and Iraq. For example, improving water service provision in Iraq helped increase trust in government as long as the benefits of improvement were "equitably distributed."[42] Moreover, the importance of providing water and sanitation has also been integrated into military policy as part of the United States' counterinsurgency strategy.[43]

The SDGs have a strong focus on peace and building peaceful societies (SDG 16). For SDG 6 to be relevant for mitigating conflict and promoting effective U.S. foreign policy in conflict-affected states, implementation may require fundamentally different solutions that are context-specific so as to leverage water to build peaceful societies.[44,45]

Further complicating water management in conflict-affected countries is the absence of baseline data, owing to years of war. In Afghanistan, for example, data on river flows and groundwater tables have not been collected because of 30 years of continued conflict.[46] For fragile/conflict-affected countries, fostering water cooperation requires investing in rebuilding data collection systems and ensuring that countries share water data. Negotiations led by the UN Environment Programme (UNEP) to foster cooperation between Afghanistan and Iran over the Sistan wetlands to cope with drought, mismanagement, and population displacement between 2005 and 2006 focused on bolstering information sharing; ultimately, negotiations collapsed because of increasing political insecurity in the region.[47] Thus, to implement many of the water targets (including those connected to other SDGs), U.S. development assistance will need to invest in rebuilding water information systems.

As the SDGs aim to take an integrated approach to development, they can be used as a solid starting point for linking state fragility, governance, water management, and economic development. Research on civil conflict illustrates that less-developed countries are more likely to experience civil conflict.[48] Sound water management can provide an avenue to increase economic development because "water insecurity acts as a drag on global economic growth,"[49] which emphasizes "the importance of investment in water security for development."[50] Because water security can increase economic development, and more economic development is associated with a lower conflict risk, we underscore the importance of providing new insights and evidence about fragile states, water access/management, and how fragile and conflict-affected states are approaching the SDG challenges and trade-offs.

The dynamic between fragile states and water access has been a key concern of U.S. overseas programs. The SDGs provide an opportunity to engage more holistically in water management. To date, much of U.S. development assistance has focused directly on access to water and sanitation services. For example, in 2013, USAID "invested" about 58% of its budget in WASH activities, with the remainder allocated to water resources management, water productivity, and water-related disaster risk reduction.[51] Because the SDGs take an integrated approach to development and environment, U.S. development assistance programs will need to use a more comprehensive approach to water management issues, shifting away from short-term project-level access goals to investment in long-term sustainability, governance, and financing.

Climate Change

Climate change is one of the greatest threats to the successful implementation of many of the SDGs, and if not addressed will not only undermine the ability of countries to meet SDG 6 specifically, but could also exacerbate conflict where water is scarce. Globally, climate change is found to shift temperature and precipitation patterns, creating greater variability and extremes, especially affecting the hydrological cycle, and thus affecting livelihoods. Climate change impacts on water will make implementation of SDG 6 and other related SDGs even more difficult, especially since nearly 80% of the world's population already faces serious threats to water security that climate change will exacerbate.[52]

Consider the case of Ethiopia, which in 2016 was struck by severe drought, reducing agricultural output and affecting nearly 10 million people.[53] Managing water in Ethiopia across SDG 6 requires taking action on climate change (SDG 13) to build resilience; yet, despite implementing food aid programs and an early warning system in response to prior droughts and famine, the magnitude of change in precipitation requires planning for surprises such as the 2016 drought that put additional stress on financial and institutional resources in Ethiopia.

The DNI's 2012 assessment of global water security noted that over the next decade, countries vital to U.S. national security would likely experience water shortages, floods, and poor water quality owing to climate change impacts on water availability; combined, these effects are likely to increase instability, state failure, and regional tensions.[54] The confluence of climate change impacts on water availability and quality (i.e., protracted drought) and instability is likely to affect implementation of SDG 6 in the Middle East, for example. At the same time that Israel has expanded its use of desalination as a source of drinking water, thereby reducing its vulnerability to climate change,[55] populations and villages within the West Bank found their water supplies severed in summer 2016 owing to the combination of a heat wave and increased consumption, compounded by Israeli restrictions on the West Bank's ability to develop water resources and infrastructure.[56] Given the growing water crisis in the West Bank and Gaza, implementing SDG 6 may require not only technical interventions but also new international efforts to elevate water negotiations in the Middle East between Israel and Palestine.

Further highlighting the linkages between individual SDGs is the connection between cities, climate change, and water. The World Bank notes that by 2030 the urban population may increase by 2.5 billion people, a majority of whom will be in Africa and Asia.[57] The combination of population growth and climate change is expected to increase the number of people who will experience seasonal water shortages to 1.9 billion by 2050. Urbanization is also expected to increase the demand for water by 50% to 70% over the next 30 years.[58] The increased demand, compounded by climate change effects on water supply, creates "fears of a perfect storm in which water shortages combine with periodic climate disasters to produce major social and economic disruptions."[59] The increased demand for water and the effects of climate change on decreasing supply can thus be constraints on meeting other SDGs, such as SDG 11, which aims to "make cities and human settlements inclusive, safe, resilient, and sustainable."[60]

In September 2016, the intersection of urban growth, increasing demand, and shrinking water supplies resulted in violent protests over the Indian Supreme Court's decision to implement a water-sharing plan that would require the Indian state of Karnataka to release a greater share of the Cauvery River to the neighboring state of Tamil Nadu; protests erupted in Bengaluru, aimed largely at the Tamil population and their businesses.[61] Devising effective U.S. foreign policy to address the rise of such intrastate water disputes will require recognizing the linkages between increasing urbanization, climate change, and water access as outlined in the SDGs.

Taking into account these crucial interlinkages with other SDGs, including the goals on cities and climate change, will also be vital to advance the health and education of populations in these growing urban environments.

Global Conflict and Cooperation

Addressing SDG 6 will also require that U.S. engagements situate programming within larger global water frameworks and connect to a wide range of intergovernmental environmental agreements. Many of the challenges facing the global community, from climate change to biodiversity protection to desertification, all have a water component. SDG 6 calls for implementing IWRM at all levels, including through transboundary cooperation as appropriate. In many ways, SDG 6 will require new architectures for linking global conventions and agreements so as to reinforce the SDGs rather than undercut their implementation.

Whereas the MDG on water and sanitation heavily focused on WASH, the SDGs also recognize the need to situate action within global conventions. For water, this means national water management decisions must engage the UN Watercourses Convention. It should not go unnoticed that this is the first time ever that transboundary cooperation on water has been mentioned; it was unthinkable even five years ago. The SDGs provide an opportunity to mitigate water conflict by elevating the importance of transboundary water cooperation and creating opportunities to use water as a tool for cooperation and peacebuilding, which UN Deputy Secretary-General Jan Eliasson discussed at a General Assembly High Level side event,[62] especially in basins prone to conflict. Although the U.S. has been hesitant to sign international treaties, it has supported efforts to enhance transboundary cooperation (e.g., on the Mekong River). The U.S. will need to consider whether it will continue to engage globally in this area, as increasingly (and despite a legacy of institutionalization over water basins) new conflicts are coming to the fore—e.g., in South Asia, Central Asia, the Middle East, and East Africa.

The SDGs thus offer a renewed opportunity to engage water cooperation in Central Asia, as water cooperation over the Amu and Syr Darya rivers has never been just about water, but also about energy and agriculture.[63] Through emphasizing the linkages across the SDGs, opportunities may arise not only to revitalize the water cooperation that faltered in the early 2000s,[64] but to cooperate on broader issues such as ecosystem management, climate change, and infrastructure development.

Most recently, the Indus Water Treaty, which has survived two wars between India and Pakistan, has been under threat. India has chosen to "suspend" "meetings between the Indus commissioners of both countries,"[65] a significant move given the stability that the Indus Water Treaty has provided over the past six decades. Suspending the treaty would give India full control of the Indus River, which would undoubtedly affect Pakistan's water security. As new conflicts arise in transboundary basins, the emphasis that the water SDG places on cooperatively managing transboundary rivers creates an opportunity to reevaluate the extent to which former river basin organizations are resilient and the extent to which climate change will place additional stressors on them.

The DNI considers transboundary river management in its assessment, focusing on the Indus, Brahmaputra, Jordan, Mekong, Nile, Tigris-Euphrates, and Amu Darya rivers. It assesses the management capacity in these river basins, which ranges from moderate to inadequate, and cites expected impacts such as a reduction in food security as products of poor water management, among other causes. Despite these problems, the assessment concludes that water-related conflict between countries "is unlikely during the next 10 years."[66] While this is a significant conclusion, it is important to recognize that riparian countries in many of these transboundary basins have experienced civil conflict, and although they have been classified as post-conflict countries—e.g., Democratic Republic of the Congo in the Nile basin and Iraq in the Tigris-Euphrates basin—there is a risk of conflict relapse.[67,68] Iraq offers an example of conflict recurrence that is relevant to U.S. foreign policy. Significantly, the Islamic State has captured "key upper reaches of the Tigris and Euphrates . . . on which all Iraq . . . depends for food, water and industry."[69] The group has also used water as a weapon of war.[70]

Additionally, while there are many water-related treaties (e.g., Ramsar Convention, UN Convention to Combat Desertification, Convention on Biodiversity), very few mechanisms exist for coordination between them, and their mandates may not be sufficient to address the crosscutting nature of the SDGs. The vertical hierarchy of the current convention approach to managing the global environment has already been noted to stifle coordination on biodiversity and climate change issues, for one. At World Water Week in 2015, members were pointing to the need for measures addressing climate change to be more integrated with water management in the preparations for Paris. While scholars and practitioners all understand these connections, the design of agreements with specific secretariats provides another obstacle to building linkages across issues.

Overall, there is great need for member states to design and agree on an intergovernmental mechanism to discuss and review SDG 6 and other water-related targets at a global scale. Further implementing Target 17.14, which calls for improving policy coherence, is more likely to occur if there is a global coordination mechanism that includes stakeholders from all sectors—that is, government, civil society, and the private sector.

Partnerships and Innovative Financing

The SDGs are designed to create new partnerships to address the three dimensions of sustainable development—social, economic, and environmental—and governance. Yet, owing to the complexity of the agenda, this will require organizations to move beyond their traditional ways of providing assistance; coordination, consultation, and partnerships will be required.

Partnerships will be central to SDG implementation, and the potential to forge partnerships around these goals could be beneficial for achieving universal access to water and sanitation because the responsibility to fulfill the goal need not wholly fall on states. Public-public and public-private partnerships can provide additional sources of funding, which is especially important because water and sanitation infrastructure is costly to implement. Partnerships with nongovernmental organizations can reduce the financial and bureaucratic pressure on states to deliver services. The U.S. recognizes the importance of partnerships for progress, as evidenced by USAID's Development Innovation Ventures, which creates "partnerships with innovators and social entrepreneurs."[71] One example of this is a $17 million collaboration with the Gates Foundation "called WASH for Life to promote cost-effective and scalable WASH services."[72] Another example is the U.S. Water Partnership, which "will allow USAID to increase donor focus and coordination, pool resources and collaborate on solutions, and increase project sustainability through improved information sharing, monitoring, and evaluation."[73]

For the SDGs to achieve their aspirations, it will be critical for the U.S. to foster partnerships across government agencies that not only address energy, water, and climate, but also deal with state fragility, housing, and infrastructure, as well as to partner with stakeholders outside of government in the non-state and private sectors. U.S. policy-makers in the water sector should continue to deepen partnerships with regional water institutions such as the African Minister's Council on Water and the Lower Mekong Initiative.[74]

There is widespread recognition that financing will no longer come from governments only but must be leveraged from domestic and private sector investment. Yet, at the same time that U.S. foreign policy in the water sector will need to leverage private sector investment, research on post-conflict countries finds that investment by the private sector in the water sector lags behind other areas of investment in post-conflict situations, owing to the perceived political risks to achieving a financial return on these investments.[75] In these situations, donors and development banks will need to partner with national governments to rebuild war-torn infrastructure that is essential for basic services and livelihoods.

Furthermore, ensuring that measures addressing SDG 6 targets do not negatively impact other targets will require the design of programs and projects to take place across sectors: for example, drinking water programs have to be based on joint criteria for potable water and water resource protection. As a consequence, governance structures will have to be redesigned for both management and accountability to facilitate such partnerships. Donors will have to be motivated to agree to "cross" funding that allows for different actions to be taken and results to be measured in terms of output and outcomes.

The Importance of U.S. Involvement in Supporting SDG 6

The water SDG provides an excellent opportunity to highlight ways that the U.S. government can use the SDGs to cut across and influence other goals, notably energy, food, climate change, gender equality, and building peaceful societies—all of which are vital for U.S. foreign policy.

Improving access to water and sanitation is in the "strategic interest" of the United States because "[e]very $1 invested in safe water and sanitation yields an economic return of between $3 and $34."[76] As the largest bilateral donor in the water and sanitation sector, contributing $523.8 million in FY 2013,[77] the U.S. government recognizes the importance of investing in WASH and water management. The Senator Paul Simon Water for the Poor Act (WPA) of 2005 and follow-up legislation in the form of the Senator Paul Simon Water for the World Act (WWA) of 2014 both "ma[ke] water, sanitation, and hygiene (WASH) a U.S. foreign policy priority," which "sets principles for WASH projects to achieve maximum impact by focusing on" countries with the "greatest need and opportunity."[78] While the emphasis in the WPA and the WWA is on prioritizing countries based on need, USAID includes a category of strategic priority countries, "places in which USAID anticipates continued WASH programs due to a combination of strategic considerations and development needs."[79] The six strategic priority countries are the West Bank/

Gaza, Jordan, Lebanon, Yemen, Afghanistan, and Pakistan.[80] In January 2016, USAID released its list of priority countries to receive development assistance for WASH activities in FY 2016. Of the 13 countries selected, four are on the strategic priority countries list (Afghanistan, Jordan, Lebanon, and the West Bank/Gaza).[81]

Integrating the SDGs into a U.S. water strategy will, however, require orienting current U.S. overseas water engagements away from the predominant focus on WASH activities and toward investing in integrative water policies that account for water quality, river basin management (subnational and transboundary), and ecosystem health. This will require communication across all U.S. government agencies involved in water activities, including but not limited to USAID, MCC, NASA, USGS, DOE, DOT, and the EPA. Addressing climate change impacts on water and water management in different political contexts will, furthermore, require a more holistic and cross-sectoral approach to water management that is inclusive and participatory. Ultimately, demonstrating a commitment to implementing the water SDGs at home and abroad contributes to U.S. prosperity and security interests.

Notes

1 United Nations General Assembly (UNGA). 2015. *Transforming Our World: The 2030 Agenda for Sustainable Development A/RES/70/1*. Available at https://sustainabledevelopment.un.org/post2015/transformingourworld.
2 UNGA 2015.
3 Defense Intelligence Agency. 2012. *Global Water Security*. Office of the Director of National Intelligence, p. 4. Available at www.dni.gov/files/documents/Special%20Report_ICA%20Global%20Water%20Security.pdf.
4 Palmer-Moloney, L. J. 2013. "Water's Role in Measuring Security and Stability in Helmand Province, Afghanistan." In *Water and Post-Conflict Peacebuilding*, eds. E. Weinthal, J. Troell, and M. Nakayama. London: Earthscan, pp. 211–235.
5 United Nations. 2009. *The Millennium Development Goals Report 2009*. New York: United Nations.
6 UN-Water. 2016. *Water and Sanitation Interlinkages Across the 2030 Agenda for Sustainable Development*. Geneva.
7 UN 2009, p. 45.
8 WHO/UNICEF. 2012. "Millennium Development Goal Drinking Water Target Met—Sanitation Target Still Lagging Far Behind," Joint News Release, March 6. Available at www.who.int/mediacentre/news/releases/2012/drinking_water_20120306/en/.
9 Zawahri, N., J. Sowers, and E. Weinthal. 2011. "The Politics of Assessment: Water and Sanitation MDGs in the Middle East." *Development & Change* 42: 1153–1178.
10 USAID. 2012. *Global Waters*. Vol. 3, Issue 2.
11 United Nations Economic and Social Council (ECOSOC). 2016. *Progress Towards the Sustainable Development Goals Report of the Secretary-General*, p. 10. July 24. Available at www.un.org/ga/search/view_doc.asp?symbol=E/2016/75&Lang=E.
12 AquaFed. Undated. "Safe Drinking Water." Available at www.aquafed.org/Water Issues/Entry/item/safe-drinking-water--1.sls.
13 United Nations Department of Economic and Social Affairs (UN-DESA). Undated. "Open Working Group on Sustainable Development Goals." Available at https://sustainabledevelopment.un.org/owg.html.

14 UN-DESA. Undated. "Water and Sanitation." Available at https://sustainabledevel opment.un.org/index.php?page=view&type=9502&menu=1565&nr=3.
15 High Level Panel on Water (HLPW). 2016. *Action Plan.* September 21.
16 UNGA 2015.
17 UNESCO. 2015. *The United Nations World Water Development Report 2015: Water for a Sustainable World.* Paris: United Nations Educational, Scientific and Cultural Organization (UNESCO). Available at http://unesdoc.unesco.org/images/0023/002318/231823E.pdf.
18 The ILO concluded in the World Water Development Report 2016 that of all paid jobs in the energy and water sector, only 17% are filled by women. See www.unesco.org/new/en/natural-sciences/environment/water/wwap/wwdr/2016-water-and-jobs/.
19 Schuster-Wallace, C. J., K. Cave, A. Bourman-Dentener, and F. Holle. 2015. *Women, WaSH, and the Water for Life Decade.* Hamilton, Ontario, Canada: United Nations University Institute for Water, Environment and Health and the Women for Water Partnership.
20 *The Dublin Statement on Water and Sustainable Development.* Available at www.wmo.int/pages/prog/hwrp/documents/english/icwedece.html.
21 Brabeck-Letmathe, P. 2013. "Water Scarcity and Food Security the Role of Virtual Water Trade." Available at www.water-challenge.com/posts/water-scarcity-and-food-security-the-role-of-virtual-water-trade.
22 Adams, B. and K. Judd. 2016. *Silos or System? The 2030 Agenda Requires an Integrated Approach to Sustainable Development.* Global Policy Watch #12. September 23. Available at www.globalpolicywatch.org/wp-content/uploads/2016/09/GPW12_2016_09_23.pdf.
23 UN-Water 2016, p. 11
24 "IWRM is a process which promotes the coordinated development and management of water, land, and related resources in order to maximise the resultant economic and social welfare in an equitable manner without compromising the sustainability of vital ecosystems." See Global Water Partnership: www.gwp.org/Global/The%20 Challenge/Resource%20material/IWRM%20at%20a%20glance.pdf.
25 UN-Water 2016.
26 IWRM is an empirical concept that was built from the on-the-ground experience of practitioners. With its roots in the first global water conference in Mar del Plata in 1977, it was not until after Agenda 21, the World Summit on Sustainable Development (WSSD) in 1992 in Rio, and the WSSD in Johannesburg in 2002 that the concept was made the object of extensive discussions as to what it means in practice.
27 UN-Water 2016.
28 Adams and Judd 2016.
29 See for example the report of the Special Rapporteur on the human right to safe drinking water and sanitation, Sept. 2016 http://ap.ohchr.org/documents/dpage_e.aspx?si=A/HRC/33/49.
30 UN-Water 2016.
31 UNDP. 2010. "Armed Violence Threatens Progress on Millennium Development Goals." Available at www.undp.org/content/undp/en/home/presscenter/pressreleases/2010/05/11/armed-violence-threatens-progress-on-millennium-development-goals.html.
32 Troell, J. and E. Weinthal. 2014. "Harnessing Water Management for More Effective Peacebuilding: Lessons Learned." In *Water and Post-Conflict Peacebuilding* (pp. 405–469), eds. E. Weinthal, J. Troell, and M. Nakayama. London: Earthscan.
33 International Finance Corporation. 2015. *IFC in Fragile and Conflict Situations.* Available at www.ifc.org/wps/wcm/connect/aad96f804f36e47f9be0df032730e94e/AM2014_IFC_Issue_Brief_FCS+.pdf?MOD=AJPERES.
34 g7+. 2016. "New Deal Document." Available at www.g7plus.org/en/new-deal/document.

35 Troell and Weinthal 2014.
36 Collier, P., L. Elliot, H. Hegre, A. Hoeffler, M. Reynal-Querol, and N. Sambanis. 2003. *Breaking the Conflict Trap: Civil War and Development Policy*. Washington, DC: World Bank and Oxford University.
37 World Health Organization and United Nations Children's Fund Joint Monitoring Programme for Water Supply and Sanitation (JMP). 2008. *Progress on Drinking Water and Sanitation: Special Focus on Sanitation*. UNICEF, New York and WHO, Geneva.
38 World Health Organization and United Nations Children's Fund Joint Monitoring Programme for Water Supply and Sanitation (JMP). 2010. *Progress on Sanitation and Drinking-Water 2010 Update*. Available at www.unicef.org/media/files/JMP-2010Final.pdf.
39 JMP. 2016. "Data & Estimates." Available at www.wssinfo.org/data-estimates/tables/.
40 Vidal, J. 2016. "Water Supplies in Syria Deteriorating Fast Due to Conflict, Experts Warn." *The Guardian*, September 7. Available at www.theguardian.com/environment/2016/sep/07/water-supplies-in-syria-deteriorating-fast-due-to-conflict-experts-warn.
41 *Al Jazeera*. 2015. "Red Cross: Water Being Used as Weapon of War in Syria." September 2. Available at www.aljazeera.com/news/2015/09/red-cross-water-weapon-war-syria-150902114347090.html.
42 Brinkerhoff, D. W., A. Wetterberg, and S. Dunn. 2012. "Service Delivery and Legitimacy in Fragile and Conflict-Affected States." *Public Management Review* 14: 273–293.
43 Palmer-Moloney 2013.
44 Troell and Weinthal 2014.
45 OECD. 2015. *State of Fragility 2015: Meeting Post-2015 Ambitions*. Paris: OECD Publishing, http://dx.doi.org/10.1787/9789264227699-en.
46 Dehgan, A., L. J. Palmer-Moloney, and M. Mirzaeei. 2014. "Water Security and Scarcity: Potential Destabilization in Western Afghanistan and Iranian Sistan and Baluchestan Due to Transboundary Water Conflicts." In *Water and Post-Conflict Peacebuilding* (pp. 305–326), ed. E. Weinthal, J. Troell, and M. Nakayama. London: Earthscan.
47 UNEP. 2009. *From Conflict to Peacebuilding: The Role of Natural Resources and the Environment*. Nairobi: UNEP.
48 Collier, P. and A. Hoeffler. 2004. "Greed and Grievance in Civil War." *Oxford Economic Papers* 56: 563–595.
49 Sadoff, C. W., J. W. Hall, D. Grey, J. C. J. H. Aerts, M. Ait-Kadi, C. Brown, A. Cox, S. Dadson, D. Garrick, J. Kelman, P. McCornick, C. Ringler, M. Rosegrant, D. Whittington, and D. Wiberg. 2015. *Securing Water, Sustaining Growth: Report of the GWP/OECD Task Force on Water Security and Sustainable Growth*. Oxford: University of Oxford, p. 58.
50 Sadoff et al. 2015, p. 19.
51 U.S. Department of State. 2014. *Annual Report to Congress: Senator Paul Simon Water for the Poor Act P.L. 109–121; Sec. 6 (g)(2)*. Available at www.state.gov/documents/organization/229278.pdf.
52 IPCC 2014, pp. 248–249.
53 Albers, L., J. van Roosmalen, and A. K. Tura, 2016. "Climate Change and Neonatal Survival: The Case of Ethiopia." *The Lancet Global Health* 4, e236.
54 Defense Intelligence Agency 2012.
55 Feitelson, E., A. Tamimi, and G. Rosenthal. 2012. "Climate Change and Security in the Israeli—Palestinian Context." *Journal of Peace Research* 49: 241–257.
56 Hass, A. 2016. Israel Admits Cutting West Bank Water Supply, but Blames Palestinian Authority. *Haaretz*. June 21.
57 World Bank. 2016. *High and Dry: Climate Change, Water, and the Economy*. Washington, DC: World Bank, p. 28.

58 World Bank 2016, p. 27.
59 World Bank 2016, p. 27.
60 United Nations. Undated. "Sustainable Development Goal 11: Make Cities and Human Settlements Inclusive, Safe, Resilient and Sustainable." Available at https://sustainabledevelopment.un.org/sdg11.
61 Safi, M. and V. Doshi. 2016. "Angry Clashes in Karnataka as India's Water Wars Run Deep." *The Guardian*, September 15. Available at www.theguardian.com/global-development/2016/sep/15/india-angry-clashes-karnataka-water-wars-run-deep-tamil-nadu.
62 Eliasson, J. 2016. "Amid Growing Demand, Cooperation over Finite Water Resources Key to Ending Conflict, Deputy Secretary-General Tells General Assembly Event." Available at www.un.org/press/en/2016/dsgsm1009.doc.htm.
63 Weinthal, E. 2002. *State Making and Environmental Cooperation: Linking Domestic and International Politics in Central Asia*. Cambridge, MA: MIT Press.
64 Bernauer, T. and T. Siegfried. 2012. "Climate Change and International Water Conflict in Central Asia." *Journal of Peace Research* 49: 227–239.
65 Kugelman, M. 2016. "Why the India-Pakistan War Over Water Is So Dangerous." *Foreign Policy*, September 30. Available at http://foreignpolicy.com/2016/09/30/why-the-india-pakistan-war-over-water-is-so-dangerous-indus-waters-treaty/.
66 DNI 2012, p. iii.
67 Collier, P. and N. Sambanis. 2002. "Understanding Civil War: A New Agenda." *The Journal of Conflict Resolution* 46: 3–12.
68 Walter, B. F. 2010. *Conflict Relapse and the Sustainability of Post-Conflict Peace*. World Development Report 2011 Background Paper.
69 Vidal, J. 2014. "Water Supply Key to Outcome of Conflicts in Iraq and Syria, Experts Warn." *The Guardian*, July 2. Available at www.theguardian.com/environment/2014/jul/02/water-key-conflict-iraq-syria-isis.
70 Vidal 2014.
71 USAID. 2013. "Millennium Development Goals: On the Crest of Success, USAID Focuses on Challenges." Available at www.usaid.gov/global-waters/june-2012/millennium-development-goals.
72 USAID 2013, www.usaid.gov/global-waters/june-2012/millennium-development-goals.
73 USAID 2013, www.usaid.gov/global-waters/june-2012/millennium-development-goals.
74 U.S. Department of State. Undated. "Programs & Partnerships." Available at www.state.gov/e/oes/water/programspartnerships/index.htm.
75 Schwartz, J., and P. Halkyard. 2006. *Post-Conflict Infrastructure: Trends in Aid and Investment Flows*. Public Policy for the Private Sector Note No. 305. Washington, DC: World Bank.
76 U.S. Department of State. 2009. *Senator Paul Simon Water for the Poor Act: Report to Congress*. Available at www.state.gov/documents/organization/125643.pdf.
77 U.S. Department of State 2014.
78 USAID. Undated. *Senator Paul Simon Water for the World Act of 2014*. Available at /www.usaid.gov/sites/default/files/documents/1865/WfW_fact%20sheet_2.27.TH_.pdf.
79 USAID. 2014. *Water and Development Strategy: Implementation Field Guide*. Available at www.usaid.gov/sites/default/files/documents/1865/Strategy_Implementation_Guide_web.pdf.
80 USAID 2014.
81 Dyer, B. 2016. "USAID Changed Its Water and Sanitation Priorities and It Makes a Lot of Sense." *Global Citizen*, January 26. Available at www.globalcitizen.org/en/content/usaid-announces-priority-countries-for-water-and-s/.

3 Climate Variability, Water and Security in El Salvador

Herman Rosa

Introduction

El Salvador has experienced unprecedented record-breaking extreme weather events since 2009. Intense- or heavy-precipitation extreme events during 2009–2011 were followed by drought in 2012–2015. Drought has impacted the livelihoods of vulnerable rural families and has had significant impacts on the agricultural sector. Extreme precipitation also had a severe impact on road infrastructure sectors, with broader impacts on the national economy. Looking ahead, rising temperatures and even more severe extreme weather events will further threaten staple food crops such as corn and beans, thus threatening food security; export crops, particularly coffee; and costly transport infrastructure.

While extreme weather–related loss and damage have been significant and have further compounded the security issues faced by El Salvador, these losses will most likely worsen and increase unless El Salvador advances very quickly in restoring its highly degraded environment. Large-scale landscape restoration under a homegrown approach, if adequately supported, can not only achieve that goal but also help advance other critical goals, such as rebuilding the country's torn social fabric and developing climate-resilient communities and an economy that provides opportunities for all. The alternative path toward permanent economic crisis, social breakdown, insecurity, and continued large-scale out-migration to the United States is not at all desirable. Therefore, achieving those goals should be a common endeavor for both countries.

Extreme Weather: Evolution, Trends, and Impacts

In the past decade, El Salvador suffered from successive extreme weather events that triggered intense or heavy precipitation during 2009–2011 and drought during 2012–2015. The 2015–2016 El Niño, one of the strongest on record since the last century (Figure 3.1), exacerbated the impacts of the drought, particularly in the eastern part of the country.

During the 2012 rainy season, the spatial variation in precipitation for July was so extreme (Figure 3.2a) that it varied between two points separated by just 217 kilometers (135 miles) from 564 millimeters in a coffee-growing

El Salvador: Climate, Water, Security 57

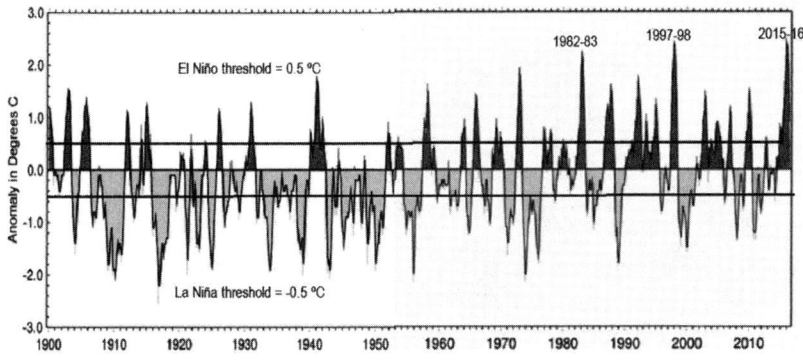

Figure 3.1 SST Anomaly in Niño 3.4 Region (5N–5S, 120–170W)

Note: El Niño (La Niña) is a phenomenon in the equatorial Pacific Ocean characterized by sea surface temperature (SST) anomalies in the Niño 3.4 region (5°N—5°S, 120°W—170°W) that is above (below) the threshold of +0.5°C (-0.5°C).

Sources: National Centers for Environmental Information/National Environmental Satellite, Data, and Information Service/National Oceanic and Atmospheric Administration

region in the west (Juayúa) to just 5 millimeters in La Unión, the eastern port in the Gulf of Fonseca (Ministerio de Medio Ambiente y Recursos Naturales [MARN], 2016a).

During the 2013, 2014, and 2015 rainy seasons, the eastern part of El Salvador again experienced many consecutive days without rain. In 2014 and 2015, other parts of the country were also significantly affected (Figure 3.2b). Average national precipitation for July 2014 was just 95 millimeters, well below the 1981–2010 average of 291 millimeters and the lowest since 1971 for the month of July. In 2015, national precipitation for the first quarter of the rainy season (May through July) was 523 millimeters, also the lowest since 1971 (MARN, 2016a).

These prolonged dry spells depleted soil moisture, negatively affecting maize crops from small producers. During the 2014 drought, according to the UN Office for the Coordination of Humanitarian Affairs (OCHA), almost two-thirds of basic grain producers registered crop losses, with 82% of those affected in eastern El Salvador reporting having completely lost their crops (OCHA, 2014). In 2014 and 2015, Guatemala and Honduras also suffered from drought. As shown in Table 3.1, this left 3.3 and 3.5 million people in 2014 and 2015, respectively, in need of humanitarian assistance in the Northern Triangle countries of Central America (OCHA, 2014, 2016).[1]

Given those impacts, members of many rural families have migrated as a coping strategy.[2] In El Salvador, the drier municipalities, which more frequently suffer the impacts of drought, have also been areas with significant out-migration and where a large percentage of households are recipients of remittances. As remittances became an important source of income and agriculture a less

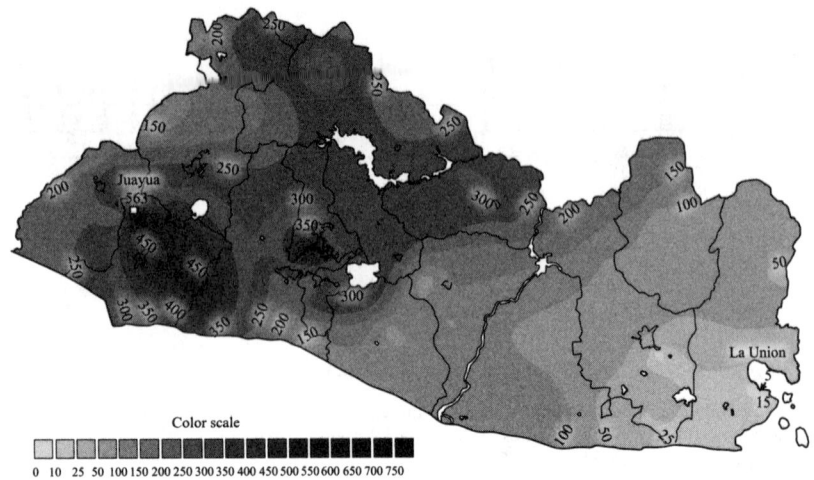

Figure 3.2a Variation in Precipitation, July 2012

Source: MARN, 2016a

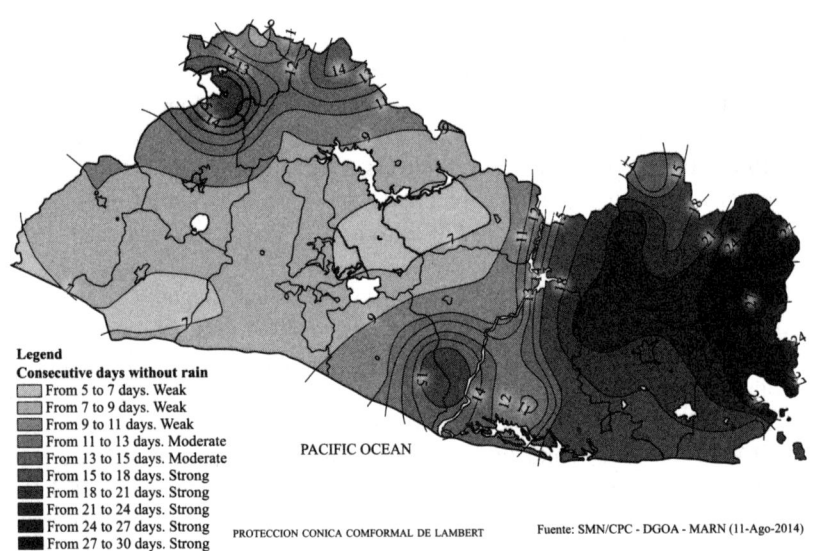

Figure 3.2b Consecutive Days Without Rain, July 4–August 3, 2014

Source: MARN, 2016a

Table 3.1 Drought-Affected Population in El Salvador, Guatemala, and Honduras Requiring Humanitarian Assistance, 2014 and 2015 (million)

	2014	2015
El Salvador	0.8	0.7
Guatemala	1.5	1.5
Honduras	1.0	1.3
Total	**3.3**	**3.5**

Source: OCHA (2014, 2016)

attractive proposition, some of these municipalities turned into reforestation hotspots (see Figure 3.3). According to Redo et al. (2012), woody vegetation in El Salvador increased 586 square kilometers from 2001 to 2010, or almost 3% of the surface of the country.

In contrast, from 1990 to 2000, in wetter areas where migration levels have been lower and shaded coffee—the largest tree cover in El Salvador—is grown, 247 square kilometers of shaded coffee were cleared. Urbanization was responsible for most of the clearing, particularly in the west and central regions (Blackman et al., 2006).

Low coffee prices and large losses in production due to irregular or extreme precipitation, as well as the coffee rust epidemics, could be accelerating the deforestation of shaded coffee areas. During the 2009/10 crop year, coffee production fell 26% in comparison with the previous year (Figure 3.4), in part due to intense precipitation in a coffee-growing area in the middle of the country when a low-pressure system (Low E96) triggered such intense precipitation that it established a six-hour record of almost 300 millimeters (12 inches) at the San Vicente volcano (Figure 3.5a). Although production recovered in the following year, it fell 38% during the 2011/12 crop year when coffee-growing areas were impacted by heavy precipitation under tropical depression 12E. Over 10 days (October 10–19), 762 millimeters of rain, or 42% of average annual precipitation for the period 1971–2000, fell over the country, and one location in a coffee-growing area registered 1,513 millimeters (almost 5 feet), setting in both cases all-time records (Figure 3.5b). Production hardly recovered in the following year and then fell again by 41% during the 2013/14 crop year due to the coffee rust epidemic and irregular rainfall patterns. During the 2015/16 crop year, production was only 30% of the production achieved in the 2010/11 crop year (Figure 3.4).

The successive shocks brought about by extreme precipitation, drought, and the coffee rust epidemic have resulted in a massive breakdown of rural livelihoods that has most likely fueled migration from areas that traditionally had lower migration levels. Extreme precipitation events have also destroyed assets and livelihoods in urban areas, thus likely also fueling international migration from urban centers.

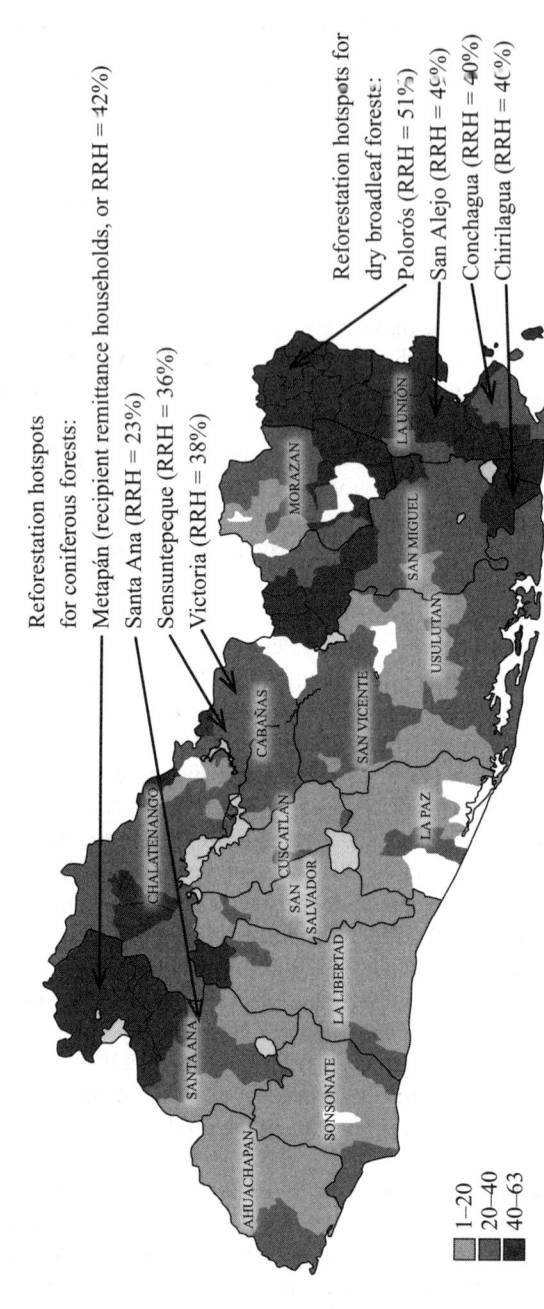

Figure 3.3 El Salvador-Recipient Remittance Households by Municipality in 2004 (% of total households) and Reforestation Hotspots Between 2001 and 2010

Sources: Rosa, Kandel, and Cuéllar (2005) for remittances and Redo et al. (2012) for reforestation hotspots

El Salvador: Climate, Water, Security 61

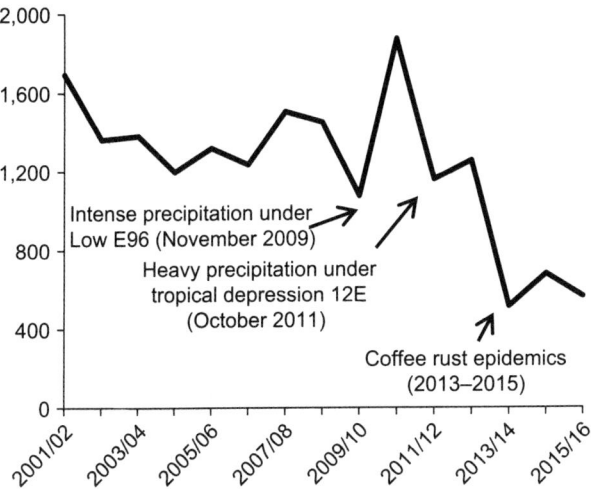

Figure 3.4 El Salvador Coffee Production, Crop Rears 2001/02 to 2015/16 (thousand 60-kilogram bags)
Source: International Coffee Organization

Extreme precipitation events—those that trigger more than 100 millimeters of rain in 24 hours and 350 millimeters in 72 hours—increased in frequency from one in the 1960s and 1970s to four in the 1990s and eight in the 2000s. In addition, events triggered by tropical cyclones and low-pressure systems from the Pacific Ocean that directly affect the largest urban concentrations in El Salvador increased from one in the 1980s (Hurricane Paul in 1982) to four in the 2000s, including Low E96 in 2009 and Tropical Storm Agatha in 2010. The most destructive event—tropical depression 12E in 2011—also formed in the Pacific Ocean. The temporal distribution of these events has also expanded from just September until December mid-1980s to almost every month in the May–November rainy season, except August, since the mid-1990s (Figure 3.6).

Recent extreme precipitation events had severe economic impacts (GOES-CEPAL, 2011):

- In November 2009, intense precipitation under Low E96 saturated the soil on the slopes of the San Vicente volcano and triggered a major landslide that buried dozens of families. The sudden, massive surge of water in rivers downstream washed away a settlement in the city of San Vicente, collapsed 24 bridges, and damaged another 55 bridges. The death toll reached 198, and 122,000 people were affected. Economic loss and damage reached $315 million, or 1.4% of gross domestic product (GDP).
- In May 2010, Tropical Storm Agatha established a 24-hour record for precipitation (483 millimeters, or 19 inches) on the border with Guatemala, destroying an international bridge over the Paz River. In addition, it

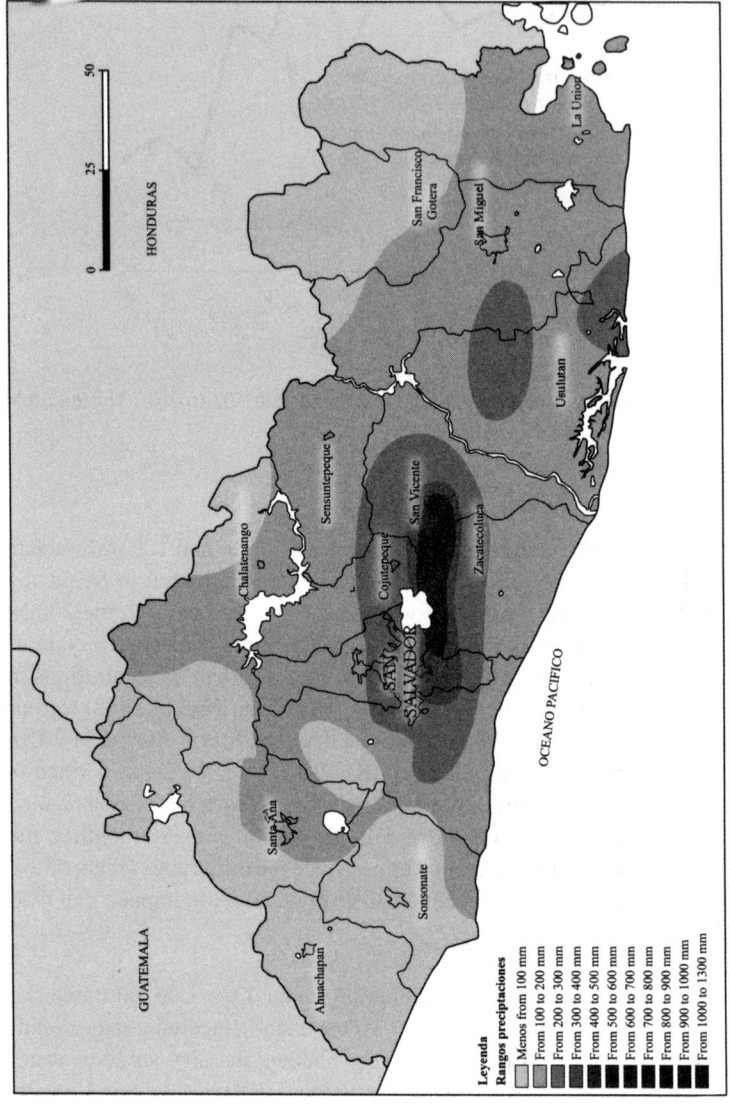

Figure 3.5a Spatial Distribution of Precipitation During Low E96 (November 7–9, 2009)
Source: Ministry of the Environment and Natural Resources, government of El Salvador

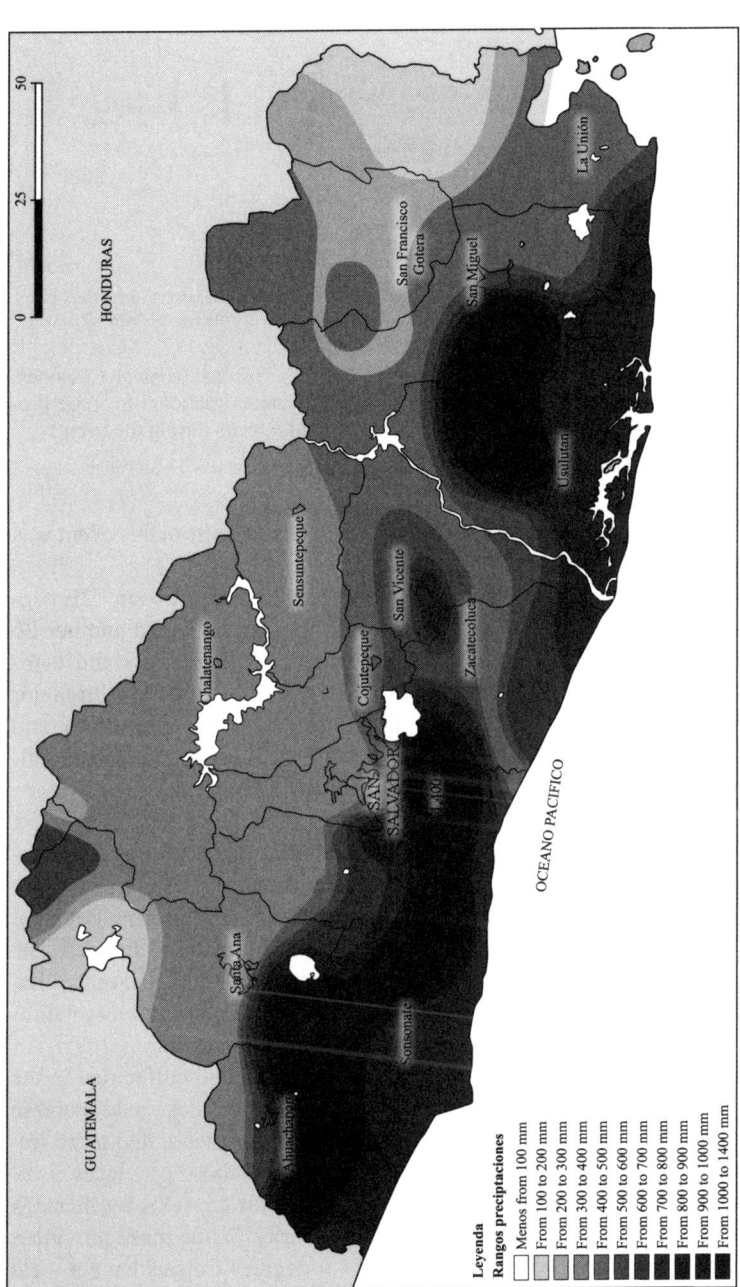

Figure 3.5b Spatial Distribution of Precipitation During TD12E (October 10–19, 2011)

Source: Ministry of the Environment and Natural Resources, government of El Salvador

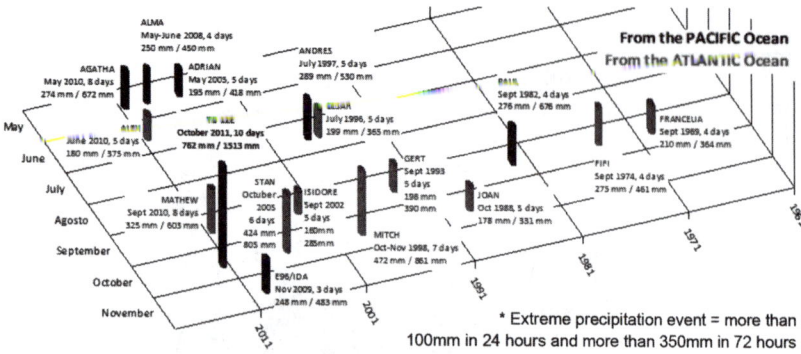

Figure 3.6 El Salvador: Extreme Precipitation Events* Due to Tropical Cyclones, Tropical Depressions, and Low-Pressure Systems (national average precipitation/peak precipitation registered in millimeters during the event)

Source: Ministry of the Environment and Natural Resources, government of El Salvador

damaged 24 other bridges. Economic loss and damage from this event was estimated at $112 million.

- In October 2011, heavy precipitation under tropical depression 12E triggered major flooding, collapsed eight bridges, and damaged another 26 bridges as well as 40% of the road network. Economic loss and damage was estimated at $902 million—more than 4% of GDP—with major impacts in infrastructure ($244 million), agriculture ($174 million), and housing ($141 million). The impact on women was estimated at $162 million due to the destruction of family assets and livelihoods.

For families, the destruction of assets and livelihoods increases their vulnerability. At a macro level, economic losses from frequent disasters in El Salvador add to fiscal pressures and constrain wealth accumulation, lowering potential growth (Calvo-Gonzalez and Lopez, 2015). In the aftermath of disasters, the government has diverted resources to fund emergency response and rehabilitation efforts. Reconstruction investments have been usually focused on rebuilding what has been lost rather than creating new stocks of capital.

While the impacts of extreme precipitation, drought, and coffee rust in the past decade have been severe, the impacts in the next decades could increase due to rising temperatures and a tendency toward more intense and more frequent extreme weather events. For instance, currently those provinces in El Salvador with higher mean temperatures have much lower yields for maize, a staple food for Salvadorans and vital for food security. While those provinces with mean temperatures between 22.5 and 24.0 degrees Celsius have a yield between 2.9 and 3.3 tons per hectare, those provinces with mean temperatures between 26 and 28 degrees Celsius have yields below 2.0 tons per hectare (Caballero, 2014). Thus, with current agricultural practices, maize yields will tend to decrease with temperature rise.

Coffee is also very sensitive to temperature rise. Although its contribution to foreign exchange has been overshadowed by remittances, coffee is still critical for rural livelihoods and for the environment. Many smallholders grow coffee on lots smaller than 5 hectares each; in addition, many families depend on the crop for employment as farm hands and laborers along the coffee processing and export chain (Läderach et al., 2013).

In El Salvador, shaded coffee is by far the largest tree cover; it protects the soil and biodiversity, and it is also critical for water infiltration and regulation. However, with a projected temperature increase of 2.0–2.25 degrees Celsius by 2050, 78% of the current coffee-growing area would be impacted (Läderach et al., 2013). While moving to higher altitudes can compensate for the increased temperature, that is not possible in El Salvador because there are no longer higher altitudes where coffee can be grown. Therefore, an absolute reduction of coffee-growing area is the most likely scenario (International Center for Tropical Agriculture [in Spanish, Centro Internacional de Agricultura Tropical, or CIAT], 2012).

A more immediate issue for coffee is the changing temporal patterns in rainfall. Earlier-than-expected precipitation with dry days afterward can lead to early flowering that is not sustained, thus leading to large loss in production. Likewise, very heavy precipitation before the fruit is collected can also lead to significant loss. Another problem is the outbreak of coffee rust, which led in recent years to massive losses in production, as discussed before. It is yet to be determined if coffee rust is related to a changing climate.

The Migration Option

Migration of Salvadorans to the U.S. skyrocketed from the late 1970s as the country descended into the 1980s civil war (Figure 3.7). It continued in the 1990s, as the economic growth under structural adjustment reforms that began in 1990 and post-war reconstruction after the 1992 peace accords fizzled out and there began a long period of economic stagnation. Migration acquired a new impetus into the present century as Salvadorans were expelled by high levels of violence, a stagnating economy, and weather extremes, or sought to reunite their families in the U.S.

It is extremely difficult to disentangle how those factors come into play. El Salvador's homicide rates have been the highest or the second-highest in the world during most years since 2005. In 2012 and 2013, a truce among the main gangs led to a sharp decrease in homicide rates (Figure 3.8), but still they were the fourth- and third-highest in those years, respectively. In 2014, El Salvador was again the country with the second-highest homicide rate in the world.

According to the U.S. Census Bureau, in 2008 there were approximately 1.1 million foreign-born Salvadorans in the U.S. (Figure 3.9). From 2009 until 2011—when the homicide rates in El Salvador were the highest—there was a sharp rise in that population; it stabilized in 2012–2013, coinciding with the sharp fall in homicide rates, and rose again in 2014, when the homicide rates also increased and the number of Salvadoran immigrants in the U.S. reached

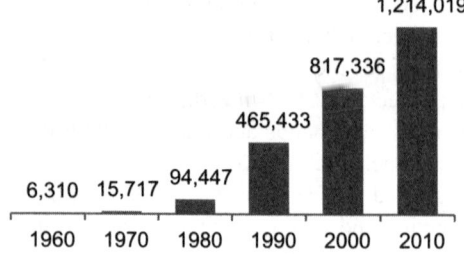

Figure 3.7 U.S. Immigrants From El Salvador, 1960–2010
Source: Migration Policy Institute

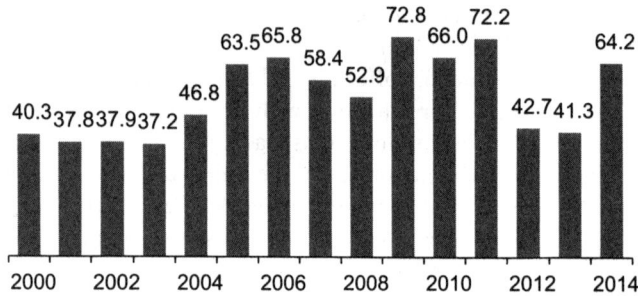

Figure 3.8 El Salvador: Homicide Rate per 100,000 Inhabitants, 2000–2014
Source: United Nations Office on Drugs and Crime

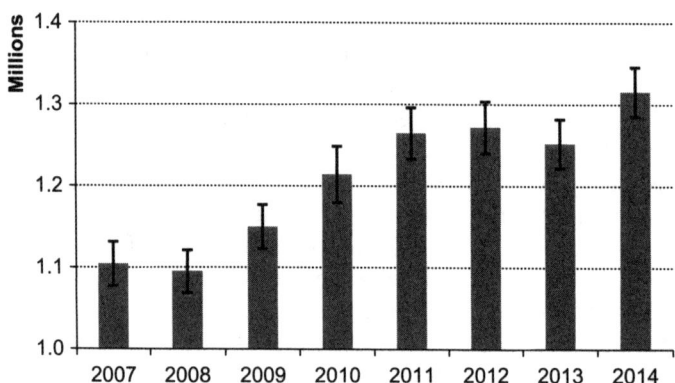

Figure 3.9 Foreign-Born People of Salvadoran Origin Residing in the U.S., 2007–2014
Source: U.S. Census Bureau

around 1.3 million, an increase of almost 20% in just six years. Thus, the level of violence seems to be related to the trends in migration of Salvadorans to the U.S.

However, it should be noted that the highest homicide rates and increases in migration between 2009 and 2011 also coincided with extreme precipitation events that brought havoc and destruction to El Salvador. Likewise, the increase in homicides and migration from 2013 to 2014 coincided with a drought that had a severe impact. Thus, in addition to violence levels, extreme weather events could be playing a role in the migration dynamics.

Economic contraction during the war and economic stagnation afterward have been major push factors for many years. During the 12-year civil war (1979–1991), per-capita GDP contracted 25% (Calvo-Gonzalez and Lopez, 2015). It rebounded to almost 4% per year over 1990–1995 under economic stabilization and structural adjustment reforms, and post-war reconstruction. However, those levels of economic growth were short-lived: The average growth rate over 1996–2001 was less than 1%, and since 2000, annual per-capita GDP has averaged 1.5%, well below the 3.9% growth rates of countries that share characteristics similar to El Salvador's (Calvo-Gonzalez and Lopez, 2015). These low per-capita GDP growth rates have occurred under rapidly declining population growth rates. While the drop in fertility rates from 6.6 in 1965 to 3.4 in 1995 was the most determining factor in population growth in earlier decades, international migration has been the most important factor in the reduction in population growth in El Salvador in the past two decades, to the extent that the growth of the population during 2005–2010 and 2010–2015 was only 1.5% (Figure 3.10).

According to Calvo-Gonzalez and Lopez (2015), low economic growth, violence, and migration reinforce each other. Low growth limits income and

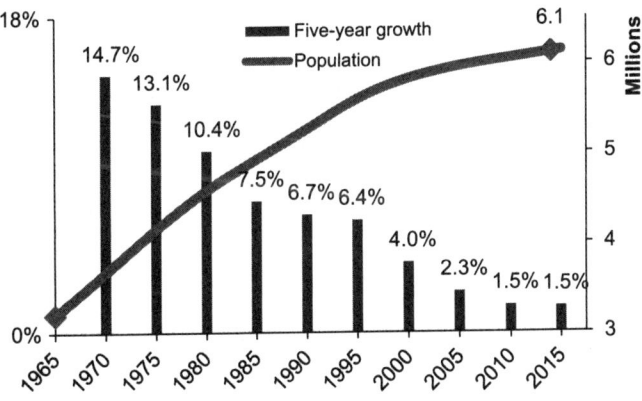

Figure 3.10 El Salvador Population and Fertility Rates, 1965–2015
Source: World Bank

opportunities, therefore creating incentives for some individuals to join a gang or narco group, which may offer significant short-term earning opportunities. In turn, high levels of violence increase security costs and deter investment, lowering productivity growth. The interaction of low growth and violence pushes many Salvadorans to migrate in search of better opportunities, often following relatives who previously migrated for similar reasons, leading in turn to a massive inflow of remittances, amounting to 16% of GDP in 2012. Remittances increase reservation wages and reduce labor participation rates, while low education outcomes prevent the country from moving to high-value-added tradable sectors to offset those Dutch disease forces. Together, all those effects lower the competitiveness of the economy and, ultimately, growth.

Reducing Vulnerability Through Ecological Landscape Restoration in El Salvador

The impacts from the unprecedented extreme weather events during the past decade in El Salvador have been magnified by the severe soil degradation in most watersheds. Because the soils lack the infiltration capacity to regulate water runoff, intense precipitation quickly translates into flooding and infrastructure-destroying water surges in rivers. Lack of infiltration also reduces aquifer recharge and dry season base water flows. With little organic matter and low water retention, crops quickly shrivel even during brief drought periods.

El Salvador is the most deforested country in the Central American isthmus (Table 3.2), and further deforestation and degradation of soils would increase its vulnerability to a changing climate, as well as to negative impacts from climate change. While out-migration may be reversing deforestation in some areas where natural reforestation is taking place, the impacts so far are not significant enough to substantially change the overall dynamics of environmental degradation. At the same time, as mentioned before, El Salvador is rapidly losing its largest tree cover—shaded coffee—and with it, humus-rich soils that are critical for the retention and regulation of water.

Table 3.2 Central American Isthmus Forest Cover by Country, 2010

	%
El Salvador	21%
Nicaragua	29%
Guatemala	37%
Honduras	41%
Panama	45%
Costa Rica	46%
Belize	63%

Source: Redo et al. (2012)

El Salvador: Climate, Water, Security 69

Given the trend toward more intense and frequent weather extremes, increased climate variability, and temperature rise, El Salvador urgently needs to reverse those dynamics of degradation through a massive landscape restoration effort that allows for a rapid increase in permanent vegetation cover and soil organic matter. Since agroecosystems represent three-quarters of the land use in El Salvador (Figure 3.11), the restoration effort has to focus on radically transforming the country's agriculture and on restoring and conserving critical ecosystems, such as mangrove and riparian forests.

The technical groundwork for landscape restoration is well advanced. The Ministry of the Environment of El Salvador designed a national landscape and ecosystem restoration program with a goal to restore degraded landscapes and ecosystems in 1 million hectares, half the territory of El Salvador.

The main focus of the National Landscape and Ecosystem Restoration Program is to establish a more climate-resilient and biodiversity-friendly agricultural system, recognizing that agriculture is heavily based on unsustainable practices such as full tillage, burning off fallow growth, and intensive use of agrochemicals. These practices deplete and contaminate soil and water sources, destroy biodiversity, and make the rural landscape highly vulnerable to climate variability and climate change. Through the expansion of agroforestry, soil and water conservation practices, reduced use of agrochemicals, and the enhancement of pastures and partial stabling of livestock, the expected outcomes are improved soil fertility and water retention, infiltration, and regulation of the water cycle; a significant increase in permanent vegetation cover; enhanced conservation of biodiversity and agrobiodiversity; expanded mitigation of climate change through the sequestration of carbon in soil and permanent vegetation; and the protection of infrastructure such as hydroelectric dams, bridges, and seaports through the reduction of water surges and sedimentation. Complementarily, the program seeks to restore and conserve critical ecosystems,

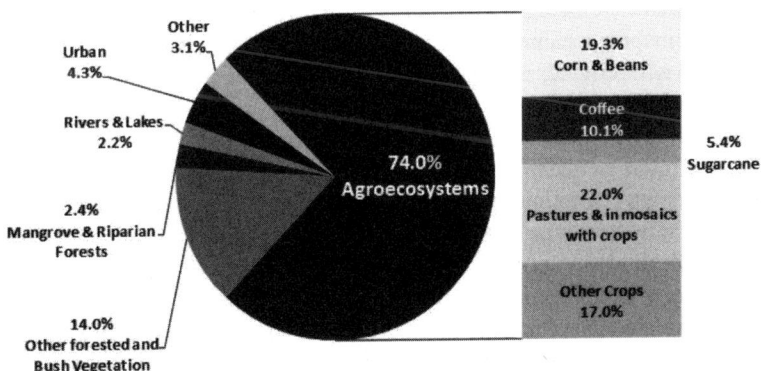

Figure 3.11 Land Use in El Salvador, 2010
Source: MARN (2013)

recognizing that natural forests and agroforests, mangrove forests, and wetlands provide essential ecosystem services for the national economy and local communities. The program recognizes that the restoration and conservation of these critical ecosystems can be sustained only if they are rooted in local practices and inclusive management schemes that link up local communities and governments with national public institutions.

Under the same vision, El Salvador designed its national REDD+ program with financial support from the Forest Carbon Partnership Fund. Given the low percentage of forest cover in the country and pressure from expanding agriculture, cattle farming, and urbanization, El Salvador took a different approach to REDD+ than have forested countries. Adaptation to reduce the adverse impacts of climate change is the primary national goal, and large-scale landscape restoration is the central task. This means reversing widespread degradation, restoring riparian and mangrove forests, and embarking upon large-scale efforts to expand agroforestry systems like shade coffee and cacao, and to reorient cattle farming. In that process, it is expected that carbon sinks will be enhanced and carbon stocks will increase, thus contributing to the global climate change mitigation effort. It is noteworthy that this is the first REDD+ program in the world with an "adaptation-based mitigation" approach.[3] In other words, it is a mitigation program, but its entry point is adaptation. Thus, the REDD+ strategy includes the design and establishment of a monitoring, reporting, and verification (MRV) system to evaluate and quantify both the adaptation and the mitigation outcomes.

This MRV system, which will be much more comprehensive than that required in REDD+ strategies that simply focus on forests, will complement the impressive monitoring capacities that El Salvador has put in place since 2010 to respond to extreme precipitation and drought.[4] El Salvador also recently completed the processing of high-resolution data for the whole country, including a lidar-based digital elevation model with a 1-meter resolution and multispectral photographs with a 40-centimeter resolution, both in 1-square-kilometer mosaics. These data are being used to build a high-resolution functional soil map using soil-mapping technology developed by Purdue University. The functional map will provide, among other data, categorized information such as the highest- and lowest-yielding areas, how much water the soil would store after a rainfall event, and how fast a farmer could expect runoff. While this map can help all farmers better manage their land and crops, and optimize the use of inputs, the high resolution of the El Salvador map makes it an appropriate tool for small farmers who grow their crops on very small plots.

Although the Landscape and Ecosystem Restoration Program responds to a global reality—a changing climate—and has a national vision, it is being implemented with a keen awareness that it has to ground itself in local realities and rely on local leadership and municipal governments to target the most pressing issues identified by local actors and communities. It also takes into account the lessons learned from concrete experiences in Central America.

A particularly relevant experience occurred in the 1990s in one of the poorest and most isolated regions of Honduras, close to the border with El Salvador, where farmers cultivating small plots of land (of 1 to 5 hectares) on hilly terrain adopted an indigenous agroforestry system, called *Quesungual* after the village in which it was first developed. This system combines grain crops with naturally regenerated trees and shrubs, and with high-value/multipurpose timber and fruit trees. Maize is intercropped with sorghum and beans using zero-tillage, mulching, and direct-sowing technologies.

As discussed by Alvarez Welchez and Cherret (2002), under this system erosion nearly stopped, and with it the loss of nutrients, which was estimated to be more than 10 times lower in the Quesungual system than in the slash-and-burn system. The system also retained 15% more water in the soil in the driest month (April) than did the slash-and-burn system, a difference equivalent to 20 millimeters of rainfall, which meant that crops could be sustained 20 more days without rainfall. As a result, during the dry period of El Niño in 1997, Quesungual farmers didn't experience a total loss in maize production as did those who continued the old system of slash and burn. In the following year, many of the latter lost their crops for a second time due to the excessive rainfall produced by Hurricane Mitch, whereas Quesungual farmers produced more or less the same quantity as the year before.

According to Hruska and Welchez (2012), the key to the large-scale adoption of this agroforestry system in southern Honduras was "the role played by local farmers' organizations and support from local government. Once the farmers and their organizations owned the process, they were able to exert pressure on their neighbors in ways that no outside organization can achieve in a sustainable manner" (p. 322).

The Importance of Greater U.S. Involvement in Supporting Landscape Restoration

The U.S. became heavily involved in El Salvador during the 1980s' civil war, and many Salvadorans sought refuge in the U.S. That migration flow, as we have seen, has continued unimpeded until today, and migrants' remittances sustain the economy and the survival of many families in El Salvador. Given the trends discussed earlier, migration to the U.S. will continue to play an important role for a long time to come. However, the forced nature of that process could be reduced if greater efforts are supported in El Salvador to revitalize its economy, rebuild its torn social fabric, and reduce its extreme vulnerability to a changing climate. Taking into account the strong linkages between El Salvador and the U.S., achieving those goals is strategically important for both countries and should be addressed as a common endeavor.

While the outsize role of the U.S. in El Salvador has diminished since the 1980s and early 1990s, this country is still an important partner in current U.S. initiatives. El Salvador is one of the four countries in the world where the

Partnership for Growth Initiative is being supported (the other three countries are the Philippines, Ghana, and Tanzania). El Salvador also benefited from a $461 million compact during 2007–2012 from the Millennium Challenge Corporation and is currently executing a second $277 million compact for 2014–2019. In addition, El Salvador is a key player under the Central American Regional Security Initiative. Cumulatively, since 1980, the U.S. has committed around $10 billion (2013 dollars) to El Salvador in economic and military aid (Figure 3.12).

Therefore, the issue has not been lack of interest, involvement, or even resources. What has been sorely missing is a comprehensive vision that integrates environmental concerns with the economic and security issues that dominate the U.S. agenda toward El Salvador.

Large-scale landscape restoration in El Salvador needs to be seen by the U.S. not simply as an environmental strategy but as a critical intervention to address the economic and security issues faced by El Salvador, with significant spillover effects in the U.S. itself. Beyond its environmental outcomes, such a strategy, if adequately supported, can help rebuild the torn social fabric of the country at local and national levels, and develop climate-resilient communities and economic opportunities for all, thus addressing the root causes of violence, insecurity, and massive out-migration to the U.S.

While extreme weather–related loss and damage have been significant and have further compounded the security issues faced by El Salvador, these problems will most likely worsen and increase unless El Salvador advances very quickly in restoring its highly degraded environment. El Salvador is already moving ahead to showcase its commitment to ecological landscape restoration; however, it is doing so with meager resources and, therefore, far too slowly to really make a difference.

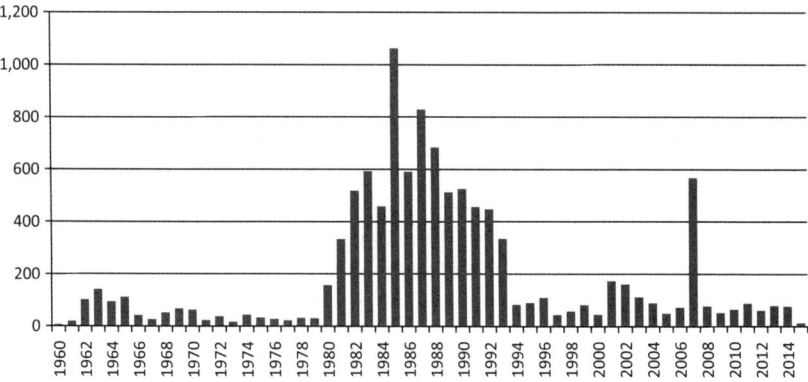

Figure 3.12 Obligated U.S. Economic and Military Aid to El Salvador in Real Terms, FY 1960–2015 (millions of constant 2013 dollars)

Source: USAID Explorer

El Salvador: Climate, Water, Security 73

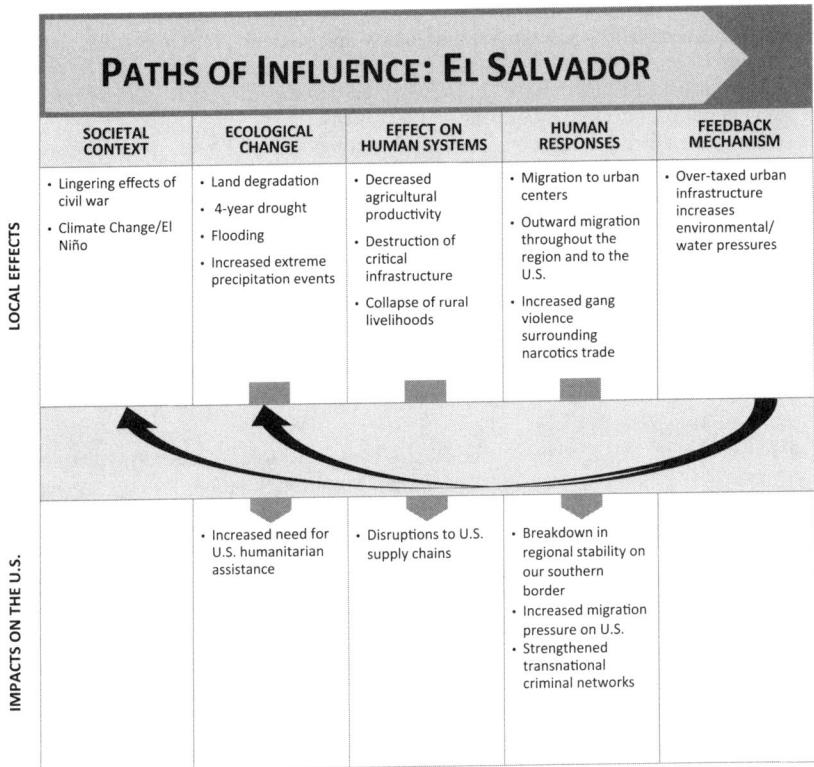

Figure 3.13 Paths of Influence: El Salvador

Note: This graphic reflects the analysis provided by the author. While it may be reductive, it highlights key pathways that link ecological change to broad security challenges that affect U.S. interests.

Given the crisis in public finances, skyrocketing public debt, and the large amounts of resources that are being used to address the more immediate security crisis, the scope for expanding national funding for ecological restoration is small. Therefore, the mobilization of concessional external funding at an adequate scale is vital to keep and significantly expand the momentum of the efforts already under way, and this calls for a far greater role in this area from the U.S.

Notes

1 The Famine Early Warning Systems Network (2016) determined that the number of people facing a food insecurity crisis (Integrated Food Security Phase Classification [IPC] Phase 3) would increase in 2016 until the primera harvests in August 2016, particularly in southern Honduras, eastern El Salvador, and northwestern Nicaragua. The phases under the IPC are 1. minimal; 2. stressed; 3. crisis; 4. emergency; and 5. famine.

2 The 2015 report "Hunger Without Borders" by the World Food Program, the International Organization for Migration, and the London School of Economics established a positive correlation between food insecurity and migration for Guatemala, Honduras, and El Salvador.
3 This phrase was coined by the author when he was the minister of the environment and natural resources of El Salvador (2009–2014).
4 More than a hundred automatic weather stations monitor rain, river levels, and other meteorological variables. Eight local weather radars monitor rainfall patterns throughout the territory, and 17 automatic weather stations that are equipped with soil- and leaf-moisture sensors monitor the evolution of droughts.

Bibliography

Alvarez Welchez, L. and Cherret, I., 2002, "The Quesungual Systems in Honduras: Alternative to Slash-and-Burn," *LEISA* (Low External Input and Sustainable Agriculture) *Magazine*.

Avelino, J. et al., 2015, "The Coffee Rust Crises in Colombia and Central America (2008–2013): Impacts, Plausible Causes and Proposed Solutions," *Food Security 7*, no. 2: 303–21.

Blackman, A. et al., 2006, "Tree Cover Loss in El Salvador's Shade Coffee Areas," July 5, www.rff.org/research/publications/tree-cover-loss-el-salvadors-shade-coffee-areas-0.

Caballero, K., 2014, "Evaluación Del Impacto Del Cambio Climático En La Biodiversidad Salvadoreña," Report for the Ministry of the Environment of El Salvador prepared under the "Economics of Climate Change in Central America" project (CEPAL, CCAD/SICA, UKAID, and DANIDA).

Calvo-Gonzalez, O. and Lopez, J., 2015, "El Salvador: Building on Strengths for a New Generation. Systematic Country Diagnostic" (Washington, DC: World Bank).

CIAT (International Center for Tropical Agriculture), 2012, "Escenarios Del Impacto Del Clima Futuro En Áreas de Cultivo de Café En El Salvador" (Cali, Colombia: Author).

Famine Early Warning Systems Network, 2016, January, "Central America and Caribbean—Key Message Update: Sun, 2016-01-31,"*FEWS NET*, www.fews.net/central-america-and-caribbean/food-security-outlook-update/january-2016-0.

GOES-CEPAL, 2011, October, "Evaluación de Daños Y Pérdidas En El Salvador Ocasionados Por La Depresión Tropical 12E," *Informe Preliminar*.

Hruska, A. and Welchez, L. Á., 2012, December, "Adoption of Conservation Agriculture by Small-Scale Farmers in Southern Honduras," in the *3rd International Conference* (p. 322).

Laderach, Peter; Haggar, Jeremy P.; Lau, Charlotte; Eitzinger, Anton; Ovalle, Oriana; Baca, Maria; Jarvis, Andrew; Lundy, Mark. 2010. Mesoamerican coffee: building a climate change adaptation strategy. Centro Internacional de Agricultura Tropical (CIAT), Cali, CO. 2 p. (CIAT Policy Brief No. 2).

López, G., "Hispanics of Salvadoran Origin in the United States, 2013" (Washington, DC: Pew Research Center, September 2015).

MARN, 2013, "Estrategia Nacional de Biodiversidad" (San Salvador: Gobierno de El Salvador, 2013).

MARN, 2016a, "Cuatro Años Continuos de Sequía en El Salvador" (San Salvador: Gobierno de El Salvador).

MARN, 2016b, "Propuesta de Estrategia de Restauración de Ecosistemas y Paisajes. REDD+ MbA El Salvador 2016–2017" (San Salvador: Gobierno de El Salvador).

Nawrotzki, R., Riosmena, F. and Hunter, L. M., 2013, "Do Rainfall Deficits Predict US-Bound Migration from Rural Mexico? Evidence from the Mexican Census," *Population Research and Policy Review 32*, no. 1: 129–58.

OCHA (UN Office for the Coordination of Humanitarian Affairs), 2014, December 10, "Drought in Central America: Situation Report No. 01," http://reliefweb.int/sites/reliefweb.int/files/resources/Drought%20in%20Central%20America.pdf.

OCHA, 2016, January, "Humanitarian Needs Overview: Central America Sub-Regional Analysis," https://www.humanitarianresponse.info/system/files/documents/files/2016-hno-centralamerica-7jan.pdf

Redo, D. J. et al., 2012, "Asymmetric Forest Transition Driven by the Interaction of Socioeconomic Development and Environmental Heterogeneity in Central America," *Proceedings of the National Academy of Sciences 109*, no. 23: 8839–44.

Rosa, H., Kandel, S. and Cuéllar, N., 2005, "Dinámica Migratoria, Medios de Vida Rurales y Manejos de Recursos Naturales en El Salvador" (San Salvador: Programa Salvadoreño de Investigación sobre Desarrollo y Medio Ambiente).

World Bank, 2015, "El Salvador. Building on Strengths for a New Generation. Systematic Country Diagnostic" (Washington, DC: Author).

World Food Programme, the International Organization for Migration, and the London School of Economics, 2015, "Hunger Without Borders," http://documents.wfp.org/stellent/groups/public/documents/liaison_offices/wfp277544.pdf?_ga=1.115899843.477836633.1491418542

4 Panama

Water Security and Social Conflicts within the Context of Climate Change

Ariel Cuschnir

Introduction

The political and socioeconomic stability of Panama has been critical to U.S. national interests for more than a century, and increasingly so since the canal was returned to Panama on December 31, 1999. Panama plays a central strategic role in the region, particularly regarding world maritime commerce and the war on drugs and terrorism in the Darien area on the border with Colombia. The strategic importance of the Panama Canal for the transport/passage of U.S. commercial goods and U.S. military vessels provides additional examples of the importance of reducing the risk of conflicts in the country. This situation has been recently accentuated with the commercial expansion of China's interests in Central America, including its plans for the potential construction of the "Grand Canal of Nicaragua." The recently completed expansion of the Panama Canal will double its shipping capacity because larger ships and freighters will be accommodated. This expanded traffic has also triggered port expansion activities across the Americas (e.g., in the United States, Colombia, Peru, Brazil) to receive the expanded maritime commercial activity. For Panama, water is a driver of success not only for canal operations but also nationwide, and for the United States, it is a resource that the country must help Panama protect and manage.

Although the stability of the canal and its watershed is of central strategic importance for Panama and for U.S. interests, the Panama Canal Authority has well-developed short- and long-term water management plans and is constantly implementing adaptive management procedures. However, it is actually the rest of the country's water availability, management, and quality and readiness to confront climate change that must be taken into consideration, since these things can have a direct effect on the stability of the region and, as a consequence, indirectly affect the Panama Canal.

The present and future availability of water for human and ecosystem use in Panama is probably less dependent on resource availability (even conservative models predict an abundant water supply for this country) and more dependent on how its population manages its water resources and prepares for unpredictable variations associated with climate change.

Panama's current water resource capacity is well beyond the current demand. The country utilizes only 25% of the available fresh water, yet in vast areas of the country (particularly the northwestern provinces, the famous Arco Seco, and also the eastern portion, such as Darien), the majority of the population lacks access to clean, potable water and sanitation services. Additionally, there is often competition among watershed users (e.g., hydroelectric generation versus diverse downstream users). In the past, this competition has led to notable social conflict and instability. This is of particular importance in view of the unpredictable effects of climate change, which can exacerbate and have a direct effect on regional variations in rainfall across the country, potentially acting as a catalytic driver in creating significant social conflicts.

In contrast to other countries in Latin America, throughout the past 20 years Panama has invested significant resources and efforts in building water resource management (WRM) capabilities. It has been progressive in its preparation for upcoming climate change events, both in human capacity and in technology. More recently, there has been a national effort (development of a National Water Security Plan) to ensure sufficient water quality and quantity for all of Panama's citizens through proposals to heavily invest in new water and sanitation infrastructure across the country.

In addition, Panama has forged collaboration agreements with a broad range of international entities—financial institutions and multilaterals, the U.S. and other governments, non-governmental organizations (NGOs), etc.—detailing adaptation programs to address climate change and WRM. These agreements have helped improve both the capacity of the national ministries in charge of managing these resources and the capacity of the citizenry at large to participate in this process. However, these national and international efforts still confront important implementation obstacles and deficiencies that could present risk for a country of such strategic value to the United States. In turn, this could undermine the positive results already achieved. Such is the case, for example, with illegal logging and heavy deforestation occurring in Darien, and the continued deterioration of water infrastructure in other areas of the country. Helping reduce the risk of water-related conflicts is of critical importance to help maintain stability in Panama, and to reduce the need for further U.S. intervention.

The Importance of Panama's Water Security

Although Panama as a whole enjoys a high abundance of water resources, this condition is variable across the country, and even in areas with abundant water resources, deficient water management and supply, and a decline in water quality, present a challenge to the authorities and a potential risk of conflict among users. Interestingly, it is also the urban areas of modern Panama City that are experiencing water supply problems, primarily associated with systemic losses and frequent interruptions in supply. The Instituto de Acueductos y Alcantarillados Nacionales (IDAAN) is the national entity responsible for water supply, and one of the major problems that this entity and Panama City as a whole face today is the deterioration of water infrastructure. It is estimated that approximately 40% of the water in the system is lost to leaks.

It is important to mention that although many regions of Panama could provide sufficient water to satisfy the demand (for diverse economic and public sectors), the quality of this water is in many cases below levels that can ensure its viability. The MiAmbiente (or Ministry of Environment in English and formerly the Autoridad Nacional del Ambiente, or ANAM) is the ministry in charge of monitoring the quality of water in more than 90 rivers nationwide.[1] According to this entity, water quality in the country is influenced by several anthropogenic sources of impacts that include, among others, chemical pollution, accidental and intentional hydrocarbon spills, agrochemicals, and untreated water sewage.

Other areas of the country, like the Arco Seco (covering the provinces of Veraguas, Herrera, Los Santos, and Cochle), are among the driest in the country, with water resources often insufficient for all users. Recently, this region has experienced social conflicts over this issue. Urban and rural residents of Panama, as well as indigenous populations, are only some of the users demanding better quantity and quality of water. The Chiriqui Viejo River is a prime example of competition among multiple stakeholders (hydropower energy providers, farmers, etc.) generating conflict and, occasionally, violent confrontations in pursuit of the same water resources. Generally, hydroelectric energy in Panama has encountered strong opposition from a variety of stakeholders that include, among others, indigenous peoples, farmers living in areas to be inundated by power plant reservoirs, and communities that live downstream of these dams and fear the impacts these changes will have on their livelihoods and/or water supply.

The concerns of the population in the central part of the country regarding the effects of mining activities on water availability and quality, and regarding the deforestation activities in the eastern Darien Province, have also created conflict among stakeholders who either are direct users or are simply alarmed by the potential impact on water resources and biodiversity. In the Darien area, populated by the Wounaan and Embera tribes, water quality has been impacted by landscape modification and soil degradation due to high levels of illegal deforestation activity. Loss of control over the land and water resources in the Darien region has been the focus of deadly conflicts between local indigenous peoples and "outsiders" (loggers, farmers, guerrilla groups from Colombia, etc.).[2] These conflicts and risks for renewed armed action in the Darien area have implications dating back to the Colombian civil war of the 1990s, which led to a huge refugee movement involving thousands of people and conflicts with local indigenous populations.[3] Conflicts between local populations, refugees, paramilitary groups, illegal loggers, and drug traffickers have created a lawless atmosphere, with negative consequences for the environment and the region's water resources. The U.S. government's Agency for International Development and the Inter-American Development Bank (IADB), among others, have been active in the region, funding diverse programs that promote increased economic opportunity, improved infrastructure and agriculture, and ecotourism, as well as cooperation with indigenous communities (the Embera and Wounaan tribes, the Kuna tribe, and also Afro-Darien tribes). These programs have achieved significant progress. However, more change is needed and security problems still persist.

Climate Change as a Destabilizing Social Driver

Climate change and associated variations in precipitation trends (exacerbated by recent El Niño events), combined with a need for further implementation of water management policies at the national and regional levels, poor sanitation in rural areas, and past and existing conflicts among users,[4] are generating conditions that must be reversed or improved in order to avoid future conflicts over water in Panama. Analyzing future U.S. involvement is vital since a future crisis could have an impact on the level of effort and resources the U.S. government may need to invest in order to help correct the situation versus the level of investment needed to prevent it.

Additionally, WRM in the context of climate change is of importance regarding one of the main economic sectors of the country: the Panama Canal watershed and its influence on global maritime commerce. Gatun Lake and the canal watershed constitute the main water sources for canal operations, Panama City, and some small tribal groups living on the lake's shores (such as the Wounaan and Embera tribes), as well as for the farmers who live on the shores of Gatun Lake and depend on these water bodies for their livelihoods. Due to recent El Niño events, during early 2016 the lake's level was rapidly descending. (In February 2016, it was already more than one meter below its average height for that time of the year. The 2016 rainy season was insufficient, and in August of the same year, the lake was still almost 50 centimeters below its normal levels.) During 2014–2016, the canal had to restrict the passage of certain ship sizes and even suspended electricity generation at the Gatun hydroelectric plant (located on the dam that feeds Gatun Lake) for more than a month so it could use all of Gatun Lake's water supply for maritime passages. Comparatively, past climatic variations also created flood conditions such as the ones that occurred during 2010, when the Panama Canal was required to discharge huge volumes of water from the system due to flooding and record-high water levels.

It is clear that impacts to water resources in Panama associated with climate change variability can have negative social consequences. These could lead to migration processes and increased competition for land, wildfires and/or floods, socioeconomic degradation, and weakening infrastructure and water supply. The crisis could affect not only the way the Panama Canal will operate, but also the steady flow of fresh water for other uses (energy, transport, human consumption, recreation and tourism, natural features, ecosystem functioning, etc.) at various watersheds across the country.

Panama's Advances in Water Resource Management, Climate Change Adaptation, and Climate Resilience Building

Panama's government, both past and present, has invested major effort and resources into developing agencies and programs to confront the challenges of managing a national effort for climate change risk reduction and to better manage the country's water resources. The risk of a potential water-related crisis that could extend into other areas of the country has led the government of Panama, through an interministerial committee, and with the support of the United Nations Development Program (UNDP) as external consultants, to develop the National Water Security Plan with a vision that extends into the year 2050.

The Inter-Ministerial National Water Security Committee is composed of MiAmbiente; the Ministries of Agriculture, Commerce and Industry, Education, and Public Works; the Foreign Ministry; the State Comptroller; the Ministry of Economy and Finance; the Panama Canal Authority; the Public Services Authority; the national energy secretary; the national science secretary; the Electric Transmission Company, S.A.; and additional contributors including the Universidad Tecnologica de Panama, UNDP, and others.

In August 2016, the Panamanian government approved a resolution adopting the National Water Security Plan (PNSH in Spanish) and establishing the National Water Council (CONAGUA) and the technical secretary for water security, and will implement an aggressive action plan across the country with measures that, one way or another, will impact 52 watersheds, including 498 rivers and small streams. The interministerial committee does recognize the importance of establishing clear goals and objectives that include, among others, universal access to quality water, basic sanitary conditions, inclusive economic growth, healthy watersheds, and sustainable hydric conditions.

Although the government has correctly identified water security and water infrastructure as central drivers of the country's development, only the adoption of harmonizing, sustainable measures to achieve these goals will ensure that other problems will be avoided (e.g., social conflicts and impacts on ecosystem services due to unstable hydric systems).

As of today, PNSH includes numerous infrastructure projects and subprojects (water distribution pipelines, water treatment plants, aqueducts, water system controls, wastewater treatment, etc.) aimed at reducing losses and increasing capacity and distribution. The plan also includes a component for the integrated management of national watersheds, as well as watershed restoration and reforestation. However, few of the proposed actions focus on important socioenvironmental elements of adaptation to climate change.

This focus on infrastructure is probably due to the fact that individual ministries (e.g., MiAmbiente) are already involved in climate change–related programs. However, since the actions proposed in the PNSH do not address climate change and WRM in an integrated manner, engineering solutions might not fully address future water demand. There are only a few programs in the plan that address vulnerability and resilience to climate change (such as building data reservoirs about landslides, floods, and droughts). Additional projects address the need for increased technical capacity in local communities and land management education. Building a link between these highly important infrastructure plans of the PNSH and water management resilience to climate change is crucial for reducing risks.

Similarly, the efforts on climate change included, among others, restructuring Panama's own institutions and programs as well as collaborating with international interest holders on a variety of bilateral programs. Some examples include the development of the National Climate Change Program,[5] the transformation of this program into the Climate Change Unit (UCC in Spanish) within the MiAmbiente,[6] the 2011 efforts to fight desertification caused by climate change,[7] the creation of the Proyecto de Agua y Saneamiento en Panama Program of the Ministry of Health (MINSA) to ensure access to clean water in urban and rural communities,[8] climate change adaptation action plans for the conflict-ridden Chiriqui Viejo River,[9] and the REDD+ office within the UCC.

These government entities and others within the Panamanian government have worked to build a national conscience and awareness about the risks of climate change. They've attempted to educate the public on climate change's effect on citizens' way of life and the need to develop adaptation measures and climate resilience-building in communities across the country. For this purpose, they have developed national strategies and action plans, and have reached collaboration agreements with national and international entities that have also invested efforts and resources in their plans. In 2013, the government of Panama reached a very important memorandum of understanding between the ANAM (currently MiAmbiente), the COONAPIP (the national coordination unit for the indigenous peoples of Panama), and other governance entities of indigenous peoples for the improvement of natural environments and WRM in their territories (the *Comarcas*).[10] In 2014, MiAmbiente (then the Autoridad Nacional del Ambiente), together with the UN-REDD+ program, and financed by the World Bank's Forest Carbon Partnership Facility program, provided capacity-training workshops on greenhouse gas inventory and reforestation to representatives of the indigenous peoples, public and private entities, representatives of rural communities, and others.

To strengthen the actions and processes previously mentioned, in 2015, MiAmbiente developed the National Strategy for Climate Change (ENCCP in Spanish) as part of its Strategic Government Plan 2015–2019. The ENCCP constitutes a very important and ambitious effort to address a broad range of climate change issues, including water security and resilience building. Diverse additional programs and national commitments are examples of the country's concern for its future and readiness for action plans to better manage its water resources and to build resilience to climate change.

The ENCCP also addresses issues such as the promotion of low-emission technologies, the development of concepts and plans to increase resilience and adaptation to climate change, the promotion of reforestation programs, and capacity training. However, institutional fragmentation of responsibilities (e.g., Panamanian law dictates that the Ministry of Environment be in charge of national watersheds, while MINSA is responsible for drinking water and sanitation for communities of 1,500 or fewer people and IDAAN is responsible for urban areas larger than 1,500 people), coupled with limited economic resources, prevents this National Strategy for Climate Change from being properly implemented.

These national efforts to better manage and supply fresh water to the population are obviously not sufficient and would benefit from additional U.S. support. The Panamanian government, already aware of the risks for increased social conflicts associated with water availability and quality, has developed a joint effort with the UNDP to address WRM in communities living in areas of the country that have already experienced social unrest.[11] These areas include the Tabasara Watershed, inhabited in the upstream reaches of the river by the Ngabe-Bugle people, and the middle and downstream portions of the same watershed in the provinces of Chiriqui and Veraguas. This joint program also addresses a highly conflicted area of Panama on the border with Colombia, on the Chucunaque River in Darien Province, inhabited by the Embera, Wounaan, Kuna, and other indigenous groups. Activities in these regions included strengthening local organizations, building local capacity to develop local water safety plans, and improving water infrastructure. The IADB has also been heavily involved in Panama as part of its sustainable cities program and recently submitted an action plan for the Panama City metropolitan area that includes, among others, actions to improve water and sanitation conditions in the area. Panama has signed on to the Sustainable Development Goals and the 2030 Agenda of the United Nations Sustainable Development Summit, and has committed to increasing its reforestation efforts along the Panama Canal Watershed and in other areas of the country. However, it is important to emphasize that the positive efforts and plans developed by Panamanian government entities cannot be fully implemented if they are not supported also by sufficient monetary and human resources. As an example, reforestation programs cannot come to fruition if there are no sufficient providers of plantings, seedlings, and other needed components of the program.

The country also developed the National Integrated Water Resources Management Plan of the Republic of Panama 2010–2013, which was complemented by the recently developed PNSH. These plans, together with other national initiatives (e.g., the Ministry of Environment's climate change and adaptation plans included in the National Climate Change Strategy 2015–2019[12]), could be an important initial institutional platform to which U.S.-funded water programs can be linked. The ENCCP includes water management measures such as the development of flood infrastructure controls and reforestation programs (Alliance for the Reforestation of a Million Hectares). It also includes sustainability drivers: the correlation between water, societal development, climate change adaptation, and institutional improvement.

Other positive efforts in improving ecosystem and WRM for the benefit of the environment and for local communities include the Global Water Partnership (GWP) to improve the management of the Rio Indio Watershed.[13] This program succeeded in organizing a community-/municipal-based watershed management program that included technical and financial support for the locals, capacity training, and participation of all age groups within the communities. According to the GWP, there have been four regional hydrogeological studies in the country that attempted to address the groundwater cycle. However, the available information is still limited and the same study noted that Panama's daily per-capita water consumption is among the highest in the world,[14] hinting at a need for better management also on an individual-use level.

Similarly, other WRM programs such as a recent proposal by Fundacion Natura in Panama to obtain funding for the implementation of a climate change adaptation program in the Chiriquí Viejo River and Santa Maria River watersheds either concur with or complement a few of the proposed water management recommendations in the present policy paper (e.g., a multi-sectoral approach for effective solutions to climate change—related risks, finding synergies with national-scale planning processes and plans, and strengthening adaptive capacity at all levels of Panamanian society).

Although what the PNSH previously discussed is of crucial importance and a major step forward, the question remains: will the expected results of this national plan and the selected indicators truly help achieve water security in a manner that will reduce future conflict among users and increase sustainable environmental conditions in a dynamic environment affected by climate change?

Opportunities for U.S. Government Involvement

Panama's water availability, management, and quality, and the country's readiness to confront climate change and its effects on water resources, are factors that must be taken into consideration as part of the regional and global geopolitical stability and U.S. foreign policy. Despite the tremendous efforts by the government of Panama to advance its institutions and population toward a more climate change–resilient environment, the risks previously described still persist, and it would be in the best interest of the U.S. government to increase its support for water programs that can reduce instability and security risks. Helping Panama build resilience to the risks posed by future climate change–related effects on water is vital and in the best interest of the U.S. government. As previously mentioned, a U.S. strategy, and subsequently a policy, oriented toward this goal will also aid in reducing the need for costly climate change impact interventions.

The Panama Canal Treaty of 1977 required Panama to provide sufficient water to operate the canal. This requirement translated into several U.S. Agency for International Development (USAID)–funded activities, which included the creation of national parks for soil and water management, the promotion of sustainable development activities across the canal watershed, and the development and implementation of an Integrated Water Resource Management Program.[15] Since the return of the canal to Panama, the U.S. government has been extremely active, directly and indirectly, in promoting important programs that include antidrug initiatives, military support, trade and development assistance, and others.[16]

In addition, there are several national and international organizations working in Panama on community-based initiatives related to water resource management, climate change adaptation, and reduction of social conflicts. The U.S. government would benefit from finding synergies with some of these organizations (e.g., Fundacion Natura, government of Panama), in order to maximize the effect of U.S.-funded programs.

A U.S. policy that focuses on sustainable development and the inclusion of visible climate change considerations in Panama's new national effort for improved WRM would be a major contribution toward the goal of reducing the frequency of potential conflicts. Providing complementary support for these and other national plans targeting WRM and climate change resilience should be a top U.S. priority in the region. The following recommendations could facilitate this process. Bridging between U.S.-funded programs in water security, the Panama government's Water Security Strategy at the national level, and the existing and planned local programs being implemented by the civil society will go a long way in reducing the potential threats to water security.

Conclusions, Recommendations, and Opportunities for Collaboration

The development and adoption of the PNSH in Panama, as well as the multiple national and international efforts in the country, give the U.S. government a tremendous new opportunity to find compatibility with established entities so it can more safely funnel portions of its development assistance to programs that emphasize long-term sustainability. U.S. assistance must also emphasize resilience to climate change and programs that enhance water security and improve social stability. This U.S. support should be built on a framework of broad water management assistance that can be designed to address the needs of all sectors of the population. Action plans must be implemented across all platforms, from government to the communities, by bridging specific objectives, the infrastructure work proposed in national plans, and the readiness of end users/receptors to execute those plans.

Several documents offering adaptation recommendations have been produced around the world (e.g., the United Nations Framework Convention on Climate Change, or FCCC,[17] United Nations Economic Commission for Europe, or UNECE[18]), but there is no existing standard that can address site-specific problems. Generally, the search for adaptation solutions follows two main approaches: bottom up and top down. As summarized by the FCCC, bottom-up approaches are based on learned experiences from past and present climatic events and recommended adaptation measures that are appropriately scaled. Top-down solutions use forecasted models and scale them down to the conditions found at the site-specific/regional level. This policy paper proposes a combined approach for developing solutions to potential climate change–driven conflicts associated with water resources, a risk management approach that will take into consideration past experiences and forecasting models, but from a risk perspective. That is, it will assess the probability of occurrence of a climatic event and its potential effects on water resources and develop a range of potential outcomes. Then, starting with the worst-case scenario and scaling down, it will develop policies and recommendations for a range of short- and long-term scenarios. Assessing this range of solutions will allow government and decision-makers to adopt measures that are appropriate for a constantly changing area-specific vulnerability (intervention prioritization). This process should be monitored, and proposed interventions modified if needed. In addition, it is necessary to establish a framework for decisions at the local level (needed regulations, etc.) and to adopt integrated risk management into the monitoring of climate impacts.

It is important to validate risk predictions at the site level. Validation of existing threats will take into consideration historical information (bottom-up approach) combined with predictions for the region (top-down approach). This risk-based combination of approaches can aid in developing more stable adaptation/risk-mitigation solutions by overlapping historical threats/events with projected climate change threats to assess the risk that past events might strengthen in the future.

Climate change risk management is already being adopted by U.S. government agencies working abroad. Building on these existing strategies can produce more effective and rapid results. Such is the case with USAID, which in 2015 developed its own strategy on funding efforts,[19] opening opportunities for additional contributions to risk reduction on water-related issues. This strategy also includes the recommendation to conduct screenings utilizing USAID's climate risk-screening tool.[20] USAID has developed a mandatory reference on climate risk management at the project and activity levels for its missions across the world, in order to provide directions on how to screen these funding activities and include considerations for climate change mitigation.[21] In this strategic document, the agency recognizes the potential of climate variability to undermine development promotion by the U.S. government and increase risks and insecurity in developing countries. Addressing this variability in the early stages of project development can provide a strong foundation for increasing the resilience of these USAID-funded projects to the effects of climate change (e.g., changes in rainfall patterns and incidence of vector-borne diseases; effects on agricultural crops, with subsequent impacts on food and water security and livelihoods; changes in water supply; and increased competition among users).

The USAID Mandatory Reference, effective October 2015, requires that climate risk analysis and implementation of appropriate measures be part of the development of all new USAID strategies worldwide (including country strategies). This policy constitutes an important step toward ensuring that nations receiving USAID aid also take into consideration climate risk analysis in their own policies and programs. The USAID strategy uses a risk-based approach in which once-high climate risk elements are identified and must be addressed, by adopting an overall shift in direction if necessary. As an example, if the U.S. government is considering supporting hydroelectric energy development in a region/country considered to be at high risk for altered riverine flow streams, then the funded activities are also considered to be at risk. This proposed risk-screening process is then translated into a strategic level of policy support. The new USAID strategy also recommends not only a climate change risk analysis but also the adoption of climate resilience-building considerations. The document's initial recommendations include adopting clean energy programs and strengthening sustainable landscapes (reducing deforestation or doing compensatory reforestation, adapting to build resilience to water scarcity, conflict, and political instability, etc.).

As previously mentioned, the proposed climate change and water resources management framework should be a risk-based program that would include multiple action items (see the following).

Examples of Potential Action Items in the Proposed U.S. Climate Change Framework for Panama

Work with national entities to improve cooperation

- Work with the government of Panama to help reduce institutional fragmentation and/or to build more effective operational links.
- Significantly expand the role of climate change variability and adaptation measures in all future infrastructure planning and programs in Panama, particularly in areas identified as "high conflict risk." This should include promoting public-private partnerships (PPPs) in conjunction with international financial institutions at the regional and local levels and with local stakeholders.
- The Panama Canal Authority (ACP in Spanish) has taken important steps forward to build up the resilience of the Panama Canal Watershed to climate change by creating the Panama Canal Green Route program, which focuses on the protection of the existing forested areas along the Panama Canal Watershed (including proper land management and water management procedures), reforestation programs for altered ecosystems, and commercial community-based reforestation, which have important positive environmental and social effects. This authority also plays a key role at the national level in activities that go beyond the operational needs of

the canal. Working with ACP as a local partner will ensure a higher probability of success for any adopted U.S. program in Panama.

Work with the private sector, civil society, and NGOs

- Address past disagreements to build trust in the process. The focus of the proposed programs should be on areas/regions either that have experienced conflicts among stakeholders or that the risk analysis determines have a high probability for conflicts to occur. These potential conflicts are, for example, those associated with competition between renewable energy companies and downstream users, and inappropriate government intervention.
- Proactively fund programs that co-manage water resources through public participation in (community-based) government programs supported by international funding.
- Offer solutions that reduce investment risks to the private sector if they help strengthening PPPs.
- Develop and strengthen a common vision with local stakeholders by establishing a clear participatory process (enhance bottom-up decisions), with full inclusion of the various stakeholders in the decision-making process, as a condition to ensure the long-term success of the proposed water and sanitation infrastructure projects.
- Elevate the quality of information that people receive and encourage participation in the management of water resources ("owning the process," "larger involvement of the civil society"). Building community resilience to climate change and water security means also ensuring that the community receives a constant (updated) flow of information on forecasts, predicted risks, threats, and potential solutions.
- Increase people's awareness on the detrimental environmental impacts of their own actions (unsustainable environmental and water resources practices).
- For certain communities, access to water is also directly related to their ways of life and cultural traditions. This means that water supply must take into consideration cultural values as well as ecosystem services provided by neighboring water bodies.

Incentivize Panamanian companies to expand investment in updated U.S. green technologies

- Panamanian society as a whole (both the private and public sectors) has been influenced by U.S. culture, technical advances, and technologies for decades; and Panamanians are notably tuned in to the development of new green technologies, climate change information, and approaches to climate change adaptation and resilience. However, economic conditions and government budget constraints have often limited the ability of national and local institutions to adopt the broad range of green technologies available. Providing economic support to enhance Panama's purchasing capabilities in green technologies (e.g., renewable sources of energy, new wastewater treatment facilities) and smart grids for small electricity sources will contribute to the common goal of achieving water security.

Work to promote commitments to best practices that are linked to the aid received

- Bilateral collaboration between the United States and Panama has existed for many years; consequently, the appropriate channels of communication, as well as existing procedures and frameworks, can form the foundation of an expanded U.S. role in promoting the adoption of climate change resilience measures across Panama. In addition to existing bilateral frameworks, it is important to promote the adoption of international sustainability frameworks and standards that can aid in reducing the risk of conflicts among water users. These include, among others, the Equator Principles and International Finance Corporation Performance Standards (for management/reduction of social conflict risks and impacts to ecosystem services).

U.S. government emergency response to water resources problems created by climate change

- Planning and aiding in the development of adaptation measures and resilience-building for unpredicted climatic variations in rainfall and water supply are good investments for the future. However, U.S. foreign aid to those programs must also be supplemented by the development of contingency plans for emergency conditions that were not previously identified and that could lead to conflicts among water users. This recommendation has already been developed by David Reed,[22] and it should be revisited if it has not yet been implemented by U.S. agencies and departments. In his work, Reed correctly recommends that the U.S. government help the various stakeholders/water users (e.g., mining companies, farmers, water providers) discuss the costs and benefits of each potential water use alternative. In addition, Reed also proposes measures that have been already discussed in the present policy paper but that might have not been implemented yet in Panama.

Build an environment of trust and collaboration

- All the previous recommendations assume that the multiple stakeholders are ready for a collaboration framework that will help them achieve the ultimate goal of building resilience to future climatic variations in rainfall and, subsequently, decreasing competition for water resources. Such is the case for most areas of Panama. However, in order to achieve that goal, it is also important to help reduce existing social conflicts in certain areas of the country (e.g., Darien Province, watersheds in the western part of the country). Without this effort, it will be extremely difficult to build the social interaction and collaboration needed to jointly confront future problems. The U.S. government should expand its involvement in this area by working to reduce tensions between stakeholders, e.g., in Darien Province, where decades of military conflict, unsustainable use of natural resources (e.g., illegal logging), and conflicts with indigenous groups have created animosity that, if left unresolved, will interfere with any plans for climate change adaptation and climate resilience building, as well as national and international plans to ensure water security.

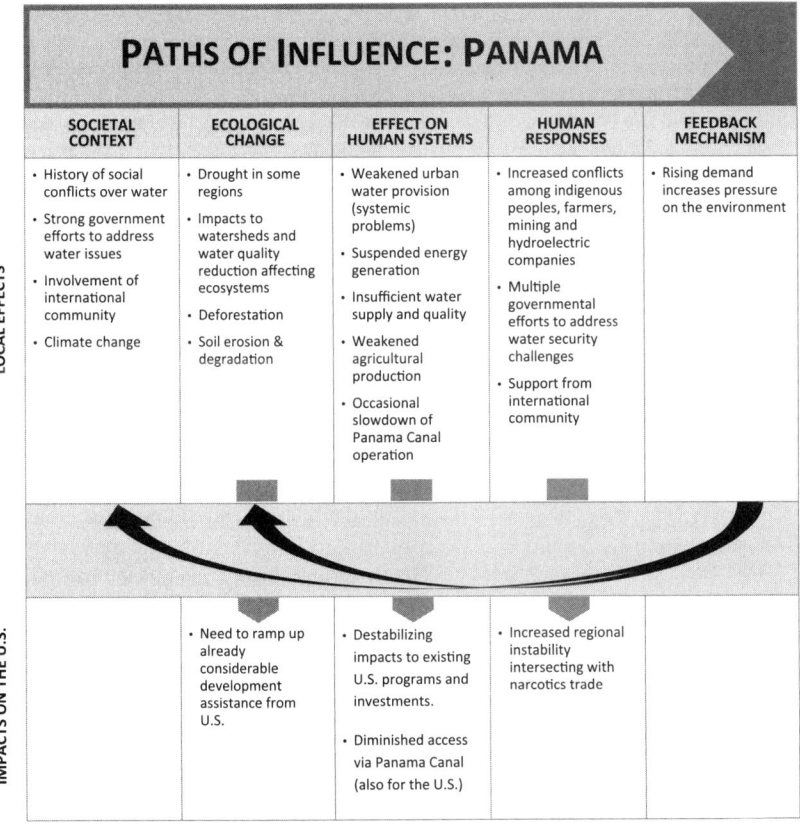

Figure 4.1 Paths of Influence: Panama

Notes

1 MiAmbiente. 2015. Ministry of Environment. www.miambiente.gob.pa/index.php/departamentos/laboratorio-de-calidadambiental/informes-de-la-calidad-ambiental.
2 Kristen French. 2016. The Wounaan people are deploying drones and using GPS technology to get evidence of logging in their customary lands. america.aljazeera.com/articles/2016/2/25/new-weapons-for-panamanian-tribes-in-old-fight-to-save-forests.html.
3 Sara Trab Nielsen. 2006. ICE Case Studies #198. The spillover effect of the Colombian conflict: Ecological Damage in the Darien Gap.
4 In the past, the Panama Canal attempted to create new reservoirs in the western region of the canal watershed to increase its operational capacity and encountered stiff opposition from local stakeholders. This led to the cancellations of the plans and the development of water-recycling basins at the new locks.
5 National resolution No AG-0583–2002.
6 Resolution No. AG-0280 gave the UCC the authority to coordinate issues related to climate change and energy resources. http://miambiente.gob.pa.
7 This was accomplished by creating the Integrated Management Unit of National Watersheds.
8 www.minsa.gob.pa/proyecto/proyecto-de-agua-y-saneamiento-en-panama-pasap.
9 "Programa conjunto de incorporación de medidas de adaptación y mitigación del cambio climático en la gestión integrada de los recursos naturales en cuencas prioritarias de panamá." UNDP/Panamá Joint Program. www.miambiente.gob.pa/images/stories/documentos_fomento/Brochure_Cambio_Climatico_Rio_Chiriqui.pdf.
10 Memorando de entendimiento entre la Autoridad Nacional del Ambiente (ANAM), la coordinadora nacional de los pueblos indígenas de Panamá (COONAPIP), congresos y consejos de las estructuras tradicionales indígenas. 2013. 6pp.
11 UNDP/Panamá Joint Program. "Programa conjunto de incorporación de medidas de adaptación y mitigación del cambio climático en la gestión integrada de los recursos naturales en cuencas prioritarias de panamá." www.pnuma.org/cuencas/.
12 MiAmbiente. 2015. Ministry of Environment. Estrategia Nacional de Cambio Climático para Panamá.
13 Global Water Partnership. 2011. Experiencias Municipales en GIRH: Municipio de Chagres, Panamá. 18pp.
14 Global Water Partnership—Central America. Situación de los Recursos Hídricos en Centroamérica. Panama. GWP, 2015. 38pp.
15 Guillermo H. Castro. Panama–The Management of the Panama Canal Watershed, GWP. Case #5. 11pp.
16 U.S. Government Foreign Assistance Report, fiscal years 1946–2014. United States Agency for International Development (USAID), Green book reports. www.explorer.usaid.gov/reports-greenbook.html.
17 UN. FCCC/TP/2011/5. Water and climate change impacts and adaptation strategies. Technical paper.
18 UNECE. 2009. Guidance on Water and Adaptation to Climate Change. www.unece.org/index.php?id=11658.
19 USAID. 2015. Climate Change in USAID Strategies. A Mandatory Reference for Automated Directive Systems (ADS) Chapter 201. GC/A&A.
20 https://www.usaid.gov/sites/default/files/documents/1876/201mat.pdf
21 USAID. 2015. Climate Change in USAID Strategies. A Mandatory Reference for Automated Directive Systems (ADS) Chapter 201. 23pp.
22 Reed, David. 2015. In Pursuit of Prosperity. U.S. foreign policy in an era of natural resource scarcity. Ed. David Reed. 362pp.

5 Mexico's Pursuit of Water Security

Román Gómez González Cosío

I. The Challenge: A Looming National Water Security Crisis

Mexico has experienced two critical junctures over the past 75 years when it has attempted to address the steadily growing water challenges facing the country. In each moment, different actors tried to reshape Mexico's water policy and institutions in response to new pressures, drivers, and opportunities. With each successive response, a new constellation of strategies and institutions was put in place, bringing, on one hand, major changes in the country's water management regime while, on the other hand, locking the country into institutional arrangements, policies, and political processes that are no longer capable of meeting either new conditions or demands.

Today, Mexico is on the cusp of the third critical juncture in its approach to water management. In reality, conditions caused by extensive overuse of water resources, inappropriate institutional design, adverse macroeconomic conditions, and rising impacts of climate change leave absolutely no latitude in implementing new reforms far more extensive and unswerving than those implemented during previous efforts. If the country fails to respond in a comprehensive, democratic, and socially inclusive manner, we will face direct threats to our prosperity, political stability, and social order. In this sense, the pending water security crisis cannot be considered a "black swan"—that is, a deviant event with major disruptive effects that was difficult to predict. The dire consequences for Mexico and, quite directly, for American interests are predictable, if not already upon us.

As I reconstruct some elements of the historical path that Mexico has followed and explore the current crisis before us, I will use two terms to explain our experience. This first is the concept of *water polity*. I use the term to indicate an integrated constellation of public, private, and social actors organized under a set of water management institutions and conducted through formal and informal political processes. I also use the term to indicate the comprehensive, inclusive set of institutional arrangements, sociopolitical relations, and political culture for managing water. The second term I use is *critical junctures*, by which I mean moments of major change in Mexican water polity wherein new conditions, new actors, and new threats demand transformative changes in both institutions and political processes used to manage water resources. Implied in the use of the two terms is the notion that once institutions and political processes are changed, self-reinforcing dynamics set in within the polity, often with a host of unforeseen and unintended consequences, that make reversing, or even changing, the prescribed management approach very difficult (i.e., path dependencies).

II. The Genesis of the Water Security Crisis

1. Modern State-Building in the Water Sector: The 1950s

Mexico's underlying institutional approach for water management was set in the 1950s, when its first modern water policies, planning approaches, and regulatory institutions were put in place. This period corresponded to a stage in Mexican economic development when all planning and implementation mechanisms, for water and for many other sectors of the economy, were concentrated in the Mexican federal government. Initially, centralized water planning was accompanied by provision of ample financial resources that incentivized the construction of a wide range of water infrastructure across the country. As such, water management was considered central to economic growth in the agricultural, urban, and industrial sectors. The downside of this initial approach for both water and the broader economy was that economic and political centralization in governmental functions stifled economic and institutional dynamism, locking the polity into policy rigidities and sociopolitical relations of privilege and exclusion (i.e., corporatism, clientelism, and political patronage) that later were eclipsed by the needs of the expanding economy, achieving greater social development, and addressing environmental problems.

2. The Second Critical Juncture: Seeking to Implement Integrated Water Resource Management—The 1990s

During the 1990s, the Mexican water polity experienced another critical juncture that was driven by macroeconomic restructuring that shifted the economy from a protectionist, import substitution model to an export-oriented one shaped by the prevailing global neoliberal economic policies of the period. In turn, the Mexican state also abandoned many of its centralizing features, embracing a neoliberal model that required far greater openness to external pressures and interests and great responsiveness to the dynamics of a market-driven economy, as well as greater pressures for decentralization and democratization. I should note that by the turn of the 1990s, Mexico's population was around 81.2 million inhabitants, of which 71.3% lived in urban areas and 28.7% lived in rural communities. Mexico City had 8.1 million inhabitants.

The National Water Commission (Comisión Nacional del Agua, or CONAGUA) was created in 1989 as the apex government institution for water resources management, becoming in 1994 a government body under the Secretary of the Environment, Natural Resources, and Fisheries (Secretaría de Medio Ambiente, Recursos Naturales y Pesca). The purpose of this elevation of CONAGUA's influence was to break the technocratic management regime that so far had focused on providing engineering solutions for agricultural and industrial needs. In its place, the government sought to promote sustainable and environmental protection as crosscutting policy objective. The succession of institutional reforms that followed were impressive.

Between 1995 and 1998, CONAGUA launched its first administrative-territorial organization of the water polity by creating 13 new Hydrological Administrative Regions (or Regiones Hidrológico-Adminstrativas) managed by River Basin Organisms (or Organismos de Cuenca). These bodies were entrusted with water resources management and planning functions, among the most important the development of regional water programs and the planning and programming of essential regional infrastructure. Throughout 1992–1994 the Water Property Rights Registry (Registro de Propiedad de Derechos de Agua, or REPDA) was created, in order to begin to develop one of the core functions of CONAGUA: the management of water resources. At the same time, CONAGUA initiated the irrigation management decentralization process and transfer to water user associations at the level of irrigation districts and units. Also, there was a further push to decentralize municipal water supply and sanitation services in a way that effectively decoupled responsibility for such services from the federal government. Fifth, the water sector was opened to private-sector involvement, giving way to the organization of some service contracts and "build, operate, and transfer" operations—especially in water treatment and utilities' commercial operations.

Sixth, and of central importance, efforts were made to implement an integrated water resources management (IWRM) strategy through the establishment of multi-stakeholder platforms (MSPs) for water resources management. The goal was to strengthen social participation and interinstitutional coordination, including the river basin councils and their auxiliary bodies, the river micro-basin commissions and groundwater management technical committees (Comités Técnicos de Aguas Subterráneas, or COTAS.[1] In 2000, the Water Advisory Council was created as another MSP to provide an institutional venue for organized civil society to participate in the water-policy process at the highest level. This transformation pathway gradually sought to build a new and semi-decentralized water governance system in a context of a water polity now populated by innovative and more politically active local actors.

It is important to highlight that Mexico, as a country, was also starting to undergo a democratic transition after 70 years of one-party rule—by the Partido Revolucionario Insitucional, or PRI—to the first competitive elections in 2006.

3. The Third Critical Juncture: Striving for Water Security Under Adverse Macroeconomic Conditions—2017 and into the future

Today, despite the progress made so far, the Mexican water polity faces another critical juncture, albeit this time with a stronger sense of urgency, a realization that things are going wrong under a business-as-usual approach. This situation has prompted the Mexican government to treat water resources—at least at the level of discourse—as a matter of national security. This critical juncture is the product of a number of drivers and factors that have grown in complexity and size due to strong path dependencies. Other drivers and factors, such as climate change and a severe macroeconomic-budgetary crisis, are relatively new or particular to the present critical juncture. A particular feature of this juncture is that water challenges and problems can now be characterized as risks or threats that demand a new, risk-based policy approach, more adaptive institutional arrangements, and more broadly, a different role for the state, and new collaborative relations between the state and civil society. Next, I highlight some of the most important challenges and national security threats as we enter this new moment in the water polity's evolution.

a) Challenge/Threat 1: Dwindling Water Resource Endowments and a Growing Water Gap

Today, population and socioeconomic activities are concentrated in the regions with the least water availability. This outcome is not surprising given the decades of lack of institutional coordination between the urban and regional planning bodies, the agricultural and livestock sectors, and the water resources management agency. The situation has been equally aggravated by a disregard for the limits imposed by the country's water and other natural resources endowments. Paradoxically, on the one hand the Central region—the highlands (the North and Northeast territories), which have 32% of the national water resources availability—is where 77% of the population has settled and 79% of gross domestic product (GDP) is being produced. On the other hand, in the South and Southeast regions—home to 68% of the national water resources availability—are where 23% of the population lives and 21% of GDP is produced (see Figure 5.1).

Moreover, water availability per capita in Mexico dwindled from an average of 18,035 cubic meters per inhabitant per year in 1950 to 3,982 cubic meters per inhabitant per year in 2015. It is estimated that currently 35 million Mexicans (of a total population of 119,530) live under constant water scarcity conditions (CONAGUA, 2014).

An analysis conducted by CONAGUA determined that the gap between available water resources and water demand by the year 2030 will be 23 billion cubic meters of water per year (see Figure 5.3; CONAGUA, 2014). If water resources continue to be managed under a business-as-usual scenario, it is possible to envisage a number of extremely negative socioeconomic consequences

Figure 5.1 Complex Water Resources Endowments
Source: CONAGUA

such as local and interstate water conflicts; the migration of industries to water-abundant regions of the country—or in some cases, their repatriation to the United States; land grabbing for agricultural activities in the South and Southeast regions; the dislocation of local economies and communities in the Central, North, and Northeast regions; an accelerated rural-urban migration, as well as migration to the United States; and a water crisis in the Valley of Mexico.

b) Challenge/Threat 2: Severe Groundwater Overexploitation

Due to the strategic nature of groundwater resources for water security in Mexico, the World Bank has considered sustainable groundwater management a

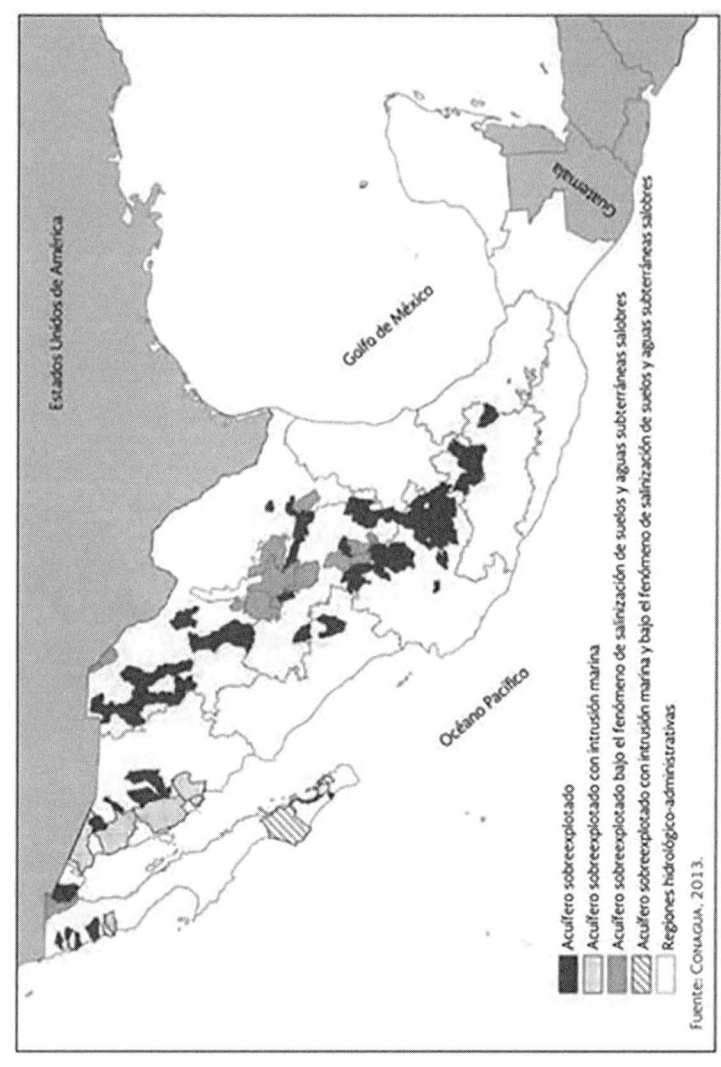

Figure 5.2 Overexploited and Compromised Aquifers
Source: CONAGUA

matter of national security (World Bank, 2006). CONAGUA has identified and demarcated a total of 653 aquifers across the national territory. It has been estimated that these aquifers provide 37% of the total volume of water allocated to consumptive use and 65% of the overall demand of cities and industries throughout the country (CONAGUA, 2014). In certain regions of the country, mostly in the North and Northeast regions and in the "el Bajio" territory—the states of Aguascalientes, Querétaro, Guanajuato, and San Luis Potosi, and parts of Jalisco—there are cities, industries, and agricultural lands that depend on up to 95% on groundwater for their source of water. Currently, it is estimated that 106 aquifers are being rapidly overdrawn—including aquifers in the aforementioned territories. This situation threatens to hinder future economic growth and stall social development in regions that support most of the country's GDP production. Furthermore, groundwater depletion has a direct impact on the livelihoods of poor and marginalized rural, small-scale producers and the *Ejidos*.[2]

The case of Mexico City's aquifer is perhaps one of the most dramatic examples of groundwater overexploitation. Groundwater has allowed Mexico City, with a population of 9 million inhabitants, and also its greater metropolitan area, with a population of 29 million inhabitants, to satisfy unceasing demands for water resources.[3] This untenable situation has negatively transformed the aquifer's balance, estimated to have a deficit of 111 million cubic meters per year. In the 1960s and 1970s, wells were drilled at depths of 100 or 150 meters; today, it is necessary to reach depths of approximately 300 or 450 meters. Due to this unsustainable abstraction rate, the city is experiencing significant ground subsidence that, in some places, can be up to 30 centimeters per year. In the city center, the accumulated subsiding in 60 years totals some 10 meters. The economic cost of this subsiding process due to damage to buildings and infrastructure is incalculable. Estimates indicate that at this rate Mexico City's aquifer could have a life span of only 30 or 40 years (Sistema de Aguas de la Ciudad de México, or SACMEX, 2015).

Already many inhabitants of Mexico City experience water insecurity. In Chalco, Netzahualcoytl, and Iztapalapa, hundreds of thousands of people suffer from discontinuous and unreliable water provision (*tandeos,* in Spanish). In some parts of these regions, inhabitants receive water only two or three times a week and mostly by water trucks. This form of water poverty and inequality is a constant source of social-political conflict and also of migration from these affected places, mainly to the United States, as explained by experts in a recent symposium in Mexico.[4] Mexico City's extreme water challenges require urgent changes. Mexico City is simply "too big to fail," and the political system needs to harness the necessary political will and financial resources to treat this situation as a matter of national security.

c) Challenge/Threat 3: Urban and Rural Water Insecurity

Today, 92.4% of people in Mexico have access to drinking water and 91% have access to sanitation. In terms of the urban population, the numbers are

95.1% and 96.3%, respectively, and for the rural population, 82.9% and 72.8%, respectively. The urban-rural divide is significant—as in most parts of Latin America. But these aggregate numbers hide some important problems and inequalities. For example, across the country, deprived peri-urban regions in cities have very unreliable water provision. Dispersed rural communities are also experiencing rising water insecurity. There is no simple solution to these two problems, which require very innovative financial, institutional, and technical approaches designed for these specific contexts. Generally speaking, the performance of Mexico's water utilities is below international standards. Because the utilities are always struggling for financial resources for operation and maintenance activities, they lack adequate and capable personnel, and also become the bounty of local political motivations and corruption due to fragile accountability and transparency mechanisms. The challenging situation of water utilities acquires a strategic and national security dimension if we consider that it is expected that by 2030 approximately 80% of the Mexican population will live in cities. About 56.9% of Mexicans live in 59 metropolitan areas across the country (CONAGUA, 2016) and it is expected that by 2030 there will be an increase in the urban population of 13.42 million inhabitants. Furthermore, Mexico is following an important urbanization process that creates severe inter-institutional coordination processes across the urban, housing, and water policy sectors, a situation that is also driven by the institutional atomization created by decentralization/municipalization policies. In addition, the present federal budgetary cutbacks of some 70% to programs supporting water utilities will definitely complicate matters.

d) Challenge/Threat 4: Unsustainable Agricultural Water Management

Approximately 76.7% of offstream water resources are dedicated to agriculture across a complex network of agri-water infrastructure. As a result of important investments in irrigation over decades, Mexico occupies seventh place worldwide in terms of irrigation infrastructure, with 6.4 million hectares (CONAGUA, 2016). The irrigation transfer management process embarked upon during the second critical juncture has been only partially successful because many irrigation districts lack the financial and technical capabilities to maintain local infrastructure. Not surprisingly, their water productivity ratios are very low. Approximately 35% of the water allocated to agriculture is groundwater, and, due to highly subsidized electricity tariffs groundwater is not used efficiently. This subsidy, by the way, benefits mostly the richest producers, who take full advantage of it. This situation creates perverse water efficiency incentives, spillage, severe groundwater overexploitation, and enormous fiscal strain.[5]

e) Challenge/Threat 5: Climate Change

Mexico is extremely exposed to the impacts of climate change, and this exposure is increasing fast. For example, the socioeconomic damage from extreme hydro-meteorological events—floods and droughts—has spiked from an average of US$36 million per year in the period 1980–1999 to US$1,095 million per year for the period 2000–2012 (CONAGUA, 2014). This is also accompanied by great rainfall uncertainty. For example, in 2009, Mexico experienced the worst drought in 60 years; 2010 was the rainiest year on record; and in 2011, 40% of the territory experienced the worst drought in 70 years. This uncertainty has created a very complex landscape for decision-making and can portend massive socioeconomic loss.

Unfortunately, due to weak land-use and urban planning systems, a great number of people live in high-risk areas, under very high vulnerability conditions, including on hillsides, in ravines, and near rivers. This situation is worsened by the rapid destruction of forests and wetlands. The lack of coordination between the agricultural and water-policy sectors also affects the possibility of embracing rapid and efficient adaptation measures in the agricultural and livestock sectors, a situation that has important implications for producers as well as for an already weak food-security system. The Mexican water polity is ill-prepared to adequately face climate change, but some adaptation efforts are under way.

f) Challenge/Threat 6: Adverse Macroeconomic Conditions and Massive Budgetary Cuts

Mexico's new macroeconomic conditions, driven by the crisis in oil prices, are severely affecting government spending, and budgetary cuts are being applied throughout the public administration. CONAGUA has been hit considerably in its budget allocation, receiving in 2014 MXN47.3 thousand million pesos, in 2015 MXN40.1 thousand million pesos and in 2016 MXN36.1 thousand million pesos. This has important consequences for continuing capital investments in strategic infrastructure and maintenance work, as well as for supporting an already frail and ineffective governance framework. As part of the austerity measures, approximately 1,500 employees have been laid off, affecting CONAGUA's institutional capability. Almost all of the subsidy-based programs for water utilities operations and modernization, as well as for irrigation infrastructure development and maintenance, have been severely curtailed for the 2017 federal budget cycle (up to a 72% decrease for the water utilities and up to a 40% decrease for agricultural water).

The decentralization gains achieved over more than two decades are being reversed by the dissolution of River Basin Organizations' planning departments and a reduction of spending in institutional development. Actually, there seems to be an ongoing de facto recentralization process that goes against international best practices. The budgets for supporting governance frameworks—including the MSPs—are extremely low, evidencing a statehood orientation that still invests heavily in infrastructure and not so much in information and institutions. In terms of Mexico City, on September 14, the Secretary of the Exchequer (Secretaria de Hacienda y Crédito Público) announced a 70% decrease in the federal budget to Mexico City's water system (SACMEX).

SACMEX's director commented to the media that such a decrease is "little shorter than suicidal."[6]

III. Urgently Needed Responses to Pursue Water Security

In my opinion, the looming water security crisis urgently demands a range of responses that should be considered and implemented steadfastly.

a) Water security should be integrated as a central policy priority of the entire political system and should be considered a national security concern. Accordingly, the new General Water Law, currently still under negotiation, should have a strong focus on water security, integrated water resources management, and water resiliency. Furthermore, it should be fully supportive of broader poverty alleviation, social inclusion, and democratization goals. The new General Water Law should be considered an opportunity to forge a new "social pact" for water, and so its design process should allow for authentic multi-stakeholder dialogue and consensus building. Regional water security plans should be designed and water resiliency investment portfolios should be prepared.

b) Severe groundwater overexploitation problems demand that Mexico strengthen the REPDA, enable a flexible and formal water market system, curb groundwater black markets across the country, support the institutional development of the groundwater management committees (to foster multi-stakeholder participation, collective action, and even self-regulation), invest in green infrastructure in aquifer recharge areas, implement a nationwide circular economy initiative (to reuse served waters in agriculture), and curb corruption and illegal groundwater exploitation.

c) Urban water security requires a national utility turnaround program that could address the political, legal, institutional, financial, technical, and human resources deadlocks that have been hampering their institutional development for decades now. It is imperative to establish autonomous state-level regulatory agencies and create a more enabling environment for private-sector participation and the corporatization of private utilities. Large metropolitan regions should have metropolitan water utilities, instead of the atomized institutional landscape that prevails. In response to the massive budgetary cuts, an emergency financial fund could be set up to support distressed water utilities at risk of failing to provide essential services, a threat with potentially very severe consequences for public health. This could be organized with the support of international lending institutions.

d) There is no simple path to address the challenges in the agri-water management sector. Reliance on old policies and institutions is deeply entrenched, and it will require great political will and an incredibly arduous consensus-building process to change the situation in rural Mexico, because it is extremely risky in sociopolitical terms. Still, efforts should be made to review Tariff 09—perhaps through a targeting based on socioeconomic conditions—to support agricultural reconversion and irrigation modernization (more crop per drop), to boost commodity chains, to reduce food waste (from crop to fork), to harness greater private capital in the agri-water sector through public-private partnerships (PPPs) and other financial mechanisms, and to implement a water-food-energy nexus approach.

e) To address climate change, CONAGUA is implementing an innovative PPP scheme to modernize meteorological and forecasting capabilities. Modernization is urgent. There is also a special-purpose program to address the impacts of droughts, but the program is mostly reactive. As such, the Mexican water polity is only beginning to internalize climate change as a priority concern, and the speed at which it is going is too slow. Some important steps could entail the development of climate-proofing water infrastructure planning, the institutionalization of water resiliency planning in high-risk areas—most importantly the Cutzamala System that feeds water to Mexico City—the development of water markets with a focus on water security (to optimize scarce water resources), and more broadly, the development of a water-food-energy nexus approach.

f) Massive budgetary cuts are aggravating an already bad situation. It seems that in order to achieve the greatest economic efficiency with the use of public funds it is necessary to strengthen the planning and programming processes. This reform should support the design of capital investment prioritization systems oriented by water security and resiliency considerations—and the creation of a more enabling environment for PPP formation that could bring more financing from the private sector and other national and international institutional investors.

IV. Establishing the Grounds for U.S. Foreign Policy Concern in the Mexican Water Polity: The Interdependence and Integration of Mexico and the United States

Since the United States, Canada, and Mexico entered into the North American Free Trade Agreement (NAFTA) in January 1994, trade liberalization fundamentally reshaped the regional political economy, driving unprecedented integration and interdependence of the countries. In general terms, what seems relevant to mention is that regional trade sharply increased from US$290 million in 1993 to US$1.1 trillion in 2016 (Council on Foreign Relations, 2016). Supporters of NAFTA say that U.S. foreign direct investment (FDI) contributes to job creation abroad—14 million jobs rely on trade with Mexico and Canada—and also complements investments and job creation in the United States. In the same tenor, the Council on Foreign Relations says that offshoring has strengthened the competitiveness of the United States' outward investors, consolidating the country's role in driving regional economic growth.[7]

Currently, Mexico is the United States' third-largest goods trading partner. U.S. goods imports from Mexico totaled US$295 billion in 2015, in these top categories: vehicles (US$74 billion), electrical machinery (US$63 billion), other machinery (US$49 billion), mineral fuels (US$14 billion), and optical/medical instruments (US$12 billion). U.S. imports of agricultural products from Mexico totaled (US$21 billion), corresponding to the second-largest agricultural imports, including fresh vegetables (US$4.8 billion), wine and beer (US$2.7 billion), snack foods (US$1.7 billion), and processed fruits and vegetables (US$1.4 billion).[8] All these activities and exchanges rely on stable long-term water security.

FDI in Mexico in 2015 was US$28 billion, 53% of which came from the United States. The Mexican states with the most FDI are the Distrito Federal (US$4.8 billion), Nuevo Leon (US$2.6 billion), Jalisco (US$2.5 billion), Chihuahua (US$2.5 billion), and Veracruz (US$1.4 billion).[9] U.S. firms with affiliates in Mexico operate in a variety of industries and, not surprisingly, given the extent of offshoring by the U.S. automobile companies, the largest of these is transportation equipment, which includes automobile manufacturing. Most of these automobile companies are established in the North and Northeast regions (in the states of Coahuila, Sonora, and Nuevo Leon) and in the Central and el Bajio regions (in the states of San Luis Potosi, Guanajuato, and Aguascalientes). A study by the Colegio de la Frontera Norte (2010) highlights that 66% of multinational corporations (MNCs) have established operations in the Central and el Bajio regions—and 34% in the Northern border states—corresponding to the *maquiladora* industry.

These statistics, when decoupled from the conditions and trends of Mexico's water resource endowments, and the challenges faced by the water polity, seem very positive and promising. However, when we overlay them with information on water stress and water insecurity conditions, things start to look a bit less optimistic. The maps presented depict this situation. Map No. 1 (upper left corner) describes the country as subdivided into two main regions where MNCs have chosen to locate: the Northern region (where 34% of MNCs are) and the Central Southern region (where 66% of MNCs are). Map No. 2 (upper right corner) describes the availability of renewable water resources in 2015, whereby the Central/Northern/Northwest region shows a dwindling availability of 1,604 cubic meters per inhabitant per year and the Southeast shows an availability of 10,852 cubic meters per inhabitant per year. Paradoxically, most MNCs are located in a region with dwindling renewable water resources. Map No. 3 (lower left corner) describes the location and presence-intensity of MNCs per state. Map No. 4 shows the location of aquifers with severe groundwater overexploitation rates.

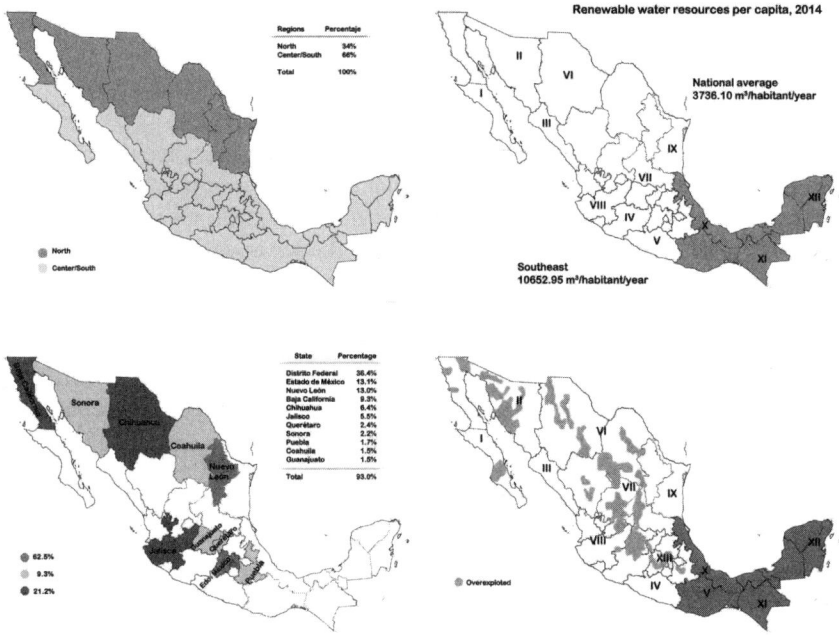

Figure 5.3 Location Map
Source: CONAGUA

As such, most MNCs are located in regions with overexploited aquifers and groundwater abstraction bans. Furthermore, a very large percentage of MNCs are located in Mexico City and the Valley of Mexico (the Distrito Federal), where water security is at great risk. In particular, most of the U.S. automobile industry—Chrysler, Ford Motor Co., and General Motors—is located in either the North/Northwest regions (Coahuila and Sonora) or the Central (Estado de Mexico) and el Bajio regions, where there is severe and very severe water stress and very high reliance on overexploited aquifers. The argumentation presented in the preceding sections shows that there are grounds for U.S. foreign policy concern over the future development path of the Mexican water polity.

V. Key Entry Points for U.S. Involvement in the Mexican Water Polity

The following are some prospective entry points for policy involvement in the Mexican water polity to support its pursuit of water security.

a) U.S. Foreign Policy Must Be More Directly Concerned With the Performance and Future of the Water Development Path of the Mexican Water Polity in Light of the Relevance That Water Security Has for U.S. Economic, Commercial, Migratory, and Security Concerns

So far, it seems that CONAGUA has no clear strategy to pursue water security in a structured and comprehensive manner, and elevating water security to the level of a national security issue has not translated into more strategic political, institutional, and financial support to the water sector. Therefore, it is extremely relevant to make water security concerns a legitimate and warranted item in U.S. policy dialogues with Mexico in an attempt to foster greater cooperation between the countries to support the Mexican water polity's efforts to embark on a water security transformation pathway. At the center of this foreign policy dialogue could be the design and implementation of a comprehensive institutional development and technical-assistance program on various themes and with different components.

b) Both Countries Should Devote Effort to Developing a Comprehensive Institutional Development and Technical-Assistance Program to Support Water Security

In this context, the role of the U.S. Water Partnership is extremely important for coordinating and harnessing a vast repertoire of human, technical-technological, and financial resources to support technical assistance and capacity-building efforts.[10] It is urgent to open a constant dialogue between representatives of the U.S. Water Partnership (and its member organizations) and CONAGUA, state and local authorities, the Consejo Consultivo del Agua, academia, and other organizations.

c) In Order to Support the Institutional Development and Capacity-Building of Strategic Actors in the Mexican Water Polity, Technical Assistance Seems Particularly Relevant in the Following Areas

- *The Implementation of IWRM, With a Focus on Water Security and Resiliency*

Technical assistance in the implementation of IWRM is paramount, including aspects such as multilevel and multi-stakeholder water governance, interinstitutional coordination, water security portfolio integration, investment prioritization, sustainable and participatory groundwater governance, water resiliency planning, gray and green infrastructure portfolio investment, utility turnaround, utility regulation, metropolitan water planning, sustainable agri-water management, the water-food-energy nexus approach, circular economy in the water sector, climate change scenario-making and decision-making tools, climate-proofing infrastructure evaluation, and new financial mechanisms such as green bonds, climate-ready water utilities approaches, and risk mitigation and insurance and reinsurance schemes, among other important themes.

- *The Full Modernization of the Registry for Public Water Rights Under the Framework of Water Security as a Main Policy Priority*

Of particular importance for the water polity is to devote effort to modernizing the REPDA, but with due consideration for national, regional, and local water-security considerations. Technical advice to this purpose is highly warranted, including on the latest satellite remote-sensing and GIS technologies for surface- and groundwater-resources management.

- *PPP and Cooperation*

There are many forms of PPP and cooperation that can be exploited between U.S.-based private corporations with legitimate interests in the performance of the Mexican water polity's development path and the CONAGUA, as well as other strategic actors.

d) Due Diligence and Water Risk Integration

U.S.-based corporations conducting business—or expecting to—in Mexico should be liable for conducting all the due diligence necessary to respect environmental safeguards and safety protocols established by the Mexican authorities and international treaties. Furthermore, they should implement internationally accepted water-risk integration protocols to guarantee that water security conditions are met for their offshoring operations in Mexico.

e) Participation in Multi-Stakeholder Platforms

It is important for U.S. foreign policy to entice U.S.-based corporations and other institutional investors in Mexico to form part of the existing MSPs in the Mexican water polity, and hence to participate proactively in multi-stakeholder governance processes, including the relevant river basin councils, groundwater management committees, and the Consejo Conultivo del Agua. These are legitimate institutional spaces to exert voice and influence over water policymaking. This participation, in turn, also supports the pursuit of other democratic values, such as democratic decision-making, transparency, accountability, and social inclusion.

f) Support the Activities of Relevant Organizations of Civil Society and Nongovernmental Organizations Working in the Field of Sustainable Water Resources Management and Water Security

The U.S. has an important history of associative activity that has supported the development of democratic culture and that has served as a power check and balance to government institutions. In Mexico, this associative activity is still somewhat underdeveloped. Still, there are a number of important and legitimate organizations of civil society and nongovernmental organizations working in the field of sustainable water resources management and local water security. U.S. foreign policy should devote effort to mapping these organizations and understanding their roles and contributions to water security, so as to explore possibilities for allocating direct or indirect support for their activities. This also helps create a more democratic water governance.

VI Outside the Water Box

Mexico is an extremely unequal society, and social exclusion and marginalization abound. Approximately 45% of its people live under the poverty line.

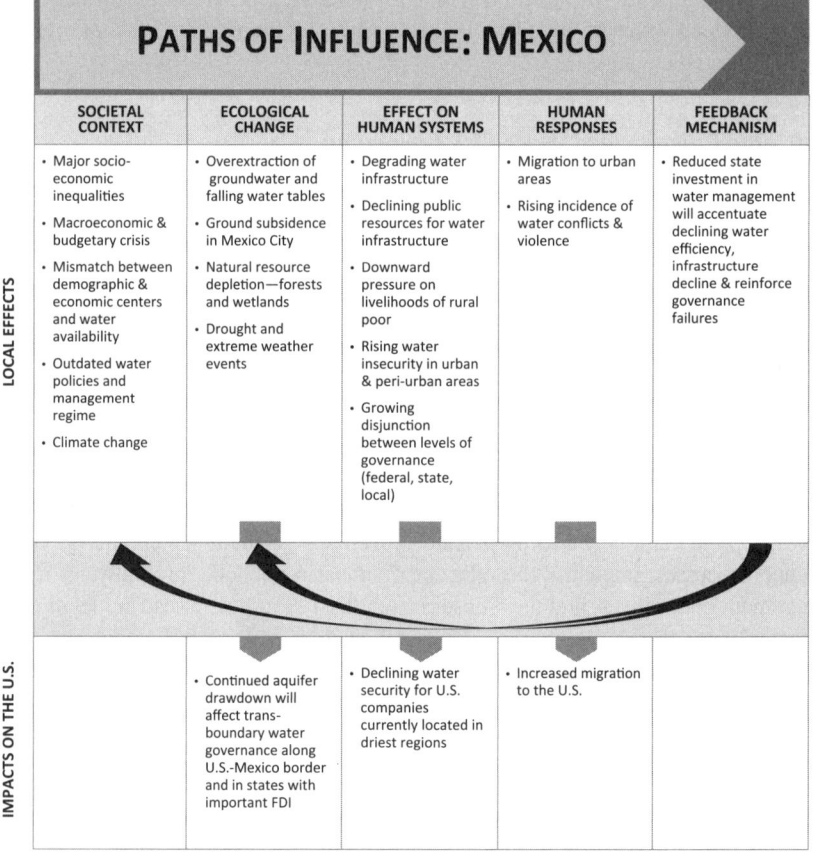

Figure 5.4 Paths of Influence: Mexico

Therefore, foreign policy efforts to implement broader poverty alleviation and social inclusion initiatives are extremely relevant. Also, democratic polities rely on a well-educated, informed, and proactive citizenship; thus, supporting water stewardship education in schools and universities seems like a critical entry point for U.S. foreign policy involvement.

Finally, the challenges presented in this chapter will test the resolve and ability of Mexico's political system to address the looming water security crisis in a democratic, socially inclusive, and consensus-driven way. In contrast, the alternative is to pursue a more centralized, autocratic approach with increased exclusion of local stakeholders and communities and greater power exercised by technocrats operating from a national security state perspective. It will require several years, if not decades, to assess which approach current and future administrations will follow.

Notes

1 The MSPs bring together government and social actors in a deliberative arena to discuss the most pressing challenges and threats affecting the river basin/aquifer, and to enable greater social participation, consensus building, and conflict resolution.
2 As water tables lower, it becomes increasingly expensive to invest in the construction, operation, and maintenance of water wells and the pumping of groundwater. Poor *Ejidos*—the peasants' organizational unit for agricultural production—and small-scale producers cannot bear these costs. This situation has triggered peasants' outward migration to cities and also to the United States, as agriculture no longer represents a viable economic activity for younger generations.
3 It is important to highlight that mean water availability in Mexico City is only a staggering 152 cubic meters per inhabitant per year, while the mean water availability in the country is 3,983 cubic meters per inhabitant per year.
4 Please see: *La Jornada*, "El Desabasto de Agua ha Influido en la Migración a Estados Unidos," Saturday, June 18, 2016, www.jornada.unam.mx/2016/06/18/sociedad/029n1soc.
5 The contribution from the agricultural sector to Mexico's GDP is still only 3.5% (World Bank, 2016).
6 *La Jornada*, Wednesday 14, 2016, p. 32.
7 Council on Foreign Relations: www.cfr.org/trade/naftas-economic-impact/p15790.
8 Office of the U.S. Trade Representative: https://ustr.gov/countries-regions/americas/mexico.
9 Wilson Center: www.wilsoncenter.org/article/infographic-foreign-direct-investment-mexico.
10 For more information on the U.S. Water Partnership, please see www.uswaterpartnership.org.

Bibliography

Aboites, L. (1998), *El Agua de la Nación, Una Historia Política de México (1888–1946)*, CIESAS, México D.F., Mexico.

——— (2009), *La Decadencia del Agua de la Nación: Estudio sobre la desigualdad social y cambio político en México*, El Colegio de México, México D.F., Mexico.

Acemoglu, D. and J. Robinson (2012), *Why Nations Fail: The Origins of Power, Prosperity and Poverty*, Profile Books, London, UK.

Aguilar-Barajas, I., Mahlknecht, J. Kaledin, J. Kjellen, M., and A. Mejia-Betancourt (2015), *Water and Cities in Latin America*, Earthscan, London, UK.

Aguilar Villanueva, L. (2006), *Gobernanza y Gestión Pública*, Fondo de Cultura Económica, México D.F., Mexico.

Backstrand, K., Khan, J., Kronsell, J., and E. Lovbrand (eds.) (2010), *Environmental Politics and Deliberative Democracy*, Edward Elgar, Cheltenham, UK.

Barry, J. and R. Eckersley (2005), *The State and the Global Ecological Crisis*, MIT Press, Cambridge, MA, USA.

Bell, S. and A. Hindmoor (2009), *Rethinking Governance, The Centrality of the State in Modern Society*, Cambridge University Press, Cambridge, UK.

Carillo, J. and R. Gomis (2010), *Corporaciones Multi-Nacionales en México, Un Primer Mapeo*, Colegio de la Frontera Norte, Tijuana, Mexico.

Comisión Nacional del Agua (2010), *Documentos Básicos de los Consejos de Cuenca*, CONAGUA, México City, Mexico.

Comisión Nacional del Agua (2014), *Programa Nacional Hidraúlico*, CONAGUA, México City, Mexico.

Comisión Nacional del Agua (2016), *Estadísticas del Agua*, Edición 2015, CONAGUA, México City, Mexico.

―――― (2011), *Agenda 2030*, CONAGUA, Mexico City, Mexico.

―――― (2015), *Water Statistics 2015*, CONAGUA, Mexico City, Mexico.

Council of Foreign Relations (2016), *NAFTA's Economic Impact*, www.cfr.org/trade/naftas-economic-impact/p15790.

Dryzek, J. (2000), *Deliberative Democracy and Beyond, Liberals, Critics and Contestations*, Oxford University Press, Oxford, UK.

Endelnbos, J., Bresssers, N., and P. Scholten, *Water Governance as Connective Capacity*, Ashgate, Surrey, UK.

Floyd, R. and R. A. Matthew (eds.) (2015), *Environmental Security: Approaches and Issues*, Routledge, Oxford, UK.

Freyman, M., Collings, S., and B. Barton (2015), *An Investor Handbook for Water Risk Integration*, A CERES Report, Boston, USA.

Fung, A. and E. O. Wright (2003), *Deepening Democracy, Institutional Innovations in Empowered Participatory Governance*, Verso, London, UK.

Guy Peters, B. (2000), *Institutional Theory in Political Science, the New Institutionalism*, Continuum, London, UK.

Hajer, M. (1995), *The Politics of Environmental Discourse, Ecological Modernisation and the Policy Process*, Oxford University Press, Oxford, UK.

Hajer M. and H. Wagenaar (eds.) (2003), *Deliberative Policy Analysis, Understanding Governance in a Network Society*, Cambridge University Press, Cambridge, UK.

Harvey, D. (1996), *Justice, Nature and the Geography of Difference*, Blackwell, Oxford, UK.

Hough, P. (ed.) (2014), *Environmental Security: An Introduction*, Routledge, Oxford, UK.

Hufbauer, G. and J. Schott (2005), *NAFTA Revisited*, Institute for International Economics, Washington, USA.

Hussain, I. (2010), *The Impacts of NAFTA on North America*, Palgrave, New York, USA.

Jimenez, B., Torregrosa, M., and L. Aboites (eds.) (2010), *El Agua en México: Cauces y Encauces*, Aosociación Mexicana de Ciencias, México D.F., Mexico.

Kemper, K., Blomquist, W., and A. Dinar (2010), *Integrated River Basin Management through Decentralisation*, World Bank, Washington, USA.

Mahoney, J. and K. Thelen (2010), *Explaining Institutional Change, Ambiguity, Agency and Power*, Cambridge University Press, Cambridge, UK.

Moulaert, F. (ed.) (2014), *The International Handbook on Social Innovation: Collective Action, Social Learning and Trans-Disciplinary Research*, Edgar Elgar, Cheltenham, UK.

OECD (2011), *Water Governance in OECD Countries*, OECD, Paris, France.

Oswald, S. (ed.) (2011), *Water Resources in Mexico: Scarcity, Degradation, Stress, Conflict, Management and Policy*, Springer, London, UK.

Pierre, J. and B. Guy Peters (2000), *Governance, Politics and the State*, Palgrave, Basingstoke, UK.

Pierson, P. (2004), *Politics in Time, History, Institutions and Social Analysis*, Princeton University Press, New Jersey, USA.

Reed, D. (ed.) (2015), *In Pursuit of Prosperity: U.S. Foreign Policy in an Era of Natural Resource Scarcity*, Routledge, London, UK.

Sadoff, C. W., Hall, J. W., Grey, D., Aerts, J. C. J. H., Ait-Kadi, M., Brown, C., Cox, A., Dadson, S., Garrick, D., Kelman, J., McCornick, P., Ringler, C., Rosegrant, M., Whittington, D., and Wiberg, D. (2015), *Securing Water, Sustaining Growth: Report of the GWP/OECD Task Force on Water Security and Sustainable Growth*, University of Oxford, Oxford, UK.

Scoones, I., Leach, M., and P. Newell (2015), *The Politics of Green Transformation*, Earthscan, London, UK.

Sistema de Aguas de la Ciudad de México (SACMEX) (2015), *El Gran Reto del Agua en la Ciudad de México*, SACMEX, México City, Mexico.

Torfing, J., Guy Peter, B., Pierre, J., and E. Sorensen (2012), *Interactive Governance*, Oxford University Press, Oxford, UK.

Warner, J. (ed.) (2016), *Multi-Stakeholder Platforms for Integrated Water Management*, Routledge, Oxford, UK.

Weidner, H. and M. Janicke (2010), *Capacity-Building for Environmental Policy: A Comparative Study of Environmental Governance*, Springer, Berlin, Germany.

World Bank (2006), *Mexico Water Public Expenditure Review*, Document, World Bank, Washington, USA.

―――― (2012), *Integrated Urban Water Management: A Summary Note*, World Bank, Washington, USA.

―――― (2016), *High and Dry, Climate Change, Water and the Economy*, World Bank, Washington, USA.

World Bank-GW Mate (2004), *México-Los COTAS: Avances en la Gestión Participativa del Agua Subterránea en Guanajuato*, World Bank, Washington, USA.

Wright, E. O. (2010), *Envisioning Real Utopias*, Verso, London, UK.

6 Who Stole the Water
Water, Security, and U.S. Foreign Policy in Guatemala

Eduardo Stein with Lilian Marquez

Introduction

In April 2016, thousands of Guatemalans set out from various locations in the country toward the capital city to arrive on April 22, Earth Day, and demonstrate peacefully for water as a human right, denouncing attempts at its privatization and highlighting its uncontrolled pollution and the lack of adequate water access suffered by most rural and impoverished Guatemalans.

Along their 11-day walk, demonstrators were greeted by locals, received food and shelter, and were joined by many. On Earth Day, the "Walk for the Water, Mother Earth, Land, and Life" reached the capital city, demanding leadership and action to ensure that every Guatemalan have adequate access to water. Despite its transcendental significance, "the Walk" was scantily covered by media, and although noticed by the government, it was easily dismissed days later.

This peaceful demonstration came only a few months after Guatemalans had taken to the streets to demand the resignation of the country's vice president and president for corruption scandals. The "Guatemalan Spring" renewed civil society's faith in their power to demand ethical and fair governance from their leaders. However, as Earth Day ended and demonstrators headed home, the many challenges to ensure fair and adequate water governance remained unaddressed.[1]

Background

Colin Kahl, former Vice President Joseph Biden's foreign policy advisor, argues that "demographically and environmentally induced conflicts are most likely to occur in countries that are deeply split along ethnic, religious, regional or class lines, and which have highly exclusive and discriminatory political systems."[2] Guatemala's history of occupation and civil war, flagrant inequalities, high poverty, and significantly weak governance, coupled with a large indigenous and marginalized population, make it a textbook case for increased conflict.

Located on the Central American isthmus, Guatemala once boasted the center of the Mayan civilization. Along with other Latin American and Caribbean countries, it endured Spanish invasion and colonization, and in the postcolonial period it has undergone several periods of instability and civil unrest. Nevertheless, Guatemala is the largest economy in the region, with a US$63.794 billion GDP (2015).[3]

Guatemala's population is also the largest in the isthmus, with more than 16.1 million people, half of whom are indigenous and rural, and survive on subsistence agriculture. Nonindigenous Guatemalans, known as *ladinos*, concentrate in urban centers and, more important, disproportionately control the country's wealth and land, and control the government.

The roots of poverty stem from historical land and wealth divisions dating back to the invading and occupying Spaniards, who appropriated and distributed indigenous lands and enslaved indigenous peoples. The oppressive and unjust treatment of indigenous peoples began with the Spanish conquest but carried on into independence and remains unabated up to the present day.

Despite Guatemala's strong economy, the country suffers from high inequality and poverty. The 2015 UN Human Development Index ranks it as the 125th country, and its Gini coefficient is 55.9 (2013).[4] Poverty in Guatemala can be characterized as extensive, rural, landless, and primarily indigenous—extensive and rural given that in 2014, the rural population was 50.5% of the total population, the vast majority of whom were poor. Poverty is landless because the majority of rural poor do not own land, making a livelihood out of subsistence agriculture and substandard agricultural employment. Land distribution is highly unequal, with the largest 1.86% of farms controlling 51.88% of agricultural land, with a farmland distribution with a Gini coefficient of 0.84.[5] Poverty is indigenous in that Guatemala's indigenous population suffers poverty disproportionately. The latest national livelihoods survey (2014) reports a significant increase in the percentage of indigenous people living in extreme poverty, rising 12 percentile points to 39.8% in 2014, from 27.1% in 2000. Overall poverty is on the rise in Guatemala, with the nonindigenous population's poverty rate also climbing, from 7.8% in 2000 to 12.8% in 2014. The survey determined that poverty has risen to 59.3% of the population, with 23.4% of the population living in extreme poverty. These figures are significantly higher than those reported for the year 2000; overall poverty was then 56.4% and extreme poverty was 15.3%.[6]

Guatemala's rural poor are also food insecure and malnourished. Almost half of all Guatemalan children suffer from chronic malnutrition (47%, 17% of whom have severe malnutrition). The children whose mothers are uneducated are more likely to suffer from chronic malnutrition (67%); similarly, children living in highly indigenous areas of the country are more likely to be chronically malnourished.[7] Globally, Guatemala ranks number 51 for child malnutrition and has the highest percentage of stunting in children from birth to age five in Latin America and the Caribbean, ranking eighth in the world for this indicator.⁸ The situation is compounded by the lack of investments in food security. Currently, investments are less than US$43 per rural capita, with a ranking of number 18 out of 21 Latin American and Caribbean countries. In contrast, Mexico invests US$266, and Nicaragua US$77, per rural capita.⁹

Drugs, Violence, and Weak Governance

Increasing violence has made the already-vulnerable rural poor more vulnerable, pushing them to migrate from rural areas to urban centers and from Guatemala into Mexico and ultimately into the United States. According to the U.S. Department of State, "Guatemala's worrisome murder rate is driven by four key factors: increased drug trafficking, growing gang-related violence, a heavily armed population (more than 60% possess firearms), and a police/judicial system unable and unwilling to hold most criminals accountable."[10]

Violence has been a chronic problem for the region, with the latest wave starting in the early 2000s. Together, the countries of Guatemala, El Salvador, and Honduras have the fourth-highest murder rate in the world, with 56 murders per 100,000 inhabitants in 2010.[11] In Central America, 77% of all murders are committed with a firearm,[12] and youths are the most vulnerable to violence. In 2014, UNICEF estimated that Guatemala and El Salvador have the highest rate in the world of homicide of children from birth to age 19.[13]

In 2006, Mexico's offensive against drug trafficking effectively shifted regional drug transit south of the border. Central America was transformed from a refueling stop into the territory through which a significant amount of drugs flows on its way north. Increased drug trafficking is partly to blame for violence and insecurity, but the problem is complex. Traffickers compete to control drug-smuggling routes, while street gangs compete to control the drug retail market and extortion rackets.[14] The most affected by the violence and insecurity are the poor, who find themselves caught between conflicting territorial groups or within the domain of street gangs and their predatory control over entire neighborhoods, where all are expected to "pay for security" (i.e., extortion). The young are expected to join the gang world.

To compound the problem, the country has been unable and unwilling to address the chronic weaknesses of its judicial and law-enforcement institutions. In 2016, the International Police Science Association presented the first iteration of the World Internal Security and Police Index. This composite index measures the ability of security institutions to maintain security, the effectiveness of those services, the public's trust in rendered services, and police operations and activities. Out of 127 countries, Guatemala ranked among the worst at number 107, El Salvador at 86, Honduras at 116, and Mexico at 118.[15] Inadequate budgets, weakened capacities, corruption, and powerful groups interested in profiting from a crippled justice system render Guatemala unable to control violence and organized crime. In 2015, Guatemala's political crisis showcased how rampant corruption has reached every level of government. The country's president and vice president had to resign and are now facing multiple charges for embezzlement and money laundering because they led a sophisticated network of bribery, obscure business arrangements, and deception.[16]

The surge in violence and insecurity has changed migration demographics. In 2014, nearly 52,000 unaccompanied minors were apprehended at the U.S. border with Mexico—more than twice the number from the previous year.[17] The surge began in 2012, and by 2014 the main demographics of people crossing illegally had shifted from single men and Mexican families to minors from El Salvador, Guatemala, and Honduras. The influx of minors from Central America's "Northern Triangle" stems from a complex array of factors, which include "push" factors such as high levels of violence and poverty in Central America and "pull" factors such as the desire to reunite with family members living in the U.S. and perceptions of U.S. immigration policies. President Barack Obama declared it a humanitarian crisis,[18] and in response Congress approved a US$750 million appropriations bill spearheaded by former Vice President Joseph Biden. The bill had the stated aim of assisting the governments of Central America's Northern Triangle "to improve security, promote peace, and tackle gang violence."[19]

In 2015, apprehension of minors near the southwestern U.S. border decreased by 45% but increased again in the first five months of 2016.[20] Unless insecurity and socioeconomic conditions change in these countries, the influx of migrants seeking refuge and economic opportunities will continue.[21] Women and children are fleeing violence, extortion, rape, and abuse inflicted upon them by organized criminal networks but also within their families.[22] They are also fleeing poverty and a government unable and unwilling to offer any protection or alternative. If fleeing such a dire situation were not enough, the road toward a safer and more stable arrangement up north is now as dangerous as what they endure every day at home. Migration into the U.S. across the Mexican border has become deadlier than ever, with organized crime controlling human trafficking and pushing the price up to US$7,000 to smuggle one person across the border.[23]

Migration is a response not only to violence and poverty but also to increased food insecurity.[24] While an unprecedented number of minors were attempting to cross into the U.S., the World Food Programme was providing food aid to El Salvador, Guatemala, and Honduras in response to a two-year drought. In Guatemala alone, 1.5 million subsistence agriculture farmers had depleted their food stocks and 75% of maize crops had been lost.[25] El Niño oscillation has extended the dry conditions, and at the end of 2016, the Dry Corridor of Central America experienced its worst drought in decades. Those most affected have already endured two to three years of hardship, and their ability to cope is greatly compromised.[26]

Water—Who Steals it?

In this insecure, unequal, poor, racist country facing age-old unresolved underdevelopment problems, a key unifying element is also at risk. Water, the essential resource, is another victim of Guatemala's inability to govern its people and resources. Guatemala is a moderately water-endowed nation, with 15.8% of the water resources in the Central American region.[27] Water quality is poor, flows are highly erratic, and access to water is grossly unequal.[28] To date, no comprehensive water policy exists. Sanitation and potable water provision are severely limited and wastewater treatment is nonexistent. The country lacks a national water authority, while municipalities manage water as they see fit. In 2010, only 10% of municipalities chlorinated water, despite the legal mandate to do so. Rampant water extraction and pollution profit from the absence of adequate legislation and enforcement.[29] The inequitable access to water that most Guatemalans face is a violation of human rights.

Water for Agriculture and Food Security

Agriculture production represents 13.6% of the GDP and employs more than 31% of the labor force.[30] Agriculture accounts for the largest share of per-capita water extraction (145 cubic meters per year), compared with domestic use (33 cubic meters per year) and industry use (36 cubic meters per year). In 2000, only 5% of land with irrigation potential was irrigated, with the private sector accounting for the majority of irrigation infrastructure.[31] In coastal areas, commercial farmers are known to divert rivers to irrigate plantations, literally stealing the water from downstream residents and farmers.

Diverting rivers for commercial agriculture is "standard procedure." Although it is illegal, rivers are diverted toward banana, oil palm, and sugarcane plantations. In 2016, in response to grassroots groups' denunciations, including the "Walk for the Water," the Ministry of Environment found more than 50 dams diverting river flows. In May 2016, the ministry proceeded to denounce seven agribusinesses, requesting an investigation from the Public Prosecutor's Office.[32] However, by July 2016, the minister stated that the office would not be pressing charges but would dialogue with violators to resolve the situation.[33] The "Walk for the Water" made a few specific requests to the government, including one to make diverting a river a crime. Congress was to vote on this request in early May, but 70 congresspeople voted against considering the motion.[34] These outcomes confirm the unwillingness of government authorities to exercise the rule of law over powerful interest groups. Meanwhile, communities downriver lack water.

Water and Energy

Energy-wise, Guatemala is chronically poor; its installed capacity is a fraction of its potential; and a significant number of its inhabitants lack access to electricity, relying on biomass (firewood) for most of their energy. The sector is largely undiversified, consisting mostly of hydropower (36%) and fossil fuels (48%).[35] Efforts to increase electricity generation capacity have focused on additional hydropower, which has met with significant opposition from civil society due to concerns of environmental degradation and the impact on indigenous peoples and territories. Most hydropower projects in the country have not carried out adequate consultation processes with those groups whose livelihoods and territories are directly impacted by dams, nor have they offered adequate and fair compensation. For most affected groups, a hydropower plant does not mean new stable sources of employment, access to education, or other strategies to break out of poverty—it does not even guarantee access to electricity. Hydrological projects also impact the land, locals' historical attachment to it, and the natural resources on which they rely for their livelihoods.[36] In a country plagued with injustice and inequalities, the land and its resources are not only a source of historical pride and roots to ancestors, but also the only resources on which people are able to rely for sheer survival.

Water and Mining

Although mining's contribution to Guatemala's economy is negligible, constituting less than 1% of Guatemala's GDP, mining is at the heart of many conflicts that privilege the elites over the rural and poor. Existing laws are deficient, especially in regard to social and environmental concerns. Mining companies are not required to pay more than 1% in royalties to the government. Mining royalties and taxes paid account for 0.1% of government fiscal revenue.[37] Mining is a significant source neither of jobs nor of income, but it is a significant source of social conflict.[38]

On March 2, 2012, resisting communities of La Puya began a peaceful, 24-hour blockade at the entrance to the El Tambor mine in San Pedro Ayampuc, just outside Guatemala City. Their concerns were for the health and environmental impacts of the mine. A major concern was water because their communities were already water deprived, receiving water only two days a week while a mining operation can use in one day what a family with adequate water access will use in a year. The communities were also protesting for their right to be consulted about a project that would directly affect their lives and livelihoods. In four years of continued and peaceful resistance, they endured repeated intimidation, violence, prosecution, and even an attempted murder. On July 15, 2015, the nonviolent movement of La Puya had a major victory when a Guatemalan court ruled in its favor, ordering the mine to suspend all activities and cease any construction until a community consultation was held. The mining company appealed but the court held to its decision. Despite the rulings, the mining company continued to operate. Added pressure came on October 26, 2015, from 12 members of the U.S. Congress, who sent a letter to President Alejandro Maldonado expressing their concern for the abuses related to the illegal operations of El Tambor gold mine and calling for the immediate halt of its operation. In February 2016, the Guatemalan Supreme Court ruled to provisionally suspend the mining license due to lack of prior consultation. La Puya is now a source of inspiration for many other communities protesting for their right to protect the water and land in their territories.[39]

Increasing Vulnerability Due to Climate Change

Guatemala is at a critical threshold where it must strengthen its agriculture and energy sectors and invest in food security while addressing sources of conflict and violence. It must do so, taking into account the compounding effects and risks associated with climate change.

Central America is already hard-hit by the effects of climate change. Guatemala is experiencing severe drought and increasingly frequent hurricanes, mudslides, and flash floods, most notably in 1998, 2005, and 2010. According to the World Bank, Guatemala ranks fifth among countries with the highest economic risk exposure to three or more climate change–related hazards, and 83.3% of Guatemala's GDP is in risk areas.[40] The Global Climate Risk Index identifies Guatemala as one of the 10 countries currently experiencing the effects of climate change most intensely.[41]

Climate change has serious consequences for Guatemala. Decreased precipitation and increasing temperatures are significant threats to agriculture and food security. Coffee, one of Guatemala's top export crops, is now hit with coffee rust, a pest that was previously controlled. Higher temperatures led to a surge of the pest and, more important, to an expansion of its area of influence because of the increase in temperature experienced at higher elevations where prime coffee is grown.[42]

Subsistence farmers are highly vulnerable to crop failure due to decreased precipitation and increased temperatures. In 2015–2016, the El Niño effect, compounded by the ongoing drought, caused damages and losses throughout the country. It is estimated that the crop failure and low reserves of grain impacted 2 million people.[43] Vulnerable families lost between 50% and 100% of their food reserves, mainly of maize. In addition to drought, Oxfam predicts a doubling of food prices by 2030 that will exacerbate food insecurity as demand continues to rise amid a collapsing resource base and markets rigged against the poor.[44]

The specter of electricity blackouts, as recently experienced in Venezuela, hangs over the country as reduced water storage for hydroelectricity generation becomes more serious. In April 2016, the Venezuelan government instituted a two-day working week for public-sector workers in order to cut energy consumption in response to a climate-induced energy crisis. Severe drought caused by El Niño decreased water levels of the main hydroelectric dam and caused widespread blackouts. These blackouts have only deepened Venezuela's economic crisis.[45]

A "Plebeian" Rebellion: The Local Control of Natural Resources

Triggered by diversion of river flows, construction of hydroelectric dams built without consultation, and poisoning of the water due to mining activities and oil exploitation, rural peasant and indigenous communities are determined to gain local control over their territories.

Rural communities argue for full control of their natural resources and that any decision as to how to take advantage of their resources should be made by them—those resources' inhabitants and keepers. Furthermore, they claim that the benefits and revenues derived from any exploitation of these resources should be invested first and foremost in the communities where the resources were sourced. Currently, all benefits are controlled by the large national or international investors, who benefit from multiple government-sanctioned incentives and care very little, if at all, for the well-being of the local communities. To add insult to injury, the royalties paid to the government are first negligible and second controlled by the central government, with no royalties devoted to local reinvestment programs.

Their rebellious claim is in direct confrontation with the "aristocracy" of the central government authorities and their environmentally unsustainable approach to economic development. While the government and corporations are concerned with short-term profits only, rural communities argue, the people who live in those communities are best suited to manage the resources sustainably. Their historical attachment to the land and their reliance on natural resources for their livelihoods have taught them how to use and protect the resources in a responsible and sustainable way.

The most recent expression of this claim is the "Walk for the Water, Mother Earth, Land, and Life," convened for Earth Day 2016. The Walk for the Water's manifesto called to peacefully demonstrate for water, Mother Earth, land and, life, making the case for the right of access to water and against its privatization. In their own words:

> "Water is not a merchandise. It is a human right."
> – Irene Barrientos, from the Committee of Peasant Unity

> "Peasants are impoverished by the lack of water, crops dry out, and drought affects several communities. That is why we support this march: to ask the government for concrete measures to stop those who use the rivers for their selfish benefit without realizing the damage they are causing."
> – Gerson de León, peasant leader

Water Security and Guatemala's Contested and Ungoverned Resource

The United Nations defines *water security* as the capacity of a population to safeguard sustainable access to adequate quantities of acceptable quality water for sustaining livelihoods, human well-being, and socioeconomic development, for ensuring protection against water-borne pollution and water-related disasters, and for preserving ecosystems in a climate of peace and political stability.[46]

Water security captures the complex interconnectedness that makes water the unifying resource necessary for security, sustainability, development and human well-being. Unfortunately, Guatemala's water management is weak institutionally, technically, and financially. It is the ungoverned resource as the country lacks a national authority in charge of its governance; it is the contested resource since it is the source of nationwide and local divisive conflicts. This section outlines the key elements of Guatemala's ungoverned and contested water scenario and analyzes what is required to resolve this fundamental problem once and for all.

Guatemalans can be grouped into three interest groups in regards to water management and governance:

a   Rural water users organized in favor of water rights for all
b   Large-scale water users who are against water-use regulations
c   Politicians and public decision-makers who

- avoid their responsibility to develop and enact water-related policies and legislation, or
- choose to favor powerful groups interested in the absence of regulations

Previous attempts at developing and enacting water management legislation have ended in divisiveness between these groups and failure to move forward. Any effective strategy in support of water management and governance must facilitate an open and representative dialogue with these groups.

For rural populations, the problem is compounded by a perception that someone is "stealing" or "poisoning" the water while the government looks the other way. This perception has its foundations in specific events in which rivers are diverted or water bodies are polluted, while entire communities lack access to water. Such a perception is a logical consequence of rural peoples' experienced difficulties and disasters, including the following elements:

a   The belief that water was an ever-present and inexhaustible resource generously provided by Mother Nature, the people's duty being to defend it against the devastation brought on by modern agriculture and industry.
b   The experience of having water stolen from them through a river being diverted, a mine taking over a water body, or the water being dammed. These communities not only lose their water but also are disrespected and dismissed when they dare raise their voices against the injustice.
c   The evidence of rampant pollution of rivers and lakes, with total disregard for existing legislation.

d   The lack of trust in public officials and institutions to enforce existing legislation because they either are bribed or are simply incapable of designing and enforcing sound legislation and regulation.

All of these elements stem from the absence of adequate water governance and management. Guatemalan authorities suffer from an endemic failure to arrive at solid and sustainable political agreements on how to manage public goods. They are also incapable of building institutional scaffolding, providing necessary technical know-how, and managing budgets. In regard to water, this has resulted in the total absence of institutions to manage such a basic and indispensable resource.

The Road Towards Water Security

Despite the difficult circumstances in which water policy must be developed and enacted, some of the key factors are already in place. An adequate assessment and analysis of the water situation in Guatemala and what needs to be done already exists. Key members of society, including scientists and academics, reputable think tanks, environmental organizations, a few public authorities, and some influential figures in the private sector are in agreement on the elements of Guatemala's water problem that must be addressed. There is a general agreement on the following:

a   Water is a human right and a public good.
b   Guatemala is water stressed and climate change will aggravate it.
c   Water bodies are being polluted and depleted.
d   The country's water-management and governance crisis is characterized by the following:
    - Lack of reliable, systematic, and comparable data on water
    - Lack of adequate water-related infrastructure
    - Lack of competent water-related government institutions
    - Lack of adequate government budgets for operating government institutions
    - Lack of interinstitutional coordination for water management
    - Lack of political will of congressional representatives to confront powerful interest groups lobbying against water governance
    - Lack of interest and capacity to enforce existing regulations against violators
    - Lack of fair and equitable access to water

Furthermore, there is also consensus on the existence of latent conflicts that can arise from lack of access to water, with water-deprived groups migrating to water-rich areas, creating territorial conflicts. This type of conflict could lead to internal struggles for water-rich territories but could also spill over into

Honduras and El Salvador. These conflicts could then result in an added layer of insecurity and violence, which may trigger even more migration beyond the region toward Mexico and the U.S.

These are all elements of a complex web of intertwined problems related to underdevelopment and political shortcomings. These problems are already producing alarming conflicts that can lead to more serious internal and regional governance crises. In the words of Jorge Cabrera, one of the most respected environmentalists in the country:

> Mining and electricity-generation projects, land use, border conflicts, and social demands for education, healthcare, and reductions in the cost of electricity have all been identified by the National System for Dialogue as sources of conflict: some have deep historical roots; others, a more presently cyclical nature. But all are impossible to avoid.[47]

Guatemala has lacked adequate water governance and management arrangements for too long. It now must resolve this once and for all, and it must do so with the added threats of climate change. An integrated and systemic approach to water security must be built. The approach will require the following:

a   Consensus on the need to address the problem with a water security focus, taking into account access to water, water quality and quantity, and its role in environmental systems

b   A renewed water policy with coordinated implementation strategies across relevant sectors

c   A legal and regulatory framework: both a general Water Law and regulatory instruments with graduated sanctions according to the seriousness and scale (national, subregional, local) of infractions

d   Water management institutions capable of coordinating with other relevant institutions

e   A plan with an adequate and secured budget to develop and maintain water management infrastructure

f   Agreements for the responsible management of transboundary water basins with neighboring countries

g   Agreements among the different sectors of society on sharing responsibilities for fair and equitable water access and management

An innovative, all-encompassing legal framework for water governance is long overdue. Although in early 2016 the Guatemalan Congress commissioned a draft for a General Water Law, the probability that it would issue an actual law was slim. The Congressional Commission on Water Resources, in charge of assisting Congress in the drafting and approval, has moved at a glacial pace, and the newly elected Speaker of the House has not demonstrated any of the former Speaker's interest in the water agenda.

Guatemala must find a way to overcome its political inability and unwillingness to address water governance. The dire scenario the country is facing will only get worse and will lead to even-more-difficult environmental, social, and economic hardships. A profound, structured, national dialogue on water could demand an adequate response from Congress. Guatemala's citizens could also revive the 2015 massive mobilizations against corruption, redirecting them to demand from Congress a participatory and transparent process to once and for all issue water-related legislation.

The country's civil society has been able to rely on support from the international diplomatic and development aid community working in Guatemala. Now, more than ever, it is necessary to join forces and push for a comprehensive government response to this potentially devastating crisis.

The U.S. is in a particularly advantageous position to support efforts for water security given the recently deployed Alliance for Prosperity Program. Through this program, the U.S. could support a transparent dialogue between government and civil society sectors to address this issue, whose repercussions for national security are unquestionable.

A Proposed Agenda for the U.S. Government

Water security can be a privileged means for promoting democratic governance, strengthening food security and environmental resilience, and reducing agricultural and energy risks. Moreover, water governance and management must be considered fundamental elements of a security strategy. Guatemala is sitting on a time bomb, and the clock is already ticking.

Aside from the support the U.S. can offer to address Guatemala's water security, the U.S. could consider the following three proposals:

a   The U.S. must address the regional security crisis through an integrated strategy that includes both Mexico and Central America's Northern Triangle. The security challenges the region faces are not relevant only to the Central American Northern Triangle; Mexico is directly responsible for the most recent peak in violence in the region. Mexico's increased offensive against drug trafficking and organized crime pushed the violence southward, and Central American migrants endure terrible and dangerous perils in Mexico trying to reach the U.S. southwestern border. The region must be viewed as a dynamic, integrated whole in a concerted strategy to eliminate violence and organized crime.

b   The U.S. government should strongly take into account that increased enforcement against illegal migration has not stopped the flow of immigrants, but has driven them further into the shadows. If the Trump administration fulfills its promise of building a wall along the border with Mexico, the consequences for both migrants and the economies of countries like Guatemala will be significant. Guatemala relies heavily on the remittances that migrants send home from the U.S. As those migrants face more obstacles to carrying on productive lives in the U.S., incoming remittances will diminish, affecting Guatemala's overall economy and especially the poor. Without alternatives, more of the poor will migrate, despite vulnerabilities to human trafficking, extortion, and abuse.

The U.S. could support poverty-reduction strategies in Guatemala including linking local food producers with relevant markets in the U.S. Supporting export-oriented cooperatives and medium-sized companies as they search for U.S. markets could offer employment and local economic strategies for many Guatemalans currently unable to reap a decent livelihood. This perspective would be quickly picked up by U.S. companies and is compatible with the Alliance for Prosperity and USAID's economic development priorities for Guatemala. This strategy would have the added value of contributing to strengthening food security and water sustainability.

c As the new U.S. administration takes office, it should not lose sight of the strategic importance that the Alliance for Prosperity and the Central

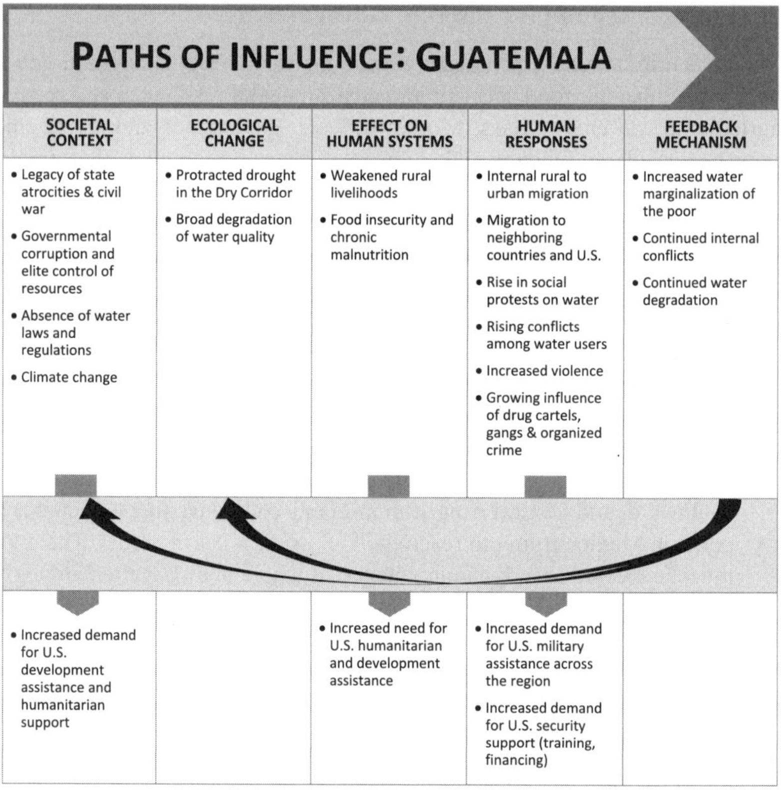

PATHS OF INFLUENCE: GUATEMALA

	SOCIETAL CONTEXT	ECOLOGICAL CHANGE	EFFECT ON HUMAN SYSTEMS	HUMAN RESPONSES	FEEDBACK MECHANISM
LOCAL EFFECTS	• Legacy of state atrocities & civil war • Governmental corruption and elite control of resources • Absence of water laws and regulations • Climate change	• Protracted drought in the Dry Corridor • Broad degradation of water quality	• Weakened rural livelihoods • Food insecurity and chronic malnutrition	• Internal rural to urban migration • Migration to neighboring countries and U.S. • Rise in social protests on water • Rising conflicts among water users • Increased violence • Growing influence of drug cartels, gangs & organized crime	• Increased water marginalization of the poor • Continued internal conflicts • Continued water degradation
IMPACTS ON THE U.S.	• Increased demand for U.S. development assistance and humanitarian support		• Increased need for U.S. humanitarian and development assistance	• Increased demand for U.S. military assistance across the region • Increased demand for U.S. security support (training, financing)	

Figure 6.1 Paths of Influence: Guatemala

American Regional Security Initiative (CARSI) have for both the U.S. and the region. The Alliance requires active and continued support, and the new administration must carry it forward with the same intensity that former Vice President Biden gave it. Security concerns ought to be addressed both through institutional reform (an ongoing major focus of the Alliance) and through economic development and productivity, wherein the private sector is vital. The new U.S. administration must remember that the Alliance's effectiveness relies on a combined approach to address security concerns (reduced violence and organized crime), as well as the other main causes of the crisis: poverty, food insecurity, and environmental vulnerability.

The crossroads at which Guatemala finds itself has serious and dangerous repercussions for the entire North American continent, and requires an integrated and thorough response. Both Guatemala and the U.S. government should face this crisis with their full strength to prevent Guatemala from falling into uncontrollable violence and abject poverty.

Notes

1 Muñoz, Geldi and Mammuel Hernández, Diputados comprometidos en aprobar que desvío de ríos sea delito, *Prensa Libre*, 22 April 2016, accessed on 10 December 2016, www.prensalibre.com/guatemala/comunitario/caminata-llega-hoy-a-la-plaza-central; Rey Rosa, Magaly, La Marcha por el Agua Fue Tremenda y los 70 Diputados que votaron a favor del Desvío de Ríos, *Plaza Pública*, 4 May 2016, accessed 10 December 2016, www.plazapublica.com.gt/content/la-marcha-por-el-agua-fue-tremenda-y-70-diputados-que-votaron-favor-del-desvio-de-rios.
2 Kahl, Colin, *States, Scarcity and Civil Strife in the Developing World* (Princeton: Princeton University Press, 2006).
3 "World Bank Open Data," accessed 30 November 2016, http://data.worldbank.org/country/guatemala.
4 "United Nations Human Development Index," accessed 30 November 2016, http://hdr.undp.org/en/content/income-gini-coefficient.
5 "Food and Agriculture Organization Livestock Census" data from *Instituto Nacional de Estadística, VI Censo Nacional Agropecuario, Características Generales de las Fincas Censales y de Productoras y Productores Agropecuarios, Resultados Definitivos, Vol. I., January 2004.* (Rome: United Nations Food and Agriculture Organization, 2004), 78–79, accessed 30 November 2016, www.fao.org/docrep/013/i1595e/i1595e01.pdf; Barry, Bridget, "Guatemala and Integrated Rural Development: Towards Inclusive Growth in the Rural Sector." *The International Policy Centre for Inclusive Growth*, Research Brief No. 37 (2012), accessed 30 November 2016, www.ipc-undp.org/publication/26606.
6 Bolaños, Rosa María and ACAN EFE. "Pobreza sube a 59.3 %: son 9.6 millones de guatemaltecos los afectados" *Prensa Libre*, 10 December 2015, accessed 30 November 2016, www.prensalibre.com/economia/se-dispara-a-593-la-pobreza-96-millones-de-guatemaltecos-viven-en-pobreza; Instituto Nacional de Estadística, "Encuesta Nacional de Condiciones de Vida 2014" (Guatemala: Instituto Nacional de Estadística, 2015), accessed 30 November 2016, www.ine.gob.gt/sistema/uploads/2016/02/03/bWC7f6t7aSbEI4wmuExoNR0oScpSHKyB.pdf.
7 The national survey on mother and child health (2014–2015) found the highest rates of chronic malnutrition in the highly indigenous departments of Totonicapán

(70%), Quiché (69%), Huehuetenango (68%), and Sololá (66%); Ministerio de Salud Pública y Asistencia Social, Instituto Nacional de Estadística y Secretaria General de Planificación. "Encuesta Nacional de Salud Materno Infantil 2014–2015" (Guatemala: MSPAS, INE, SEGEPLAN, November 2015), accessed 29 November 2016, http://pdf.usaid.gov/pdf_docs/PBAAD728.pdf.

8 "Ending Rural Hunger." Country rankings on child malnutrition and under-five stunting for all regions and for Latin America and the Caribbean (Washington D.C.: Brookings Institution), accessed November 29, 2016, https://endingruralhunger.org/data/rankings/; "UNICEF Statistics at a Glance: Guatemala" (New York: United Nations Children's Fund), accessed November 29, 2016, www.unicef.org/infobycountry/guatemala_statistics.html.

9 "Ending Rural Hunger." Country rankings on resources and public and private investments in Latin America and the Caribbean. The Brookings Institute. Accessed November 29, 2016, https://endingruralhunger.org/data/rankings/.

10 Overseas Security Advisory Council, "Guatemala 2015 Crime and Safety Report" (Washington, DC: U.S. Department of State Bureau of Diplomatic Security), accessed 30 November 2016, www.osac.gov/pages/ContentReportDetails.aspx?cid=17785.

11 For the year 2010, 56 murders per 100,000 inhabitants is the average murder rate for El Salvador, Honduras, and Guatemala together. The Honduras murder rate is 82, El Salvador's is 65, and Guatemala's is 41. The top three murder rates are: Honduras (82), El Salvador (65), and Côte d'Ivoire (57), followed by the Central American Northern Triangle rate of 56 in the number four spot. Guatemala is ranked at number nine. United Nations Office on Drugs and Crime, "Transnational Organized Crime in Central America and the Caribbean: A Threats Assessment" (Vienna: United Nations Office on Drugs and Crime (UNODC), 2012), accessed 30 November 2016, www.unodc.org/documents/data-and-analysis/Studies/TOC_Central_America_and_the_Caribbean_english.pdf.

12 United Nations Office on Drugs and Crime, "Transnational Organized Crime in Central America and the Caribbean: A Threats Assessment."

13 United Nations Children's Fund, "Hidden in Plain Sight: A Statistical Analysis of Violence Against Children" (New York: United Nations International Children Fund, 2014), Figures 3.10B and 3.10C, accessed 1 December 2016, http://files.unicef.org/publications/files/Hidden_in_plain_sight_statistical_analysis_EN_3_Sept_2014.pdf.

14 United Nations Office on Drugs and Crime, "Transnational Organized Crime in Central America and the Caribbean: A Threats Assessment."

15 Abdelmottlep, Mamdooh A. "World Internal Security and Police Index" (1st edition) (Land o' Lakes: International Police Science Association, 2016), accessed November 24, 2016, http://wispindex.org/sites/default/files/downloadables/WISPIpercent20Report_EN_WEB_0.pdf; EFE, "Guatemala entre los peores calificados en Seguridad y Policía" *Prensa Libre,* accessed November 24, 2016. www.prensalibre.com/guatemala/comunitario/guatemala-entre-los-peor-calificados-en-seguridad-y-policia.

16 Lohmuller, Michael, "Guatemala's Government Corruption Scandals Explained," *Insight Crime,* June 21, 2016, accessed November 15, 2016, www.insightcrime.org/news-analysis/guatemala-s-government-corruption-scandals-explained.

17 Meyer, Peter J., Rhoda Margesson, Clare Ribando Seelke, Maureen Taft-Morales, "Unaccompanied Children from Central America: Foreign Policy Considerations" (Washington D.C.: Congressional Research Service, 11 April 2016), fig. 1, accessed 24 November 2016, www.fas.org/sgp/crs/homesec/R43702.pdf.

18 United States Office of the Press Secretary, the White House, "Presidential Memorandum—Response to the Influx of Unaccompanied Alien Children across the Southwest Border," 2 June 2014, accessed 30 November 2016, www.white

house.gov/the-press-office/2014/06/02/presidential-memorandum-response-influx-unaccompanied-alien-children-acr.
19 U.S. Office of the Press Secretary, the White House, "FACT SHEET: The United States and Central America: Honoring Our Commitments," 14 January 2016, accessed 30 November 2016, www.whitehouse.gov/the-press-office/2016/01/15/fact-sheet-united-states-and-central-america-honoring-our-commitments.
20 Meyer, Peter J. et al., "Unaccompanied Children from Central America: Foreign Policy Considerations"; Jonas, Susanne, "Guatemalan Migration in Times of Civil War and Post-War Challenges" (Washington, DC: Migration Policy Institute, March 27, 2013), accessed 30 November 2016, www.migrationpolicy.org/article/guatemalan-migration-times-civil-war-and-post-war-challenges.
21 Rosenblum, Marc R. and Isabel Ball. "Trends in Unaccompanied Child and Family Migration from Central America," Factsheet (Washington, DC: Migration Policy Institute, January 2016), accessed 30 November 2016, www.migrationpolicy.org/research/trends-unaccompanied-child-and-family-migration-central-america; Meyer, Peter J. et al., "Unaccompanied Children from Central America: Foreign Policy Considerations."
22 United Nations High Commissioner for Refugees, "Women on the Run: A first account of refugees fleeing El Salvador, Honduras, Guatemala and Mexico" (Geneva, United Nations High Commissioner for Refugees (UNHCR), 2016), accessed 30 November 2016, www.unhcr.org/publications/operations/5630f24c6/women-run.html.
23 "Havocscope Global Black Market Information," accessed 30 November 2016, www.havocscope.com/black-market-prices/human-smuggling-fees/.
24 World Food Program and International Organization for Migration "Hunger Without Borders: The Hidden Links Between Food Insecurity, Violence and Migration in the Northern Triangle of Central America" (Panama City: World Food Program (WFP) and International Organization for Migration (IOM), 2015), accessed 30 November 2016, documents.wfp.org/stellent/groups/public/documents/liaison_offices/wfp277544.pdf?_ga=1.26677821.1852681487.1479920693.
25 Office for the Coordination of Humanitarian Affairs, "Drought in Central America: Situation Report as of December 10 2014" (Panama City, Office for the Coordination of Humanitarian Affairs (OCHA), 2014), accessed 30 November 2016, http://reliefweb.int/sites/reliefweb.int/files/resources/Droughtpercent20inpercent20Centralpercent20America.pdf.
26 Office for the Coordination of Humanitarian Affairs, "El Nino: Overview of Impact, Projected Humanitarian Needs and Response as of 02 June 2016" (Panama City, Office for the Coordination of Humanitarian Affairs -OCHA, 2014), accessed 30 November 2016, http://reliefweb.int/sites/reliefweb.int/files/resources/OCHA_ElNino_Monthly_Report_2Jun2016.pdf.
27 Ballestero, Maureen, Virginia Reyes, and Yamileth Astorga, "Groundwater in Central America: Its Importance, Development and Use With Particular Reference to Its Role in Irrigated Agriculture," in *The Agricultural Groundwater Revolution Opportunities and Threats to Development*, edited by Mark Giordano and Karen Villholt. 2007. Colombo: International Water Management Institute. Chapter 6, accessed 30 November 2016, www.iwmi.cgiar.org/Publications/CABI_Publications/CA_CABI_Series/Ground_Water/protected/Giordano_1845931726-Chapter6.pdf?galog=no.
28 U.S. Army Corps of Engineers, "Water Resources Assessment of Guatemala." Mobile District and Topographic Engineering Center (Washington, DC: U.S. Army Corps of Engineers, June 2000), accessed 30 November 2016, www.sam.usace.army.mil/Portals/46/docs/military/engineering/docs/WRA/Guatemala/Guatemalapercent20WRApercent20English.pdf.

29 "Global Water Partnership Country Water Partnerships: Guatemala," accessed 30 November 2016 www.gwp.org/en/About-GWP/Country-Water-Partnerships/Guatemala-/.
30 "U.S. Central Intelligence Agency the World Factbook: Guatemala," accessed 30 November 2016, www.cia.gov/library/publications/the-world-factbook/geos/gt.html.
31 Ballestero, Maureen et al., "Groundwater in Central America: Its Importance, Development and Use With Particular Reference to Its Role in Irrigated Agriculture."
32 Contreras, Geovanni, "MARN denuncia a empresas por desvío de ríos," *Prensa Libre*, 9 May 2016, accessed 10 December 2016, www.prensalibre.com/guatemala/denuncian-a-empresas-por-desvio-de-rios.
33 Alvarez, Carlos, "Ambiente ya no denunciará por desvío de ríos," *Prensa Libre*, 12 July 12, 2016, accessed 10 December 2016, www.prensalibre.com/guatemala/comunitario/ambiente-ya-no-denuncia.
34 Rey Rosa, Magaly, La Marcha por el Agua Fue Tremenda y los 70 Diputados que votaron a favor del Desvío de Ríos.
35 "Climate Scope Guatemala," accessed 1 December 2016, http://global-climatescope.org/en/country/guatemala/#/details; Koberle, Alex, "Reality Check for Guatemala's Energy Plans," International Rivers Review. Vol 27/No. 2, June 7 2012, accessed 1 December 2016, www.internationalrivers.org/resources/reality-check-for-guatemalapercentE2percent80percent99s-energy-plans-7504; U.S. Army Corps of Engineers, "Water Resources Assessment of Guatemala."
36 Hirsh, Cecilie and Miguel Utreras. "Power to the People: Hydropower, Indigenous People's Rights and Popular Resistance in Guatemala" (Oslo: Association for International Water Studies (FIVAS), September 2010), accessed 1 December 2016, http://fivas.org/wp-content/uploads/2015/05/Power-to-the-people_FIVAS_2010.pdf.
37 United Nations Economic Commission for Latin America and the Caribbean, "Pactos para la Igualdad, Hacia un futuro sostenible" (Santiago: UN Economic Commission for Latin America and the Caribbean -ECLAC, 2014), Table VI.2 page 289, accessed 12 December 2016, www.cepal.org/es/publicaciones/36692-pactos-la-igualdad-un-futuro-sostenible.
38 Escalon, Sebastian and Angelica Medinilla, "Los muchos favores del estado a la minería," *Plaza Pública*, 31 October 2016, accessed 10 December 2016, www.plazapublica.com.gt/content/los-muchos-favores-del-estado-la-mineria.
39 Guatemalan Human Rights Commission, "La Puya Environmental Movement" (Washington, DC: Guatemalan Human Rights Commission, 2016) accessed 10 December 2016, www.ghrc-usa.org/our-work/current-cases/lapuya/#puyarecent updates.
40 "World Bank Guatemala Dashboard: Risk Screening Overview," accessed 1 December 2016, http://sdwebx.worldbank.org/climateportal/countryprofile/home.cfm?page=country_profile&CCode=GTM&ThisTab=RiskOverview.
41 Kreft, Sönke, David Eckstein and Inga Melchior. "Global Climate Risk 2017: Who Suffers Most From Extreme Weather Events? Weather-Related Loss Events in 2015 and 1996 to 2015" (Bonn: Germanwatch, November 2016), accessed 1 December 2016, https://germanwatch.org/en/download/16411.pdf.
42 Malkin, Elizabeth, "A Coffee Crop Withers, Fungus Cripples Coffee Production Across Central America," *New York Times*, May 15, 2014, accessed 10 December 2016, www.nytimes.com/2014/05/06/business/international/fungus-cripples-coffee-production-across-central-america.html?_r=0.
43 World Food Program, "Guatemala Country Brief September 2016" (Rome: World Food Program, 2016), accessed 1 December 2016, http://documents.wfp.org/stellent/groups/public/documents/ep/wfp272220.pdf?_ga=1.243236098.1852681487.1479920693.

44 Oxfam, "Extreme Weather, Extreme Prices: The Costs of Feeding a Warming World" (Oxfam: Oxford, September 2012), accessed 1 December 2016, www.oxfam.org/sites/www.oxfam.org/files/20120905-ib-extreme-weather-extreme-prices-en.pdf; Oxfam. "Entering Uncharted Waters: El Niño and the Threat to Food Security" (Oxfam: Oxford, 1 October 2015), accessed 1 December 2016, www.oxfam.org/sites/www.oxfam.org/files/file_attachments/mb-el-nino-uncharted-waters_1.pdf.
45 "Venezuela Introduces Two-Day Work Week to Deal With Energy Crisis," *BBC*, 27 April 2016, accessed 1 December 2016, www.bbc.com/news/world-latin-america-36145184.
46 United Nations Water, "Water Security and the Global Water Agenda: A UN Water Analytical Brief" (Hamilton: United Nations University, 2013), accessed December 1 2016, www.unwater.org/downloads/watersecurity_analyticalbrief.pdf.
47 Jorge Cabrera (Guatemalan environmentalist) in discussion with Eduardo Stein, 2016.

7 Water Stress, Instability, and Violent Extremism in Nigeria

Marcus King

Introduction

Nigeria is a rich case study for understanding the security-related dynamics and consequences of water stress due to the simultaneous existence of multiple water challenges, geographic and climatic variability, and myriad social cleavages and economic divisions. Rarely, if ever, have so many factors conspired to reduce societal stability within a sovereign territory important to U.S. security interests. There is a chain of causality between water stress and conflict in Nigeria, but the exact nature of the relationship is poorly understood.

This chapter takes a different approach than did previous assessments. Previous analyses of water and conflict in Nigeria have focused exclusively on the possibility of interstate disputes among the 10 nations of the Niger River Delta, or on one particular region within the Nigerian state. While allocation of waters in the Niger River Basin has the potential to cause disagreements between riparian states, the possibility of interstate violence is relatively remote.

The focus in this chapter is on water-driven security problems with implications for internal stability in a country ranked among the most fragile on the globe.[1] Instability is accentuated by the symbiotic relationship between water stress and violence perpetrated by three sub-state militant groups analyzed herein: Boko Haram; Niger Delta Vigilante (NDV) Group and its predecessor, Movement for the Emancipation of the Niger Delta (MEND); and militant Hausa-Fulani herdsmen.

Nigerian policies that encourage inefficient or exclusionary allocation of natural resources led to a civil war that raged from 1967 to 1970. New grievances, stemming from similar policies, are leading to reactions that threaten to undermine Nigerian stability. Today, these grievances must be evaluated in the context of increasing water stress, demographic pressure, and accelerating climate change. The stakes are high. The unraveling of Nigeria would undermine the security architecture of West Africa, a region vital to U.S. economic and security interests.

Therefore, this chapter examines the conditions of water resources in Nigeria and trends that conspire to create instability. Next, it provides a snapshot of how water stress relates to distinct and growing internal conflicts in three ecoregions. Finally, it provides recommendations for more effective interventions by the U.S. and the Nigerian governments in order to forestall some of the worst consequences of water stress and promote policies that prevent or mitigate conflict.

Background

Nigeria as a Geostrategic Power

Nigeria is a critical actor on the world stage and wields heavy influence in international and regional institutions. It is Africa's most populous country, with the eighth-largest population in the world.[2] Nigeria also contributes a large number of troops to peacekeeping operations in Africa.[3]

Nigeria is also a regional economic power. It is the largest oil producer in Africa and was the world's fourth-largest exporter of LNG in 2015.[4] Oil and natural gas revenue is the country's main source of foreign exchange, making up more than 95 percent of total exports in 2014.[5] This condition has brought along with it a "resource curse" prevalent in countries whose wealth derives from a single commodity. The fluctuations of commodity prices inhibit investments that would otherwise lead to a more stable economy that would promote human development. Furthermore, extractive industries tend to employ fewer

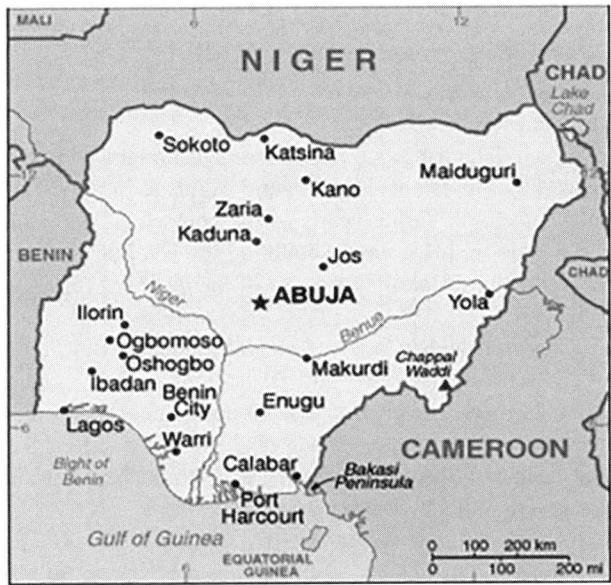

Figure 7.1 Political Map of Nigeria
Source: CIA World Factbook

people than sectors like agriculture, concentrating the wealth in the hands of the political elite and leading to governance that is relatively more authoritarian, corrupt, and prone to conflict.⁶ Inequality in Nigeria is extreme and getting worse; in 2010, 61 percent of Nigerians lived on less than a dollar a day.⁷

As of 2016, the country has been hit hard by slumping oil prices.⁸ In addition to oil, Nigeria's economic security is also dependent on the agricultural sector, which accounts for more than 40 percent of GDP and roughly 70 percent of employment.⁹ However, despite this dependence, only 40 percent of available land is cultivated.¹⁰ As a response to declining oil prices, the Nigerian government implemented the Agricultural Transformation Agenda (ATA), shifting investment "to create over 3.5 million jobs" to increase employment among youth and women especially. The ATA prioritizes cultivation of rice, sorghum, cassava, cotton, cocoa, and oil palm as well as animal husbandry and fishing.¹¹

Internal conflicts are at a boiling point, and if they are not brought under control soon, the Nigerian government will lack the political and economic capacity to complete the ATA initiative. Implementation requires a stable security environment in the Niger River Basin, where much of the farming is based. Meanwhile, northeastern Nigeria is undergoing a humanitarian crisis, with famine-levels of insecurity and thousands of people displaced, placing demand on already scarce food resources.

Three Conflict Flash Points

Three internal conflicts in Nigeria have rough geographical bases and feature distinct subnational militant groups as the primary aggressors, although the Government of Nigeria (GON) has retaliated mercilessly in most cases. The three conflicts are generally worsening, and water stress plays a variable but significant role in each.

First, the militant Islamist group Boko Haram has waged a violent insurgent campaign since 2010 to establish an Islamic caliphate in the country's north and northeast. Attacks in various parts of the country are estimated to have killed 11,000 people in 2015, nearly doubling the level of violence from the year before, and making Boko Haram one of the deadliest militant groups in the world.¹²

Second, in a region widely referred to as the Middle Belt, a conflict pits members of the semi-nomadic Muslim Hausa-Fulani tribe against predominantly Christian farmers. The primary issue is access to grazing land. This land has been ravaged by climate change and drought. The resulting violence has left thousands dead over the past five years. In 2016 the herdsmen killed more Nigerians than were killed by Boko Haram.¹³

Third, 2016 has seen renewed attacks by a handful of militant groups in the oil-rich southern Niger Delta. These groups are waging guerilla warfare with the primary tactic of bringing the government to its knees by destroying oil infrastructure. Their grievances stem in part from the central government's oligarchical monopolization of oil resources without transferring benefits to local communities, and ecological destruction caused by oil spills. Economic grievances and ethnic, cultural, and religious tensions led to the secession of the southeastern provinces in the Biafran War, leading to an estimated 1 million to 3 million deaths, mostly from starvation.[14]

Geophysical and Ecological Trends

Status of Water Resources in Nigeria

Water resources vary greatly in Nigeria. The country spans six vegetation zones, from mangrove-saltwater swamps to grassland and desert. Weather and precipitation patterns vary widely, and altitudes range from 10 to 3,000 feet above sea level.[15] Three diverse water systems, including ground aquifers, rivers, and wetlands, provide services for irrigation, human consumption, hydroelectric power generation, and navigation.

Despite sometimes-copious water resources, Nigeria faces nationally pervasive challenges with both availability and quality of water. Water for human consumption is at a premium, with less than 65 percent of the population having access to improved water supply.[16,17] Only 8 percent of homes nationwide have treated pipe-borne water.[18] While some regions suffer from physical water scarcity, economic water scarcity is a greater overall trend. Economic water scarcity occurs when government services and relevant technical and financial capacity are insufficient to access available water supplies.[19] The level of water conservation is also insufficient: the United Nations Food and Agriculture Organization rates Nigeria's practices poor by international and African standards.

Water and energy poverty converge in Nigeria. Despite the existence of hydropower dams on the upper reaches of the Niger River, evaporation and abstraction of water for irrigation conspire to diminish generation capacity. The unpredictability of water flows in times of drought, and unusually heavy rainfall associated with climate change are contributory variables. These conditions commonly lead to power failures, forcing citizens to rely on privately owned generators.

Lack of water availability for sanitation and hygiene affects approximately two-thirds of Nigeria's people.[20] Sanitation infrastructure such as piped sewerage is almost nonexistent, and installation rates are declining, according to reports by the World Health Organization. The government struggles to provide adequate water for drinking and hygiene, with the rising demand of burgeoning populations in urban areas.[21] Poor sanitation and hygiene have led to high rates of diarrhea and pneumonia and are associated with other diseases, such as trachoma and worm-related illnesses.[22] Diarrhea is the second-largest direct cause of child mortality in Nigeria.[23]

Climate Change Impacts

Climate change has a significant impact on water resources. However, there is no national-level analysis that paints a clear picture of climate vulnerabilities, so a full understanding relies heavily on models for the world or West Africa.[24] Generally, the Intergovernmental Panel on Climate Change (IPCC) predicts that climate change will amplify existing water stress in Africa.[25] Water stress is the most visceral manifestation of climate change, meaning that its effects will be felt sooner and more acutely than others.

Temperatures are rising on average across the country. Parts of the country, especially the arid north, are facing more heat, accompanied by less precipitation. This is consistent with general IPCC predictions that West Africa generally will see 10 percent less rainfall by 2100. Parts of Nigeria's northern Sahel area, where the Sahara Desert is encroaching to the south, now receive 10 inches of rain annually, a full 25 percent less than 30 years ago.

More severe weather is also predicted.[26] Government figures show torrential rains and windstorms becoming more frequent and intense across Nigeria: over the past 40 years, recorded volumes of torrential rains increased 20 percent across various southern states, some of which already see up to 160 inches of rainfall a year.[27] Finally, sea levels could rise from 1.5 to 3 feet by century's end along the southern coastline, causing saltwater to intrude into the potable water supply.[28]

The Notre Dame Global Adaptation Index (ND-GAIN) is a useful data-driven tool used to assess a country's exposure, sensitivity, and ability to adapt to the negative impact of climate change. The strength of six life-supporting sectors—food, water, health, ecosystem services, human habitat, and infrastructure—is factored into the index. Nigeria is ranked as the 12th-least-ready country to deal with climate change impacts, due largely to weaknesses with agricultural yield, political stability, and governance.[29]

Demographic Challenges

Nearly half of Nigeria's people live in cities, and the urban population is expected to grow by 3.75 percent annually due to migration.[30] Rural-to-urban migration on a large scale has increased economic inequality in large cities like Lagos and, in equal measure, has increased pressure on water service delivery.

The median age in Nigeria is 19 years, and 62 percent of the population is under the age of 25.[31] Social scientists concur that populations with a disproportionately large idle youth population have a higher potential for destabilization and violence, especially in states where the government and economy are incapable of incorporating youth into the formal employment sector. Under such conditions, youths become susceptible to recruitment into violent organizations.[32]

Demographic diversity is another significant contributor to tensions within Nigeria. Divisions are based on class, religion, and tribal affiliation, among other factors too numerous to list in detail. Religious tensions are especially acute between the Muslim communities in the north/northeast and the predominantly Christian communities in the south.[33] Nigeria is also home to 250 ethnic groups.[34] These cleavages would create challenging conditions in nations even with the best governance structures.[35]

Governance Challenges

The Federal Ministry of Water Resources has overarching responsibility for management and control of water resources in Nigeria. It is charged with assuring the supply of underground and surface water resources for all domestic and agricultural uses. In addition, the Ministry sustains national water security by developing and supporting irrigation and other methods to reduce agricultural vulnerability to irregular rainfall.[36] The federal government administers 12 river basin development authorities.[37]

Water governance is further devolved to 37 state water agencies and 774 local government authorities responsible for maintaining rural water and sanitation facilities in conjunction with beneficiary communities. However, only a few have the resources and skills to carry out their tasks.[38]

Climate change presents another policy challenge. The current water management practices in Nigeria are assessed to be insufficient to cope with the impacts of climate change in the areas of water supply reliability, flood risk, health, agriculture, energy, and aquatic ecosystems.[39] In addition, good policy decisions have been hampered by the lack of robust environmental and social impact assessments.[40]

Water allocation is also a significant problem. Diffuse government responsibility is a contributing factor. The country's byzantine land-use systems and institutional governance of land use complicate allocation schemes. Years of improper operation and maintenance due to poor planning and unsustainable public investment in water management infrastructures are additional contributing factors.[41]

Conflict Linkages

Geophysical, ecological, and demographic factors have created water stress in many regions. Social scientists observe a correlation between environmental scarcity and internal conflict.[42] This observation is consistent with the fact that 40 percent of all intrastate conflicts in the past six decades involved disputes over limited natural resources.[43]

A copious and contentious academic literature attempts to further explain this correlation. The theories of eco-violence explain that the diminished quality and quantity of natural resources coupled with poor resource management and societal cleavages can ignite a competitive quest for resources that can take the form of violence.[44]

Climate change has been linked to such violence. A groundbreaking study examined 60 examples of human conflict, ranging from the individual to the national level, and situated them relative to climate-related events. The study found that with warmer average temperatures and more extreme rainfall, the estimate of the frequency of intergroup conflict—i.e., civil war—rose by 14 percent.[45] Other research, presented in the Social Conflict Analysis Database, indicates that conflicts in Africa are more prevalent in extremely wet or dry years.[46] Other scholarship has found that the coincidence of weather variations and war is dramatically higher in countries with inter-ethnic tensions. This study found a stronger correlation between ethnic division and weather disasters than with poverty or income inequality, or even past propensity for conflict.[47]

Any attempt to fully explain connections between water stress and conflict will fall short due to the difficulty of isolating water stress from the myriad of other variables that contribute to violent conflict in most cases. However, the following sections provide a systematic approach to tracing a chain of causality in the impacts of water stress on Nigeria's three major ongoing conflicts.

Regional Analysis of Water Stress and Conflict in Nigeria

The North and Northeast Region

The World Wildlife Fund categorizes areas with similar ecology as ecoregions, defined as a "large unit of land or water containing a geographically distinct assemblage of species, natural communities, and environmental conditions."[48] Northern Nigeria includes four ecoregions: Sahelian Acacia Savanna (or the Sahel), West Sudanian Savanna, East Sudanian Savanna, and Mandara Plateau Mosaic. Broadly, these ecoregions consist predominantly of grassland, savanna, and shrubland.

Droughts, leading to desertification, are taking a heavy toll. Rainfall over the Sahel has experienced an overall reduction over the course of the 20th century.[49] The Sahel is expanding southward, with roughly 3,500 square kilometers of land turning to desert each year and significantly reducing the land area available for grazing and farming, contributing to the abandonment of as many as 200 villages.[50] The use of fuel wood as a main energy source in the Sahel has led to further desertification. The trees that are cut down for wood and charcoal are needed to protect topsoil and replenish aquifers.

The impacts of droughts are especially "troubling when government data show rural households harvest rain for more than half their total water consumption, and northern groundwater tables have dropped sharply over the last half century, owing partly to less rain."[51] In many areas, wells have already run dry due to dropping groundwater tables and unsustainable withdrawals; wells dug even deeper will still be exhausted without adequate hydrological replenishment of water supplies.[52]

The northern region has also been ravaged by floods. Especially severe flooding in 2010 tested the government's ability to build shelter and relocate populations.[53] The prevailing ecological conditions imperil livelihoods and create increased strain on institutions and communities that are already resource challenged.

It is under these environmental conditions in the north that the militant Islamist group Boko Haram was formed in 2002 with the aim of declaring Islamic Sharia law as a means of eliminating Western influence and government corruption. The term *Boko Haram* is translated from the Hausa language as "Western Education Is Sinful."[54]

The bombing of the UN headquarters in Abuja, Nigeria's capital, marked a surge of violence in 2011. Since then, Boko Haram has staged continual attacks, such as planting bombs in public places and churches, as well as conducting a traditional guerilla war against the Nigerian military.[55] Boko Haram has used general guerilla tactics, but there were warnings from the Nigerian military in late 2015 that Boko Haram could be attempting to sabotage water supplies.[56] According to *The Economist*, Boko Haram has pillaged food, stolen cattle, and poisoned water.[57]

While hatred toward the West and Christianity provides nominal motivation for Boko Haram, it is worth taking a closer look at its motivations. According to John Campbell, former U.S. ambassador to Nigeria, "Boko Haram 'writ large' is a movement of grassroots anger among northern people at the continuing depravation [sic] and poverty in the north."[58] However, the group pledged allegiance to the Islamic State in March 2015, demonstrating a clearer threat to U.S. allies and interests.[59]

Boko Haram's insurgency—driven in part by environmental degradation—has damaged the economy, diverting national resources toward combating terrorist activity that could have otherwise been directed toward development and environmental remediation. Lack of resources has diminished economic and political capacity to initiate and maintain new water infrastructure projects

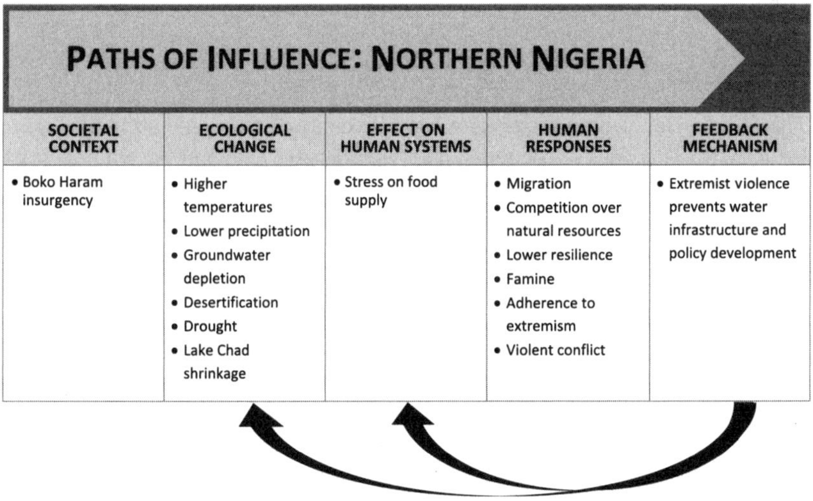

PATHS OF INFLUENCE: NORTHERN NIGERIA

SOCIETAL CONTEXT	ECOLOGICAL CHANGE	EFFECT ON HUMAN SYSTEMS	HUMAN RESPONSES	FEEDBACK MECHANISM
• Boko Haram insurgency	• Higher temperatures • Lower precipitation • Groundwater depletion • Desertification • Drought • Lake Chad shrinkage	• Stress on food supply	• Migration • Competition over natural resources • Lower resilience • Famine • Adherence to extremism • Violent conflict	• Extremist violence prevents water infrastructure and policy development

Figure 7.2 Water Stress Paths of Influence: Northern Nigeria

that could improve conditions in a region dependent on agriculture. Desperate conditions diminish the population's capacity to resist Boko Haram's attacks and may encourage recruitment efforts. Lack of national resources has also inhibited a response to a UN-declared humanitarian crisis based on the dislocation of thousands of refugees who have fled the violence.

The Lake Chad Crisis

As of the summer of 2016, a humanitarian catastrophe has been growing in the region of Lake Chad—which originally straddled Nigeria, Niger, Chad, and Cameroon—where approximately 2.6 million people have been uprooted or displaced.[60] The existing water crisis has grown more extreme: in Nigeria alone, 4.4 million people are in need of urgent food assistance, and nearly a quarter of a million children are severely malnourished.[61,62]

The region has traditionally been host to a large community of fishermen, pastoralists (or herders), and farmers. Since droughts in the 1970s, the lake started to recede at an unprecedented pace, and surface water has shrunk to 5 percent of its previous levels, receding completely from Nigeria and moving toward Cameroon and Chad. This problem has been compounded by mismanagement and pollution. Fishermen are catching only one-fourth of their previous catches. The fishermen who depended on Lake Chad are now farming land that was once covered in water.

This environmental catastrophe has been exacerbated by increasing droughts and scarcer rainfalls, population growth, and the pumping of the lake's waters for the water-intensive exploitation of uranium in Niger. National irrigation projects conducted by the riparian states have all accelerated the recession of the lake.[63] A vast expanse of once-arable land is uninhabitable and out of production.

Middle Belt

The Middle Belt is home to great natural diversity, forming the transition space between the extremes of the temperate Niger River Delta and the country's dry northeast. The Middle Belt's ecoregions are predominantly tropical and subtropical, and include savanna, grasslands, shrubland, and some forest.[64] Due to the geophysical diversity of this region, the impacts of climate change vary widely.[65]

The Middle Belt is on the traditional migration route of herders from northern Nigeria and beyond belonging to the Muslim Hausa-Fulani, a large ethnic group spread across a number of countries in Africa, including Niger, Cameroon, Chad, Nigeria, Senegal, and Mali.[66] The herders' annual search for arable land for cattle grazing encroaches on farming communities of several other tribes, including the Christian Yorubas, who are also trying to survive on the decreasingly fertile land, a condition that threatens their security and destroys their crops.[67,68] Due to changing weather patterns, Hausa-Fulani herders are staying longer in the Middle Belt, where the rainy season lasts longer.

Hausa-Fulani herders sometimes attempt to settle. However, the farmers generally believe they have the right to the land, citing inheritance rights and land tenure laws that give special recognition and social services through local and state law to generational landowners, also known as *indigenes*.[69] Both state and local governments favor indigenes, a contributing factor to the conflict.[70]

When cattle consume or trample their crops, the farmers retaliate against the pastoralists by destroying their livestock, through physical attacks or by poisoning the water or grass. These actions, combined with dissatisfaction caused by sparse economic opportunities for either of the belligerent parties, have created a conflict spiral and led to thousands of deaths.

The conflict in the Middle Belt is an understudied and growing crisis. The 2015 Global Terrorism Index reported that "Fulani militants" were the fourth-most deadly terrorist group in the world, responsible for the deaths of 1,229 people in 2014.[71] Furthermore, there are concerning linkages between the conflicts in the northern region and the Middle Belt. Sources within the Nigerian military allege that their advances have caused Boko Haram members to infiltrate Hausa-Fulani communities in an attempt to flee.[72]

Nigerian authorities' attempts to mitigate and manage the conflict have been largely ineffective. Local authorities expelled pastoralists from a number of northeastern states in an attempt to mitigate retaliatory killings. By expelling one group, local governments have depicted pastoralists as aggressors and created the perception of taking sides, rather than remaining a neutral party.[73] In the Middle Belt, like the northern region, ongoing conflict hampers the implementation of more effective water management policies that could reduce the competition for arable lands.

PATHS OF INFLUENCE: NIGERIA'S MIDDLE BELT

SOCIETAL CONTEXT	ECOLOGICAL CHANGE	EFFECT ON HUMAN SYSTEMS	HUMAN RESPONSES	FEEDBACK MECHANISM
• Pastoralists and herders are engaged in civil conflict	• Increased temperature variability with an upward trend • Less predictable precipitation • Groundwater depletion • Desertification • Drought	• Stress on food security (farming and animal husbandry) • Alteration of ancestral migration patterns	• Herdsmen encroach on pastoralists • Competition over resources • Violent conflict	• Violence prevents effective water management and land tenure policy

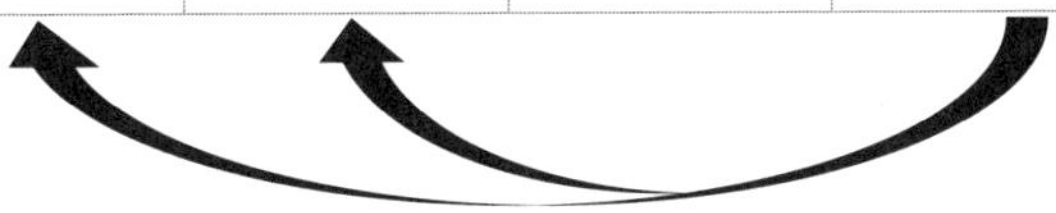

Figure 7.3 Water Stress Paths of Influence: Middle Belt

The Niger Delta

The Niger Delta is characterized by a high level of biodiversity, an intense rainy season, and high human population density, and encompasses several forest and mangrove ecoregions. The expected impacts of climate change on the Delta are severe. Sea level rise is expected to cause coastal flooding and storm surges in this network of estuaries, rivers, creeks, and streams that sits especially low, as does Nigeria's largest city, Lagos.[74] Flooding contaminates freshwater aquifers and rivers, leaving them with high salinity and polluted with sediment and sewage.[75] Polluted water is unsuitable for hygiene and causes deaths from communicable diseases such as cholera.

The Delta region is rich in oil and natural gas reserves. Fifty years of extractive processes—largely undertaken by Shell, Exxon, and their subsidiaries—have resulted in hundreds of oil spills each year, causing tremendous water pollution that has diminished water availability for human consumption and hygiene, destroyed and degraded agricultural lands, and destroyed coastal wetlands and mangroves where fish breed.

Water is used in extractive processes such as oil drilling and mining. In oil drilling, water is mixed with solvents that pollute the ecosystem. This use of water in drilling and mining has led to conflicts between the national government and local villages over water rights. Intentional oil theft by puncturing pipelines—known as *bunkering*—has resulted in spillage into surrounding swampland and waterways.[76] In addition, mining operations near rivers create significant downstream pollution of heavy metals and other dangerous substances that affect fish stocks.

The best available estimates indicate that as many as 13 million barrels of crude have escaped into the environment—roughly equal to one Exxon Valdez spill per year.[77] The United Nations Environmental Program found that the damage to one area of the Delta, known as Ogoniland, could take up to 30 years to remediate.[78] Collectively, the oil spills have been found to have led to "a 60 percent reduction in household food security" and "a 24 percent increase in the prevalence of childhood malnutrition."[79]

A history of grievances in the Delta, motivated by environmental degradation and the lack of benefits to the local communities, intensifies current widespread popular unhappiness over unequal distribution of water and the government's disregard for water pollution. The failure of the GON to protect its population from the ravages of environmental destruction has led communities to turn to UK courts to demand reparations that they have so far been unable to obtain through the Nigerian system.[80]

Instead, the GON's efforts to suppress an earlier environmental protest movement resulted in approximately 2,000 fatalities between 1993 and 1999 alone.[81] Ken Saro-Wiwa, a protest movement leader who promoted nonviolent resistance, was executed in 1995, sparking international outrage and attention to the Delta's plight.

Today, several insurgent groups operating in the Delta have struck at oil company personnel and targets in an attempt to bring the government to its knees. The activities of these factions, described next, are also linked to water stress.

- The Movement for the Emancipation of the Niger Delta (MEND) is the original and most well known of these insurgent groups. MEND is composed mainly of fighters who live in small villages adjacent to the Niger River.[82] It has used terrorist tactics, including sabotage of oil rigs. The group has called on the central government to invest more in basic services, including clean water.[83]
- The Niger Delta People's Volunteer Force (NDPVF), founded in 2004, called for ownership of lands and waterways to be returned to local control. The NDPVF sees all national control, including the imposition of environmental protection laws, as an intrusion.[84]
- In 2016, The Niger Delta Avengers (NDA) emerged from the violent milieu of the Delta with the expiration of an amnesty program extended to militants seven years previously.[85] The NDA took responsibility for a major attack resulting in the destruction of an underwater pipeline owned by Royal Dutch Shell. This and other attacks are seriously affecting oil production and have dealt a huge blow to government revenue.[86] The group is social media–savvy and maintains a website.

There is a detrimental feedback effect similar to that seen in the other two conflicts in Nigeria. As the Delta-based insurgent groups destroy oil infrastructure, these actions in turn cause grave damage to the water resources of the Niger River Delta—through leaking pipelines, for example—destroying the livelihoods of the local populace they are seeking to defend against environmental destruction.

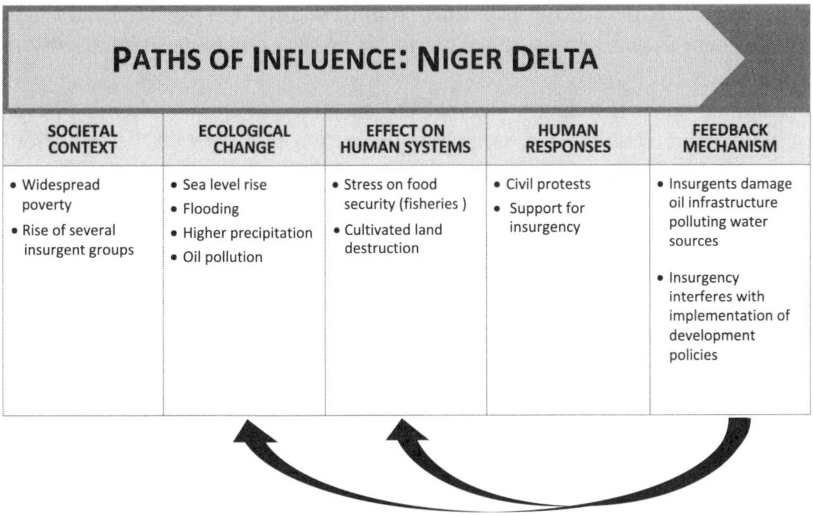

PATHS OF INFLUENCE: NIGER DELTA

SOCIETAL CONTEXT	ECOLOGICAL CHANGE	EFFECT ON HUMAN SYSTEMS	HUMAN RESPONSES	FEEDBACK MECHANISM
• Widespread poverty • Rise of several insurgent groups	• Sea level rise • Flooding • Higher precipitation • Oil pollution	• Stress on food security (fisheries) • Cultivated land destruction	• Civil protests • Support for insurgency	• Insurgents damage oil infrastructure polluting water sources • Insurgency interferes with implementation of development policies

Figure 7.4 Water Stress: Paths of Influence in the Niger Delta

The Bonga Oil Spill

A devastating offshore oil spill was reported on December 20, 2011, when a flexible oil transfer tube ruptured at Bonga, the site of Nigeria's first deepwater discovery well.[87] The exact amount of crude oil released into the ocean due to the spill is unknown, but 40,000 barrels is commonly cited.[88] The massive spill occurred in one of the most biodiverse areas of the world.[89]

The spill compounded long-term degradation from oil and gas extraction.[90] Just prior to the Bonga spill, the UN Environment Programme (UNEP) published an environmental assessment that identified extremely high levels of water contamination in adjacent Ogoniland, including contamination of well water with 900 times the allowable levels of benzene.[91] The spill also resulted in extensive ecological damage that put the health and livelihoods of local populations at risk. Environmental analyses show that marine life, as well as water and soil quality, was decimated, profoundly impacting human security for the local populations. Traditionally, local groups such as the Bodo and Itsekiri communities are fishers and farmers. However, much of the land can no longer sustain vegetation due to oil slicks moving up the Delta, and the marine life that has survived is too toxic for consumption.[92]

Global Implications of Instability in Nigeria

While Nigeria is a rising power due to its geography, population size, relative (if uneven) wealth, and oil production, its government faces many challenges, including war, poverty, and increasing environmental degradation. The resulting instability and potential for state failure carry global repercussions. The volatility of the Nigerian oil sector is an example. While world oil supply exceeds demand at this time, supply disruptions for Nigerian crude oil have put upward pressure on world oil prices and created uncertainty in the market.

Developments in Nigeria warrant the sustained attention of U.S. foreign policy-makers. The country is a strategic partner that the U.S. can ill afford to lose as an ally in the struggle against extremism. Nigeria's military participates in regional peacekeeping missions even as it struggles to manage multiple conflicts within its borders. The United States is spearheading efforts to stem the influence of Al Qaeda in the Islamic Maghreb (AQIM), which has already gained influence in neighboring countries including Niger, Mali, and Mauritania. Although AQIM has not established itself in Nigeria, where Boko Haram, a group with a similar outlook, already operates, it is incumbent upon the United States to limit fertile ground for the spread of extremism on the African continent. Finally, outward migration from Nigeria is already putting pressure on important U.S. allies in Europe, as Nigerians are joining a massive wave of migrants from nations such as Syria seeking passage across the Mediterranean.

Recommendations

For the GON

Effective policies to address water stress problems identified in each of the three ecoregions will be site-specific and must be undertaken primarily by organizations at various levels within the GON. Major factors that make solving the water stress challenges identified in this chapter a monumental undertaking include ill-defined and often conflicting responsibilities for water governance at the state, local, and tribal levels as well as endemic corruption. A full assessment of all policy options is beyond the scope of this chapter, but some potential courses of action are identified.

In light of increasing climate impacts, developing and planting drought-tolerant seeds of maize, an important food staple, would increase yields in Nigeria's north and Middle Belt. Therefore, the GON should encourage ongoing research and deployment of drought-tolerant seed varieties. Scientists at the International Institute of Tropical Agriculture (IITA), in Ibadan, Nigeria, have developed and released a number of high-yielding and drought-tolerant varieties, and they are also carrying out research and developing more varieties.[93] Cooperation with the United States has already been established. The U.S. Agency for International Development (USAID) is also working across 18 Nigerian states to facilitate the use and marketing of drought-resistant seeds under the Feed the Future Program.[94] Greater maize crop productivity would ease emerging food insecurity and reduce contention between pastoralists and herders in the Middle Belt.

Also in the Middle Belt, key elements of a solution could be the establishment of grazing reserves for livestock, increasing funding for communities affected by the clashes, and improving security at conflict hotspots. Therefore, the federal government in cooperation with other levels of Nigerian government should delineate and regulate clear and enforced cattle migration routes and establish grazing reserves to preempt clashes over land encroachment.[95] In order to promote a more sustainable peace, these reserves should contain clinics, schools, and other basic infrastructure to support the Hausa-Fulani herders during the time that they reside there.

Developing solutions to the conflict in northern Nigeria and the Middle Belt has been made more difficult by lack of complete information from the physical and social sciences. Better information could be used to create detailed state-specific maps that plot overlaying elements such as desertification patterns, land tenure holdings, grazing routes and reserves, and past locations of violence and conflict. Creation of these visual tools, which would be possible with the assistance of international organizations, would enable the GON to design interventions that guide policies that build resilience and prevent or resolve emerging farmer–pastoralist conflicts.

These maps should be coupled with information from downscaled regional climate models in order to assess the range of possibilities for the worsening impacts of climate change in some regions. These maps can also be used as advocacy tools for local stakeholders to encourage sometimes reluctant state and federal organizations to take action and design effective water infrastructure and water management systems that take all risks into account.

For the U.S. Government (USG)

There are some situations where direct interventions by the USG could also contribute to sustainable solutions to problems created by water stress, using the tools of defense, development, and diplomacy—the 3Ds, traditionally wielded by the U.S. Department of State, USAID, and the U.S. Department of Defense (DoD). These are the tools that provide the foundation for promoting and protecting U.S. national security interests.

The purpose of U.S. defense engagement in the framework of conflict prevention is to cut the chain of causality from environmental stressors to violence early enough to maintain peace and forestall a situation that will breed extremism. The United States' defense engagement with Nigeria has been limited due to the GON's unease about U.S. strategic intentions in the Gulf of Guinea. Furthermore, the GON has expressed vocal opposition to U.S. aspirations to find a location for U.S. Africa Command's headquarters on the continent.[96]

Capacities resident in the U.S. military also include the U.S. Army Corps of Engineers (USACE). The USACE can cooperate with civilian agencies such as the U.S. Geological Survey to build early warning capabilities through application of remote sensing technologies that advance situational awareness for more effective policy-making and response to emerging crises. The USACE also provides international partners with analysis of emerging water resources trends and issues, and state-of-the-art planning and hydrologic engineering methods, models, and training.[97]

The U.S. DoD should ramp up existing efforts of the U.S. Special Operations Forces. Deployment of these and the transfer of equipment support the GON's overall military capacity and battlefield effectiveness to combat Boko Haram.[98] U.S.-supplied surveillance drones could supply intelligence for environmental missions, such as pinpointing the locations and effects of droughts.[99]

In the Niger Delta's Ogoniland region, the GON set in motion a $1 billion cleanup and restoration program in the summer of 2016. This environmental remediation is likely to be the world's largest terrestrial cleanup ever seen, yet the environmental destruction in Ogoniland represents only a small percentage of the overall damage to the Delta.[100] A USG strategy should support the Nigerian government's efforts but focus on the other aspects of the problem.

Improving human security in the Niger Delta will require environmental remediation, adaptation, and redistribution of wealth. Therefore, the USG must encourage the implementation of more accountable, transparent governance structures. A history of corruption indicates that the USG would have the best chances of success working directly with local communities to provide support such as technical assistance, investments in water infrastructure, and vocational training. A development strategy of direct investment at this level allows for clearer monitoring and evaluation of allocated funds. U.S. federal budgets are inconsistent from one year to the next, so short-term projects are easier to implement than long-term projects. Finally, the USG should use diplomacy to encourage multinational oil companies with assets in the United States and Nigeria to reach financial settlements for addressing minority grievances in the Delta.

Conclusion

The analysis presented in this chapter illustrates paths of influence linking water stress to other factors that diminish societal resilience. It demonstrates that Nigeria's water challenges are significant and vary greatly among the three ecoregions. Therefore, the potential set of solutions is large.

Effective governance is the cornerstone of water stress management. The three major conflicts described in this section are connected to water stress through various mechanisms. A common denominator in each chain is a feedback mechanism whereby each conflict impedes good governance.[101] Weak governance has combined with existing ethnic and religious cleavages to foster

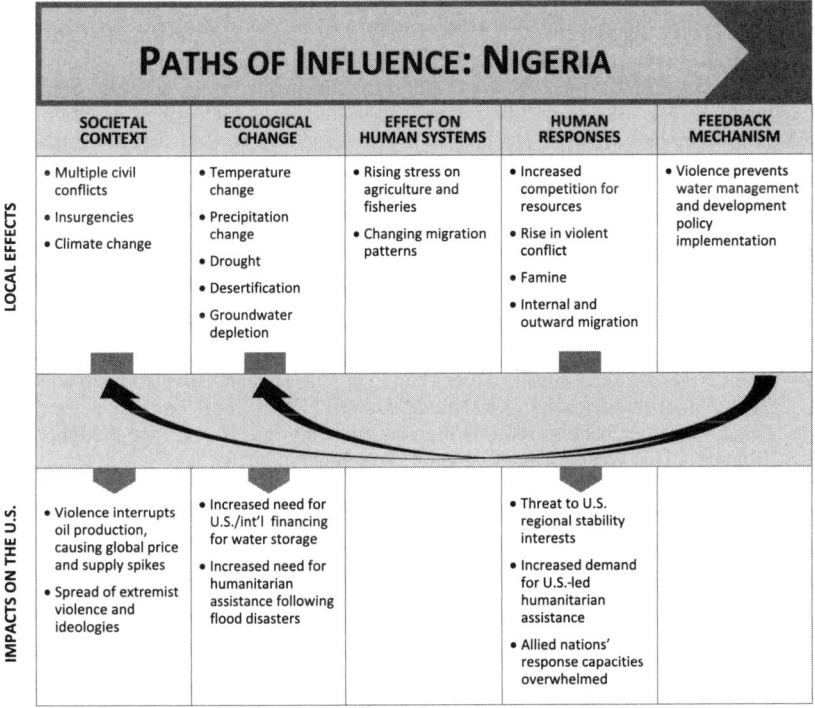

Figure 7.5 Water Stress: Paths of Influence for Whole of Nigeria

organized violence. However, improved governance is a goal that seems unobtainable when Nigerian policy-makers are hamstrung by the drain on resources created by three simultaneously intensifying conflicts.

It is therefore advisable that a comprehensive U.S. policy to assist the GON to address water challenges starts with support to prevent or mitigate existing conflict. An integrated approach incorporating diplomatic and development strategies as well as direct defense assistance will go the longest way toward this goal and jump-start needed development and technical-assistance initiatives. While the outlook for stability in Nigeria is uncertain at best, the United States must double down on attempts to assist one of the most economic and geo-strategically important nations in the world.

Addressing water stress in Nigeria is a test case for U.S. engagement with complex environmental problems in fragile states that are important to U.S. national security interests. It is fortunate that foreign assistance for water projects in developing nations enjoys relatively high support among U.S. policy-makers.[102] Therefore, systematic prioritization of global water challenges is an important and politically feasible basis of a renewed U.S. foreign policy that foregrounds human dignity while combating extremism and supporting the security of future generations.

Notes

1 "The Fragile States Index 2016," *The US Fund for Peace*, 2016, accessed August 10, 2016, fsi.fundforpeace.org.
2 "Nigeria," *CIA World Factbook* (Central Intelligence Agency), accessed September 6, 2016, www.cia.gov/library/publications/the-world-factbook/geos/ni.html.
3 John Campbell. *Nigeria, Dancing on the Brink* (Lanham: Rowman & Littlefield, 2013).
4 U.S. Energy Information Administration. "Nigeria: Analysis-Energy Sector Highlights" (U.S. Department of Energy, 2016), www.eia.gov/beta/international/country.cfm?iso=NGA.
5 U.S. Energy Information Administration. "Nigeria" (U.S. Department of Energy, 2016), www.eia.gov/beta/international/analysis.cfm?iso=NGA.
6 M. L. Ross. "Does Oil Hinder Democracy?" *World Politics* 53, no. 3 (2001): 325–361.
7 G. P. "Nigeria's Economy Is Bigger Than Everyone Thought: Give Yourself an 89% Raise," Baobab: Africa (blog), last modified April 7, 2014, *The Economist*, www.econ omist.com/blogs/baobab/2014/04/nigerias-economy-bigger-everyone-thought.
8 "Business This Week," *The Economist* (September 3, 2016): 8.
9 Nwajiuba, Chinedum. "Nigeria's Agriculture and Food Security Challenges," *Green Deal Nigeria* (Heinrich Boell Foundation 2013), accessed at www.boell.org/downloads/4_Green_Deal_Nigeria_AGRICULTURE.pdf.
10 Dinfin Mulupi. "Agriculture Should Be a Business, Not a Development Activity," *How We Made It in Africa*, last modified May 27, 2013, www.howwemadei tinafrica.com/agriculture-should-be-a-business-not-a-development-activity/.
11 Nigerian Federal Ministry of Agriculture and Rural Development (FMARD). *Agricultural Transformation Agenda Support Program—Phase 1 (Atasp-1): Strategic Environmental and Social Assessment* (African Development Bank Group, July 2013).

12 Alexander Smith. "ISIS Owns Headlines, but Nigeria's Boko Haram Kills More Than Ever," *NBC News* (January 1, 2016), www.nbcnews.com/storyline/2015-year-in-review/isis-owns-headlines-nigeria-s-boko-haram-kills-more-ever-n480986.
13 John Campbell, ed. "Nigeria Security Tracker: Mapping Violence in Nigeria," *Council on Foreign Relations blog*, last modified 2012, www.cfr.org/nigeria/nigeria-security-tracker/p29483.
14 "Biafra War," *Global Security*, accessed January 15, 2015, www.globalsecurity.org/military/world/war/biafra.htm.
15 Aaron Sayne. "Climate Change Adaptation and Conflict in Nigeria," special report no. 274 (Washington DC: United States Institute of Peace, 2011): 3, accessed at www.usip.org/publications/climate-change-adaptation-and-conflict-in-nigeria.
16 Ignatius A. Madu. "Spatial Vulnerability of Rural Households to Climate Change in Nigeria: Implications for International Security," The Robert S. Strauss Center for International Security and Law, Working Paper No. 2 (May 2012): 13.
17 Sayne. "Climate Change Adaptation and Conflict," 4.
18 Sayne. "Climate Change Adaptation and Conflict," 4.
19 "What's Really Causing Water Scarcity in Africa South of the Sahara?" *International Food Policy Research Institute Blog*, last modified August 13, 2013, International Food Policy Research Institute, www.ifpri.org/blog/what%e2%80%99s-really-causing-water-scarcity-africa-south-sahara.
20 "Nigeria," *WaterAid*, accessed August 1, 2016, www.wateraid.org/ng.
21 O. B. Akpor and M. Muchie. "Challenges in Meeting the MDGs: The Nigerian Drinking Water Supply and Distribution Sector," *Journal of Environmental Science and Technology* 4, no. 5 (2011): 480–489.
22 "Progress for Children: a Report Card on Water and Sanitation," *UNICEF* (2006): 4, www.unicef.org/publications/files/Progress_for_Children_No._5_-_English.pdf.
23 "Sanitation Fact Sheet: Nigeria," *UNICEF* (2008), www.unicef.org/french/wash/files/NigIYSFact.pdf.
24 Sayne. "Climate Change Adaptation and Conflict," 2.
25 I. Niang, O. C. Ruppel, M. A. Abdrabo, A. Essel, C. Lennard, J. Padgham, and P. Urquhart. "Executive Summary, Chapter 22 Africa," IPCC 5th Assessment Report (Intergovernmental Panel on Climate Change, 2014), accessed September 5, 2016, at http://ipcc-wg2.gov/AR5/images/uploads/WGIIAR5-Chap22_FINAL.pdf, 1202.
26 Sayne. "Climate Change Adaptation and Conflict," 3.
27 P. Odjugo. "An Analysis of Rainfall Pattern in Nigeria," *Global Journal of Environmental Science* 4, no. 2 (2005): 139–145.
28 See Federal Ministry of Environment, "Nigeria and Climate Change: Road to Cop15" (Abuja, Federal Ministry of Environment, 2009).
29 "Nigeria," *Notre Dame Global Adaptation Index*, last modified 2014, http://index.gain.org/country/nigeria.
30 "Nigeria," *Notre Dame*.
31 Michael Werz and Laura Conley. "Climate Change, Migration, and Conflict in Northwest Africa: Rising Dangers and Policy Options Across the Arc of Tension," Center for American Progress (April 2012): 21.
32 Lionel Beehner. "The Effects of 'Youth Bulge' on Civil Conflicts." *Council on Foreign Relations* (April 27, 2007).
33 Moses Ochonu. "The Roots of Nigeria's Religious and Ethnic Conflict," *Public Radio International* (March 10, 2014), www.pri.org/stories/2014-03-10/roots-nigerias-religious-and-ethnic-conflict.
34 "Nigeria," *CIA Factbook*.
35 Jeffrey Haynes. "Conflict, Conflict Resolution and Peace-Building: The Role of Religion in Mozambique, Nigeria and Cambodia," *Commonwealth & Comparative Politics* 47, no. 1 (2009): 52–75, http://dx.doi.org/10.1080/14662040802659033.

36 Ministry website: www.waterresources.gov.ng/.
37 State Water Agencies in Nigeria, Chapter 2,"Water Sector Institutions and Governance."
38 Abel Bove, Alexander Danilenko, Berta Macheve, L. Joe Moffitt, and Roohi Abdullah. *State Water Agencies in Nigeria: A Performance Assessment*, The World Bank, September 2015,
39 Senator Ireogbu. "Expert Proffers Solution to Fulani Herdsmen, Farmers Clashes," *This Day Live*, www.thisdaylive.com/index.php/2016/07/09/expert-proffers-solution-to-fulani-herdsmen-farmers-clashes-2/.
40 Yunana Magaji Bature, Abubakar Aliyu Sanni, and Francis Ojo Adebayo. "Analysis of Impact of National Fadama Development Projects on Beneficiaries Income and Wealth in FCT, Nigeria." *Journal of Economics and Sustainable Development* 4, no. 17 (2013): 11–23.
41 O. B. Akpor and M. Muchie. "Challenges in Meeting the MDGs," 480–489.
42 Thomas F. Homer-Dixon. *Environment, Scarcity, and Violence* (Princeton: University Press, 1999).
43 Sayne. "Climate Change Adaptation and Conflict," 5.
44 Homer-Dixon. *Environment, Scarcity, and Violence*, 30–31; Al Chukwuma Okoli and Atelje George. "Nomads Against Natives: A Political Ecology of Herder/Farmer Conflict in Nasarawa State, Nigeria," *American International Journal of Contemporary Research* 4, no. 2 (February 2014): 79.
45 Solomon M. Hsiang, Marshall Burke, and Edward Miguel. "Quantifying the Influence of Climate on Human Conflict," *Science* 13 (September 2013): 341, doi:10.1126/science.1235367.
46 Idean Salehyan, Cullen S. Hendrix, Jesse Hamner, Christina Case, Christopher Linebarger, Emily Stull, and Jennifer Williams. "Social conflict in Africa: A new database." *International Interactions* 38, no. 4 (2012): 503–511.
47 Carl-Friedrich Schleussner, Jonathan F. Donges, Reik V. Donner, and Hans Joachim and Schellnhuber. "Armed-Conflict Risks Enhanced by Climate-Related Disasters in Ethnically Fractionalized Countries." *Proceedings of the National Academy of Sciences of the United States* (May 20, 2016), accessed July 28, 2016, at www.pnas.org/content/113/33/9216.
48 World Wildlife Fund, "Biomes," accessed August 1, 2016, www.worldwildlife.org/biomes.
49 I. Niang et al. "Executive Summary," 1209.
50 Werz and Conley. "Climate Change," 15.
51 Sayne. "Climate Change Adaptation and Conflict," 4.
52 Sayne. "Climate Change Adaptation and Conflict," 4.
53 Werz and Conley. "Climate Change," 24.
54 "A Look at the Nigerian Extremist Group Boko Haram and Its Major Attacks in Recent Years," *Associated Press*, last modified April 14, 2014, http://hosted.ap.org/dynamic/stories/a/af_nigeria_explosions_glance?site=ap&section=home&templat e=default&ctime=2014-04-14-12-54-09.
55 Andrew Walker. "What Is Boko Haram?" Special report no. 308 (Washington, DC: United States Institute for Peace, 2012):1, accessed at www.usip.org/publications/what-boko-haram.
56 Michael Kaplan. "Nigeria's Boko Haram to Poison Water? Army Urges Northeast to Stock Up Amid Fears of a Terrorist Plot," last modified October 26, 2015, accessed October 22, 2016, at www.ibtimes.com/nigerias-boko-haram-poison-water-army-urges-northeast-stock-amid-fears-terrorist-plot-2155999.
57 "Nigeria's Food Crisis: Hunger Games," *The Economist*, accessed September 3, 2016, at www.economist.com/news/middle-east-and-africa/21706261-famine-looms-areas-devastated-boko-haram-hunger-games.
58 Walker. "What Is Boko Haram?" 9.

59 "Nigeria's Boko Haram Pledges Allegiance to Islamic State," *BBC*, accessed September 1, 2016, at www.bbc.com/news/world-africa-31784538.
60 "Quick Facts About the Lake Chad Crisis," Mercy Corps, accessed July 15, 2016, at www.mercycorps.org/articles/niger-nigeria/quick-facts-about-lake-chad-crisis.
61 Cullen Hendrix. "Water and Security in Niger and the Sahel," *Climate Change and African Political Stability* no. 24 (2014), www.strausscenter.org/ccaps-news/water-insecurity-in-the-sahel.html.
62 Kieran Guilbert. "Lake Chad 'Forgotten' as World Focuses on Boko Haram-Hit Nigeria." *Reuters*, last modified August 4, 2016, www.reuters.com/article/us-chad-boko-haram-idUSKCN10F1Z3.
63 "Lake Chad's Unseen Crisis: Voices of Refugees and Internally Displaced People from Niger and Nigeria," Oxfam (2016), accessed at www.oxfam.org/sites/www.oxfam.org/files/file_attachments/bn-lake-chad-refugees-idps-190816-en.pdf.
64 The World Wildlife Fund, "Biomes."
65 Peter Akpodiogaga-a Ovuyovwiroye Odjugo. "General Overview of Climate Change Impacts in Nigeria." *Journal of Human Ecology* 29, no. 1 (January 2010): 47–55. http://search.proquest.com/docview/742950318?accountid=11243.
66 "Bandit attacks displace Nigeria herders," *IRIN*, last modified June 19, 2013, www.irinnews.org/news/2013/06/19/bandit-attacks-displace-northern-nigeria-herders.
67 Werz and Conley. "Climate Change," 22.
68 Al Chukwuma Okoli and Atelje George. "Nomads Against Natives," 82.
69 Al Chukwuma Okoli and Atelje George. "Nomads Against Natives," 82.
70 Aaron Sayne. "Rethinking Nigeria's Indigene-Settler Conflicts," special report no. 311 (Washington DC: United States Institute of Peace, 2012): 2, accessed at www.usip.org/publications/rethinking-nigeria-s-indigene-settler-conflicts.
71 "Nigeria: Investigate Massacre, Step Up Patrols: Hundreds Killed by Mobs in Villages in Central Nigeria," Human Rights Watch, last modified March 8, 2010, www.hrw.org/news/2010/03/08/nigeria-investigate-massacre-step-patrols.
72 Chika Oduah. "Nigeria: Deadly Nomad-Versus-Farmer Conflict Escalates," *Al Jazeera*, last modified July 6, 2016, www.aljazeera.com/news/2016/07/nigeria-deadly-nomad-farmer-conflict-escalates-160704043119561.html.
73 Nate Haken (Fund for Peace), interview by Lourdes Eliacin Mars, Alyssa Gomes, and Maya Jacobs. "Environmental Peacebuilding: Nigeria" report, Elliott School of International Affairs, March 13, 2015.
74 Sayne. "Climate Change Adaptation and Conflict," 4.
75 Sayne. "Climate Change Adaptation and Conflict," 4.
76 Abosede Babatunde. "Environmental Politics and the Politics of Oil in the Oil-Bearing Areas of Nigeria's Niger Delta," *Peace and Conflict Review* 5, no. 1 (2010): 1–12.
77 Nigerian Conservation Foundation, et al. "Niger Delta Natural Resources Damage Assessment and Restoration Project, Phase I Scoping Report" (Abuja: Federal Ministry of Environment, 2006).
78 "Executive Summary," UNEP Environmental Assessment of Ogoniland (United Nations Environment Programme, 2011): 12–15, accessed at http://postconflict.unep.ch/publications/OEA/01_fwd_es_ch01_UNEP_OEA.pdf.
79 Best Ordinioha and Seiyefa Brisibe. "The Human Health Implications of Crude Oil Spills in the Niger Delta, Nigeria: An Interpretation of Published Studies," *Nigerian Medical Journal: Journal of the Nigeria Medical Association* 54.1 (2013): 10–16. PMC, accessed September 23, 2016, at www.ncbi.nlm.nih.gov/pmc/articles/PMC3644738/.
80 "Oil spill: Monarchs seek Buhari's intervention in Niger Delta," *Today Nigeria*, last modified May 27, 2016, www.today.ng/news/nigeria/128037/oil-spill-monarchs-seek-buharis-intervention-niger-delta.
81 "The Curse of Oil in Ogoniland," accessed September 2, 2016, at www.umich.edu/~snre492/cases_03–04/Ogoni/Ogoni_case_study.htm.

82 Stephanie Hanson. "MEND: The Niger Delta's Umbrella Militant Group" Council on Foreign Relations (March 2007).
83 Caroline Duffield. "Who Are Nigeria's Mend Oil Militants?" *BBC*, accessed September 2, 2016, www.bbc.com/news/world-africa-11467394.
84 Judith Asuni. Working Paper "Understanding the Armed Groups of the Niger Delta," Council on Foreign Relations, September 2009.
85 Ewokor. "Niger Delta Avengers."
86 Ewokor. "Niger Delta Avengers."
87 Newsdesk. "Oil Spill at Shell Deepwater Field off Nigeria," *Public Radio International*, December 21, 2011.
88 Joe Brock and Camillus Eboh. "Shell Faces $5 Billion Fine Over Nigeria Bonga Oil Spill," *Reuters*, July 17, 2012.
89 Ebeku, Kaniye S. A. "Biodiversity Conservation in Nigeria: An Appraisal of the Legal Regime in Relation to the Niger Delta Area of the Country," *Journal of Environmental Law* 16, no. 3 (2004): 361–375.
90 Kayinwaye Omorede Christiana. "Assessment of the Impact of Oil and Gas Resource Exploration on the Environment of Selected Communities in Delta State, Nigeria," *International Journal of Management, Economics and Social Sciences* 3, no. 2 (2014):6, accessed October 22, 2016, at http://proxygw.wrlc.org/login?url=http://search.proquest.com/docview/1610119064?accountid=11243.
91 "Environmental Assessment of Ogoniland," United Nations Environmental Programme, 2011.
92 Kayinwaye Omorede Christiana. "Assessment of the Impact of Oil and Gas Resource Exploration on the Environment of Selected Communities in Delta State, Nigeria."
93 Ojoma Akor. "How drought-tolerant maize variety enhances production," February 13, 2014, accessed October 12, 2016, at www.dailytrust.com.ng/daily/agriculture/16699-how-drought-tolerant-maize-variety-enhances-production#hMRcwyfC9bwKyRJs.99.
94 USAID. *Maximizing Agricultural Revenue and Key Enterprises in Targeted Sites. Mid-Term Update April 2012—March 2016* (N.D.), accessed September 1, 2016, at www.chemonics.com/OurWork/OurProjects/Documents/MARKETS2_MidTermUpdate_PrintVersion_Dollars.pdf.
95 Al Chukwuma Okoli and Atelje George. "Nomads Against Natives," 79.
96 John Campbell. *Nigeria Dancing on the Brink*.
97 "Mission and Vision," *Institute for Water Resources*, U.S. Army Corps of Engineers, accessed July 2, 2016, at www.iwr.usace.army.mil/About/Mission-and-Vision/.
98 Temitayo Famutimi. "U.S. donates 24 armored personnel carriers to Nigeria," U.S. Consulate General, Office of Public Affairs, Monday, January 11, 2016, www.africom.mil/NewsByCategory/article/27874/u-s-donates-24-armored-personnel-carriers-to-nigeria; Eric Schmitt and Dionne Searchy, "U.S. Plans to Put Advisers on Front Lines of Nigeria's War on Boko Haram," *The New York Times*, February 25, 2016, accessed at www.nytimes.com/2016/02/26/world/africa/us-plans-to-help-nigeria-in-war-on-boko-haram-terrorists.html?_r=0.
99 Sinead O'Sullivan. "Drones for Climate and Security," The Center for Climate and Security (2016), https://climateandsecurity.org/2016/08/10/drones-for-climate-and-security/.
100 UN Environmental Program. "UN Environment's Ogoniland Assessment Back in Spotlight," August 4, 2016, accessed September 1, 2016, at www.unep.org/newscentre/Default.aspx?DocumentID=27081&ArticleID=36239&l=e.
101 Marcus DuBois King and Jay Gulledge. "The Climate Change and Energy Security Nexus," *The Fletcher Forum of World Affairs* (2013).
102 Marcus DuBois King. "Water, U.S. Foreign Policy and American Leadership," Elliott School of International Affairs, October 15, 2013.

8 Water Resources, Climate Change, and the Destabilization of Modern Mesopotamia

Peter Gleick

Introduction

The links between water and political tensions and conflict have been the subject of extensive analysis for several decades, beginning with the development of the literature on "environmental security" in the 1980s. But the history of actual conflicts over water resources goes back literally millennia to some of the earliest organized civilizations in the ancient Middle East, with a reported violent dispute over irrigation canals between the ancient cities of Umma and Lagash around 2500 BCE. Disputes over access to and use of water in the arid river basins of the Nile, Tigris, Euphrates, and Jordan rivers have continued to the present day, as populations and economies grow and compete for scarce water resources.

New factors, including human-induced climate change and growing political uncertainties throughout the region, make it increasingly urgent that solutions to water tensions be found and implemented. These solutions are not solely technical or scientific; they include diplomatic, economic, and management approaches as well. This chapter reviews the water resource issues facing modern Mesopotamia, the new challenges facing the region, and strategies for moving toward more sustainable water management and use in order to reduce the risks of water-related violence.

Fresh water is vital to all human economic and social activities, from the production of food and energy to the maintenance of natural ecosystems that provide basic services for us. Yet freshwater resources are limited, unevenly distributed in space and time, increasingly contaminated or overused, and poorly managed. These constraints are compounded by growing populations and economies, which mean increased demand for limited water resources. Even in regions where natural water resources were previously considered abundant, water management increasingly has political and security implications.

These concerns are not new. As the Cold War waned in the late 1970s and early 1980s, researchers concerned about international security and conflict began to shift their focus from realpolitik and superpower politics to an evaluation of other threats to national and international stability, including environmental threats such as energy security and oil transfers, transboundary

150 Peter Gleick

environmental pollution, and conflicts over water resources and the potential impacts of climate changes.[1]

The fundamental concept, now widely accepted, is that political instability and violence, especially at the local or regional level, do not have purely political roots but are extensively influenced by economic, demographic, and social factors that are themselves sensitive to resource and environmental conditions.[2] This will be considered further in this chapter, but first, an overview of the current regional context regarding Mesopotamia's demographics and hydrology. This will be followed by a discussion of changes currently posing challenges for water security in the region, as well as for modeling of near-future changes. Finally, linkages between climate, water, and security will be explored, along with some recommendations for addressing regional challenges.

Modern Mesopotamia: The Region, Hydrology, and Demographics

The region known as Mesopotamia—literally translated as "the land between rivers"—encompasses roughly the area of the Tigris and Euphrates river basins. Mesopotamia has been nicknamed the "cradle of civilization" in the western world, having supported some of the earliest organized cultures of the Bronze Age, Sumer, Babylon, Assyria, and more. It is the region archaeologists credit with the first organized planting of cereal crops, key developments in mathematics and astronomy, and the earliest organized irrigation of agriculture. It is also the area where the very first reported water conflict was recorded, around 2500 BCE, when Urlama, the king of Lagash, diverted water from boundary irrigation canals between the Tigris and Euphrates rivers to deprive a competing region, Umma, of water.[3] Today, this region corresponds to parts of modern-day Iraq, Syria, southern Turkey, and some of the northern Arabian/Persian Gulf countries.

The Rivers

The Euphrates and Tigris rivers are the only renewable water resources of any note in the area, though some modest groundwater resources are also tapped in parts of the region. Both rivers originate in the mountains of southern Turkey, flow south through Syria, join together in Iraq in the Shatt al-Arab waterway, and discharge in the Arabian/Persian Gulf. The two rivers together drain a watershed of around 789,000 square kilometers (Table 8.1).

Data on the water discharged by the two rivers are extremely hard to come by and of mixed reliability. Massive infrastructure development in the form of large dams and extensive irrigation systems, especially on the Euphrates River, has led to a substantial decrease in flows over the past few decades. Figure 8.1 shows measured flows at Jarabulus in northern Syria near the Turkish border from 1937 to 2010. Flows have been steadily declining, with a significant drop after 1990 due to major Turkish irrigation withdrawals associated with the completion of the Ataturk Dam.

Table 8.1 Tigris-Euphrates/Shatt al-Arab Watershed

Total watershed area (km²): 789,000

Country	Watershed Area (km²)	(Percentage of Watershed Area)
Iraq	319,400	40%
Turkey	195,700	25%
Iran	155,400	20%
Syria	116,300	15%
Jordan	2,000	<1%
Saudi Arabia	80	<1%

Source: International River Basin Register (2012)

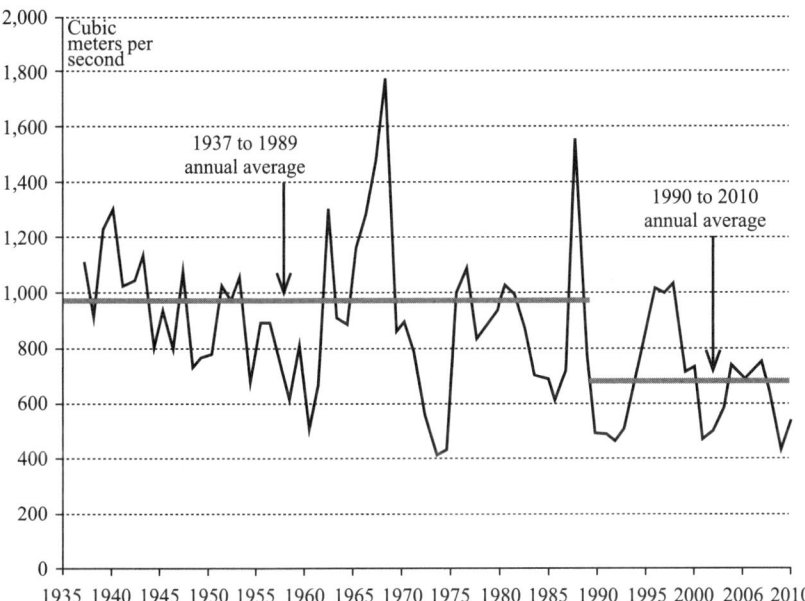

Figure 8.1 Flows in the Euphrates River From 1937 to 2010 (in cubic meters per second) Measured at Jarabulus, Syria

Note: Long-term averages from 1937 to 1989 and from 1990 to 2010 are marked by gray bars.
Source: UN-ESCWA (2013)

Water Use: Turkey, Syria, and Iraq

Figure 8.2 shows water-use estimates in billion cubic meters per year for Turkey, Syria, and Iraq from the UN Food and Agriculture Organization's AQUASTAT database.[4] As of the mid-2000s, all three countries used 75%–85% of water for agricultural production. Note that these data are prior to the current Syrian civil war, which has completely disrupted agriculture and urban water systems and prevented the collection of more recent data.

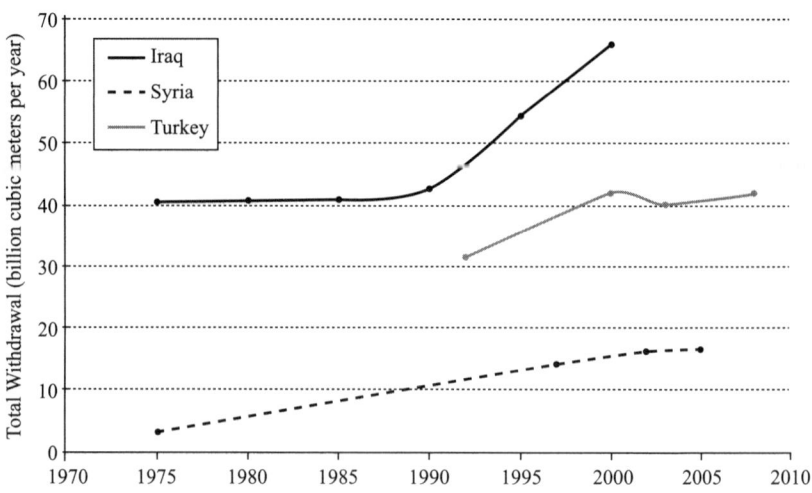

Figure 8.2 Water Withdrawals in Billion Cubic Meters Per Year for Iraq, Syria, and Turkey

Source: UNFAO (2016a)

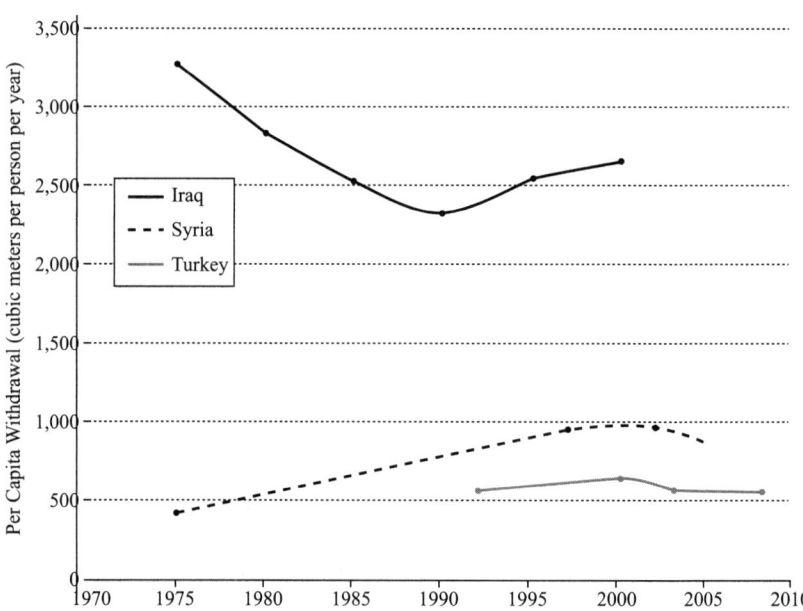

Figure 8.3 Per-Capita Water Withdrawals in Cubic Meters Per Person Per Year for Iraq, Syria, and Turkey

Source: UNFAO (2016a)

Figure 8.3 normalizes the total country withdrawals by population, showing per-capita withdrawal estimates (in cubic meters per person per year). As with total withdrawals, only limited data on water use are available in recent years, and overall, water-use data in the region are unreliable.

Key Interbasin Water Management Details

Every major river basin in the Middle East crosses a political border, and every significant watershed, including the Nile, Jordan, Tigris, Euphrates, and Orontes, is shared by two or more countries. As a result, international agreements about sustainable river management are critical, yet there have been very few successful interbasin treaties signed for the region and none for the Tigris-Euphrates river basin. Instead, a series of informal and incomplete agreements and committees have been set over the past four decades that have proved to be inadequate to address disagreements and equitable sharing of the waters. In addition, a series of dams on the rivers, in all countries, have greatly modified overall flows and led to impacts to downstream communities and (in the case of Iraq) the delta ecosystems at the mouth of the Tigris-Euphrates river. The lack of adequate international agreements has contributed to a series of political and potentially violent water-related disputes and raised political tensions over the years. A few examples from the Water Conflict Chronology[5] are described here.

In 1974, Iraq threatened to bomb the al-Thawra (Tabaqah) Dam in Syria and massed troops along the Syrian-Iraq border, alleging that the dam reduced flows of Euphrates water to Iraq. In 1975, Iraq asked the Arab League to intervene, claiming that flows reaching the Syria-Iraq border were "intolerable." Syria closed its airspace to Iraqi flights, and both countries reportedly transferred troops to their mutual border. Saudi Arabia helped mediate a peaceful resolution to that dispute.

In 1983, Turkey, Iraq, and Syria created a joint technical committee to address Tigris-Euphrates water disputes, but the group was ineffective and dissolved in 1993.[6] An informal agreement between Turkey and Syria, negotiated in the late 1980s, guaranteed a minimum flow of 500 cubic meters per second at the Turkish-Syrian border, but Syria has regularly accused Turkey of violating this standard. A separate agreement between Syria and Iraq signed in 1990 requires Syria and Iraq to share Euphrates River water on a 42%–58% basis (respectively) based on an assumed flow from Turkey,[7] but this agreement also has not been enforced.

In 1990, the flow of the Euphrates was temporarily interrupted when Turkey finished construction on the massive Ataturk Dam. Both Syria and Iraq protested the interruption. The political situation worsened when Turkish president Turgut Özal threatened to restrict water flow to Syria to force the country to withdraw support for Kurdish rebels.[8] No formal resolution on how to manage or operate the dam in the context of the overall flows of the Euphrates has ever been reached, although a joint water institute was proposed in 2008

154 Peter Gleick

to work toward developing agreements on more effective use of water. These efforts have been put on hold due to recent conflicts in the region. Turkey was one of only three countries in the world to vote against the United Nations Convention on the Law of Non-Navigational Uses of International Watercourses, which would require application of a set of principles and guidelines of equitable sharing of the waters of the Tigris and Euphrates rivers. Because Turkey is the upstream nation for both rivers, it is reluctant to acknowledge responsibility for provision of flows to downstream parties, even though such responsibility is a key principle of international water law.

The Changing Landscape

Even in a static world, conditions in the Middle East and in the core watersheds of modern Mesopotamia would continue to raise tensions over water resources. As noted previously, all the major water systems cross political borders, total current water demand already approaches the limits of renewable water availability, and few successful international water-sharing or water-management agreements are in place.

Yet the world is not static, experiencing dynamic and often rapid changes in demographics and environmental conditions. Populations are growing rapidly, economies are expanding or changing their focus, and the climate is undergoing increasingly severe shifts due to rising concentrations of greenhouse gases.[9]

Identifying causal links among complex social, political, and environmental conditions is not a science, especially when it comes to evaluating the contribution of these conditions to the field of security and conflict. There are no "control groups," few opportunities for genuine comparative studies, limited high-quality data, and large numbers of confounding factors. Nevertheless, it is increasingly possible to trace links between environmental factors, economic and social conditions, and political disruption. In the following sections, some of these factors are addressed. While there are opportunities for fundamental changes in policy and strategy that can reduce water-related tensions (and these are discussed later), most of the current changes appear to be worsening these tensions rather than alleviating them.

Demographic Change

Populations are growing very rapidly in many Middle Eastern countries, including the countries of the Tigris-Euphrates river basin. Figure 8.4 shows population trends since the 1960s, and today's populations are higher still. The only minor exception to this trend has been the intervention of the severe civil war in Syria and the massive diaspora of refugees, which has led to a precipitous population drop there (more recent data would show a more significant drop than shown in Figure 8.4). Even so, Syria saw a fourfold increase in population from around 5 million in 1962 to approximately 20 million at the start of

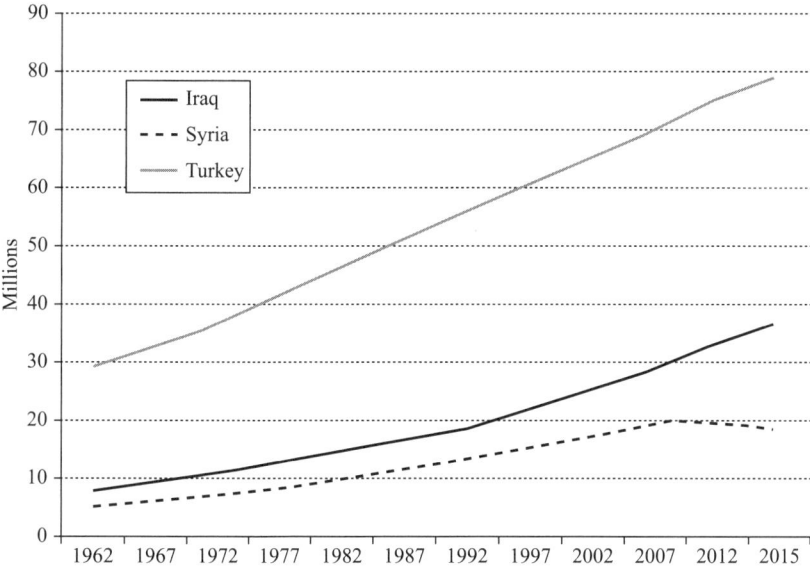

Figure 8.4 Population (in millions) of Iraq, Syria, and Turkey, 1962–2015
Source: UNFAO (2016a)

the civil war. Iraq's population rose from around 8 million to nearly 40 million. Turkey's rose from just under 30 million to around 80 million in 2015. These population increases put far greater pressures on fixed water resources and infrastructure, and they are likely to continue to grow in coming years.

Climate Change

The climate of the planet is changing because of human activities. In the language of the most comprehensive assessment of the scientific evidence (emphasis in original):

> Warming of the climate system is unequivocal, and since the 1950s, many of the observed changes are unprecedented over decades to millennia. The atmosphere and ocean have warmed, the amounts of snow and ice have diminished, sea level has risen, and the concentrations of greenhouse gases have increased.
>
> Human influence on the climate system is clear. This is evident from the increasing greenhouse gas concentrations in the atmosphere, positive radiative forcing, observed warming, and understanding of the climate system.
>
> Human influence has been detected in warming of the atmosphere and the ocean, in changes in the global water cycle, in reductions in snow and ice, in global mean sea level rise, and in changes in some climate

extremes. It is *extremely likely* that human influence has been the dominant cause of the observed warming since the mid-20th century.[10]

Climate model projections indicate that impacts on water resources will include changes in precipitation patterns, altered snowfall and snowmelt dynamics, increased water loss from higher evaporation driven by rising temperatures, and effects on water quality.

A review of climate scenarios for the eastern Mediterranean and Tigris-Euphrates basin area indicated strong warming of between 3.5 and 7 degrees Celsius above late-20th-century levels by the latter part of the 21st century.[11] Extremely hot summer conditions were projected to become more common, with a related increase in evapotranspiration water demand. These models also projected a decrease in annual precipitation. Combined, these effects led to worsening water scarcity for southern Turkey, Syria, and Iraq.

Another assessment tested large-scale general circulation models for the Tigris-Euphrates river basin under a range of emissions scenarios.[12] Looking at the impacts of climate changes on the region's hydrology, all the scenarios resulted in higher temperatures and evaporative water losses. In all the model simulations, rising temperatures led to a loss of snowpack in the upper watersheds, consistent with findings from a wide range of studies of other regions. Statistically significant reductions in runoff were seen along with a shift in the seasonality of runoff—again, consistent with other hydrologic assessments of climate change. The authors noted that these runoff changes

> suggest that the territories of Turkey and Syria within the basin are most vulnerable to climate change as they will experience significant decreases in the annual surface runoff. Eventually, however, the downstream countries, especially Iraq, may suffer more as they rely primarily on the water released by the upstream countries.[13]

Many of these modeled climate impacts have already been observed. As early as 2008, observational data suggested that drought frequency and intensity in the eastern Mediterranean area had changed.[14] An evaluation of two different drought indices showed an increasing seasonal and annual intensity of drought related to a decrease in precipitation in the wet season. Romanou et al. (2010) also reported higher evaporative water demand in the region associated with higher temperatures,[15] and Hoerling et al. (2012) explicitly linked observed regional increases in drought intensity in the form of reduced winter rainfall and increased evapotranspiration to human-induced climate changes.[16] That study concluded: "The magnitude and frequency of the drying that has occurred is too great to be explained by natural variability alone."[17]

The World Bank (2013) also noted that both model projections and actual observations pointed to worsening water conditions in the region:

> According to the latest IPCC [Intergovernmental Panel on Climate Change] assessment, the climate is predicted to become even hotter and drier in

most of the MENA [Middle East and North Africa] region. Higher temperatures and reduced precipitation will result in higher frequency and severity of droughts, an effect that is already materializing in the Maghreb.[18]

The Links Among Climate, Water, and Security: Past, Present, and Future

Regional Impacts

As noted earlier, the links between resources and security concerns include both direct and indirect effects. The history of conflicts over water resources goes back literally millennia, with some of the earliest reported disputes over access to irrigation water occurring in ancient Mesopotamia.[19]

The nature of these links varies. The most difficult to assess are indirect impacts, such as where effects on water resources lead to changes in food production or other economic factors, which in turn contribute to state instability and conflict. Such intertwined causal links are subtle and complicated to analyze, but recent analyses have drawn links between changes in climate change or water resources and security issues. In Syria, for example, there are documented links between regional long-term drought, climate change, agricultural crop failures, rising rural unemployment and migration to cities, urban unemployment and social unrest, and violent conflict.

A key point in assessments of these links is that they occur in the context of other complicated issues and factors. Each of the recent analyses that have addressed the current violence in the region are careful to avoid claims that climate change or water resources directly cause war; rather, they argue that environmental conditions have increased the probability or severity of state failure, civil unrest, and violence.[20] This important but sometimes subtle distinction between "causality" and "influence" is often missed in popular discussions of the ongoing violence.

The core argument is that severe observed hydrologic disruption in the form of reduced rainfall, reduced runoff (see, for example, Figure 8.1), and falling groundwater levels lead to significant reductions in agricultural productivity and farm income. Syria provides a particularly striking example, as its northern agricultural region suffered especially severe impacts, but overall the contribution of agriculture to the national economy suffered during the drought, when wheat production collapsed and many livestock operations were wiped out. During the height of the drought, food prices for core staples including wheat, corn, rice, and livestock feed doubled.[21] These economic disruptions contributed directly to massive migration from rural areas to Syrian cities, a growth of illegal settlements and urban unemployment, and eventually political instability—links that have been previously documented.[22] Other factors that have been cited as playing a role include local and governmental corruption and slow response by the Assad government. It is not possible to know the relative influences of each of these complex factors, but the extensive literature in this area supports the influence of each.

There are also examples of direct conflicts over access to water, although these are rare and almost always subnational rather than cases of institutionalized state violence. Attacks on water systems such as dams, water treatment and distribution plants, and hydroelectric facilities are much more common and have been regularly reported in conflicts that start for other (non resource-related) reasons. In recent years in the Middle East, major dams and water infrastructure have been attacked by different sides in the Iraqi and Syrian wars.[23] In June 2015, Islamic State militants shut off and redirected water flows below Ramadi Dam on the Euphrates River in order to facilitate military movements.[24] In August 2015, Syrian rebel groups cut off water from a spring in Ain al-Fijah, reducing water output to Damascus by 90% for three days and leading to water shortages and rationing.[25]

In December 2015, Russian Federation forces reportedly bombed the al-Khafsa water treatment facility in the city of Aleppo. The treatment plant draws water for the city from the Euphrates River. According to UNICEF,

> The bombing caused severe damage and cut off piped water supplies on which approximately 3.5 million people depend. Water pumping operations have since been partially restored, but more than 1.4 million people in rural Aleppo continue to suffer interruptions to their supply.[26]

International Impacts

While almost all water issues are largely local in nature, and while almost all examples of water-related conflicts are equally constrained to local or regional impacts, broader international security concerns arise when outside political or economic actors are involved. Unfortunately, the Middle East is a region with a long history of outside political influence, involvement, and interference. In recent years, the United States, Arabian Gulf countries, Russia, France, Britain, and others have all committed economic and military resources to the region and sometimes directly contributed to the ongoing conflicts.

Because of these links, changes in water resources or climate that affect regional conflict may lead to international responses. These risks have been explicitly acknowledged in recent reports from the U.S. military and intelligence communities that analyzed national security threats associated with international impacts of climate change or water scarcity.

For example, the concept of "environmental security" was a central topic at a November 1991 symposium at the U.S. National War College: "From Globalism to Regionalism: New Perspectives on American Foreign and Defense Policies." More recently, the U.S. Defense Intelligence Agency, in conjunction with other military and intelligence agencies, released an analysis of global and regional water security issues that concluded in part:

- "water problems—when combined with poverty, social tensions, environmental degradation, ineffectual leadership, and weak political

institutions—contribute to social disruptions that can result in state failure."[27]
- "The lack of adequate water will be a destabilizing factor in some countries because they do not have the financial resources or technical ability to solve their internal water problems. In addition, some states are further stressed by a heavy dependency on river water controlled by upstream nations with unresolved water-sharing issues.[28]
- "as water shortages become more acute beyond the next 10 years, water in shared basins will increasingly be used as leverage; the use of water as a weapon or to further terrorist objectives also will become more likely beyond 10 years."[29]

This risk of resource-related security challenges also includes climate change, which was identified in the 2014 U.S. Quadrennial Defense Review (QDR) as a "threat multiplier" posing significant challenges for the United States and the world at large. The QDR argued that climate change will aggravate stressors "that can enable terrorist activities and other forms of violence."[30] As greenhouse gas emissions increase, sea levels are rising, average global temperatures are increasing, and severe weather patterns are accelerating. These changes will devastate homes, land, and infrastructure, and will be exacerbated by other global dynamics, including populations that are growing, urbanizing, and becoming more affluent, and substantial economic growth in India, China, Brazil, and other nations. Climate change may worsen water scarcity and lead to sharp increases in food costs. The pressures caused by climate change will influence resource competition while placing additional burdens on economies, societies, and governance institutions around the world. These effects will aggravate stressors abroad such as poverty, environmental degradation, political instability, and social tensions—conditions that can enable terrorist activity and other forms of violence.

Both of these recent security assessments highlighted ongoing instability and "friction points" in the Middle East—no surprise given both U.S. interests and recent tensions and violence there. For water especially, the challenges facing the region have a long history.

Similarly, the U.S. Department of Defense and the U.S. government have acknowledged the reality of climate change and the risks it poses to U.S. global interests. The National Security Strategy issued by the administration in 2015 explicitly notes that

> climate change is an urgent and growing threat to our national security, contributing to increased natural disasters, refugee flows, and conflicts over basic resources like food and water. The present day effects of climate change are being felt from the Arctic to the Midwest. Increased sea levels and storm surges threaten coastal regions, infrastructure, and property. In turn, the global economy suffers, compounding the growing costs of preparing and restoring infrastructure.[31]

In this sense, the environmental factors are "threat multipliers" that contribute to more basic underlying destabilizing impacts. The Center for Naval Analyses (CNA) Military Advisory Board raised this point a decade ago in the context of the growing likelihood that climate change would worsen drought, famine, flood, and refugee problems.[32]

Questions also arise about the security interests and concerns of the United States or other outside powers in the region. Without such interests, the political disruptions and collapse in the region would be a largely local concern, with limited ability to influence international politics. This is not the case here, however: While the 21st century has seen a reduction in the kinds of tensions associated with the classic realpolitik of the Cold War, remnants of those tensions have always remained in the Middle East. Russia sees strategic value in access to ports and resources in the Persian/Arabian Gulf region and has been a consistent, long-term ally of the Syrian regime.

In Europe, the most recent security concern relates to the massive population dislocation and refugee flows driven by the civil war in Syria, coupled with the continued economic and political unrest in Iraq and North Africa. The European Commission has linked the refugee crisis both to the overall political situation in the region and to climate change, with some media descriptions of "climate refugees" and "climate migrants."[33]

Similarly, the United States has extensive political and economic interests in the area. Originally those interests were focused on ensuring reliable access to oil resources, and the U.S. remains dependent on Middle Eastern oil. More recently, however, the U.S. has continued to be involved because of new concerns about international terrorism associated with al-Qaeda and ISIL, maintaining long-standing security ties with Israel, and addressing recent migratory pressures on European allies. Concerns about links between refugees and terrorism were even a political talking point during the 2016 U.S. presidential election.

Recommendations: Reducing Security Risks

A variety of options are available to reduce the risks of resource-related conflicts. These include improvements and modifications to water technologies and policies, agricultural reform, economic strategies to strengthen resilience to climate and water variability through improvements in management, broader political stabilization and conflict resolution approaches with a component related to water and agricultural production, and the application of diplomatic and policy initiatives and tools.

No single strategy is likely to work consistently or broadly across regions and problems, but some fundamental principles and guidelines can be useful. Most broadly, whether or not problems with water or extreme climate events are likely, strong resource management strategies provide resilience—the ability to recover from disruptions or stresses.

For water resources, key strategies include those that both improve the effective use of scarce surface and groundwater and expand nontraditional sources of supply, such as treated wastewater and improved stormwater capture. There is great potential to improve water use efficiency in agricultural and urban uses, which can cut pressure on limited water resources. More-effective water management approaches are also needed, including economic and pricing strategies, community engagement, and diplomatic tools of conflict resolution. Israel and Jordan, for example, have put considerable resources into improving water use efficiency and turning treated wastewater into an asset. Almost all of Israeli agriculture now uses treated wastewater, while a series of desalination plants along the coast are meeting a growing proportion of urban water demand.

In addition to reform in the water sector, agricultural reforms can play an important role in strengthening regional stability. If droughts and water shortages are to worsen, traditional agricultural policies will have to be modified. Reforms should include a reassessment of the crops that are grown, the irrigation methods used, and the institutional factors that support resilient agriculture in the region, such as price supports and subsidies.

A separate set of strategies related to diplomacy, law, international agreements, and security policies can also be central to risk reduction. On a global scale, efforts to develop fundamental principles for transboundary watershed management have led to the drafting, adoption, and ratification of the 1997 UN Convention on the Law of Non-Navigational Uses of International Watercourses. This convention establishes standards and principles for best practices around joint basin management, data sharing, and conflict resolution; while not universally accepted, the fundamental concepts in the convention are widely respected.

Even more effective than broad international legal guidelines are specific watershed agreements that can be negotiated between affected parties. The long history of cooperative transboundary water agreements, described extensively by Wolf and colleagues, provides tools and models for nations that want to share water resources.[34] Indeed, part of the problem in the region of modern Mesopotamia, as noted earlier, is the lack of such specific acceptable agreements on levels of withdrawals, seasonal standards for river flows, and rules to govern management of large dams and other infrastructure. When conditions permit, efforts to restart negotiations on such agreements would be valuable. The top priority should be a joint agreement on the Tigris and Euphrates rivers among all the parties.

For water resources, the combination of these approaches has been synthesized in descriptions of a "soft path for water."[35] An application of these kinds of strategies could have reduced the role that water played in the recent Syrian civil war: More-efficient agricultural water use would have permitted greater food production and the retention of rural jobs; policies to more effectively manage variable supplies could have lessened the economic costs of the drought; and modern technologies such as precision irrigation, soil-moisture monitors, desalination, distributed wastewater treatment, smart metering, and more could have also reduced the difficulties associated with sustainable water management.

162   *Peter Gleick*

The ongoing conflict makes implementing any of these policy recommendations difficult—not just for individual countries but for the region as a whole. Nevertheless, even in the absence of a complete resolution of the violence, efforts must begin to plan for reconstruction of infrastructure and rebuilding of local institutions, such as agricultural and water agencies and committees.[36] International resources are available for such efforts: Water resources expertise is available through the multiple UN agencies that make up UN Water. Agricultural organizations such as the Food and Agricultural Organization, the International Water Management Institute, and the Consultative Group on International Agricultural Research all have extensive experience with agricultural reform. Finally, major European powers, together with the United States, have a vested interest in reducing refugee pressures and rebuilding the stability in the region to a point that functioning institutions return.

Summary

Concerns about water resources have led the World Economic Forum and other global bodies to highlight the nature of water threats and the need to develop strategies to manage water more effectively. In part, this reflects the

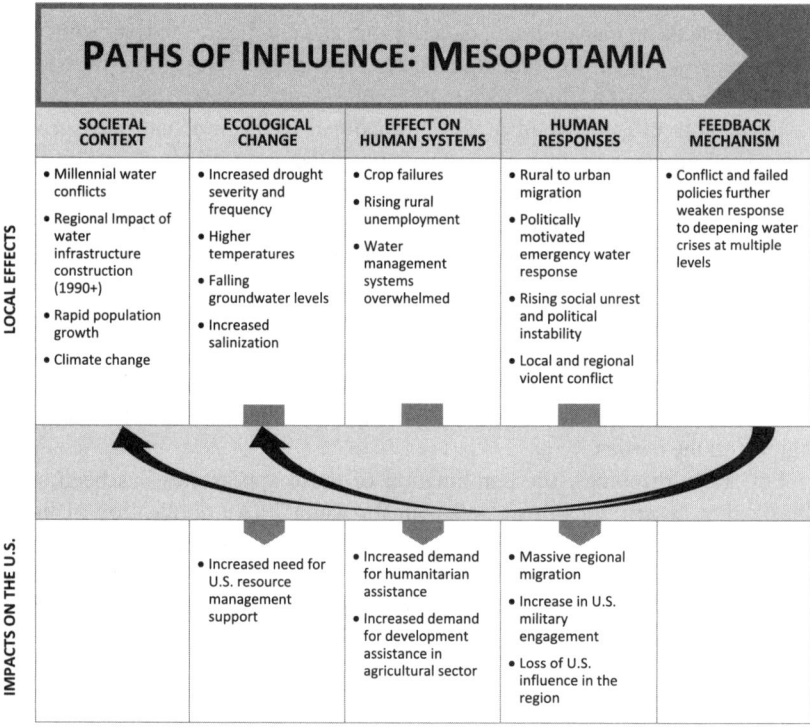

Figure 8.5 Paths of Influence: Mesopotamia

vital importance of water for human and environmental health and for the health of local and regional economies. But it also reflects a long history of political tensions and violence associated with poor water policies and management. This history goes back millennia, and much of the history revolves around the arid Middle East and the region between the Tigris and Euphrates rivers. Disputes over access and use have long been evident here and continue to manifest themselves as populations and economies grow and compete for scarce water resources.

New factors, including human-induced climate change and growing political uncertainties throughout the region, also contribute to the risks and make it increasingly urgent that solutions to water tensions be found and implemented. These solutions include diplomatic, economic, and management approaches, as well as the application of new technologies for monitoring and using water and reforming agriculture. This chapter reviewed water resource and climate issues facing modern Mesopotamia and the risks of water-related violence in the face of both old and new problems, and offered an overview of strategies for moving toward more sustainable water management.

Notes

1 Lester R. Brown, "Redefining National Security" (Washington, DC: Worldwatch Institute, 1977); Richard Ullman, "Redefining Security," *International Security* 8 (1983): 129–53; Norman Myers, "Environmental Dimensions to Security Issues," *Environmentalist* 6 (1986): 251–57; Jessica T. Mathews, "Redefining Security," *Foreign Affairs* 68, no. 2 (1989): 162–177; Peter H. Gleick, "Global Climatic Changes and Geopolitics: Pressures on Developed and Developing Countries," in A. L. Berger, S. Schneider, and J. Cl. Duplessy (eds.), *Climate and Geo-Sciences: A Challenge for Science and Society in the 21st Century* (Dordrecht, Netherlands: D. Reidel Press, 1989), 603–21; Peter H. Gleick, "The Implications of Global Climatic Changes for International Security," *Climatic Change* 15, no. 1/2 (1989): 309–25; Tad F. Homer-Dixon, "Environmental Change and Violent Conflict" (Cambridge, MA: American Academy of Arts and Sciences, 1990); Peter H. Gleick, "Environment, Resources, and International Security and Politics," in *Science and International Security: Responding to a Changing World* (Washington, DC: American Association for the Advancement of Science Press, 1990), 201–523.
2 Jon Barnett and Neil Adger, "Climate Change, Human Security and Violent Conflict," *Political Geography* 26, no. 6 (August 2007): 639–55; Solomon M. Hsiang, Marshall Burke, and Edward Miguel, "Quantifying the Influence of Climate on Human Conflict," *Science* 341, no. 6151: 1235367-1–1235367-14 (September 13, 2013), doi:10.1126/science.1235367; Blake D. Ratner and Ruth Meinzen-Dick, "Resource Conflict, Collective Action, and Resilience: An Analytical Framework," *International Journal of the Commons* 7, no. 1 (2013): 183–208.
3 Heleh Hatami and Peter H. Gleick, "Conflicts Over Water in the Myths, Legends, and Ancient History of the Middle East," *Environment* 36, no. 3 (1994): 10–11.
4 UNFAO, "AQUASTAT Main Database," 2016.
5 Peter H. Gleick, "The Water Conflict Chronology," 2016, http://worldwater.org/water-conflict/.
6 Ali Akanda, Sarah Freeman, and Maria M. Placht, "The Tigris-Euphrates River Basin: Mediating a Path Towards Regional Water Stability" (The Fletcher School,

164 *Peter Gleick*

Tufts University, 2007), http://fletcher.tufts.edu/AlNakhlah/Archives/~/media/Fletcher/Microsites/al%20Nakhlah/archives/pdfs/placht-2.pdf.
7 UNFAO, "Syrian Arab Republic Country Report."
8 Gleick, "The Water Conflict Chronology," 2016, http://worldwater.org/water-conflict/.
9 "Summary for Policymakers," Climate Change 2013: The Physical Science Basis Contribution of Working Group I to the Fifth Assessment Report of the Intergovernmental Panel on Climate Change (Cambridge, UK: IPCC, 2013).
10 Ibid., p. 4, 15, and 17.
11 Jos Lelieveld, P. Hadjinicolaou, E. Kostopoulou, J. Chenoweth, M. El Maayar, C. Giannakopoulos, C. Hannides, M. A. Lange, M. Tanarhte, E. Tyrlis, and E. Xoplaki, "Climate Change and Impacts in the Eastern Mediterranean and the Middle East," *Climatic Change* 114, no. 3–4 (2012): 667–687.
12 Deniz Bozkurt and Omer Lufti, "Climate Change Impacts in the Euphrates—Tigris Basin Based on Different Model and Scenario Simulations," *Journal of Hydrology* 480 (2013): 149–61, doi:10.1016/j.jhydrol.2012.12.021.
13 Ibid., p. 149.
14 S. Mathbout and M. Skaf, "Drought Changes Over Last Five Decades in Syria," in *Economics of Drought and Drought Preparedness in a Climate Change Context*, vol. 95, CIHEAM, 2010, 107–12.
15 A. Romanou and George Tselioudis, "Evaporation—Precipitation Variability Over the Mediterranean and the Black Seas from Satellite and Reanalysis Estimates," *Journal of Climate* 23 (2010): 5268–87, doi:10.1175/2010JCLI3525.1.
16 Martin Hoerling, Jon Eischeid, Judith Perlwitz, Xiaowei Quan, Tao Zhang, and Philip Pegion, "On the Increased Frequency of Mediterranean Drought," *Journal of Climate* 25 (2012): 2146–61, doi:10.1175/JCLI-D-11–00296.1.
17 NOAA, "Human-Caused Climate Change a Major Factor in More Frequent Mediterranean Droughts," 2012, www.noaanews.noaa.gov/stories2011/20111027_drought.html.
18 World Bank, "A Strategy to Address Climate Change in the MENA Region" (Washington, DC: Author, 2013), http://go.worldbank.org/OIZZFRJZZ0.
19 Hatami and Gleick, "Conflicts Over Water in the Myths, Legends, and Ancient History of the Middle East"; Peter H. Gleick, "Water, War and Peace in the Middle East," *Environment* 36, no. 3 (1994): 6–15; Gleick, "The Water Conflict Chronology."
20 Francesco Femia and Caitlin Werrell, "Syria: Climate Change, Drought, and Social Unrest" (Center for Climate and Security, 2013); Andrea Beck, "Drought, Dams, and Survival: Linking Water to Conflict and Cooperation in Syria's Civil War," *International Affairs* 5, no. 1 (2014); Peter H. Gleick, "Water, Drought, Climate Change, and Conflict in Syria," *Weather, Climate, and Society* 6, no. 3 (2014): 331–40; Solomon M. Hsiang, *Climatic Change* 123, no. 39: 39–55 (2014), doi:10.1007/s10584-013-0868-3.
21 Colin P. Kelley, Shahrzad Mohtadib, Mark A. Canec, Richard Seagerc, and Yochanan Kushnirc, "Climate Change in the Fertile Crescent and Implications of the Recent Syrian Drought" (Washington, DC: National Academy of Sciences, 2015).
22 Jack Goldstone, "Population and Security: How Demographic Change Can Lead to Violent Conflict," *Journal of International Affairs* 26, no. 1 (fall 2002): 3–22; Hsiang, Burke, and Miguel, "Quantifying the Influence of Climate on Human Conflict"; Gleick, "Water, Drought, Climate Change, and Conflict in Syria."
23 Gleick, "The Water Conflict Chronology."
24 Hamdi Alkhshali and Laure Smith-Spark, "Iraq: ISIS Fighters Close Ramadi Dam Gates," *CNN*, June 4, 2015, www.cnn.com/2015/06/04/middleeast/iraq-isis-ramadi/index.html.

25 Alisa Reznick, "Weaponizing Syria's Water," *Boston Review*, January 4, 2016,www.bostonreview.net/world/syria-water-alisa-reznick.
26 Hanaa Singer, "UNICEF Representative in Syria on Airstrikes against the Al-Khafsa Water Treatment Facility in Aleppo" (UNICEF, December 1, 2015), www.unicef.org/media/media_86402.html.
27 "Global Water Security," Intelligence Community Assessment (Washington, DC: Defense Intelligence Agency, February 2012), p. 3.
28 Ibid.
29 Ibid., p. 4.
30 Chuck Hagel, "Quadrennial defense review" (Washington, DC: Department of Defense, March 4, 2014), p. 8.
31 The White House, "U.S. National Security Strategy" (Washington, DC: Author, 2015), www.whitehouse.gov/the-press-office/2015/02/06/fact-sheet-2015-national-security-strategy.
32 "National Security and the Threat of Climate Change" (Alexandria, VA: Center for Naval Analyses, 2007).
33 Aryn Baker, "How Climate Change Is behind the Surge of Migrants to Europe," *Time*, September 7, 2015, http://time.com/4024210/climate-change-migrants/.
34 Aaron T. Wolf, "International Water Conflict Resolution: Lessons From Comparative Analysis," *International Journal of Water Resources Development* 13, no. 3 (September 1997); Mark Giordano, Alena Drieschova, James A. Duncan, Yoshiko Sayama, Lucia De Stefano, & Aaron T. Wolf, "A Review of the Evolution and State of Transboundary Freshwater Treaties" (Dordrecht: International Environmental Agreements: Politics, Law, and Economics, May 2013); Ashok Subramanian, Bridget Brown, and Aaron T. Wolf, "Understanding and Overcoming Risks to Cooperation along Transboundary Rivers," *Water Policy* 16, no. 5 (2014): 824.
35 Peter H. Gleick, "Soft Water Paths," *Nature* 418 (July 25, 2002): 373; Peter H. Gleick, "Soft Water Paths," *Science* 302 (November 28, 2003): 1524–28.
36 Frederick Lorenz and Edward J. Erickson, *Strategic Water: Iraq and Security Planning in the Euphrates-Tigris Basin* (Quantico, VA: Marine Corps University Press, 2013).

References

Akanda, A., S. Freeman, and M. M. Placht. 2007. "The Tigris-Euphrates River Basin: Mediating a Path Towards Regional Water Stability." The Fletcher School, Tufts University. http://fletcher.tufts.edu/AlNakhlah/Archives/~/media/Fletcher/Microsites/al%20Nakhlah/archives/pdfs/placht-2.pdf.
Alkhshali, H., and L. Smith-Spark. 2015. "Iraq: ISIS Fighters Close Ramadi Dam Gates." *CNN*, June 4. www.cnn.com/2015/06/04/middleeast/iraq-isis-ramadi/index.html.
Baker, A. 2015. "How Climate Change Is behind the Surge of Migrants to Europe." *Time* magazine, September 7. http://time.com/4024210/climate-change-migrants/.
Barnett, J., and N. Adger. 2007. "Climate Change, Human Security and Violent Conflict." *Political Geography* 26 (6): 639–55.
Beck, A. 2014. "Drought, Dams, and Survival: Linking Water to Conflict and Cooperation in Syria's Civil War." *International Affairs* 5 (1): 11–22.
Bozkurt, D., and O. Lufti. 2013. "Climate Change Impacts in the Euphrates—Tigris Basin Based on Different Model and Scenario Simulations." *Journal of Hydrology* 480: 149–61. doi:10.1016/j.jhydrol.2012.12.021.
Brown, L. R. 1977. "Redefining National Security." 14. Washington, DC: Worldwatch Institute.

Center for Naval Analyses (CAN) 2007. "National Security and the Threat of Climate Change." Alexandria, VA: Center for Naval Analyses.

Femia, F., and C. Werrell. 2013. "Syria: Climate Change, Drought, and Social Unrest." Washington, DC: Center for Climate and Security.

Giordano, M., A. Drieschova, J. A. Duncan, Y. Sayama, L. De Stefano, and A. T. Wolf. 2013. "A Review of the Evolution and State of Transboundary Freshwater Treaties." *International Environmental Agreements: Politics, Law, and Economics*.

Gleick, P. H. 1989a. "Global Climatic Changes and Geopolitics: Pressures on Developed and Developing Countries." In A. L. Berger, S. Schneider, and J. Cl. Duplessy (eds.), *Climate and Geo-Sciences: A Challenge for Science and Society in the 21st Century*, 603–21. Dordrecht, Netherlands: D. Reidel Press.

——— 1989b. "The Implications of Global Climatic Changes for International Security." *Climatic Change* 15 (1/2): 309–25.

——— 1990. "Environment, Resources, and International Security and Politics." In *Science and International Security: Responding to a Changing World*, 201–523. Washington, DC: American Association for the Advancement of Science Press.

——— 1994. "Water, War and Peace in the Middle East." *Environment* 36 (3): 6–15.

——— 2002. "Soft Water Paths." *Nature* 418 (July): 373.

——— 2003. "Soft Water Paths." *Science* 302 (November): 1524–8.

——— 2014. "Water, Drought, Climate Change, and Conflict in Syria." *Weather, Climate, and Society* 6 (3): 331–40.

——— 2016. "The Water Conflict Chronology." http://worldwater.org/water-conflict/.

"Global Water Security." 2012. ICA 2010–08. Intelligence Community Assessment. Washington, DC: Defense Intelligence Agency.

Goldstone, J. 2002. "Population and Security: How Demographic Change Can Lead to Violent Conflict." *Journal of International Affairs* 26 (1): 3–22.

Hatami, H., and P. H. Gleick. 1994. "Conflicts Over Water in the Myths, Legends, and Ancient History of the Middle East." *Environment* 36 (3): 10–11.

Hoerling, M., J. Eischeid, J. Perlwitz, X. Quan, T. Zhang, and P. Pegion. 2012. "On the Increased Frequency of Mediterranean Drought." *Journal of Climate* 25: 2146–61. doi:10.1175/JCLI-D-11-00296.1.

Homer-Dixon, T. F. 1990. "Environmental Change and Violent Conflict." 4. Cambridge, MA, USA: American Academy of Arts and Sciences.

Hsiang, S. M. 2014. "Climate, Conflict, and Social Stability: What Does the Evidence Say?" *Climatic Change* 123 (39): 39–55. doi:10.1007/s10584-013-0868-3.

Hsiang, S. M., M. Burke, and E. Miguel. 2013. "Quantifying the Influence of Climate on Human Conflict." *Science* 341 (6151): 1235367-1–1235367-14. doi:10.1126/science.1235367.

International River Basin Register. 2012. "Transboundary Freshwater Dispute Database." College of Earth, Ocean, and Atmospheric Sciences. www.transboundarywaters.orst.edu.

Kelley, C. P., S. Mohtadi, M. Cane, R. Seager, and Y. Kushnir. 2015. "Climate Change in the Fertile Crescent and Implications of the Recent Syrian Drought." Washington, DC: National Academy of Sciences.

Lelieveld, J., P. Hadjinicolaou, E. Kostopoulou, J. Chenoweth, M. El Maayar, C. Giannakopoulos, C. Hannides, et al. 2012. "Climate Change and Impacts in the Eastern Mediterranean and the Middle East." *Climatic Change* 114 (3–4): 667–87.

Lorenz, F., and E. J. Erickson. 2013. "Strategic Water: Iraq and Security Planning in the Euphrates-Tigris Basin." Quantico, VA: Marine Corps.

Mathbout, S., and M. Skaf. 2010. "Drought Changes Over Last Five Decades in Syria." *Economics of Drought and Drought Preparedness in a Climate Change Context* 95: 107–12. CIHEAM.
Mathews, J. T. 1989. "Redefining Security." *Foreign Affairs* 68 (2): 162–177.
Myers, N. 1986. "Environmental Dimensions to Security Issues." *Environmentalist* 6: 251–7.
NOAA. 2012. "Human-Caused Climate Change a Major Factor in More Frequent Mediterranean Droughts." www.noaanews.noaa.gov/stories2011/20111027_drought.html.
Omer, Tara. 2016. "FAO Country Profiles: Iraq." UN Food and Agricultural Organization. www.fao.org/ag/agp/agpc/doc/counprof/iraq/iraq.html.
Ratner, B. D., and R. Meinzen-Dick. 2013. "Resource Conflict, Collective Action, and Resilience: An Analytical Framework." *International Journal of the Commons* 7 (1): 183–208.
Reznick, A. 2016. "Weaponizing Syria's Water." Boston Review, January 4. www.bostonreview.net/world/syria-water-alisa-reznick.
Romanou, A., C. Tselioudis, C. Zerefos, J. Clayson, A. Curry, and A. Andersson. 2010. "Evaporation—Precipitation Variability Over the Mediterranean and the Black Seas from Satellite and Reanalysis Estimates." *Journal of Climate* 23: 5268–87. doi:10.1175/2010JCLI3525.1.
Singer, Hanaa. 2015. "UNICEF Representative in Syria on Airstrikes Against the Al-Khafsa Water Treatment Facility in Aleppo." UNICEF. www.unicef.org/media/media_86402.html.
Subramanian, A., B. Brown, and A. T. Wolf. 2014. "Understanding and Overcoming Risks to Cooperation Along Transboundary Rivers." *Water Policy* 16 (5): 824.
"Summary for Policymakers." 2013. Climate Change 2013: The Physical Science Basis. Contribution of Working Group I to the Fifth Assessment Report of the Intergovernmental Panel on Climate Change. Cambridge, UK: IPCC.
Ullman, Richard. 1983. "Redefining Security." *International Security* 8: 129–53.
UNFAO. 2016a. "AQUASTAT Main Database."
——— 2016b. "Syrian Arab Republic Country Report." www.fao.org/nr/water/aquastat/countries_regions/syr/index.stm.
United Nations. 1997. "UN Convention on the Law of Non-Navigational Uses of International Watercourses." www.unwatercoursesconvention.org.
United Nations Economic and Social Commission for Western Asia and the Bundesanstalt für Geowissenschaften und Rohstoffe. 2013. "Inventory of Shared Water Resources in Western Asia." Beirut, Lebanon: UN-ESCWA.
The White House. 2015. "U.S. National Security Strategy." Washington, DC. www.whitehouse.gov/the-press-office/2015/02/06/fact-sheet-2015-national-security-strategy.
Wolf, Aaron T. 1997. "International Water Conflict Resolution: Lessons From Comparative Analysis." *International Journal of Water Resources Development* 13 (3): 333–366.
World Bank. 2013. "A Strategy to Address Climate Change in the MENA Region." Washington, DC: Author. http://go.worldbank.org/OIZZFRJZZ0.

9 Iran's Impending Water Crisis

David Michel

The Islamic Republic of Iran faces a growing water crisis. Progressively mounting demands and persistent management deficiencies are imposing unsustainable pressures on the country's freshwater resources. Continuing global climate change threatens to further strain Iran's water security, diminishing the quantity, degrading the quality, and disrupting the seasonal timing and geographical distribution of available water supplies. Increasing water stresses risk impairing Iran's economic development, compromising food security, undermining public health, and potentially upsetting national and regional stability. Competition over scarce water supplies has already sparked local conflicts within Iran and clashes with its neighbors.

Dwindling water supplies increasingly preoccupy the Iranian leadership and public alike. The February 2016 elections, for instance, saw hundreds of parliamentarians sign a 15-point environmental pact calling for a national environmental plan.[1] As international sanctions have eased in the wake of the Joint Comprehensive Plan of Action (JCPOA) reached between Iran and the so-called P5 + 1 (China, France, Russia, the United Kingdom, and the United States, plus Germany) and the European Union, to monitor the Islamic Republic's nuclear program, Tehran has quickly moved to attract investment and technical collaboration in its water infrastructure, utilities, and agricultural sectors, as well as commercial opportunities abroad for its own firms, offering multiple avenues for mutual engagement.[2] The U.S. and Iran should seize this emerging opportunity together. Water offers a significant platform for data sharing, scientific exchange, and policy dialogue on which Washington and Tehran could build more cooperative and normalized relations.

"Dry, Barren, Mountainous": Iran's Freshwater Resources

"There is not in all the world that country which hath more mountains and fewer rivers," wrote Jean Chardin, a French jeweler who spent several years in Safavid Iran in the 1660s and '70s.

> The country of Persia is dry, barren, [and] mountainous. . . . This barrenness proceeds from no other Cause, than the scarcity of Water; there is want of it in most Parts of the whole Kingdom, where they are forc'd to preserve the Rain-Water, or to seek for it very deep in the Entrails of the Earth.[3]

Figure 9.1
Source: FAO Aquastat

Chardin may have overstated the realm's comparative plethora of mountains and paucity of rivers, but he was not far off the mark.

Some 90% of Iran is arid or semi-arid. More than half of the Islamic Republic's land area consists of mountains and deserts. The quantity and distribution of freshwater resources varies considerably around the country. Annual average rainfall ranges from 50 millimeters (mm) in the deserts to 2,275 mm near the southwest coast of the Caspian Sea. For Iran as a whole, average yearly precipitation amounts to 228 mm, less than one-third of the global average.

Half of annual precipitation falls in winter, outside the agricultural growing season, and two-thirds of yearly rainfall evaporates before it reaches the nation's rivers.[4]

Iran's average available renewable water resources total 137 cubic kilometers (km³) per year. Demand for these vital freshwater supplies has increased inexorably in the four decades since the 1979 Revolution. From 34.81 million people (in 1977), Iran's population has more than doubled, reaching 79.11 million in 2015. Over the same period, the economy expanded fivefold. National gross domestic product (GDP) jumped from $80.6 billion in 1977 (current U.S. dollars) to $425 billion in 2014, lifting per capita GDP from $2,315 to $5,372. Propelled by the powerful triptych of a growing population, a growing economy, and growing incomes, annual water withdrawals doubled from 45 km³ in 1975 to 93.3 km³ in 2004, the latest available figures.[5]

Water managers commonly consider that every individual requires 1,700 cubic meters (m³) of water a year to meet their needs—drinking, washing, food production, etc.—a threshold known as the Falkenmark indicator.[6] When annual renewable water availability per capita falls below this level, societies suffer "water stress." Iran's renewable water supplies have dipped from 3,935 m³ per inhabitant before the Revolution to 1,732 m³ today, just above the 1,700 m³ mark. Another widely employed metric plots total actual renewable water resources (TARWR) against demand. Water stress occurs where annual freshwater withdrawals exceed 40% of TARWR. In 2004, Iran's freshwater withdrawals topped 67% of annual renewable resources.[7]

The Falkenmark indicator and withdrawals/TARWR ratio are typically applied at the national level and on an annual time scale. Other indices attempt to capture higher geographical and temporal resolutions, reflecting that demand and availability can vary from basin to basin and fluctuate season to season. On this basis, one recent study examining the period 1996–2005 compared global monthly "bluewater" withdrawals (fresh surface water plus groundwater) with monthly bluewater availability.[8] It found fresh water use outstripped supplies during four months of the year or more across virtually all of Iran, except its northwest and northeast corners. For more than half the country, including most of the center, south, and east, annual average monthly withdrawals exceeded available resources by three to five times, or even more.

Tellingly, most of Iran's major rivers are hydrologically "closed," meaning that, under prevailing management practices, all their annual renewable water supplies are already committed to various human or environmental uses.[9] With little to no spare capacity left, these basins possess scant margin for maneuver; rising demands can be accommodated only by reducing existing withdrawals, reaching trade-offs among current users, or realizing greater water efficiencies. Many waterways are regularly overallocated. In Isfahan, the 33 arches of the city's famed 16th-century Allahverdi Khan Bridge over the Zayendeh-Rud, the "life-giving river," span dry earth much of the year.[10] According to the World Resources Institute, of the 100 most populous basins worldwide, three of the top seven, ranked by severity of water stress—the Qom, Harirud, and Helmand—lie in Iran.[11]

Iran's lakes, too, bear increasing strains. Lake Urmia in northwestern Iran once was the Middle East's largest, with a surface area varying from 5,200 to 6,000 km² in the 20th century. Dams, diversions, and drought have so diminished the 60 permanent and periodic rivers feeding into it that the lake has lost 88% of its area and 80% of its volume since 1972. With water levels now sinking by 34 centimeters (cm) annually and its volume shrinking 0.44 km²/year, Lake Urmia could disappear altogether by 2020.[12] Lake Bakhtegan, the country's second largest, has dried up almost completely, declining from 224.5 km² in 2001 to 9.32 km² in 2013.[13]

As Chardin observed in the 17th century, Iran meets many of its water needs by drawing from underground aquifers. Iran suffers extended droughts of five to six years in duration approximately once every decade, and rainfall can intermittently wither to less than 30% of the annual average, leading to increased reliance on groundwater.[14] But ballooning demands increasingly stretch this resource. By one count, the number of tube wells sunk around the country soared from 45,000–50,000 in the 1970s to 500,000 in 2006, a figure not including unregistered wells thought to equal one-third of the registered total.[15] Groundwater abstractions climbed correspondingly, from 20 km³ per year to 53 km³, and now fulfill 57% of annual water withdrawals. Iran's aquifers cannot sustain such demands.[16] Groundwater storage levels are declining 3.1–11.2 mm/year in each of Iran's six major basins.[17] At current rates of overexploitation, 12 of Iran's 31 provinces will entirely exhaust their aquifers within the next 50 years.[18]

Evaluated on a global Water Quality Index encompassing nitrogen, phosphorous, and dissolved oxygen concentrations, electrical conductivity, and pH, Iran's waters score 70.74 on a 100-point scale (100 being the highest quality), just above the global mean of 67 and comparable to countries such as Mexico and Romania.[19] Under this nationwide figure, however, several basins evince significant pollution. Nitrogen and phosphorous pollution, from agricultural runoff and domestic wastewater, exceed assimilation capacities in many southwestern rivers. Across the northern basins, nitrogen and phosphorous loads surpass levels the rivers can dilute by two- to fivefold.[20] Pressures on water quantity and quality interact. As consumers draw off water for various purposes, contaminants become more concentrated in the water remaining, eroding water quality. Decreasing water quality lowers available water quantities, as some sources become too polluted for certain uses.

Global climate change risks deepening Iran's water woes. The Intergovernmental Panel on Climate Change estimates Iran will experience average winter temperatures (December–February) 1.5–3.0°C higher by midcentury (2046–2065), while average summer temperatures (June–August) will warm 2–3°C. Autumn–winter (October–March) precipitation could increase 0–10% across northern Iran, while falling 0–10% through the middle and southern parts of the country. Spring–summer (April–September) precipitation is projected to drop 0–10% over nearly all the nation, while rising 0–10% in the southwest corner.[21]

Higher temperatures typically augment water demand—for irrigation, for cooling power plants and industrial processes, and for municipal consumption.[22] Changing precipitation patterns could make these demands harder to meet. According to some studies, diminishing rainfall and greater evapotranspiration could reduce annual recharge of Iran's water resources 22% by 2040–2050, slashing per-capita water availability to 501–750 m³ and cutting the country's TARWR by 15–19%. Between lower water availability and higher water use due to climate change, annual shortages could surpass 41% of the country's expected midcentury water needs.[23]

Agriculture would bear the brunt of the impacts. Absent robust adaptation strategies, crop production in many areas could slump substantially, and harvest failures from drought will increase.[24] Water requirement simulations for wheat production under future climate scenarios estimate water deficits during the growing season could hit 23% by 2050.[25] Locally increased precipitation may strengthen some crop yields. In other scenarios, wheat yields could plummet more than 50% and maize yields plunge more than 40% in the coming decades.[26] Hydropower, too, risks production losses, as lower precipitation will lessen river flows and impact reservoir levels. In the Karkheh basin, models anticipate annual average hydroelectricity generation could decline 15% or more by 2050.[27] Iranian government analyses project global warming will reduce river runoff, accelerate soil erosion, and aggravate flooding and dust storms.[28] So serious are the climate perils to Iran's water resources that the World Bank calculates that, even under a policy regime incentivizing more efficient water use, water scarcity will still drag down GDP growth rates by 6% in 2050.[29]

The Policy Roots of a Water Crisis

Mismanagement magnifies Iran's water challenges. Many key policies and practices affecting water supply and demand reflect primarily exogenous national priorities and purposes, before any sustainability or productivity goals, often inducing multiple inefficiencies and impairments—a picture by no means unique to the Islamic Republic.[30] Thus, for example, agriculture claims by far the lion's share of Iran's water use, accounting for 92% of total withdrawals, compared to a global average of 70%. Government policies contribute considerably to this thirst. Farming and livestock produce 10% of GDP and engage 21% of the labor force.[31] That represents a shrinking share of the economy and employment, but ensuring national food security has been an imperative regime objective since the Revolution.

To this end, Tehran pursues dual-subsidy programs supporting producers and consumers alike. Producer price supports for some 20 crops boost supply and promote self-sufficiency in staples, while bread subsidies ensure societal access to basic foodstuffs and bolster demand. In 2007, food subsidies represented 2.8% of GDP, cereal subsidies accounting for half that amount. The strategy has enhanced consumption and production, lifting Iran from the world's 16th largest wheat producer in 1999 to 12th in 2012.[32] Nevertheless, Iran still depended on imports for 29% of its cereal needs in 2014, the same percentage as in 2007–2009, and in 1990–1992.[33]

In addition to food production, Iran subsidizes agricultural water use. Groundwater is mainly private property, so it is free of charge to well owners. Regulated surface water deliveries to farmers are subsidized substantially below the cost of provision, at prices set between 1% and 3% of the value of the cultivated crop.[34] Tehran also heavily subsidizes electricity and fuel prices. Consequently, farmers pay less than 2% of the actual cost of power for pumping water from source to field.[35] Since 70% of groundwater extraction comes from deep wells averaging 90 meters (m) in depth, drafting irrigation water demands tremendous amounts of energy. In summer months, half the power used in many provinces operates tube wells.[36]

Food and energy subsidies distort the amount and geographical distribution of Iran's water demands. Growing wheat, for example, requires 1,827 m^3 of water per metric ton of crop, on average.[37] Yet as Tehran's self-sufficiency strategy expanded the area devoted to wheat by 12% from 1990 to 2010, wheat cultivation spread to cover three-fifths of Iran's arable land, extending through every one of its provinces, without regard for their diverse climatic and agro-ecologic conditions. With subsidies creating little incentive to economize, agricultural water use is highly wasteful. Irrigation efficiencies hover around 36%, with some provinces as low as 15%, far beneath the global average.[38] For crops such as maize, rice, and wheat, Iranian farmers often apply up to two to three times as much water per hectare as the world average.[39]

Iran subsidizes municipal water demand as well. Like farmers, domestic customers pay considerably less than the actual cost of provision. The average Iranian uses 250 liters of water per day. Some urban inhabitants use more than 400 liters, less than the average American's daily take of 575 liters, but well over the 150–190 liters claimed by British and German citizens, let alone the roughly 135 liters used by the average Indian.[40]

Faced with the growing fiscal burdens imposed by its subsidy system, Tehran launched major reforms in 2010. The *Targeted Subsidies Reform Act* (TSR) aimed to gradually increase energy prices to global market levels, raise electricity and water charges to cover full costs, and eliminate food subsidies, ultimately replacing the subsidy regime with nationwide cash transfers. In the first phase, prices for petroleum products, electricity, water, and bread surged four- to twenty-fold. By 2011, the International Monetary Fund calculates, water consumption had decreased 6%.[41] Nevertheless, the sharp jump in prices sparked accelerating inflation and economic contraction, even as the revenues from higher energy charges proved insufficient to cover the cash transfers committed to households. Consequently, Parliament postponed the follow-on phase of the reforms in 2012, and consumption of subsidized products rebounded.[42] The Rouhani government moved again to trim subsidies in 2014, sending petrol, electricity, and water prices skyward by 75%, 24%, and 20%, respectively, with further adjustments in 2015, but the larger TSR remains suspended.[43]

Government policy also significantly influences freshwater supply. Iran, like many countries, has adopted a highly technocratic approach to water management.⁴⁴ Sometimes termed the "hydraulic mission paradigm," this outlook conceives water resources essentially as economic goods to be dominated and distributed through engineering and technology.⁴⁵ Iran's hydraulic mission emphasizes industrializing arid regions, delivering water for hydropower, irrigation, cities, and industry via the construction of hard infrastructure—dams, canals, and groundwater pumping. Especially following the Revolution and the regime's international isolation, dam construction exemplified the nation's technological prowess. Beginning with 14 dams in 1979, many erected by foreign companies, Iran now boasts 802 large dams higher than 15 m, according to the International Commission on Large Dams.⁴⁶ With most dam-building projects executed by Khatam al-Anbia, an engineering element of Iran's Revolutionary Guards Corps, dam building also exemplifies the intertwining of technological capacity, nationalism, and security in Iran's economic development.⁴⁷

Iran's infrastructure interventions have substantially reshaped the country's water supply—but they have often done so at the cost of skewing perceptions of water availability, maximizing short-term development while ignoring growing ecosystem damages and undermining long-term sustainability. Lake Urmia illustrates these impacts. The government and water companies regularly attribute the lake's desiccation to climate change and drought. But the basin has weathered drought in the past, and statistical analyses signal no significant trend in drought or precipitation in the region as the lake has retreated from its maximum extent in 1995. What have changed markedly are water needs. Over the past three decades, the urban area in the basin spiked from 17,394 hectares in 1984 to 55,935 hectares in 2014. Irrigated agriculture leaped from 303,588 hectares to 508,660 hectares. Recent years have seen several new dams and irrigation districts. More than 200 dam and irrigation projects under construction or in final design now gird the lake. As urban and agricultural withdrawals diminish surface inflows, groundwater pumping has dropped local water tables, increasing leakage from the lake's higher water levels into the lowered surrounding aquifers. Climate factors do impact lake levels, but natural variability cannot explain Urmia's drastic drying. Human causes, rather than environmental forces, are the primary culprit in Lake Urmia's decline.⁴⁸

The Socioeconomic and Political Impacts of Iran's Water Challenges

Iran's multiple water challenges inflict significant burdens on the economy and society, threatening to sap agricultural output, curb hydropower generation, compromise public health, and potentially fuel regional frictions. World Bank assessments of environmental degradation in the Islamic Republic estimate Iran's annual economic losses and public health damages stemming from inadequate and unsafe water supplies, groundwater depletion, water-caused soil erosion, and dam sedimentation amount to 2.82% of GDP.⁴⁹ By way of comparison, 2.8% of GDP corresponds to World Bank calculations of the costs to the Iranian economy of stringent U.S. and international sanctions prior to the JCPOA.[50]

Intensifying water stress contributes to degrading the nation's land, menacing 110–118 million hectares with desertification.[51] Spreading desertification in turn exacerbates the debilitating dust storms that regularly shroud the country. Wind erosion afflicts 200,000 km² across 19 provinces, affecting 150,000 residential units, 6,300 km² of farmland, and 9,100 km of roads.[52] The salt marshes—called *Hamouns*—of the Sistan basin in Iran's southeast, nourished by the Hirmand/Helmand and other smaller rivers entering from Afghanistan, exemplify the problem. Irrigation demands and development of the Chah Nimeh Reservoir have curbed flows into the lakes. Water levels in the Hamouns have plunged, uncovering growing expanses of arid lakebed. Winds that previously rippled shallow saltwater now sweep parched salt flats, carrying off clouds of salt-laden silt. Satellite and ground data show area dust storms rising as the Hamouns recede. Storms around the city of Zabol, a regional capital, can contain 250 kilograms of dust per cubic meter of air, and the region endures up to 80 storms a year.[53]

Billowing dust clouds close roads, railways, and airports; choke crops and livestock; contaminate soils; and clog machinery. Carrying salts and sediments into rivers and canals, the storms pollute area water supplies.[54] Dust storms cost Zabol $100 million in lost economic activity and physical damage between 2000 and 2005.[55] More troubling, dust storms jeopardize public health. Fine particles can penetrate the lungs, causing infections, respiratory difficulties, and cardiovascular problems. A study of hospitals in Kermanshah Province found that every 10% rise in dust concentration swelled the number of cardiac patients by 10%, respiratory patients by 5%, and deaths from heart disease by 3%.[56]

Groundwater depletion also undermines agricultural productivity and urban development. Many of Iran's aquifers are naturally brackish or sodic. Overirrigation with saline groundwater accumulates salts in the soil, compromising its fertility, especially in arid regions where low precipitation prevents the washing of salts out of the soil. In coastal areas, falling groundwater tables can enable saltwater intrusion from the sea, salinizing aquifers. Approximately 50–75% of Iran's irrigated lands manifest varying degrees of salinity. Average yield losses from salt-affected soils may reach 50%, and annual economic losses range from $1 billion to $1.8 billion.[57] Excessive groundwater withdrawals also trigger land subsidence. As aquifer levels fall, the overlaying layers of earth can compact or collapse, disrupting farming and damaging buildings. In Tehran, groundwater levels dropped 11.65 m from 1984 to 2012, occasioning subsidence up to 20 cm per year in some areas.[58]

Lower water availability poses other risks as well. In 2014–2015, feeble rains and supply strains decreased total storage volume in Tehran's four main reservoirs by 40%. By August 2015, the Energy Ministry announced that 60% of the reservoirs of major dams were empty. Power generation halted or plummeted at Iran's 50-odd hydroelectric plants, and hydropower tumbled from 14% to under 5% of the national electricity mix.[59]

Rising water challenges also contribute to growing social pressures. Iran is rapidly urbanizing. Between 1980 and 2015, the urban population rocketed from 19.3 to 58.3 million. In 1980, 49.7% of Iranians lived in cities. In 2015, 73.4% did. Rural-to-urban migration, together with urban absorption and development of rural areas, feed Iran's expanding cities. The rural population is projected to dive 23.8% by 2050, even as the national population increases 26.6%.[60] Multiple influences drive internal population movements, including the "pull" factors of employment opportunities, education, and family networks available in cities. Rural underdevelopment represents the key corresponding "push" factor, and rising water stress constitutes a critical constraint on agricultural livelihoods, limiting economic possibilities and fueling rural outmigration.[61]

Soaring urban populations strain municipal infrastructure and water supplies. Many cities face prospective water rationing.[62] Iran's burgeoning urban areas are geographically ill-distributed relative to its freshwater resources. Nearly half of Iran's renewable water lies in the southern coastal basin, which occupies only a quarter of the national territory. The Central Plateau, covering over half the country, including metropolitan Tehran, holds less than one-third of its water supplies, with annual renewable water availability under 1,200 m^3 per capita. To compensate for this uneven distribution, Iran has developed numerous water transfer projects, using tunnels and canals to move water from one river basin to another.[63]

Interbasin transfers can help alleviate supply disparities, but risk significant negative side effects, creating "fixes that backfire."[64] In central Iran, several tunnels import water from the Karun and Dez rivers into the Zayendeh-Rud basin—more than doubling the river's natural flow—to nourish irrigated agriculture and the city of Isfahan. Several other projects export water outside the basin to Yazd, Kashan, and other cities. Yet this supply-side shuffling of resources between rivers provides only temporary solutions, while starving downstream ecosystems that then receive less water. Without efforts to moderate rising (and often inefficient) water use, demand merely grows to absorb and exceed the transferred supplies, ultimately begetting long-run shortages more severe than had the interbasin projects never been built.[65]

Iran's technocratic hydraulic mission, embodied by dams, diversions, and interbasin transfers, is sowing political pressures as well as hydrological ones. Politics, in Harold Lasswell's classic formulation, is the question of who gets what, when, and how.[66] "Pushing rivers around" to determine who gets water, when, and how can enable development, but it can also engender disputes among contending claimants.[67] In the Zayendeh-Rud basin, state-centric infrastructure interventions have effectively reapportioned benefits to favored constituencies, frequently motivated by political patronage, overturning traditional allocation rights and concentrating negative impacts on politically weaker communities and the environment. Violent opposition has resulted. In the 1980s, demonstrations against the Khamiran Dam caused a number of fatalities. In 2013, farmers outside Isfahan destroyed pumps carrying water from the local river to urban Yazd 270 km away, spurring armed clashes with police that reportedly killed five.[68]

Regional discord over water distribution regularly pits rural demands against municipal consumers, and ethnic minority communities in Iran's border provinces against the Persian majority. The Islamic Republic's water governance structures often exacerbate such conflicts. Reforms enacted under President Mahmoud Ahmadinejad reassigned water management responsibilities along provincial political boundaries rather than along the hydrological boundaries of watersheds, accentuating interprovincial competition and the tendency of water to "flow toward power."[69] In Iran, power resides with Tehran. Through successive regimes, the overriding trend has been toward centralization under Persian dominance and political subordination of minorities, rather than decentralization or self-government.[70]

Since Iran's rivers and lakes are national property, and power to manage resources resides with the central government, many minorities in peripheral provinces resent policies they regard as degrading or dispossessing their water supplies.[71] Arabs in southwestern Khuzestan contest the channeling of their rivers to distant Isfahan and Yazd, and excoriate expropriation of their irrigation waters to benefit the sugarcane industry rather than local farmers; this has sparked several riots. In the northwest, movements to protect Lake Urmia increasingly stir Azeri nationalism. Demonstrations in mid-2011 witnessed banners declaring "Lake Urmia is drying up; Iran has ordered its execution," while conspiracy theories circulate denouncing Tehran's deliberate designs to strangle the lake and render Azeri provinces salty deserts. In eastern Sistan and Baluchistan Province, perceptions of government inaction and indifference to rural poverty, desertification, and inadequate access to clean water aggravate the ethnic and religious grievances that stoke Baluch separatist insurgency.[72]

Transboundary Water Tensions

Increasingly, Iran's water challenges also hold international repercussions. Iran shares 10 major river systems with surrounding countries. Absent robust collective arrangements, rows over common water supplies are firing regional tensions.

In the west, several tributaries flow from Iran into the Tigris River in Iraq. Together, Iranian headwaters contribute up to 25% of the Tigris's mean annual flow of 26 km^3.[73] Baghdad and Tehran signed an accord in 1975 on their transboundary watercourses, specifying the division of minor tributaries but leaving allocation of the major rivers to a later joint commission.[74] That work was never done. Since 1975, Iran has established multiple dams in the Diyala sub-basin, the largest tributary, providing irrigation and hydropower, and supplying drinking water to Kermanshah. These works have lowered long-term median flows into Iraq by more than 50%. By 2020, an additional dam and an interbasin water transfer tunnel currently under construction are projected to cut the Diyala's flows to 22% of their natural unimpaired volume.[75]

Downstream Iraqis fear Iran's diversions of the Diyala will thwart their own water development, potentially throttling agriculture, fishing, and hydropower production. Farmers have blocked cross-border traffic in protest. Demonstrations erupted in 2008 and in 2011–2012. Iraqi officials, political parties, and non-governmental organizations have contested the Iranian projects, but the dispute remains unresolved. Troubled regional politics complicate the transboundary hostilities. From Iran, the Diyala first runs through the Kurdistan Region of Iraq, whose people the river's waning flows most imperil. But the Kurdistan Regional Government (KRG) is not party to the 1975 treaty and has no standing to negotiate with Iran, while the KRG's own dam-building plans on the Diyala set it at odds with Baghdad.[76]

Farther south, the Karkheh and Karun rivers empty from Iran into Iraq's Shatt-al-Arab River, formed by the confluence of the Tigris and Euphrates. The Karkheh and Karun supply an estimated 41% of the Shatt-al-Arab's historical mean annual discharge of 73.6 km^2. Yet Iran's upstream exploitation increasingly utilizes nearly all of the tributaries' waters. On the Karkheh, dams and irrigation have drastically curtailed downstream flows, shrinking the Hoor-Alazim (Haweizeh) wetlands on the Iran-Iraq border by two-thirds from 1991 to 2009. Dams and irrigation projects have also pared mean annual discharge on the Karun, while mounting salinization and agricultural, industrial, and municipal pollution render the river's water unfit for drinking much of the year. In 2004, Iraq lodged formal complaints against Iran's development activities, leading to the creation of a joint technical committee for shared water issues, but as yet no common management measures have resulted.[77]

From Afghanistan, two major rivers, the Helmand/Hirmand and the Harirud, flow into Iran. The Afghan reaches supply 96% of the Hirmand's flow into Iran. Similarly, 61% of the Harirud's flow in Iran originates from Afghanistan.[78] As Afghanistan struggles to emerge from four decades of war, water development plays a crucial role in the nation's stabilization and reconstruction. But Iran perceives Afghanistan's growing water demand and expanding infrastructure as threatening its own increasingly water-stressed eastern provinces, engendering ongoing tensions.

The Helmand conflict's roots lie in 19th-century state formation and territorial confrontations over the region then known as the Sistan and its only surface water source. In 1973, the two countries reached a treaty agreement whereby Afghanistan guaranteed Iran average annual flows of 26 m^3/second (0.82 km^3/year) in a "normal" year, with provisions for lesser deliveries in years of lower flow.[79] Due to the 1973 coup in Afghanistan, the Iranian Revolution, the 1979 Soviet invasion of Afghanistan, and subsequent civil turmoil, the treaty was never fully implemented. From 1999 to 2002, the entire region experienced intense drought. In the midst of the disaster, the Taliban government closed the Kajaki Dam, cutting off the Helmand's flow to Iran for two years—in Iranians' eyes, dramatically exacerbating a catastrophe that saw hundreds of villages on both sides of the border abandoned to sandstorms and lack of potable water.[80]

Since the fall of the Taliban, Afghanistan has moved with U.S. and international assistance to rebuild its water infrastructure, declaring dams for hydropower and irrigation a "strategic objective" for the country. Many of these projects portend considerable changes in downstream water availability. The Kamal Khan Dam under construction on the Helmand mainstream and the Bakhashabad Dam on the smaller Farah River are projected to decrease annual flows into Iran and the Hamouns by 0.73–1.29 km². Together with rehabilitation of the Kajaki Dam, damaged during the war, along with planned enlargement of the Dahla Dam and new reservoirs, the Helmand's flows could drop by 2.7 km², equal to half of Iran's entire water demand in the Sistan basin. On the Harirud, the recently inaugurated Salma Dam will enable irrigation withdrawals projected to reduce flows to Iran by 62–76%.[81]

In response, Iran has engaged paradoxically conflicting strategies. On the one hand, Iran has repeatedly endeavored to negotiate developments on its shared rivers, in 2005 putting in place the Helmand Commission prescribed by the 1973 treaty, cooperating closely with Kabul on a UN-led initiative to restore the Hamouns, calling for a joint commission on the Harirud, allocating over $3 million for reconstruction of Afghanistan's water sector, and furnishing technical assistance to the Afghan Ministry of Water and Energy. At the same time, however, Iran has reportedly prosecuted a violent proxy campaign, allegedly providing support to various Taliban factions targeting Afghanistan's water infrastructure. Several major attacks by antigovernment militias struck the Kajaki Dam in 2007. In 2009, Afghan forces found Iranian-made explosives and ammunition cached around the Bakhashabad Dam. In 2011, a captured Taliban commander publicly averred he had received money and training in Iran to destroy the Kamal Khan Dam. In addition, multiple ambushes and assassinations killed dozens of security personnel at the Salma Dam.[82] At times, Iran has taken action overtly. In 2011, Iranian border guards crossed into Afghanistan to release water from a 30-km irrigation canal that diverts flow out of the Helmand River before it enters Iran, exchanging fire with Afghan forces.[83] Since the completion of the JCPOA, shared rivers have risen high on Tehran and Kabul's common agenda. Both governments profess the desire to settle their differences, but both also continue to pronounce largely uncompromising positions.[84]

Tehran entertains more cordial water relations with its northern neighbors. Turkmenistan and Iran equally share water resources from their jointly constructed and operated Dosti (Friendship) Dam and Shirtappeh Dam on the Harirud. A joint protocol with Turkey provides Iran a guaranteed minimum flow on the Sarisu River. Nevertheless, where they exist at all, cooperative arrangements on Iran's transboundary rivers are limited in scope, partial in their coverage of all riparians, and underdeveloped in their institutional capacities.[85] In the assessment of the United Nations Environment Programme, weak legal frameworks and pervasive hydro-political frictions, combined with rising demand, increasing pollution, and environmental pressures, render most of Iran's international rivers particular "hot spots" for worsening water stress, socioeconomic impacts, and transboundary tensions.[86]

Policy Reforms and Cooperative Opportunities

Iranian policymakers are acutely aware of the nation's water challenges. "The main problem that threatens us," warned former Agriculture Minister Issa Kalantari in 2013, "more dangerous than Israel, America or political fighting . . . , is that the Iranian plateau is becoming uninhabitable," as lakes dry and groundwater dwindles.[87] President Hassan Rouhani characterizes water management as a "historic" issue, the most basic challenge facing the government, and has called for a national water-conservation plan.[88]

Some reforms have moved in this direction. In numerous districts, further well construction has been prohibited to aide aquifer recovery, and President Rouhani has quietly scaled back Iran's fervid dam construction.[89] But government planning still privileges large supply-side infrastructure proposals—such as interbasin transfers from as far afield as Tajikistan and pumping desalinated water hundreds of kilometers from the Gulf of Oman and the Caspian Sea to the interior—over demand-management efficiencies.[90] Similarly, Iran aims to shave agricultural water use from 92% to 87% of withdrawals, and Tehran's main agricultural policy frameworks nominally emphasize sustainable development. Yet the logic of food security still animates the Five-Year National Economic, Social, and Cultural Development Plan, continuing the pursuit of self-sufficiency in essential crops—with the consequent demands on water resources.[91] The national water-conservation strategy has yet to materialize.

To ensure the country's future water security, more fundamental measures will be needed. Continued subsidy reforms should further align water and fuel charges with their true cost to incentivize more efficient use. Food subsidy reforms, such as structuring guaranteed prices for specific crops in specific regions to reflect resource availabilities, could promote new cropping patterns more appropriate to land and water conditions.[92] Savings realized from subsidy reform could in turn be redirected to support agricultural investment and innovation. Analyses of cereals production, for example, suggest that shifting cultivation away from water-scarce regions, together with improved irrigation efficiencies, could considerably enhance national yields while simultaneously reducing water demand.[93] Technological developments can play an important role. Iran possesses substantial solar energy potential. Solar desalination could furnish more sustainable desalination options than the current hydrocarbon-powered plants.[94] Many emerging desalination filtration technologies are also applicable to removing other contaminants, making local, distributed solar desalination systems potentially suitable for treating brackish groundwater sources and for the reuse and recycling of agricultural and municipal wastewater.[95]

Water also provides a significant platform on which Iran and the U.S. (and the larger international community) could build more cooperative and normalized relations. The governments and policy and scientific communities in both nations have recognized that such "engagement can not only contribute to solutions of global problems but can also help improve understanding of each country's society and politics, with attendant benefits for bilateral political relations."[96] Iran and America confront many of the same water management issues, from addressing long-term drought to improving groundwater governance and adapting to climate change. Collaboration could thus take place around common challenges and experiences, rather than on the potentially more politically fraught terrain of "lessons" and "best practices" delivered from one state to another. Indeed, expert exchanges along these lines have already begun, looking at initiatives to save Lake Urmia in light of U.S. experiences with California's salt lakes.[97]

Productive subjects for potential cooperation include water information, incentives, infrastructure, and institutions. In the first category, water cannot be efficiently managed if it cannot be effectively measured. In key areas, Iran lacks adequate data. Ground-gauged precipitation data is patchy, for instance, and many groundwater aquifers remain under-characterized in their extent, capacity, flow, etc.[98] U.S. agencies such as NASA, NOAA, and the USGS could coordinate with Iranian bodies to better gather and share such information through remote sensing and in situ measures.

Under the rubric of incentives, dialogues could consider water pricing strategies and the creation and regulation of water markets. Given the importance of agricultural water demand in both countries, exchange of policy experiences in agricultural extension services could prove valuable. Studies in the U.S. and Iran alike show farmers do not necessarily readily adopt available water conservation practices or advanced irrigation technologies, for example.[99] Analyses of the incentives and interests determining farmer behaviors in the two nations could help improve the dissemination of sustainable water strategies in both.

Iran and the U.S. face important infrastructure challenges, such as refurbishing aging municipal water systems, as well as opportunities for innovation, such as desalination technologies and wastewater treatment and reuse.[100] Both areas present possibilities for mutually beneficial trade, investment, and research and development. The U.S. could take practical steps to encourage such developments by easing impediments it imposes to lending to Iran by the multilateral development banks and private sector.[101]

Fruitful exchanges might also examine the two countries' institutional structures and practices for water governance, drawing on their respective experiences with water users' associations, river commissions, etc. Iran's international agreement with Turkmenistan on the Harirud, for example, resembles in some respects the U.S.-Canada Columbia River Treaty, while Iran also maintains an underutilized memorandum of understanding on environmental cooperation with Turkey not so dissimilar from the U.S.-Mexico Border Environment Cooperation Commission, suggesting that Iran and America might usefully engage each other on transboundary water management approaches.[102]

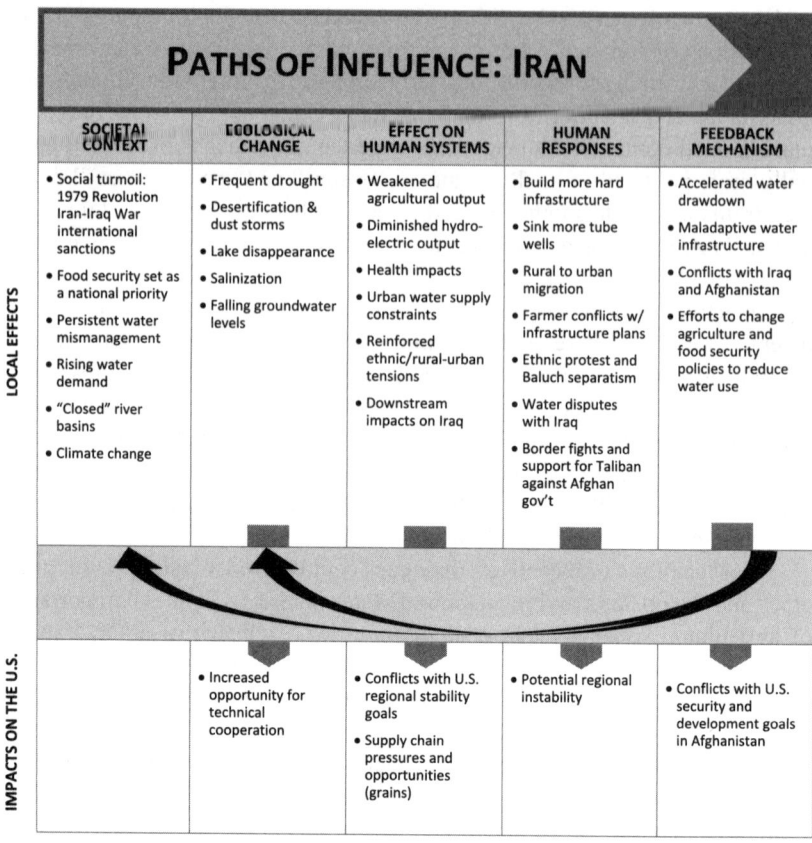

Figure 9.2 Paths of Influence: Iran

Finally, though Iran and the U.S. share no common waters, they share a common interest in water security for regional stability. This common cause is most evident in Afghanistan. Water stress constitutes a fundamental underlying driver of Afghan opium production. Two-thirds of Afghans work in agriculture. But Afghanistan is beset by persistent drought, and its irrigation systems and water infrastructure have been decimated by war. The opium poppy is both highly drought resistant, requiring a fraction of the water needed by crops like wheat, and highly profitable, offering an indispensable, if illicit, lifeline to beleaguered farmers. Armed insurgents and criminal traffickers such as the Taliban and the Haqqani Network protect and tax the poppy crop, threatening civil order in Afghanistan, destabilizing the frontier region with Iran, and fueling an opiate epidemic within the Islamic Republic.[103]

Both the U.S. and Iran have devoted significant resources to water development in Afghanistan. By committing to coordinate their efforts to rehabilitate infrastructure and irrigation systems, improve agricultural water efficiencies, increase water storage, etc., Washington and Tehran can contribute to reducing the appeal of opium cultivation, supporting viable rural economies, countering insurgency and promoting stability in Afghanistan. By the same token, water insecurity and low government capacity have rendered Kabul reluctant to negotiate with its more powerful neighbor on their shared rivers. By supporting sustainable water management and building technical capacity in Afghanistan, international engagement may help alleviate such sensitivities, defusing a destabilizing flashpoint and promoting regional cooperation on transboundary waterways.[104]

Iran has a long history of successful water management, dating back at least 2,500 years to the Persians' invention of the *qanat* system of wells still in use today.[105] Through practicable policy reforms, strong public and political commitment, and international engagement on transboundary resources, the Islamic Republic can continue to secure its water future.

Notes

1 Ali Mirchi and Kaveh Madani, "How Iran's Elections Are Going Green," *The Guardian*, February 23, 2016.
2 Toshihiro Nakanishi, "Iran seeking Japanese Investment in Autos, Water," *Nikkei Asian Review*, February 4, 2016; "Iran Seeks Foreign Investment for Water Projects," *Tehran Times*, April 6, 2016.
3 Jean Chardin, *Sir John Chardin's Travels in Persia* (New York: Cosimo Classics, 2010), 128.
4 FAO AQUASTAT, *Iran*, Water Report 34 (Rome: FAO, 2009).
5 FAO AQUASTAT Database, accessed 15 November 2016, www.fao.org/nr/water/aquastat/data/query/index.html?lang=en.
6 Nathaniel Mason and Roger Calow, *Water Security: From Abstract Concept to Meaningful Metrics*, Working Paper 357 (London: Overseas Development Institute, October 2012).
7 FAO AQUASTAT, *Iran*.
8 Mesfin M. Mekonnen and Arjen Y. Hoekstra, "Four Billion People Facing Severe Water Scarcity," *Science Advances* 2, no. 2 (2016).
9 Vladimir Smakhtin, "Basin Closure and Environmental Flow Requirements," *International Journal of Water Resources Development* 24, no. 2 (2008): 227–233.
10 Najmeh Bozorgmehr, "Iran: Dried Out," *Financial Times*, August 21, 2014.
11 Francis Gessert et al., *A Weighted Aggregation of Spatially Distinct Hydrological Indicators* (Washington, DC: World Resources Institute, December 2013), Table A3.
12 Amir AghaKouchak et al., "Aral Sea Syndrome Desiccates Lake Urmia: Call for Action," *Journal of Great Lakes Research* 41, no. 2 (2015); M. J. Tourian et al., "A Spaceborne Multisensor Approach to Monitor the Desiccation of Lake Urmia in Iran," *Remote Sensing of Environment* 156 (2015): 349–360.
13 Hadi Esandari et al., "Change Detection of Bakhtegan and Tashk Basin during 2001–2013," *International Journal of Forest, Soil, and Erosion* 6, no. 2 (2016): 67–71.
14 FAO, *Groundwater Management in Iran: Draft Synthesis Report* (Rome: FAO, 2009).
15 Poolad Karimi et al., "Reducing Carbon Emissions Through Improved Irrigation and Groundwater Management: A Case Study from Iran," *Agricultural Water Management* 108 (2012): 52–60.
16 FAO, *Groundwater Management in Iran*.

17 E. Forootan et al., "Separation of Large Scale Water Storage Patterns Over Iran Using GRACE, Altimetry and Hydrological Data," *Remote Sensing of Environment* 140 (2014): 580–595.
18 Monireh Faramarzi et al., "Modeling Wheat Yield and Crop Water Productivity in Iran: Implications of Agricultural Water Management for Wheat Production," *Agricultural Water Management* 97 (2010): 1861–1875.
19 Tanja Srebotnjak et al., "A Global Water Quality Index and Hot-Deck Imputation of Missing Data," *Ecological Indicators* 17 (2012): 108–119.
20 Cheng Liu et al., "Past and Future Trends in Grey Water Footprints of Anthropogenic Nitrogen and Phosphorous Inputs to Major World Rivers," *Ecological Indicators* 18 (2012): 42–49.
21 G. J. van Oldenbergh et al., eds., "Annex 1: Atlas of Global and Regional Climate Projections," in *Climate Change 2013: The Physical Science Basis. Contribution of Working Group I to the Fifth Assessment Report of the Intergovernmental Panel on Climate Change*, eds. T.F. Stocker et al. (Cambridge: Cambridge University Press, 2013), 1366–1369. Temperature and precipitation changes are relative to the period 1986–2005.
22 Bryson Bates et al. eds., *Climate Change and Water*, IPCC Technical Paper VI (Geneva: IPCC, 2008).
23 Walter Immerzeel et al., *Middle East and Northern Africa Water Outlook* (Wageningen: FutureWater, April 2011).
24 Alireza Gohari, "Climate Change Impacts on Crop Production in Iran's Zayandeh-Rud River Basin," *Science of the Total Environment* 442 (2013): 405–419; A. J. Challinor et al., "A Meta-Analysis of Crop Yield Under Climate Change and Adaptation," *Nature Climate Change* 4 (2014): 287–291; Sarvenaz Farhangfar et al., "Vulnerability Assessment of Wheat and Maize Production Affected by Drought and Climate Change," *International Journal of Disaster Risk Reduction* 13 (2015): 37–51.
25 Gh. R. Roshan and S. W. Grab, "Regional Climate Change Scenarios and Their Impacts on Water Requirements for Wheat Production in Iran," *International Journal of Plant Production* 6, no. 2 (2012).
26 Mohammad Bannayan and Ehsan Eyshi Rezaei, "Future Production of Rainfed Wheat in Iran (Khorasan province): Climate Change Scenario Analysis," *Mitigation and Adaptation Strategies for Global Change* 19, no. 2 (2014): 211–227; Rooholla Moradi et al., "Adaptation of Maize to Climate Change Impacts in Iran," *Mitigation and Adaptation Strategies for Global Change* 19, no. 8 (2014): 1223–1238; Milad Nouri et al., "Towards Modeling Soil Texture-Specific Sensitivity of Wheat Yield and Water Balance to Climatic Changes," *Agricultural Water Management* 177 (2016): 248–263.
27 Saeed Jamali et al., "Climate Change and Hydropower Planning in the Middle East: Implications for Iran's Karkheh Hydropower Systems," *Journal of Energy Engineering* 139 (2013): 153–160.
28 Department of Environment, *Iran Second National Communication to the UNFCCC* (Tehran: Government of the Islamic Republic of Iran, December 2010), 96ff; National Climate Change Committee, *Intended Nationally Determined Contribution* (Tehran: Department of Environment, 19 November 2015), 7–9.
29 World Bank, *High and Dry: Climate Change, Water, and the Economy* (Washington, DC: World Bank, 2016), vii.
30 Kaveh Madani, "Water Management in Iran: What Is Causing the Looming Crisis?" *Journal of Environmental Studies and Sciences* 4, no. 4 (2014): 315–328.
31 FAO, *FAO Statistical Pocketbook 2015: World Food and Agriculture* (Rome: FAO, 2015), 125.
32 Ghada Ahmed et al., *Wheat Value Chains and Food Security in the Middle East and North Africa Region* (Durham: Duke University Center on Globalization,

Governance, and Competitiveness, August 2013); FAO, *Iran: Country Factsheet on Food and Agriculture Policy Trends* (Rome: FAO, September 2014).
33 FAO, *FAO Statistical Yearbook 2013: World Food and Agriculture* (Rome: FAO, 2013), 111; FAO, *FAO Statistical Pocketbook 2015*, 124.
34 Sanaz Alasti, *Legislation on Use of Water in Agriculture: Iran* (Washington, DC: Library of Congress, October 2013), 2.
35 FAO, *Groundwater Management in Iran*, 10.
36 Karimi et al., "Reducing carbon emissions."
37 M. M. Mekonnen and A. Y. Hoekstra, "The Green, Blue and Grey Water Footprint of Crops and Derived Crop Products," *Hydrology and Earth System Sciences* 15, no. 5 (2011): 1577–1600.
38 M. Faramarzi et al., "Analysis of Intra-Country Virtual Water Trade Strategy to Alleviate Water Scarcity in Iran," *Hydrology and Earth System Sciences* 14, no. 8 (2010): 1417–1433.
39 Ghalamreza Zehtabian, "High Demand in a Land of Water Scarcity: Iran," in *Water and Sustainability in Arid Regions*, eds. Graciela Schneier-Madanes and Marie-Françoise Courel (Dordrecht: Springer, 2010), 79.
40 UNDP, *Human Development Report 2006* (New York: UNDP, 2006), 34; Madani, "Water Management in Iran," 319.
41 Carlo Sdralevich et al., *Subsidy Reforms in the Middle East and North Africa: Recent Progress and Challenges Ahead* (Washington, DC: IMF, 2014), 51ff.
42 Ozgur Demirkol et al., *Islamic Republic of Iran: Selected Issues*, IMF Country Report No. 14/94 (Washington, DC: IMF, April 2014).
43 BBC News, "Iran Petrol Prices Surge as Subsidies Cut," April 25, 2014; IMF, *Islamic Republic of Iran*, Country Report No. 15/349 (Washington, DC: IMF, December 2015).
44 Nazanin Soroush, "Political Dimensions of Iran's Water Crisis," *CRIAViews*, October 16, 2014; Madani, "Water Management in Iran."
45 François Molle et al., "Water, Politics and Development: Introducing Water Alternatives," *Water Alternatives* 1, no. 1 (2008): 1–6.
46 International Commission on Large Dams, "Register of Dams: General Synthesis—Number of Dams by Country Members," accessed October 10, 2016, www.icold-cigb.net/GB/World_register/general_synthesis.asp?IDA=206.
47 Soroush, "Political Dimensions of Iran's Water Crisis."
48 AghaKouchak et al., "Aral Sea Syndrome"; Tourian et al., "A Spaceborne Multisensor Approach"; Majid Ramezani Mehrian et al., "Investigating the Causality of Changes in the Landscape Pattern of Lake Urmia basin, Iran Using Remote Sensing and Time Series Analysis," *Environmental Monitoring and Assessment* 188 (2016): 462–474.
49 World Bank, *Islamic Republic of Iran: Cost Assessment of Environmental Degradation*, Report No. 32043-IR (Washington, DC: World Bank, 2005).
50 World Bank, *Economic Implications of Lifting Sanctions on Iran*, MENA Quarterly Economic Brief (Washington, DC: World Bank, July 2015), 2.
51 Farshad Amiraslani and Deidre Dragovich, "Combating Desertification in Iran Over the Last 50 Years: An Overview of Changing Approaches," *Journal of Environmental Management* 92, no. 1 (2011): 1–13; Islamic Republic News Agency, "Over 110mn Hectares of Land Threatened by Desertification," May 7, 2012.
52 Iraj Emadodin and Hans Rudolf Bork, "Degradation of Soils as a Result of Long-Term Human-Induced Transformation of the Environment in Iran: An Overview," *Journal of Land Use Science* 7, no. 2 (2012): 203–219
53 Eelco van Beek et al., "Limits to Agricultural Growth in the Sistan Closed Inland Delta, Iran," *Irrigation and Drainage Systems* 22 (2008): 131–143; A. Rashki et al., "Dust Storms and Their Horizontal Dust Loading in the Sistan Region, Iran," *Aeolian Research* 5 (2012): 51–62.

54 R. Sabouri et al., "Correlation Analysis of Dust Concentration and Water Quality Indicators," *International Journal of Environmental Science and Development* 2, no. 2 (2011): 91–97.
55 A. Pahlavanravi et al., "The Impacts of Different Kinds of Dust Storms in Hot and Dry Climate, A Case Study in Sistan Region," *Desert* 17 (2012): 15–25.
56 Sohrab Delangizan and Zainab Jafari Motlagh, "Dust Phenomenon Effects on Cardiovascular and Respiratory Hospitalizations and Mortality: A Case Study in Kermanshah, during March-September 2010–2011," *Iranian Journal of Health and Environment* 6, no. 1 (2013). In Persian with English Abstract.
57 Asad Sarwar Qureshi et al., *A Review of Management Strategies for Salt-Prone Land and Water Resources in Iran*, Working Paper 125 (Colombo, Sri Lanka: International Water Management Institute, 2007); Iraj Emadodin et al., "Soil degradation and agricultural sustainability: an overview from Iran," *Environment, Development and Sustainability* 14, no. 5 (2012).
58 Masoud Mahmoudpour et al., "Numerical simulation and prediction of regional land subsidence caused by groundwater exploitation in the Southwest Plain of Tehran, Iran," *Engineering Geology* 201 (2016): 6–28.
59 Arup, *Future Water Challenges: UK-Iran Workshop—Summary Report* (London: Arup, March 2016), 9.
60 UN, *World Urbanization Prospects: The 2014 Revision* (New York: UN, 2015).
61 Arezoo Soltani et al., "Poverty, Sustainability, and Household Livelihood Strategies in Zagros, Iran," *Ecological Economics* 79 (2012): 60–70; Marzieh Keshavarz et al., "The Social Experience of Drought in Iran," *Land Use Policy* 30, no. 1 (2013); Abbas Mahdi et al., "Factors Influencing Rural-Urban Migration from Mountainous Areas in Iran: A Case Study in West Esfahan," *European Online Journal of Natural and Social Sciences* 3, no. 3 (2014).
62 Jason Rezaian, "Iran's Water Crisis the Product of Decades of Bad Planning," *Washington Post*, July 2, 2014.
63 FAO AQUASTAT, *Iran*; Ahmad Abrishamchi and Massoud Tajrishy, "Interbasin Water Transfers in Iran," in Committee on U.S.-Iranian Workshop on Water Conservation and Recycling, *Water Conservation, Reuse, and Recycling: Proceedings of an Iranian-American Workshop* (Washington, DC: National Academies Press, 2005).
64 Alireza Gohari et al., "Water Transfer as a Solution to Water Shortage: A Fix That Can Backfire," *Journal of Hydrology* 491 (2013): 23–39.
65 François Molle and Alireza Mamanpoush, "Scale, Governance and the Management of River Basins: A Case Study from Central Iran," *Geoforum* 43, no. 2 (2012): 285–294.
66 Harold D. Lasswell, *Politics: Who Gets What, When, How* (New York: McGraw-Hill, 1936).
67 The phrase comes from an engineer with the U.S. Bureau of Reclamation, quoted in Donald Worster, *The Wealth of Nations: Environmental History and the Ecological Imagination* (New York: Oxford University Press, 1993), 135.
68 Molle and Mamanpoush, "Scale, Governance and the Management of River Basins"; "Water Riot Breaks Out In Iran," *Al-Monitor*, February 28, 2013.
69 Madani, "Water Management in Iran"; Azad Henareh Khalyani et al., "Water Flows Toward Power: Socioecological Degradation of Lake Urmia, Iran," *Society & Natural Resources* 27, no. 7 (2014): 759–767.
70 Nayereh Tohidi, "Ethnic and Religious Minority Politics in Iran," in *Contemporary Iran: Economy, Society, Politics*, ed. Ali Gheissari (Oxford: Oxford University Press, 2009), 299–323; Alam Saleh, *Ethnic Identity and the State in Iran* (New York: Palgrave Macmillan, 2013).
71 FAO AQUASTAT, *Iran*, 11–12; Alasti, *Legislation on Use of Water*.
72 Emil Souleimanov et al., "The Rise of Nationalism among Iranian Azerbaijanis: A Step Toward Iran's Disintegration?" *Middle East Review of International Affairs*

17, no. 1 (2013): 71–91; Zia ur Rehman, *The Baluch Insurgency: Linking Iran to Pakistan*, NOREF Report (Oslo: Norwegian Peacebuilding Resource Center, May 2014); "How Iran's Khuzestan went from wetland to wasteland," *The Guardian*, April 16, 2015.

73 UNESCWA, *Inventory of Shared Water Resources in Western Asia* (New York: UN, 2013), 100ff.

74 Government of Iran and Government of Iraq, "Agreement Between Iran and Iraq Concerning the Use of Frontier Watercourses," December 26, 1975, accessed October 10, 2016, Transboundary Freshwater Dispute Database, Oregon State University, www.transboundarywaters.orst.edu/database/interfreshtreatdata.html.

75 Furat A.M. Al-Faraj and Miklas Scholz, "Assessment of temporal hydrologic anomalies coupled with drought impact for a transboundary river flow regime: The Diyala watershed case study," *Journal of Hydrology* 517 (2014): 64–73.

76 UNESCWA, *Inventory of Shared Water Resources*; Kamal Chomani and Toon Bijnens, "The Impact of the Daryan Dam in the Kurdistan Region of Iraq," Save the Tigris and Iraqi Marshes Campaign, October 2016.

77 UNESCWA, *Inventory of Shared Water Resources*, 148ff; Nadhir Al-Ansari et al., "Present Conditions and Future Challenges of Water Resource Problems in Iraq," *Journal of Water Resources and Protection* 6 (2014): 1066–1098; Samira Fuladavand and Gholam Abbas Sayad, "The Impact of Karkheh Dam Construction on Reducing the Extent of Wetlands of Hoor-Alazim," *Journal of Water Resources and Ocean Science* 4, no. 2 (2015).

78 Vincent Thomas et al., *Developing Transboundary Water Resources: What Perspectives for Cooperation Between Afghanistan, Iran, and Pakistan?* (Kabul: Afghan Research and Evaluation Unit, 2016), 42, 53.

79 Government of Afghanistan and Government of Iran, "The Afghan-Iranian Helmand Water Treaty," Kabul, March 13, 1973.

80 Thomas et al., *Developing Transboundary Water Resources*; Alex Dehgan et al., "Water Security and Scarcity: Potential Destabilization in Western Afghanistan and Iranian Sistan and Baluchestan Due to Transboundary Water Conflicts," in *Water and Post-Conflict Peacebuilding*, Erika Weinthal et al., eds. (London: Earthscan, 2014), 305–326.

81 Renard Sexton, *Natural Resources and Conflict in Afghanistan* (Kabul: Afghanistan Watch, July 2012), 20.

82 Sexton, *Natural Resources and Conflict*; Dehgan et al., "Water Security and Scarcity"; Mathew King and Benjamin Sturtewagen, *Making the Most of Afghanistan's River Basins: Opportunities for Regional Cooperation* (New York: EastWest Institute, 2010).

83 Kerry Hutchinson, "Water Wars," *The Middle East* (January/February), 2012.

84 Fatemah Aman, *Water Dispute Escalating between Iran and Afghanistan*, Issue Brief (Washington, DC: Atlantic Council, August 2016).

85 UNECE, *Second Assessment of Transboundary Rivers, Lakes, and Groundwaters* (New York: UN, 2011); UNESCWA, *Inventory of Shared Water Resources*.

86 UNEP-DHI/UNEP, *Transboundary River Basins: Status and Trends* (Nairobi: United Nations Environment Programme, 2016), 138ff, 161ff.

87 Arash Karami, "Iran Becoming 'Uninhabitable' Says Former Agriculture Minister," *Al-Monitor Iran Pulse*, July 9, 2013.

88 Landane Nasser, "Iran's Rouhani Seeks 'National Will' to Conserve Water," *Bloomberg News*, October 30, 2013; "Management of Water Resources, Government's Basic Challenge," *Iranian Labour News Agency*, November 19, 2013.

89 FAO, *Groundwater Management in Iran*; Monavar Khalaj, "Iran's Rouhani Rolls Back on Dam Projects," *Financial Times*, March 19, 2015.

90 Bruce Pannier and Farhodi Milloh, "A Pipeline From a Land of Water to a Land of Oil," *Radio Free Europe*, August 8, 2014; "Water to Be Transferred From Persian

Gulf to Central Iran in Giant Project," *Tasnim*, March 8, 2016; "New Twist in Caspian Water Transfer Debate," *Financial Tribune*, April 17, 2016.

91 FAO, *Groundwater Management in Iran*; FAO, *Iran: Country Fact Sheet on Food and Agriculture Trends* (Rome: FAO, September 2014).

92 Madani, "Water Management in Iran," 322.

93 Faramarzi et al., "Analysis of Intra-Country Virtual Water Trade."

94 World Bank, *Renewable Energy Desalination: An Emerging Solution to Close the Water Gap in the Middle East and North Africa* (Washington, DC: World Bank, 2012).

95 Arun Subramani and Joseph G. Jacangelo, "Emerging Desalination Technologies for Water Treatment: A Critical Review," *Water Research* 75 (2015): 164–187.

96 Norman Neureiter and Glenn Schweitzer, "Engaging Iran," *Science* 319 (2008): 258.

97 Richard Stone, "Saving Iran's Great Salt Lake," *Science* 349 (2015): 1044–1047.

98 UNECE, *Second Assessment*; UNESCWA, *Inventory of Shared Water Resources*; Shiro Hishinuma et al., "Challenges of Hydrological Analysis for Water Resource Development in Semi-Arid Mountainous Regions: Case Study in Iran," *Hydrological Sciences Journal* 59, no. 9 (2014): 1718–1737.

99 Masoud Yazdanpanah et al., "Understanding Farmers' Intention and Behavior Regarding Water Conservation in the Middle-East and North Africa: A Case Study in Iran," *Journal of Environmental Management* 135 (2014): 63–72; U.S. Department of Agriculture, *2012 Census of Agriculture: Farm and Ranch Irrigation Survey (2013)*, Volume 3, Special Studies, Part 1 (Washington, DC: USDA, November 2014).

100 Committee on U.S.-Iranian Workshop, *Water Conservation*; Kaisa Korhoned-Kurki and Merc Fox, eds., *Towards New Solutions in Managing Environmental Crisis: Proceedings of the USA-Iran-Finland Environmental Workshop* (Helsinki: Helsinki University, 2010).

101 Barbara Slavin, "U.S. Is Hampering Iran's Return to the World," *Newsweek*, April 20, 2016.

102 Government of Canada and Government of the United States of America, "The Columbia Treaty," January 17, 1961; Government of the Republic of Turkey and Government of the Islamic Republic of Iran, "Memorandum of Understanding on Environment," April 28, 2010.

103 Nasser Saghafi-Ameri, "The 'Afghan Drugs' Problem—A Challenge to Iran and International Security," *Iranian Review of Foreign Affairs* 1, no. 2 (2010): 213–235; Christian Parenti, "Flower of War: An Environmental History of Opium Poppy in Afghanistan," *SAIS Review of International Affairs* 35, no. 1 (2015): 183–200.

104 United Nations Country Team in Afghanistan, *Natural Resource Management and Peacebuilding in Afghanistan* (Nairobi: United Nations Environment Programme, 2013).

105 Hassan Ahmadi et al., "The Qanat: A Living History in Iran," in eds. Graciela Schneier-Madanes and Marie-Françoise Courel, *Water and Sustainability in Arid Regions* (Dordrecht: Springer, 2010), 125–138.

10 Dammed If You Do and Damned If You Don't
Afghanistan's Water Woes

Glen Hearns

1. A Big Piece of the Puzzle

> *In many areas, farmers do not have access to irrigation water and are unable to cultivate all their land. This is why they cultivate poppy, because poppy can fulfill their expenses and needs.*
>
> —N. Ahmad[1]

As Mohammed Amin Omar wanders his fields, he looks into the opal morning sky; the sun is rising over Mushan village in Zarhi District, Kandahar Province.[2] It is that time of year again when he must consider sending his two eldest boys away to pick poppies.[3] Although he grows wheat and fruit, he is considering planting some poppies of his own next year to help offset poor yields due to the lack of water. He also knows that if he agrees to turn over part of his land to poppy cultivation he can more easily obtain a "preharvest" loan to cover household expenses during the winter months. He does not understand why there is not a more consistent water supply, as work has been done recently to rehabilitate the canal system.[4] As a farmer, Mohammad is keenly aware of the water situation. He is aware that the canals have deteriorated from decades of neglect and are in need of repair, that the governing *mirab*[5] has many issues to balance in allocating water, and that poor winter precipitation over the past few years means less water in the late growing season. He knows that people, mostly landless, have moved into the desert and are growing crops (much of which are poppies) by tapping groundwater.[6] He knows also that many of the local *kazares*[7] and shallow wells are dry sooner than they used to be. However, what he is less aware of is the dire need for large storage dams to provide continued irrigation water and help mitigate climate change, and he has no understanding of the challenges of developing storage at this scale in international rivers. What Mohammad is most conscious of, however, is the debt that he owes and the need to make changes to provide for his family's well-being in Afghanistan's fluctuating economy.

This chapter addresses the need for effective transboundary water agreements to stimulate financing and investment to increase water storage capacity in Afghanistan's water sector. The region is poised to experience water scarcity through a combination of increased human pressure and climate change.

Developing appropriate storage capacity is critical to catalyze economic development in the rural areas and enhance social stability. Yet despite being high on the government's list of priorities and years of massive international development assistance, it has remained, for the most part, an elusive goal. Without adequate irrigation and viable alternatives for income generation, rural areas will increasingly look toward poppy cultivation to hedge their bets against a worsening climate and fluctuating economy.

2. The Current Snapshot

Afghanistan is landlocked, bordered by Turkmenistan, Uzbekistan, and Tajikistan to the north, China to the northeast, Pakistan to the east, and Iran to the west. It experiences a harsh continental climate, with the winter conditions enhanced by the country's high altitude. Annual precipitation occurs from November to May, and summers are warm to very hot with little or no precipitation or streamflow, except in rivers and streams fed by melting snow and glaciers.[8]

Ninety percent of Afghanistan's surface water systems are transboundary in nature,[9] and it is predominantly an upstream country in a water-thirsty region. Afghanistan has numerous international watersheds; however, administratively there are five major basins (Figure 10.1). The Panj-Amu, Kabul, Harirod-Murghab, Northern, and Helmand basins all have international components.

The overall assessment of internally generated surface water is approximately 47.5 bcm/annum[11] and the groundwater resources are estimated to be approximately 18 bcm/annum.[12] The distribution of water resources in Afghanistan is not uniform, and 80% of the surface water resources are found in the Kabul and Panj-Amu basins, which are home to approximately 60% of the population.[13] As a consequence, while the average water resources for the country are more than 2,000 m³/cap, the Harirod-Murghab basin is below the 1,800 m³/cap mark, indicating water stress,[14] and the Northern basin is considered "water scarce," with less than 1,000 m³/cap. On average, the country uses approximately 27–32% of its water resources, and all the basins would be considered to be medium to highly stressed in terms of water use vs. water availability.[15] Water usage of 39% was estimated in 1987.[16]

Agriculture accounts for the bulk of water use in Afghanistan, where more than 95% is used for irrigation.[17] About 85% of this use is supplied through surface water;[18] however, groundwater still plays an important role in certain areas where *karezes*, springs, and shallow wells (*arhads*) are used, and is increasingly important as deeper drilling technologies are being introduced.

Afghan Development in the Regional Context: "The Catch-Up Game"

Afghanistan is highly underdeveloped in comparison with its neighbors (Table 10.1). It has the lowest Human Development Index (HDI) by a significant margin. It also has the lowest gross national income (GNI)/capita—nearly 11 times less than that of Kazakhstan. Of the countries measured, it has the

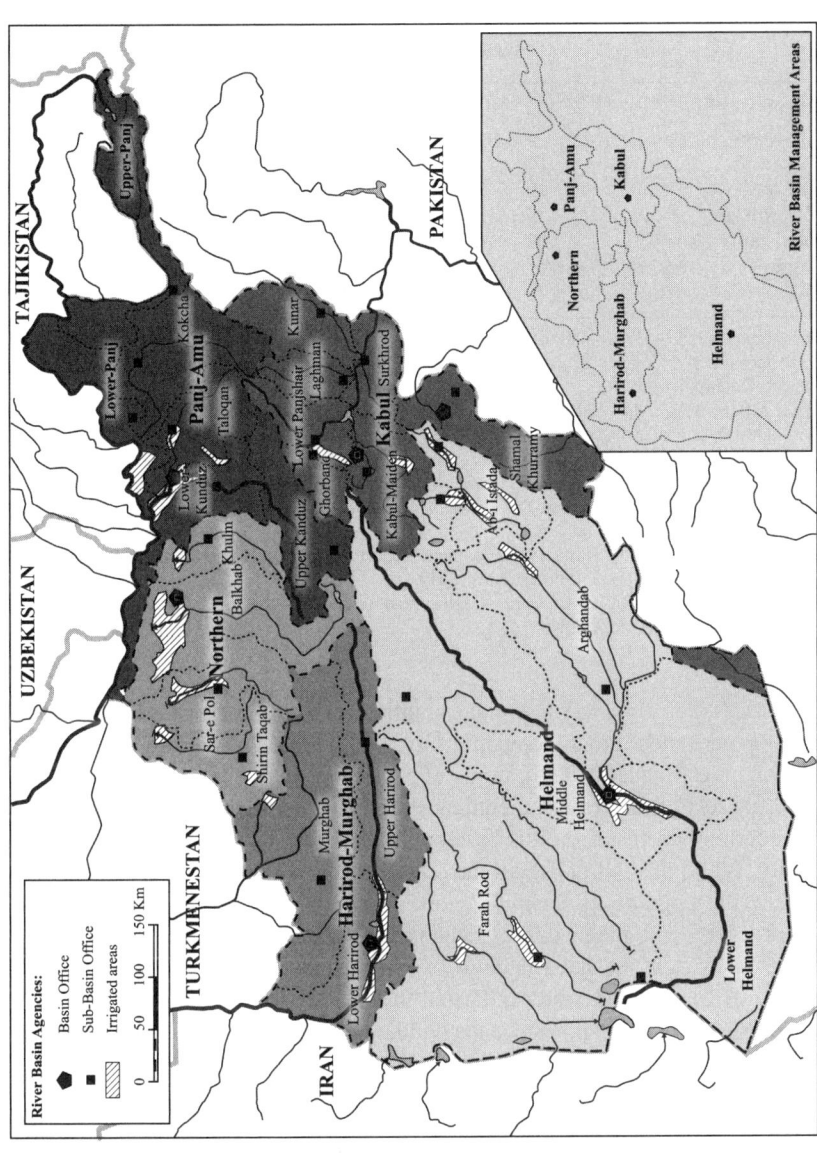

Figure 10.1 Major River Basins of Afghanistan[10]

192 Glen Hearns

Table 10.1 Regional Development Statistics

Country	Population (million)[i]	Large dams[ii]	Storage[iii] m³/cap	GNI[iv] $/cap	HDI[v] 2011	Access to Improved Qater (2015) % Pop[vi]	
						Urban	Rural
Afghanistan	33	2	135*	1,885	0.46	78	47
Kazakhstan	17.8	53	5,895	20,867	0.78	94	86
Kyrgyzstan	6	13	4,352	3,044	0.66	96	85
Tajikistan	8.6	16	4,232	2,517	0.62	93	66
Turkmenistan	5.4	3	1,240	13,066	0.68		
Uzbekistan	30	12	801	5,567	0.67	98	
Pakistan	192	?	150	4,866	0.52	93	98
Iran	80	?	396	15,439	0.78	97.7	92

* The figure is likely closer to 100–110 m³/cap.
i UN data (2016). Country Profiles—Afghanistan. Retrieved 2 August 2016 from http://data.un.org/CountryProfile.aspx.
ii FAO (2016). FAO AQUASTAT Data—Afghanistan. Retrieved 28 June 2016 from www.fao.org/nr/water/aquastat/dbase/index.stm.
iii Storage capacity from FAO (2016). FAO AQUASTAT calculated with current population figures. With updated storage figures as available.
iv GNI. See UNDP (2013), Human Development Report 2015: Work for Human Development, United Nations Development Programme, 2013.
v HDI. See UNDP (2015), Human Development Report 2015: Work for Human Development, United Nations Development Programme, 2013.
vi FAO (2016). FAO AQUASTAT Data—Afghanistan. Retrieved 28 June from www.fao.org/nr/water/aquastat/dbase/index.stm.

lowest access to drinking water and sanitation for both urban and rural populations. Where most urban populations are well above 90% access, Afghanistan is at an estimated 78%. In terms of storage capacity, it significantly lags behind the rest of the region, including Pakistan. The figure of 135 m³/cap is almost certainly inflated, as it is based on installed capacity and not effective capacity. While this is true for all the other nations, Pakistan has been able to maintain and upgrade its facilities, while Afghanistan has done so only to a limited degree. The major dams of Kajaki and Dahla were constructed more than 60 years ago and have undergone heavy sedimentation.[19] In 2013, the Ministry of Energy and Water indicated that the "real" storage capacity was between 80 and 90 m³/cap.[20] Since the building of the Salma Dam in July 2015, the potential storage capacity has increased by 22 m³/cap.[21]

The Need for Greater Storage

An estimated 80% of Afghan agriculture depends on rain-fed agriculture and cattle grazing.[22] The agricultural sector is responsible for some 30% of the gross domestic product and 78% of the labor force,[23] so increased irrigation, and hence storage, is one of the key goals of the Afghanistan National Development Strategy to stimulate the economy.[24] Unfortunately, climate change means natural storage is decreasing. Snow pack and glaciers act as nature's "water storage," releasing water more evenly as they melt; however, decreased winter precipitation along with increased temperature is making drought conditions the norm. Disputes over land and water are already the major cause of local insecurity, and the situation may further deteriorate as arable land declines. Perhaps as much as 60% has been lost since 1978.[25]

While much work has been undertaken to improve adaptability to climate change through "climate-resilient livelihoods," or improving farm water use efficiency,[26] there is little doubt that improved irrigation through increased storage is needed.

Where Are the Dams?

Over the past decade, the government has been constructing five dams (Table 10.2) and has initiated detailed studies of the Bakshabad Dam on the Farah River in the southwest.[27] These dams are all being developed through the national budget with very limited assistance from the international community.[28] The dams are all behind schedule.[29]

In addition to these projects, the government of Afghanistan has highlighted at least 19 other projects;[31] however, it lacks the budget and capacity to develop them.

Between 2003 and 2014, the international community invested at least USD56 billion in reconstructing Afghanistan.[32] Within the water sector, most has gone to improve access to clean water, sanitation programs, rehabilitating and upgrading previous irrigation areas, and improving the capacity to manage water resources. However, the only significant new storage structure that has been completed since the 1960s has been the Salma Dam on the Hari Rud River, which was constructed with bilateral funding and assistance from India, and cost USD350 million.[33] To its credit, the U.S. has committed to addressing leakages and still hopes to improve the Dahla Dam (on the Arghandab River) to compensate for sedimentation over the years.[34]

While Afghanistan has undergone a decline in its ability to regulate and utilize water resources, its neighbors have increased their use of water originating from Afghanistan.[35] This places Afghanistan in the awkward situation of any significant increase in water use for its own needs potentially having a

Table 10.2 Current Dams Under Construction[30]

Project	River/Basin	Capacity at FRL (million m³)	Diversion Flood (cumecs) (m³)	Irrigation Command Area (ha)	Power (Mwatt)
Shah-wa-Aroos	Shakardara/Kabul	9.38	66	1,500	1
Pashdan	Karukh/Harirod	45.767	1,000	12,296	2
Machalgho	Machalgho	5.28	8.5	1,824	0.4
Almar	Almar/Northern	12.817	24	2,170	0
Kamal Khan	Helmand	54	N/A	85,000	9

significant impact on its downstream neighbors. This in turn provides the international community with a paradox: *While the international community would, in theory, be eager to improve Afghanistan's storage capacity for its development goals and to address climate change, it is understandably reluctant to do so if it causes harm to downstream states.*

International law addresses this dilemma with the application of the principle of equitable and reasonable utilization of international watercourses.[36] The principle balances the right of development of one state with the right of another state not to undergo significant harm due to that development. The principle is designed to help states negotiate water rights and benefit-sharing agreements on shared watercourses to arrive at equitable and reasonable use, and thus provide for a stable investment environment for development.[37]

Lack of Water-Sharing Arrangements

Despite having multiple transboundary water basins, Afghanistan has only one water treaty, on the Helmand River with Iran.[38] The reason for the lack of treaties is primarily that when Afghanistan descended into disorder in 1978, the irrigation sector had not been well-developed and there were no potential competitions over surface water resources with its neighbors, such as exist today. For example, in the Amu Darya River, the 1992 Almaty Agreement supports the agreed water divisions between the Central Asia countries established previously while under the USSR.[39] It allocates virtually all the flow to Kazakhstan, Kyrgyzstan, Tajikistan, Turkmenistan, and Uzbekistan, leaving 3% for Afghanistan, despite it being the source of approximately 28% of the flow. Afghanistan is not a party to that agreement.

The 1973 Helmand Treaty provides a binding solution to the issue of water rights in the Sistan area that had been the site of contention between Persia and Afghanistan since the 1800s.[40] Under the treaty, Iran is allowed a maximum volume of water use even in years of excess water flow.[41] Although developed 50 years ago, it has a novel climate adaptation component whereby if there is unseasonably low water the amount provided to Iran is reduced accordingly.[42] The treaty has not been without its hiccups, but it does provide a forum for dealing with outstanding issues.[43]

Afghanistan has participated in regional dialogues since 2006; conducted bilateral meetings with Iran, Turkmenistan, Tajikistan, and Pakistan; and named the development of treaties as one of the future goals in the water sector.[44] However, the capacity of the government is limited in this area.[45]

Why Grow Poppies?

The choice of poppy cultivation is not a straightforward issue of simply higher value, but can be affected by a host of factors, including:

- Drought or environmental shocks that limit agricultural land. While poppies require consistent water at the initial phase of development, they are

drought tolerant for the latter part of their development and will offer a yield even if water is negligible.
- Preferential access to credit. Credit in the form of advance payments on their crops is possible with poppies but not easy with traditional crops.
- To subsidize new investments for other cash and sustenance crops. In places that have experienced droughts, such as Kandahar in the mid-1990s, farmers were forced to dig shallow wells to irrigate their land or lose their crops. Opium provided the funds to develop their land and maintain traditional crops. Opium continued to support other crops as irrigation systems fell into further disrepair and deep wells were developed. The concentration of tube wells and increased opium production in southern Nangarhar occurs in areas that have had problems gaining access to sufficient and consistent irrigation during the winter months.
- New technologies opening up opportunities. Deep wells and solar-powered pumping have allowed agricultural expansion into marginal desert areas (for example, in southern and southwestern Afghanistan). This provides an opportunity for many who were landless to obtain land in the irrigated valleys. However, cultivating the areas is expensive and opium allows for sufficient returns to pay off debts of development.
- Availability and cost of labor can vary across the country. Poppy harvesting and spring weeding are labor intensive and require some skill. The migration of labor affects the ability to cultivate opium.
- Economic variability. A rise in energy prices could induce less well water production and thus reduce the viability of farming in marginalized areas. Changes in local food prices might also have an effect on the choice of which crops to cultivate.
- Other social factors. Local political bargaining between powerbrokers may influence the choice to grow poppies.[46]

3. Business as Usual

Climate change predictions do not bode well for Afghanistan. Most of the past 15 years have seen floods or droughts, including the country's most severe drought ever, which lasted from 1998 to 2006. Over the next 45 years, scientists predict a decrease in rainfall and a rise in average temperatures of up to 4°C compared with 1999. Droughts are likely to be the norm by 2030, leading to land degradation and desertification.[47] The country is ill-prepared to meet this challenge and has been ranked as one of the most vulnerable countries in the world, and the most vulnerable in its region, by the Global Adaptation Index[48] (Table 10.3).

Table 10.3 Global Adaptation Index for Central Asia

Afg	Iran	Kaz	Kyrg	Pak	Taj	Turk	Uzbek
33.4	53.4	62.3	54.4	43.6	47.4	40.3	47.7

Compounding the issue of climate change on Afghanistan's water resources is the increased demand. Afghanistan's population is estimated to be approximately 32 million, with an annual increase of 2.7%.[49]

The combined population pressure and demand will further burden scarce water resources in an arid region. The problem of increased water needs will affect both surface and groundwater. Continued groundwater use through deep wells may render traditional low-investment community systems such as *karezes* and *arhads* increasingly ineffective. Even by 2008 it was estimated that 60–70% of the traditional systems were not yielding adequate water.[50] While the international community continues to invest in enhancing irrigation, it likely will not be sufficient to catalyze economic development in the way it could if combined with increased storage capacity.

Moreover, although Afghanistan is the least-capable country in the region to address climate change, the region as a whole is not well-suited (Table 10.3). Since 1991, Central Asia (including Iran) has created some six water cooperation agreements,[51] including successful agreements such as the Chu Talas between Kyrgyzstan and Kazakhstan.[52] Unfortunately, despite this effort, the region has been described as having inadequate water agreements, poor water and river basin management, degraded water quality and disruption of flows, and decreased health, all resulting in increased tensions over water.[53]

The situation for Afghanistan will become increasingly difficult as water stress shifts into water scarcity over the coming decades, likely resulting in instability from the local to the regional levels. Direct tensions between countries could well ensue, in particular in light of negative sentiments from downstream countries such as Uzbekistan over upstream development.[54] However, the more insidious threat is at the local level, where climate change increases the likelihood of further land being converted to poppy production[55] as well as creating increased disputes over land and water. Increased instability and social unrest may further provide opportunities for insurgents. Clearly, not all insurgent groups approve of the cultivation of poppies;[56] however, it is undeniable that there are groups that benefit from the lack of stability and control to help fund themselves through the production and export of opium. Indeed, some analyses suggest that the "insurgency is no longer about religion or ideology, but rather a battle for money and wealth under the guise of a broader political movement."[57]

4. Interests for the U.S.

U.S. policy emphasizes the threat of rising terrorism, crime, corruption, and instability posed by illegal narcotics production, use, and trafficking in Central Asia (and other areas).[58] It is in the U.S.'s national interest to encourage stability in Afghanistan through improved water supply to help stimulate local economies and provide viable alternatives to poppy cultivation. At the farm level, shifting politics and a variable economy challenge the ability of farmers, such as Mohammad, to provide for their families; these farmers may thus turn to opium production as a way of hedging their bets and subsidizing traditional farming. The negative spillover effects in the neighboring countries and the international community have been felt, and under the current trajectory will continue to grow.

Estimates vary, but Afghanistan is responsible for 80–90% of the heroin produced globally,[59] with an estimated Western street value of more than USD120 billion/annum.[60] While most of the heroin arriving in the United States is from Mexico and Colombia, some is also from Afghanistan.[61] But more important, as noted earlier, the U.S. is affected by the instability in Afghanistan and Central Asia—and the region is very much affected by Afghan opium production. Approximately 30% of Afghan opium is consumed in Afghanistan and its neighboring countries, which by 2009 represented some 2.6 million users.[62] Usage rates in Afghanistan might be as high as 5% of the population, representing not only a health problem but also an increasing security problem.[63] Moreover, drug-related corruption continues to undermine governance efforts both within Afghanistan and, increasingly, in the region.

The opium harvest in Afghanistan declined 48% from 2014 to 2015,[64] which may have been largely due to disease in the poppies in southwestern Afghanistan and salinization of much of the irrigated areas in the desert.[65] Nevertheless, at least some of the reduction in poppy cultivation in the Kandahar area can also be attributed to the work of the Kandahar Food Zone (see the next section).

5. Steps Toward a Solution

Creating economic and social stability in Afghanistan to counter poppy cultivation and reduce opportunities for insurgents is no simple matter, and increasing water storage capacity is not a silver bullet. Alongside improvements to infrastructure is the need for good governance at the local and national levels. Despite the fact that the country is still affected by conflict, and will likely continue to be so for the next decade, the importance of improving local government and management systems of water along with national-level capacity cannot be underscored enough. The interconnected relationship that water plays in heavily agrarian societies such as Afghanistan makes it a focal point and nexus for development. As Afghanistan continues to struggle against destabilizing forces, it is critical that the international community continue, and indeed augment, its efforts to enhance water management at local, basin-wide, and national levels. It is true that in some parts of the country, for instance the southwest, insecurity will mean that any large dam development may be impractical for the near future. Nevertheless, efforts such as the recent installation of the third turbine at Kajaki Dam on the Helmand River[66] illustrate that infrastructure development is possible and that insurgencies and instability will not stop needed growth and development for the Afghan people.

There are some arguments that improvements to the water system could translate to enhanced poppy cultivation. However, the bulk of improved irrigation, combined with education, has provided local communities with greater alternatives for developing cash crops.[67] Trying to eliminate poppy cultivation without alternatives creates the risk of undue economic hardship on the most vulnerable of the population, which in turn leads to civil unrest, as seen through several "poppy bans" in Afghanistan.[68]

As witnessed through programs such as the Kandahar Food Zone, a multi-pronged approach incorporating alternative livelihoods, training, and creation of licit economic opportunities gives farmers opportunities to make choices that do not necessitate poppy cultivation to make ends meet.[69] At the inauguration of several rehabilitated canals in March 2015, the district chief of Zarhi, Sayed Jamal Agha, noted that due to the work of the Kandahar Food Zone, in six months poppy cultivation was reduced by 65%.[70] While the reasons to cultivate poppy are complex, one is clearly the lack of water to grow other cash crops. The importance of strengthening basin authorities and local water management councils needs to be emphasized as a prerequisite for responding to the long-term pressures associated with expanded climate change impacts. To this extent, support is needed to better engage basin- and local-level authorities in developing and implementing national policy, as per the 2009 Water Law.[71]

Unfortunately, improvements to irrigation systems, developing new markets for produce, and strengthening local governance will need to be coupled with storage. A more reliable source of irrigation water is needed throughout the growing season to ensure real economic development in the agricultural sector.[72] Indeed, a comprehensive water infrastructure program (that includes large storage) is required for reconciliation, for local stability, for local economic growth, for national economic diversification, and for regional stability.

To leverage the billions of dollars necessary to develop adequate storage, a stable investment environment is required for the international community and, potentially, the private sector. This stable investment environment will occur only once: Afghanistan and its neighbors have created agreements as to water use and water management with respect to their shared rivers. In the case of the international community, donors need assurances that their interventions and assistance in Afghanistan will not undermine other developing riparian states. Bilateral donor agencies are furthermore keenly aware of the political ramifications of such assistance. Finally, the private sector requires some assurance regarding hydrological availability for any large-scale investment. Water sharing agreements would pave the way for assurances in hydrological availability as well as address political concerns regarding water development in upstream Afghanistan.

Enhancing treaty development in the region will be a key catalyst for enhancing investment opportunities in storage capacity development as well as for reducing tensions that may build as a result of uncertainty in Afghanistan's future water use. The ground is fertile for tensions as long as downstream countries are unsure of the projected upstream uses and intentions and while no existing governance or institutional structure is in place to address issues. As noted earlier, the U.S. has logical interests in promoting regional stability among Afghanistan and its neighbors. As such, the U.S. should provide support to develop capacity, facilitating and mediating water agreements that address multiple issues such as trade, water, and hydropower in the region as a whole. This, combined with improved investment in irrigation efficiency and conservation in water-thirsty downstream countries, will help the region adapt and buffer impacts associated with climate change and water development in Afghanistan.

The mechanisms for U.S. engagement should necessarily be varied to ensure advancement when opportunities open:

1   At a bilateral level, the U.S. could help support capacity building within the key ministries associated with transboundary water management. USAID was present at a recent donor coordination meeting in August 2016 and indicated several ongoing activities, such as sharing data and information from the U.S. Geological Survey. This is applauded and should be augmented. Efforts could extend to enhancing a "negotiation" environment between Afghanistan and its neighbors, in particular Pakistan, of which the U.S. has traditionally been a supporter.
2   At a formal level, the U.S. should help support regional organizations in addressing water issues, either directly or as part of economic elements. There are already a number of organizations that are involved with economic and/or water development in the region, such as the International Fund for the Aral Sea and its affiliates, the Interstate Commission for Water Coordination (ICWC) and the ICWC Scientific Information Center, the Central Asia Regional Economic Cooperation Program, and the Economic Cooperation Organization. The U.S. should assess how to best leverage these initiatives to promote agreements between Afghanistan and its neighbors.
3   At an informal level, the U.S. should consider developing a forum in the region to encourage dialogue and education among high-level universities to exchange research and conduct awareness building on water treaty development, in light of mutual benefits such as economic development and trade. Such an initiative would be complementary to the existing USAID PEER program, which encourages the exchange of information on climate change-related research. Helping develop and deliver courses on transboundary water management and the benefits of cooperation would increase awareness of the advantages of water agreements in the region.

Treaties alone will not provide a cure to the ailments of regional water management, or a solution to economic development and stability. However, without them there is little chance that tensions over well water will be reduced or that advances will be made in agricultural development. Farmers such as Mohammed Amin Omar may not understand the complexities and machinations involved with hydro-politics in the region; however, they do understand

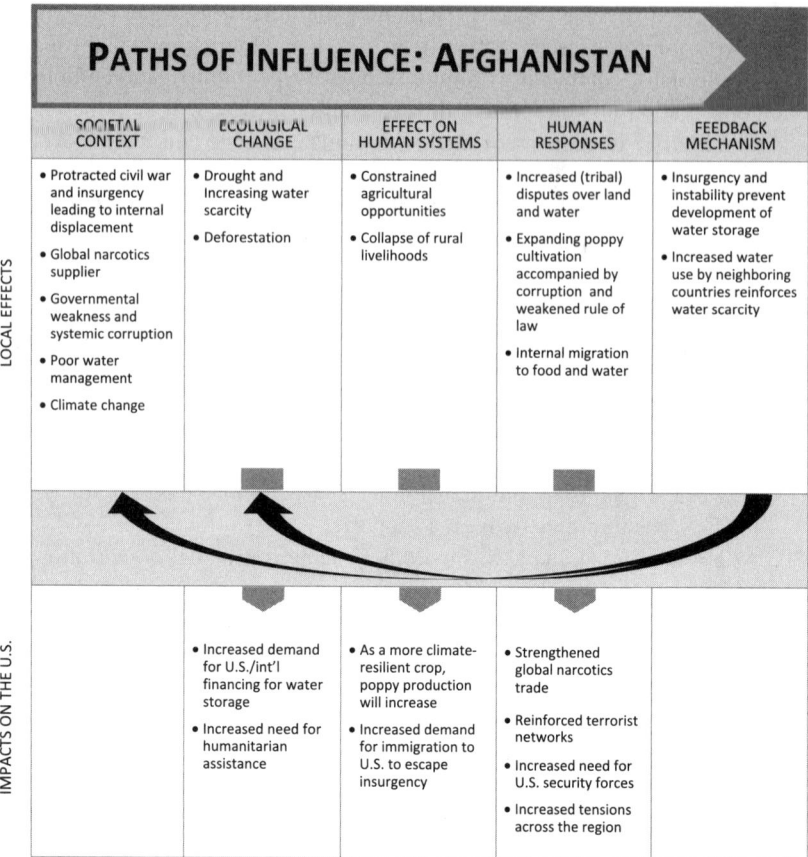

Figure 10.2 Paths of Influence: Afghanistan

that with a more reliable water supply, they will have more options available than poppy cultivation.

Notes

1 Shahwali Kot District Green House Project, Kandahar Food Zone, March 2015.
2 Mohammed Amin Omar is a fictitious character blended together from three real people in southeastern Afghanistan.
3 Mansfield, D. (2016). *A State Built on Sand: How Opium Undermined Afghanistan.* London: Hurst & Company.
4 MSI (2015). Kandahar Food Zone: Mid-term Performance Evaluation for USAID. Retrieved 23 July 2015 from http://pdf.usaid.gov/pdf_docs/PA00KDWM.pdf.
5 Local "water master" who leads decision making about local water issues. See Lee, J. (2007). The Performance of Community Water Management Systems, *Water Management, Livestock and the Opium Economy*. Case Study Series, Kabul: Afghanistan Research and Evaluation Unit, p. 3.

6 Mansfield, D., & Fishsein, P. (2016). *Moving With the Times: How Opium Poppy Cultivation Has Adapted to the Changing Environment in Afghanistan*. Watching Brief: Afghanistan Research and Evaluation Unit, p. 13.
7 Local underground water conveyance systems that depend primarily on groundwater.
8 Landell Mills (2013). Investment Plan for the Panj-Amu Basin, World Bank and Ministry of Energy and Water (Afghanistan), June 2013.
9 Favre, R., & Kamal, G. M. (2004). *Watershed Atlas of Afghanistan: Working Document for Planners*, 2 vols, Kabul: Afghanistan Information Management Services.
10 MEW (2011). Procedure on the Framework for Water Resources Management in the River Basins, Ministry of Energy and Water.
11 MEW (2014). Hydrometric Data, Directorate General of Hydrometeorology, Ministry of Energy and Water. Note: FAO (2016). FAO AQUASTAT fact sheet 2015 lists 37.5 bcm. Retrieved 2 May 2016 from www.fao.org/nr/water/aquastat/data/query/results.
12 Favre, R., and Kamal, G. M. (2004). *Watershed Atlas of Afghanistan: Working Document for Planners*, 2 vols, Kabul: Afghanistan Information Management Services; and the National Water Plan (GIZ, 2013) estimates between 10% and 12% of precipitation is groundwater and is approximately 18 bcm.
13 Basin populations are those for projected population in 2015 used in the National Water Master Plan, Ministry of Energy and Water (GIZ, 2013). The population issue is confounded by poor surveys and refugees in neighboring Iran and Pakistan, who often migrate back and forth depending on the security situation, and consequently the actual population may be less than the official numbers.
14 Rijberman, F. (2006). Water Scarcity: Fact or Fiction. *Agricultural Water Management* 80(1–3): 5–22. Note: Approximately 45% of the water availability in the Kabul basin comes from Pakistan in the Chattral-Kunar system and is not available to the majority of the basin population, as it returns to Pakistan just after joining the Kabul River.
15 Water use vs. availability of 20–40% indicates medium to high stress, and above 40% indicates severe water limitation. Vörösmarty, C., P. Green, J. Salisbury, and R. Lammers, R. (2000). Global Water Resources: Vulnerability from Climate Change and Population Growth. *Science* 289(5477): 284–288.
16 FAO (2016). FAO AQUASTAT 2015. Retrieved 16 August from www.fao.org/nr/water/aquastat/data/query/results.
17 FAO quotes 98%. FAO (2016). FAO AQUASTAT 2015. Retrieved 12 July from www.fao.org/nr/water/aquastat/data/query/results.
18 Favre and Kamal (2004) indicate an average of 15% of agricultural needs are met through groundwater. Landell Mills (2013) estimates 13% of irrigation is supplied by groundwater in the Panj-Amu Darya; national statistics indicated the use of groundwater for irrigation may exceed 15%. See Water Sector Strategy, Afghan National Development Strategy (ANDS), 21 April 2008.
19 Dowell, B. (7 April 2014). Corps of Engineers to raise Dahla Dam, provide water essential to southern Afghanistan, *U.S. Army*. Retrieved 3 June 2016 from www.army.mil/article/123474/Corps_of_Engineers_to_raise_Dahla_Dam__provide___/.
20 Sultan Mahmoodi, Director General of Hydrometeorology, Ministry of Energy and Water (Personal Communication, 15 October 2013).
21 The Salma Dam has a capacity of 633 CM (of which 560 CM is active storage). See WAPCOS (2016), Afghan-Indian Friendship Dam (Salma Dam), Water and Power Consultancy Services (India) Ltd. Retrieved 23 May 2016 from http://wapcos.gov.in/SalmaDam.aspx.
22 UNDP (2014). Climate Change Adaptation—Project Background, *Supporting Afghanistan to Deal with Climate Change*: United Nations Development Programme.

Retrieved 25 June 2016 fromwww.af.undp.org/content/afghanistan/en/home/oper ations/projects/environment_and_energy/ClimateChange.html.
23 CIA (2015). The World Factbook—Afghanistan. Retrieved 27 June from www.cia.gov/library/publications/the-world-factbook/geos/af.html.
24 Water Sector Strategy, Afghan National Development Strategy (ANDS), April 21, 2008.
25 UNDP (2014). Climate Change Adaptation—Project Background, *Supporting Afghanistan to Deal With Climate Change*: United Nations Development Programme. Retrieved 25 June 2016 fromwww.af.undp.org/content/afghanistan/en/home/operations/projects/environment_and_energy/ClimateChange.html.
26 Most major international donors and organizations such as GIZ, USAID, JICA, DIFD, UNDP, World Bank, and ADB have all participated in irrigation and agriculture improvement in Afghanistan.
27 The Ministry of Energy and Water awarded the detailed design to the Pakistani engineering company NESPAK in 2014.
28 USAID, under its EQUALS project, undertook capacity development training between 2009 and 2013 for Ministry of Energy and Water engineers responsible for managing the dams.
29 Wais A., Dam and Infrastructure, Ministry of Energy and Water, Afghanistan (Personal communication, 28 April 2016).
30 Wais Rahman Aria (2013). Building the Future: the revival of dam building in Afghanistan, conference paper Hydropower 2013 and The 3rd International Symposium on Rockfill Dams—Kunming, China (1–7 November, 2013).
31 Ibid.
32 World Bank (2016). Net official development assistance and official aid received (current US$)—Afghanistan. Retrieved 2 August 2016 from http://data.worldbank.org/indicator/DT.ODA.ALLD.CD?locations=AF.
33 *TOLOnews* (4 June 2016). Ghani, Modi Inaugurate Salma Dam, *TOLOnews*. Retrieved 4 June 2016 fromwww.tolonews.com/en/salma-dam-project/25637-ghani-modi-inaugurate-salma-dam.
34 In 2013, the U.S. Army Corps of Engineers awarded Phase 1 to repair and replace water control features to reduce water loss due to leakage. See USACE (28 February 2013), Dahla Dam Phase 1 award. Retrieved 3 August 2016 fromwww.tad.usace.army.mil/.
35 A UNEP study on water security noted a decrease in water resource availability in Central Asia between 1988 and 2008 despite a reduction in the use of cotton and application rate of water per hectare. See ENVSEC-UNEP (2011), Environment and Security in the Amu Darya Basin: UNEP, UNDP, UNECE, OSCE, REC, NATO.
36 *The United Nations Convention on the Law of the Non-Navigational Uses of International Watercourses*, 21 May 1977, New York, 36ILM 700 (1997). Entry into force 14 August 2014.
37 McCaffrey, S. (2001). *The Law of International Watercourses Non-Navigational Uses*. Oxford: Oxford University Press; Paisley, R. (2002). Adversaries Into Partners: International Water Law and the Equitable Sharing of Downstream Benefits. *Melbourne Journal of International Law* 3: 280–300; Wouters, P. (1997). *International Water Law: Selected Writings of Professor Charles B. Bourne*. The Hauge: Martinus Nijhoff Publishers; Dellapenna, J. (2003). Adapting the law of water management to global climate change and other hydro-political stresses. In J. Rodda & L. Ubertini (Eds.), *The Basis of Civilization—Water science?* (Vol. IHAS Publ 286: 291–299). Rome: International Association of Hydrological Sciences.
38 It should be mentioned that the frontier agreements conducted with the former USSR do address issues of structures that may hinder navigation or influence the flow need to be agreed upon. *The Treaty between the Government of the Union of*

Soviet Socialist Republics and the Royal Government of Afghanistan Concerning the Regime of the Soviet-Afghan State Frontier was signed in Moscow on 18 January 1958, UNTS No. 4655, 1959, vol. 321. Art. 9 (2) says "bridges, dams, and other similar structures likely to hinder navigation or influence the flow of water shall not be erected on frontier watercourses except by agreement between the two Parties." This is also supported by Art. 20 (1), which calls for special agreements to be developed for roads (bridges) and waterways intersecting the frontier line. There is also a protocol to the 1958 Agreement relating to works on the Murghab River limiting the development of dams and infrastructure.

39 See Article 2 of the *Cooperation in the Field of Joint Water Resources Management and Conservation of the Interstate Sources* (Kazakhstan-Kyrgyzstan-Uzbekistan-Tajikistan-Turkmenistan), signed 18 September 1992.

40 *The Afghan-Iranian Helmand River Water Treaty* with Protocol No. 1 and Protocol No. 2, 13 March 1973. Retrieved 13 April 2015 from www.internationalwaterlaw.org.

41 Art. 5, 1973 Helmand Treaty.

42 Art. 4, 1973 Helmand Treaty.

43 In 2001, Iran went to the UN Security Council when it believed Afghanistan was holding back water in Kajaki Dam. See *Letter dated 20 September 2001 from the Permanent Representative of the Islamic Republic of Iran to the United Nations addressed to the Secretary-General*, General Assembly Security Council, A/56/393-S/2001/896 (Iran Letter, 2001). Other outstanding issues revolve around the actual amount of water Iran is utilizing and the intake areas and groundwater resources that are linked to surface water but are not covered under the treaty.

44 National Priority Program 1: Agricultural Cluster—National Water and Natural Resource Development (2011). "The overriding principles in the context of sharing [transboundary water] resources would be equity and the just distribution of riparian resources" (p. 20).

45 The World Bank is partially addressing this issue through a capacity development program under its IRDP Project (World Bank 2016 Irrigation Restoration and Development Project). Retrieved 3 August 2016 from www.worldbank.org/projects/P122235/?lang=en&tab=overview.

46 These issues are summarized from Mansfield, D., & Fishsein, P. (2016). Moving With the Times: How Opium Poppy Cultivation Has Adapted to the Changing Environment in Afghanistan; Watching Brief: Afghanistan Research and Evaluation Unit; and Mansfield, D. (2016). *A State Built on Sand: How Opium Undermined Afghanistan*. London: Hurst & Company.

47 UNDP (2014). Climate Change Adaptation—Project Background, *Supporting Afghanistan to Deal With Climate Change*: United Nations Development Programme. Retrieved 25 June 2016 from www.af.undp.org/content/afghanistan/en/home/operations/projects/environment_and_energy/ClimateChange.html.

48 2014 data, Notre Dame Global Adaptation Index. Retrieved 24 July 2016 from index.gain.org.

49 Based off figures from UN Stat (2016), United Nations Population Information Network. Retrieved 2 August 2016 from www.un.org/popin/data.html.

50 Water Sector Strategy, ANDS (2008).

51 EU-UNDP (2011). Overview of Regional Transboundary Water Agreements, Institutions and Relevant Legal/Policy Activities in Central Asia. United Nations Development Programme 2011. In addition to the existing agreements, states are working on other agreements related to institutional coordination, use of water in modern conditions, and joint planning as part of the greater *Agreement on joint activities in addressing the Aral Sea and the zone around the Sea crisis, improving the environment, and ensuring the social and economic development of the Aral Sea region* (1993).

52 *Agreement Between the Government of the Kyrgyz Republic and the Government of the Republic of Kazakhstan on Using Water Distribution Facilities of Interstate Use*

at Chui and Talas Rivers, Bishkek, 21 January 2000, vol. 2196, reg. 38892, UNLS (2000).
53 ICA (2012). Intelligence Community Assessment, Global Water Security, ICA 2012–08.
54 On 1 August 2015, Uzbekistan's Ministry of Foreign Affairs issued an unequivocal statement reaffirming the country's objections to the Rogun project in upstream Tajikistan. See Putz, C. (4 August 2015). Uzbekistan Still Hates the Rogun Dam Project, The Diplomat. Retrieved 15 August 2016 from http://thediplomat.com/2015/08/uzbekistan-still-hates-the-rogun-dam-project/.
55 Mansfield, D., & Fishsein, P. (2016). Moving With the Times: How Opium Poppy Cultivation Has Adapted to the Changing Environment in Afghanistan. Watching Brief: Afghanistan Research and Evaluation Unit.
56 Since December 2015, areas of the upper reaches of Achin in Nangarhar province controlled by Daesh (Islamic State) are subject to an "opium ban." Mansfield, D., & Fishsein, P. (2016). Moving with the Times: How Opium Poppy Cultivation Has Adapted to the Changing Environment in Afghanistan. Watching Brief: Afghanistan Research and Evaluation Unit.
57 Eisler, D. (2012). Afghanistan's Opium Economy: Incentives, Insurgency, and International Demand. Journal of International Affairs, Online 5 December 2012. Retrieved 5 June 2016 from http://jia.sipa.columbia.edu/online-articles/afghanistans-opium-economy/. For a more in-depth analysis of how the drug economy feeds the desire for instability and lack of control of state authorities, see Pearce. J. (1990), Colombia: Inside the Labyrinth. London: Latin American Bureau.
58 Nichol, J. (2010). Central Asia's Security: Issues and Implications for U.S. Interests: Congressional Research Service. Retrieved 3 July 2016 from www.fas.org/sgp/crs/row/RL30294.pdf.
59 UNODC (2011). The Global Afghan Opium Trade: A Threat Assessment: United Nations Office on Drugs and Crime; UNODC (2015). Afghan Opium Survey 2015 Cultivation and Production (Vol. December): United Nations Office on Drugs and Crime, Ministry of Counter Narcotics Afghanistan.
60 Eisler, D. (2012). Afghanistan's Opium Economy: Incentives, Insurgency, and International Demand. Journal of International Affairs, Online 5 December 2012. Retrieved 5 June 2016 from http://jia.sipa.columbia.edu/online-articles/afghanistans-opium-economy/.
61 UNODC (2011). The Global Afghan Opium Trade: A Threat Assessment: United Nations Office on Drugs and Crime. Also see: The International Heroine Market, U.S. Office of National Drug Control Policy, available at www.whitehouse.gov/ondcp/global-heroin-market.
62 UNODC (2011). The Global Afghan Opium Trade: A Threat Assessment: United Nations Office on Drugs and Crime.
63 Goldstein, J. (20 December 2014). Kabul Residents Watch as Heroin Addiction Grows, The New York Times.
64 UNODC (2016). World Drug Report 2016. Vienna: United Nations Office on Drugs and Crime.
65 Mansfield, D., & Fishsein, P. (2016). Moving With the Times: How Opium Poppy Cultivation Has Adapted to the Changing Environment in Afghanistan. Watching Brief: Afghanistan Research and Evaluation Unit.
66 Ashaki Guyton-Blanton, Hydropower Team Lead, U.S. Agency for International Development, Kabul, Afghanistan (Personal Communication, 15 September 2016). USAID has installed the turbine at Kajaki Dam, and is scheduled for full commission by the end of November 2016.
67 UNODC (2011). The Global Afghan Opium Trade: A Threat Assessment: United Nations Office on Drugs and Crime.

68 Mansfield, D. (2016). *A State Built on Sand: How Opium Undermined Afghanistan*. London: Hurst & Company.
69 The Kandahar Food Zone is a USAID-sponsored program to improve the options of local farmers in seven Kandahar districts through training in market development, greenhouse technologies, water use efficiency as well as upgrading irrigation systems and improving the capacity of the Ministry of Counter Narcotics. MSI (2015), Kandahar Food Zone: Mid-term Performance Evaluation for USAID available at http://pdf.usaid.gov/pdf_docs/PA00KDWM.pdf.
70 Minutes of the Alternative Development Conference, Panjowai District Center, April 12 2015, report for Kandahar Food Zone.
71 The Water Law of Afghanistan outlines the structure for decentralized decision making based on River Basin Agencies and Councils. See Chapter 3 of the Water Law (Unofficial English translation of the Water Law as published in the Ministry of Justice Official Gazette No. (980), 26 April 2009).
72 Matti Monawar, Senior Advisor to the Minister, Afghan Ministry of Irrigation, Land and Livestock (Personal Communication, 14 June 2016).

11 Winter Is Coming

U.S. Strategic Interests and the Water-Energy-Agriculture Conundrum in Central Asia

Richard Kyle Paisley

Introduction

The water-energy-agriculture nexus has long been a conundrum in Central Asia (CA) that poses a direct threat to U.S. prosperity, stability, and interests. The crux of the conundrum is the conflict between upstream states wanting to release water in the winter to generate hydropower and downstream states wanting the water released in the summer for downstream agricultural purposes. Release in the winter also causes undesirable flooding downstream. The conundrum stems from the disintegration of the resource-sharing system imposed on the region by Moscow until the collapse of the Soviet Union in 1991. Under that system, Kyrgyzstan and Tajikistan provided water to Kazakhstan, Turkmenistan, and Uzbekistan in summer (largely for agricultural purposes) and received Kazakh, Turkmen, and Uzbek coal, gas, and electricity in winter. That system had largely broken down by the late 1990s. Various bilateral and regional agreements and resolutions that were reached in that decade and the following decade, including the seminal 1998 Long-Term Framework Agreement brokered by USAID, failed to fix the situation.

Significant U.S. interests in CA continue to include preventing the emergence of failed states that could become a staging ground for international terrorism; avoiding regional conflicts that could draw in neighboring powers (many of them with nuclear weapons); and limiting CA's ability to act as a transit route for transnational threats, such as drugs.

There is now fresh evidence of acute, increasing instability in CA through a perfect storm of worsening water quality and quantity issues; climate change; infrastructure deterioration; failing agricultural yields; ethnic conflicts on the borders of CA states; endemic corruption; various unsuccessful initiatives by China, Russia, and the European Union to broaden their engagements in the region; and personal animosity among the presidents of the various CA states. The situation could be further exacerbated by the additional political instability potentially flowing from the recent (September 2016) death of longtime Uzbekistan President Islam Karimov. When construction on the now-dormant Rogun and Kambarata hydroelectric power projects in Tajikistan and Kyrgyzstan resumes in earnest, a new and dangerous threshold will likely be crossed, likely requiring urgent, intensive, high-level resolution.

The highly respected International Crisis Group (ICG) has advanced two possible solutions to the CA water-energy-agriculture nexus conundrum going forward: first, a redoubling of efforts by all CA states, including Afghanistan, to develop one or more legally binding international agreements across all CA countries, or CA-wide, on water and related resources, and, second, a high-level mediation to address Uzbekistan's objections to new upstream hydro projects.

These initiatives are not a panacea for all that ails CA. However, these initiatives would clearly help focus the region on establishing a much-needed nascent international river basin organization that could eventually embrace larger water/energy governance solutions that actually address and reduce climate change impacts and risks.

Inspired leadership, from the U.S., the United Nations Regional Centre for Preventive Diplomacy for Central Asia (UNRCCA), the World Bank, the United Nations Department of Political Affairs (UN DPA) Mediation Support Unit, and like-minded others, is urgently needed to give CA states the knowledge, skills, and political space they need to successfully conclude and implement three draft CA-wide legally binding international agreements on water and related resources, as well as advance high-level mediation. The U.S. is well situated, due to long-standing interest and aptitude, to make this happen.

1.0 CA: Basic Facts[1]

Total area: 1.6 million (M) sq. mi. (larger than India); Kazakhstan: 1.1 M sq. mi.; Kyrgyzstan: 77,000 sq. mi.; Tajikistan: 55,800 sq. mi.; Turkmenistan: 190,000 sq. mi.; Uzbekistan: 174,500 sq. mi.

Total population: 64.97 M (slightly less than France); Kazakhstan: 17.74 M; Kyrgyzstan: 5.55 M; Tajikistan: 7.91 M; Turkmenistan: 5.11 M; Uzbekistan: 28.66 M[2].

Total gross domestic product: US$414.74 billion (B) in 2012 (slightly less than Belgium). Per-capita GDP is about US$6,400 (slightly less than Bhutan). There are large income disparities (and relatively large percentages of people in each country are in poverty). Kazakhstan: US$231.3 B; Kyrgyzstan: US$13.47 B; Tajikistan: US$17.72 B; Turkmenistan: US$47.55 B; Uzbekistan: US$104.7 B[3]

2.0 Problem Statement

When the CA region comes to mind, it prompts memories of the Silk Road, ancient civilizations with limitless barren lands, the sands of Karakum and Kyzylkum, the dazzling ice-covered mountains of the Pamir, Alai, Zarafshan and Hindu Kush, the Aral Sea, wild rapid rivers, and the historical centers of Samarkand, Bukhara, Termez, Sogdiana, Balkh, and Merv.[4]

The water-energy-agriculture nexus has long been a conundrum in CA that poses a direct threat to U.S. prosperity, stability, and interests.[5] The crux of the conundrum is the conflict between upstream states wanting to release

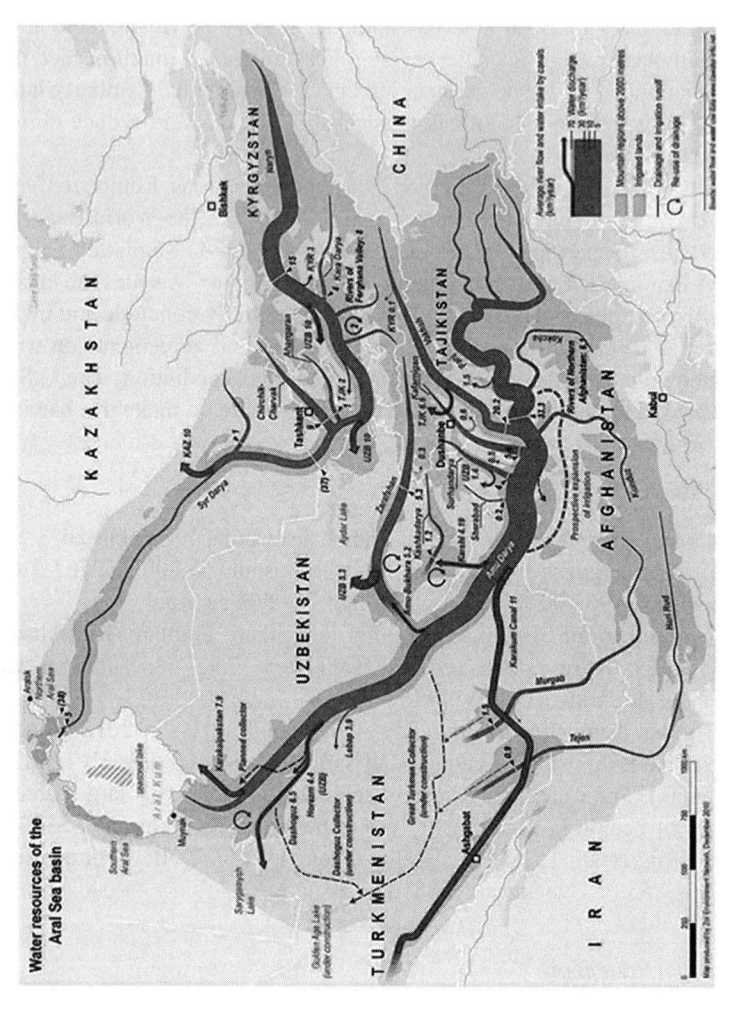

Figure 11.1 ZOI Environment

water in the winter to generate hydropower and downstream states wanting the water released in the summer for downstream agricultural purposes. Two CA states (Kyrgyzstan and Tajikistan) have a water surplus.[6] Four other CA states (Uzbekistan, Kazakhstan, Turkmenistan, and Afghanistan) say they do not get their fair share from the region's great rivers—the Syr Darya and Amu Darya—which slice across CA from the Tien Shan/Pamir Mountains and the Hindu Kush to the Aral Sea's remains.[7] According to the latest (2014) report of the highly respected ICG, pressures are mounting, especially in Kyrgyzstan, Tajikistan, and Uzbekistan.[8] The population in CA has increased by almost 10 M since 2000, and limited arable land is being depleted by overuse and outdated farming methods.[9] Extensive corruption and failing infrastructure are taking a toll, and climate change is increasingly likely to have long-term negative consequences throughout CA.[10] As CA economies become increasingly weaker and states more fragile, heightened nationalism, (perceived) religious discrimination and dogmatism, border disputes, and regional tensions complicate the search for mutually acceptable solutions to the region's water, energy, and related needs.[11]

3.0 Problem Evolution and Analysis of Impact/Consequences

The CA water-energy-agriculture nexus conundrum stems from the disintegration of the resource-sharing system imposed on the region by Moscow until the collapse of the Soviet Union in 1991.[12] Under the old system, Kyrgyzstan and Tajikistan provided water to Kazakhstan, Turkmenistan, and Uzbekistan in summer (largely for agricultural purposes) and received Kazakh, Turkmen, and Uzbek coal, gas, and electricity in winter.[13] That system had largely broken down by the late 1990s. Various bilateral and regional agreements and resolutions negotiated in that decade and the following decade, including the seminal 1998 Long-Term Framework Agreement brokered by USAID, failed to rectify the situation.[14] The 1998 Long Term Framework Agreement explicitly recognized that annual multiyear irrigation water storage has a cost that could be compensated through a barter exchange of electricity and fossil fuels or with cash.[15] However, negotiation and implementation of the various ancillary agreements necessary to make the 1998 Long Term Framework Agreement work proved unsatisfactory, particularly when the key Toktogul Reservoir in Kyrgyzstan reached a record low level of 7.5 billion cubic meters (BCM) in April 2002, eroding the multiyear regulating ability of that reservoir.[16] The shrinking of the Aral Sea, and the increasingly poor environmental conditions in the surrounding region, dramatically brought home to policy-makers that urgent action was needed to mitigate the Aral Sea's disappearance and the resulting socioeconomic disaster.[17] In the past 20 years, various international conferences and missions by national and international experts have described the deteriorating environmental and socioeconomic situation throughout CA.[18]

There is now fresh evidence of acute, increasing instability in CA, through a perfect storm of worsening water quality and quantity issues; climate change;

infrastructure deterioration; failing agricultural yields; ethnic conflicts on the borders of CA states; endemic corruption; various unsuccessful initiatives by China, Russia, and the EU to broaden their engagements in the region; and personal animosity among the presidents of the various CA states.[19]

This evidence includes but is not limited to the following.

1 A 2011 UN Environment Programme (UNEP) report notes the effects of climate change in CA are much more serious and challenging than previously thought (compared with data in the UNEP 1993 report).[20] In particular, the occurrence of severe droughts, glacier cover exhaustion, changes in rainfall patterns, increased land degradation, and pest infestation are key sources of concern.[21] The amount of salinized, degraded land has increased, as has the pressure on arable land. Structural imbalances in the regional water system of CA are stirring up tensions over use. Upstream states, limited economic alternatives, and rising demand for exporting electricity are the main factors motivating the development of large hydropower projects in upstream countries. However, these are perceived as modifying access to water resources by downstream countries, which are heavily dependent on agriculture and consequently disagree with most hydropower projects.[22] According to UNEP:

> there is little doubt that climate change will bring transboundary and multifaceted challenges to the Amu Darya basin, especially for water resources, land productivity, and sensitive biodiversity. Climate change is likely to increase the risk of extreme weather events and related natural disasters, which may become humanitarian emergencies.[23]

2 A 2012 Intelligence Community Assessment specifically identifies the Amu Darya river basin as a place with inadequate water agreements, degradation of water quality and disruption of flows, poor water management, degraded regional food security, increased regional tensions over water, decreased health of populations around the dried Aral Sea, and inadequate river basin management capacity.[24]

3 Testimony in 2014 by Director of U.S. National Intelligence James Clapper that

> the governments of the Central Asian states continue to be concerned about regional instability following the drawdown of U.S. and NATO forces in Afghanistan. . . . Central Asian militants currently harbored in Afghanistan and Pakistan would continue to pose a threat to the Central Asian region, but sources of internal regional instability would probably remain more of a threat. Such instability includes uncertain political succession contingencies, endemic corruption, weak economies, ethnic tensions, and political repression. Regional cooperation remains stymied

by personal leadership rivalries and disputes over water, borders, and energy.[25]

4   A 2015 article by Catherine Putz on the Rogun Dam in *The Diplomat Magazine* quotes the Uzbekistan Ministry of Foreign Affairs as saying that "Uzbekistan never, and under no circumstances, will provide support to this (Rogun) Project."[26]

5   A 2016 Al Jazeera report states that Uzbekistan President Islam Karimov "wasn't very subtle with his October warning that control over water resources in the republics of Central Asia may lead to full-scale war."[27] Al Jazeera also quoted international water expert Daniel Farinotti of the Swiss Federal Institute for Forest, Snow, and Landscape Research as saying, "The answer is clearly yes," to the question of whether experts think that in the coming decades an armed conflict in CA over water seems inevitable.[28]

6   A 2016 address by Miroslav Jenča, assistant secretary general for political affairs, to the United Nations Security Council in New York states that

> pressures on water and related resources have been building for some time in Central Asia—home to two great rivers, the Amu Darya and the Syr Darya, and the Aral Sea—once one of the largest lakes in the world but now "one of the worst environmental disasters in the world,'" in the words of our secretary general. Climate change is making the need to find durable and sustainable solutions all the more urgent.[29]

The availability of abundant water resources, combined with the scarcity of fossil fuels, the vulnerability of the energy sector, and the policy of reducing dependence on external energy sources are pushing Tajikistan and the Kyrgyz Republic to develop their hydropower potential, and reactivate now-dormant projects such as Rogun and Kambarata (designed under the Soviet Union) despite the risk that these projects may further reduce or modify the availability of water on which downstream users, such as Uzbekistan and Turkmenistan, are heavily dependent to irrigate crops.[30] The situation could be further exacerbated by the additional political instability potentially flowing from the recent (September 2016) death of longtime Uzbekistan President Islam Karimov.[31]

In addition, an Italian industrial group recently signed a framework agreement worth about US$3.9 B to resume construction on Rogun.[32] This same group is also the lead contractor on the huge Grand Ethiopian Renaissance Dam (GERD) project in Ethiopia.[33] Both Rogun and GERD are massive developments that look like they may be constructed without environmental and social safeguards such as those required, for example, on World Bank projects. The lack of such safeguards is known to be particularly problematic in environmentally and politically volatile regions.[34]

When construction on Rogun and Kambarata resume in earnest, a new and dangerous threshold will likely be crossed, likely requiring urgent, intensive, high-level resolution.[35]

4.0 U.S. Interests in CA/U.S. Contribution to Problem Mitigation

The United States has a number of important strategic interests in CA. According to the U.S. Congressional Research Service, these interests include securing and eliminating Soviet-era nuclear and biological weapons materials and facilities.[36] U.S. energy firms have also invested in oil and natural gas development in Kazakhstan and Turkmenistan, and successive administrations have backed diverse export routes to the West for these resources. U.S. policy toward Kyrgyzstan has long included support for its civil society. In Tajikistan, the United States focuses on developmental assistance to bolster the fragile economy and address high poverty rates. The United States and others have urged the regional states to cooperate in managing their water resources. U.S. relations with Uzbekistan—the most populous state in the heart of the region—were cool after 2005, but recently have improved. Congress has been at the forefront in advocating increased U.S. ties with CA, and in providing backing for the region for the transit of U.S. and North Atlantic Treaty Organization (NATO) equipment and supplies into and out of Afghanistan. Congress has pursued these goals through hearings and legislation on humanitarian, economic, and democratization assistance; security issues; and human rights.

Despite the closure of the Manas military base in the Kyrgyz Republic in 2014, significant U.S. interests in the CA region also include preventing the emergence of failed states that could become a staging ground for international terrorism; avoiding regional conflicts that could draw in neighboring powers (many of them with nuclear weapons); and limiting CA's ability to act as a transit route for transnational threats, such as drugs.[37]

Having a presence in CA also enhances U.S. ability to influence developments in Afghanistan and Pakistan, which both face worsening insurgencies and the growth of radical forces.[38]

According to the Foreign Military Studies Office (FMSO), conflict in CA would:[39]

> destabilize Uzbekistan and Tajikistan and have serious consequences on U.S. military efforts in Afghanistan. Central Asian countries, particularly Uzbekistan, have historically closed their borders during periods of unrest. Kazakhstan closed its border with Kyrgyzstan for a month following the April 2010 revolution in Kyrgyzstan, and Uzbekistan did the same with Kyrgyzstan during the June 2010 violence there. Any disturbance will likely disrupt regional transportation networks and hinder logistics on the Northern Distribution Network (NDN). The main line, primarily rail, of this logistics network starts in Latvia and runs through Russia, Kazakhstan, and Uzbekistan into Afghanistan. An estimated 40 percent of U.S.

and coalition nonlethal supplies are now transiting this network, and supplies are purchased along the route when possible. These supplies include construction materials and basic necessities, such as food and water. This network will continue to be an important supply line for U.S. and coalition forces in Afghanistan.[40]

5.0 Solutions

For all its complexity, the water issue in CA is thought to actually offer some opportunity for solutions.[41] The ICG has advanced two possible solutions to the CA water-energy-agriculture nexus conundrum:[42] First, a redoubling of efforts by all CA states, including Afghanistan, to develop one, or more, CA-wide legally binding international agreements on water and related resources; and, second, a high-level mediation initiative to address Uzbekistan's objections to upstream hydro projects.

5.1 CA-Wide Legally Binding International Agreements

According to the ICG:

> Kyrgyzstan, Tajikistan, and Uzbekistan (and their international backers) should act now in the border areas of the Ferghana Valley to end the annual cycle of competition and conflict over water by dividing the water issue into more manageable portions—seeking gradual, step-by-step solutions along conceptual and geographical lines rather than one all-inclusive resource settlement.[43]

The World Bank, UN DPA, UNRCCA, and like-minded others have been trying to forge such cooperation for some time. The World Bank is specifically mandated to help ensure that large infrastructure development initiatives, especially those utilizing shared international waters such as Rogun and Kambarata, are built with the full suite of World Bank environmental and social safeguards.[44] The UN Department of Political Affairs Mediation Support Unit is mandated to support the pacific settlement of disputes. The UNRCCA, based in Ashgabat, Turkmenistan, is specifically mandated to assist and support the governments of Kazakhstan, Kyrgyzstan, Tajikistan, Turkmenistan, and Uzbekistan in "building their conflict prevention capacities through enhanced dialogue, confidence building measures, and genuine partnership in order to respond to existing threats and emerging challenges in the Central Asian region."[45] At the 2009 International Fund for Saving the Aral Sea (IFAS) Summit Meeting of the CA presidents, the UNRCCA was specifically instructed to intensify efforts to assist in the resolution of transboundary international water and related resource issues.[46]

The efforts of the UNRCCA, the World Bank, the UN DPA, and like-minded others have included assembling a team of international water, energy, and conflict prevention/conflict resolution experts to progressively develop three demand-driven multilateral international shared waters-related agreements for consideration and further discussion among CA states:[47]

1. A Draft Central Asia Regional Framework Agreement (for the Sustainable Utilization and Development of the Shared International Waters, and Related, Resources) of the Syr Darya and Amu Darya River Basins (Draft CA Regional FA);
2. A Draft Agreement on a Procedure for Implementing International Legal Principles to Promote Cooperation and to Prevent and Resolve Disputes (over the Sustainable Utilization and Development of the Shared International Waters, and Related, Resources of the Syr Darya and Amu Darya River Basins) (Draft CA Procedure Agreement); and
3. A Draft Regional CA Infrastructure Development Template for the Syr Darya and Amu Darya River Basins (Draft Regional CA IDT), including mutually agreed-upon mechanisms for acknowledging the concerns of those impacted by development, encouraging joint fact-finding, offering contingent commitments to minimize impacts if they do occur, and compensating knowable but unintended impacts (including "casino" events; i.e., events with low probability of occurrence but with potentially catastrophic consequences).

The Draft CA Regional FA establishes a mutually agreed-upon framework for the conservation and management of shared international waters and related resources throughout CA that would, among other things, reflect in particular both the unique history and special arid nature of the CA region. The Draft CA Regional FA incorporates various international legal principles expressed in existing international agreements and instruments, such as the seminal 1966 Helsinki Rules, 1997 UN Watercourses Convention, and the 1992 UNECE Water Convention, while further taking into account developments since those instruments were finalized.[48] As of July 2016, the Draft CA Regional FA was in Draft 3, following four rounds of regional consultations and a regional workshop to identify customary practice in CA states. All CA states, other than Uzbekistan, have indicated they are committed to providing specific inputs on further adjusting Draft 3 of the Draft CA Regional FA, including by identifying representatives to participate in a drafting team. There is also some indication that the various CA states, other than Uzbekistan, may opt for two parallel framework agreements (one for the Amu Darya and another for the Syr Darya) rather than support one Draft CA Regional FA. Uzbekistan has so far declined to participate in consultations on a Draft CA Regional FA, taking the position that all CA states should first join Uzbekistan in ratifying the existing global water conventions (i.e., the 1997 UN Watercourses Convention and the 1992 UNECE Water Convention) before a Draft CA Regional FA can be discussed.

The Draft CA Procedure Agreement establishes a unique, mutually agreed-upon procedure for preventing and resolving conflicts that have historically arisen between CA states with regard to the use and management of shared international waters and related resources: e.g., infrastructure development and climate change, which could necessitate changes in flow regime or water allocation. Nonperformance of agreed-upon obligations has also always been a serious challenge under virtually all previous arrangements involving CA.[49] The Draft CA Procedure Agreement would help alleviate this concern by providing a systematic approach to applying the basic principles of international water law to the prevention and resolution of conflicts and potential conflicts. Importantly, a regional agreement to prevent and resolve conflict, such as the Draft CA Procedure Agreement, would still be necessary in CA even if all CA states were party to the 1997 UN Watercourses Convention and/or the 1992 UNECE Water Convention.[50] A proposed eight-step Draft CA Procedure Agreement and associated international precedents for each step were presented and critically reviewed at two workshops in Vienna, Austria, in 2014 and 2015. These workshops were attended by representatives from all CA states. As of July 2016, a revised version of the Draft CA Procedure Agreement had been consolidated into a working document designed to facilitate further discussion.

The objective of the Draft Regional CA IDT is to identify the key elements that are needed in order to secure commitments that would successfully guide the construction and operation of infrastructure projects with potential transboundary impacts. These elements could possibly include identifying contingent commitments to minimize economic, social, and/or environmental impacts, if they do occur, and compensating knowable but unintended impacts.[51] The concept of a Draft Regional CA IDT was proposed to various CA states during regional consultations in 2015, with support expressed by some CA states.

5.2 High-Level Mediation Addressing Uzbekistan's Objections to Upstream Hydro Projects

The ICG 2014 report stated the following:

> High-level mediation should be sought to address Uzbekistan's objections to upstream hydropower projects. There is no guarantee this would work, but it could give these three states (Kyrgyzstan, Tajikistan, Uzbekistan) an opportunity to modernise infrastructure and the management of water resources as well as train a new generation of technical specialists. The agreements would also set a modest precedent for other spheres in which cooperation is sorely needed and might help defuse tensions in the region, while improving the grim living conditions of most of its population.[52]

Mediation is well known to have played a critical role throughout the world in resolving conflicts over shared international waters, preventing outbreaks of violence and enhancing collaboration and cooperation between historical former adversaries.[53]

Various overarching messages, capturing lessons learned in the field, are applicable to a possible U.S.-supported, culturally appropriate mediation of Uzbekistan's objections to upstream hydropower projects in CA:[54]

1. Equal access to impartial scientific and technical information about the resource in dispute is particularly key in CA. One of the prerequisites to effective mediation processes over natural resources is for all parties to have equal access to such information.[55] This can be generated jointly by the parties themselves or by an independent third party. The very process of generating common information can also have confidence-building benefits.[56]

2. Careful attention is needed to identify the stakeholders that should be engaged in the mediation process.[57] Designers of the mediation process should think carefully about which stakeholders to involve. Inviting the participation of all stakeholders may, for example, prove too unwieldy or fragmented to produce consensus. Understanding which actors to include in mediation, and the potential political impacts of including some and excluding others, is essential. In turn, ensuring consultation with a sufficiently wide set of stakeholders is crucial to establish and maintain the legitimacy of the process. Track Two–type diplomatic initiatives have so far been conspicuous by their absence in CA. "Track Two diplomacy" refers to conflict resolution efforts by professional non-governmental conflict resolution practitioners and theorists. Track Two has as its object the reduction or resolution of conflict, within a country or between countries, by lowering the anger, tension, or fear that exists, through improved communication and a better understanding of each other's point of view.[58]

3. Mediation should aim for collaboration over shared benefits, which can in turn generate the trust needed to tackle other issues.[59] Mediators approaching any conflict in CA over shared international water and related resources should try to help parties move past zero-sum, win-lose positions.[60] Mediators should try to identify ways that stakeholders can maximize shared benefits and address common problems and challenges together. When possible, shared international water and related resources should be treated as a platform for cooperation that transcends ethnic, religious, ideological, political, or tribal differences, as initial cooperation over shared international water and related resources can sometimes be leveraged to tackle more challenging problems down the line.

4. Various mediation techniques are available to overcome critical impasses and entrenched positions.[61] Once involved in negotiations, mediators can break down impasses using a number of techniques: focusing the talks on technical issues, conducting joint information gathering, identifying and sharing multiple benefits, or using scenario-building approaches. Altering fixed or inflexible default positions can sometimes be achieved by moving parties away from questions of natural resource ownership and toward

broader issues of benefit-sharing, predictable access, and management—areas where opportunities for mutual benefit can be found. One interesting recent example with possible application to CA is the new agreement on the desalination plant in Gaza.[62] In this case, historical adversaries appeared to reach an agreement on the basis of shared benefits and basically redesigned water allocation schemes for the whole system.[63] By widening the issue (instead of focusing on one aspect), they found a workable solution.[64] By analogy, upstream states in CA may be willing to do something similar.[65] They appear to be currently focused on looking at "selling" their hydropower capacity. However, they will probably need guarantees for additional investments. A scheme to improve agriculture, including fewer problems with irrigation/rain-fed agriculture, might be an entry point. As long as an initial agreement can be reached through the efforts of technical experts, there may be a chance.[66]

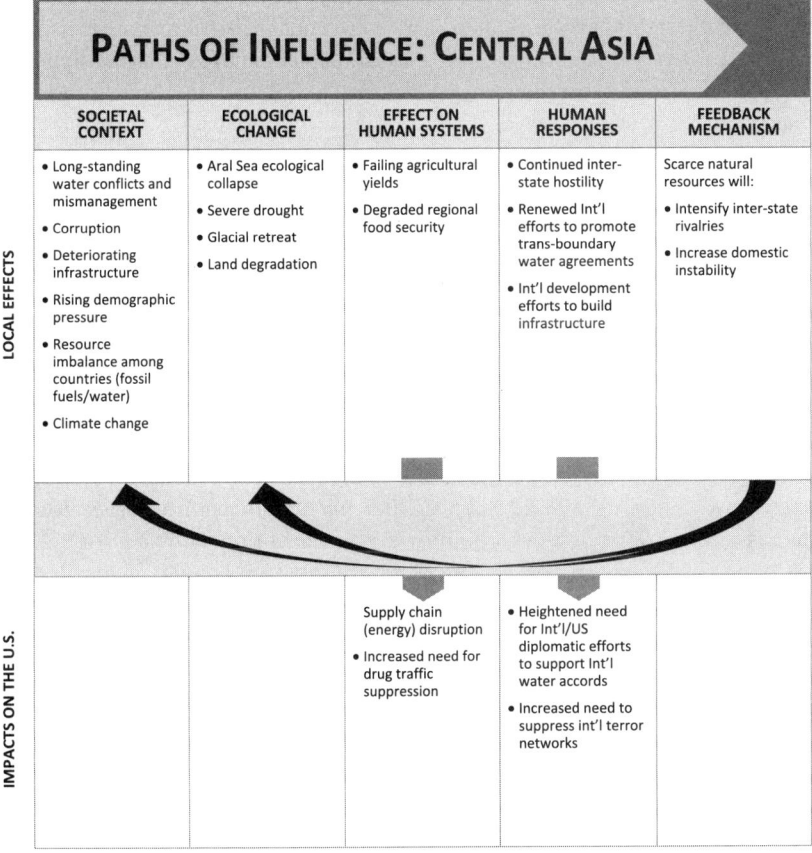

Figure 11.2 Paths of Influence: Central Asia

There are reasons for cautious optimism. First, the Rogun and Kambarata projects are increasingly likely to be built. Perhaps this will be the catalyst to move Uzbekistan into reaching an accommodation, albeit reluctantly, with upstream adversaries. The analogy is to Egypt, which has eventually been moved into working toward an accommodation, albeit reluctantly, with upstream adversaries over the GERD. Second, the Italian Consortium may be persuaded to use its considerable financial clout and international prestige to promote regional stability, if for no other reason than to protect its already significant investment in Rogun. Third, the international community has already successfully spaded the ground for cooperation in CA by negotiating the three draft agreements, suggesting that there are at least some common interests seemingly shared between the various CA states that could act as a foundation for moving forward. Fourth, the recent death of longtime President Islam Karimov of Uzbekistan may have put to rest (at least temporarily) any personal animosity between the presidents of Tajikistan and Uzbekistan.

The initiatives called for by the ICG are not a panacea for all that ails CA. However, these initiatives would clearly help focus the region on establishing a nascent international river basin organization that could eventually embrace larger water/energy governance solutions that actually address and reduce climate change impacts and risks.[67]

Inspired leadership, from the U.S., the UNRCCA, the World Bank, the UN DPA Mediation Support Unit, and like-minded others, is urgently needed to give CA states the knowledge, skills, and political space they need to successfully conclude and implement the three draft CA-wide legally binding international agreements on water and related resources, as well as advance high-level mediation. The United States is well situated, due to long-standing interest and aptitude, to make this happen.

Acknowledgments

Thank you to the three reviewers who made many constructive suggestions and to my colleague Maaria Solin Curlier for her always invaluable advice and assistance. The views or opinions expressed in this paper do not necessarily reflect the views or opinions of any of the entities (e.g., World Bank, UNRCCA, and UN DPA Mediation Support Team) the author may be, or may have been, associated with.

Notes

1 Nichol, Jim. Central Asia: Regional Developments and Implications for U.S. Interests, Congressional Research Service, 2014 at 3.
2 July 2013 estimate, *The World Factbook.[0]*
3 *The World Factbook*, purchasing power parity.
4 UNEP, GRID Arendal and Zoi Environmental Network, Environment and Security in the Amu Darya Basin. UNEP Environment and Security Initiative. 2011 at 5.

5 "Weaponizing Water: Water and Energy as Sources of Conflict Among the Central Asian Soviet Successor States." 22 *Mich. St. Int'l. L. Rev.* 409. 2013; Bichsel, Christine.
6 International Crisis Group, Water Pressures in Central Asia. *ICG Asia Report* No. 233 Osh/Brussels. 2014.
7 Ibid. at Executive Summary.
8 Ibid.
9 Ibid.
10 Ibid.
11 Miroslav, Mansur. Are "Water Wars" Imminent in Central Asia?, *Al Jazeera*, 23 March 2016.www.aljazeera.com/indepth/features/2016/03/water-wars-imminent-central-asia-160321064118684.html (27 July 2016); Bart, Jason, "Weaponizing Water: Water and Energy as Sources of Conflict Among the Central Asian Soviet Successor States." 22 *Mich. St. Int'l. L. Rev.* 409. 2013.
12 International Crisis Group, Water Pressures in Central Asia. *ICG Asia Report* No. 233 Osh/Brussels 2014 at Executive Summary.
13 Ibid.
14 World Bank, Water and Energy Nexus in Central Asia: Improving Regional Cooperation in the Syr Darya Basin. Europe and Central Asia Region. The World Bank, Washington, D.C. 2004 at v.
15 Ibid.
16 Ibid.
17 UNEP, GRID Arendal and Zoi Environmental Network, Environment and Security in the Amu Darya Basin. UNEP Environment and Security Initiative. 2011 at 5.
18 Ibid.
19 Lowe, Rebecca and Emily Silvester. Water Shortages Threaten Global Security. 68 No. 4 IBA Global Insight 42. August/September, 2014.
20 NEP, GRID Arendal and Zoi Environmental Network, Environment and Security in the Amu Darya Basin. UNEP Environment and Security Initiative. 2011 at 9.
21 Ibid.
22 Ibid.
23 Ibid.
24 Ibid. at v.; Intelligence Community Assessment, Global Water Security, ICA 2012–08, 2012 at iii.
25 Ibid. at 5.
26 Putz, Catherine. Uzbekistan Still Hates the Rogun Dam Project, *The Diplomat*, 4 August 2015. http://thediplomat.com/2015/08/uzbekistan-still-hates-the-rogun-dam-project/ (30 July 2016). The Rogun dam and power plant are planned on the Vakhsh River, upstream from the Nurek dam, currently Tajikistan's main source of electricity. Once these projects are completed, Tajikistan will be able to generate an energy surplus for export. However, to export electricity it will need to invest in extending the transmission grid. In February 2006 the energy ministers of Tajikistan, Iran, and Afghanistan signed an agreement providing for a high-voltage power line to run from Rogun and the other hydroelectric stations on the Vakhsh River to Afghanistan.
27 Miroslav, Mansur. Are "Water Wars" Imminent in Central Asia?, *Al Jazeera*, 23 March 2016. www.aljazeera.com/indepth/features/2016/03/water-wars-imminent-central-asia-160321064118684.html (27 July 2016).
28 Ibid.
29 Jenča, Miroslav. The Concept of Preventive Diplomacy and Its Application by the United Nations in Central Asia Security and Human Rights 24 (2013) at 191 and 192.
30 UNEP Environment and Security Initiative, 2011 at 49.
31 Islam Karimov: Uzbekistan President's Death Confirmed. *BBC*, 2 September 2016. www.bbc.com/news/world-asia-37260375 (12 October 2016).

220   *Richard Kyle Paisley*

32 Harris, Michael. Salini Wins Contract to Build Tajikistan's 3,000-MW Rogun Hydropower Plant. *HydroWorld*. www.hydroworld.com/articles/2016/07/salini-wins-contract-to-build-tajikistan-s-3-600-mw-rogun-hydropower-plant.html.
33 Grand Ethiopian Renaissance Dam Project. *Salini Impregilo*. www.salini-impregilo.com/en/projects/in-progress/dams-hydroelectric-plants-hydraulic-works/grand-ethiopian-renaissance-dam-project.html
34 OP/BP 7.50 is a World Bank "safeguard policy" pertaining to international waterways. On August 4, 2016, the World Bank's Board of Executive Directors approved a new Environmental and Social Framework that sets out mandatory requirements that apply to the bank in the World Bank Environmental and Social Policy for Investment Project Financing.
35 ICG 2014. See also Human Rights Brief, Energy And Forced Resettlement in Tajikistan. No. 1 Hum. Rts. Brief 21. Spring, 2015 and Lowe, Rebecca and Emily Silvester. Water Shortages Threaten Global Security. 68 No. 4 IBA Global Insight 42. August/September, 2014.
36 Nichol, Jim. Central Asia: Regional Developments and Implications for U.S. Interests, Congressional Research Service, 2014 at 2.
37 Ibid.
38 Ibid.
39 Stein, Matthew. Conflict Over Water and Related Resources in Uzbekistan and Tajikistan and Its Impact on Local Security. Foreign Military Studies Office. 2011.
40 Ibid.
41 International Crisis Group, Water Pressures in Central Asia. *ICG Asia Report* No. 233 Osh/Brussels. 2014 (ICG 2014).
42 Ibid.
43 International Crisis Group, Water Pressures in Central Asia. *ICG Asia Report* No. 233 Osh/Brussels. 2014 (ICG 2014) at ii.
44 World Bank. *Water and Energy Nexus in Central Asia: Improving Regional Cooperation in the Syr Darya Basin*. Washington, D.C.: The World Bank, 2004.
45 UNRCCA Mission Statement. http://unrcca.unmissions.org/Default.aspx?tabid=9303&language=en-US.
46 Jenča, Miroslav. The Concept of Preventive Diplomacy and Its Application by the United Nations in Central Asia Security and Human Rights 24 (2013) at 191 and 192.
47 Ibid.
48 Ibid.
49 World Bank, Water and Energy Nexus in Central Asia, 2004 at v.
50 Schwabach, Aaron. The United Nations Convention on the Law of Non-Navigational Uses of International Watercourses, Customary International Law, and the Interests of Developing Upper Riparians. 33 *Tex. Int'l L.J.* 257. 1998.
51 Susskind, Lawrence and Patrick Field. *Dealing with an Angry Public: The Mutual Gains Approach to Resolving Disputes*. New York: The Free Press, 1996.
52 ICG 2014 at ii.
53 UN DPA and UNEP. Natural Resources and Conflict: A Guide for Mediation Practitioners, UN DPA, 2015 at 54.
54 Ibid.
55 Ibid.
56 Kraska, James. Sustainable Development Is Security: The Role of Transboundary River Agreements as Confidence Building Measure (CBM) in South Asia, 28 *Yale J. Int'l L*. 465, 490 (2003).
57 UN DPA and UNEP, 2015 at 54.
58 McDonald, J. W. and Bendahmane, D. B. (Eds.). (1987). *Conflict Resolution: Track Two Diplomacy*. Washington, D.C.: Foreign Service Institute, U.S. Department of State.
59 UN DPA and UNEP, 2015 at 54. Some of the "mutual gains" to be had through cooperation in Central Asia are documented in: World Bank. *Water and Energy*

Nexus in Central Asia: Improving Regional Cooperation in the Syr Darya Basin. Washington, D.C.: The World Bank, 2004 at v and vi.
60 Grzybowski, Alex, Stephen C. McCaffrey, and Richard Kyle Paisley. Beyond International Water Law: Successfully Negotiating Mutually Beneficial Agreements for International Watercourses. *Global Business & Development Law Journal* 139 (2010).
61 UN DPA and UNEP. Natural Resources and Conflict: A Guide for Mediation Practitioners, UN DPA, 2015 at 54.
62 Personal communication, Dr. Lesha B. M. Witmer, June 2016.
63 Lieber, Dov. EU Inaugurates Gaza's First Desalinization Plant. *The Times of Israel*, 14 June 2016. www.timesofisrael.com/eu-inaugurates-gazas-first-desalinization-plant/ (31 July 2016).
64 Personal communication, Dr. Lesha B. M. Witmer, June 2016.
65 Ibid.
66 Ibid.
67 Laruelle, Marlene and Sebastien Peyrouse. *Regional Organizations in Central Asia: Patterns of Interaction, Dilemmas of Efficiency*. Working Paper Number 10, 2012, University of Central Asia, Graduate School of Development, Institute of Public Policy and Administration. According to Laruelle and Peyrouse, there are numerous historical, cultural, and political challenges to successfully establishing and maintaining regional organizations in CA.

12 The Perils of Denial
Challenges for a Water-Secure Pakistan

Ali Hasnain Sayed, Chelsea N. Spangler, and Muhammad Faizan Usman

Introduction

Pakistan's economy is dependent on water, and improper water resource management contributes to the nation's volatile socioeconomic situation. On its face, water risk might not appear as immediately hazardous to national security as the threat of terrorism—Pakistan has lost close to 100,000 people to terror attacks in the past bloody decade. However, poor water management contributes to and reinforces the conditions of weak governance and slow economic growth that provide fertile ground for terrorist insurgency. The impacts of water scarcity on agriculture lead to a loss of rural livelihoods and broad food insecurity. Waterborne diseases claim hundreds of thousands of Pakistani lives each year due to poor sanitation. As the climate changes and scarcity worsens, these effects will become more severe. Delaying improvements to water management will only increase the costs to society, and continuing system inefficiencies will likely lead to economic loss, civil unrest, and conflict.

One manifestation of the linkage between water and conflict is the current tension on the Kashmir border with India. In September 2016, Pakistan and India traded accusations about responsibility for hostile actions along the tense border. In the ensuing days, escalation included threats of imposing military action and restricting flow on the Indus River to downstream Pakistan.[1] This is not the first time that tensions have flared: shared water resources are a decades-old source of conflict between Pakistan and India, as witnessed by India's halting of the flow of the Sutlej River into Pakistan in 1948. The combination of these two factors—the inherent insecurity that comes with sharing limited water resources and attacks on both sides of the border—is contributing to increasingly frequent invocations of deploying nuclear weapons to protect national interests. Such saber rattling strains United States relations with Pakistan, which increasingly views U.S. ties with India as a threat.

The India–Pakistan water conflict is just one example of the myriad ways in which water scarcity interacts with other security concerns in mutually reinforcing ways. Water stress contributes to weakening state governance both directly—when civil unrest arises due to insufficient water access and sanitation—and indirectly as water stress contributes to food insecurity and public health crises. Karachi is the world's sixth most water-stressed city, and protests in the rapidly growing municipality have become commonplace. Residents blame the government for its failure to provide sufficient water infrastructure and to prosecute those who steal water and sell it on the black market. Weak and corrupt governance, in turn, prevents the badly needed improvements to infrastructure that could mitigate water supply and sanitation concerns. As surface and groundwater resources in Pakistan diminish, so do agricultural outputs and food security. Pakistan was scored 47.8 (out of 100) on the 2016 Global Food Security Index, and 22% of the population was undernourished in 2016.[2,3] Sanitation is so poor that around 4% of Pakistan's gross domestic product (GDP) is spent on treatment of water-related diseases (US$800 million), and nearly 300,000 infants are lost every year to waterborne diseases.[4]

Water security also has direct ramifications for Pakistan's economic well-being, and its weak economy limits its ability to address water concerns. The government has proven itself unable to tackle large water infrastructure problems, as most of its resources in the past decade have been devoted to fighting terrorism. According to the Pakistan Economic Survey, more than US$67 billion has been spent on the war on terror, directly and indirectly.[5] As a result, according to the Ministry of Planning, Development, and Reforms, Pakistan's per-capita income growth rate was a meager 2%–4% per year in this period—a rate comparable to that of other countries strongly affected by terrorism. Water stress also damages the economy by limiting agricultural productivity, which is responsible for 22% of Pakistan's GDP and employs around 43% of the workforce.[6] Of particular concern for the U.S. is cotton production, which is intensely dependent on water and constitutes the bulk of Pakistani commodities imported by the U.S.; disruptions to cotton production mean disruptions to U.S. supply chains. Cotton and its allied products make up 60% of Pakistan's foreign exchange. Additionally, as agriculture becomes less viable, rural-to-urban migration increases, placing greater strain on already-overtaxed water infrastructure in cities. Water shortages also increase competition between sectors for limited water resources.

The continued presence of terrorist groups within Pakistan makes these feedback loops even more complicated. Weak governance and economic instability provide an optimal environment for insurgency, and increased insurgency leads to sharply decreased foreign direct investment (FDI). According to the United Nations Conference on Trade and Development, Pakistan dropped in international ranking to attract FDI from number 74 in 2008 to 110 in 2010, and FDI dropped from US$5.5 billion (3.9% of GDP) in 2007 to US$0.85 billion (0.4% of GDP) in 2012.[7] This decrease in investment and aid left Pakistan with fewer resources to fight terrorism domestically or to improve the conditions that lead to insurgency in the first place. On the contrary, fighting internal terrorist threats has sapped Pakistan's economic resources and weakened the government's authority over the past decade.

In short, water stress contributes to political instability and economic struggles, which decrease Pakistan's ability to deal with terrorist threats and other forms of militancy. Increased insurgency weakens governance and the economy, and the nation is caught in a negative feedback loop. In order to expose ways to break these destructive cycles, the remainder of this chapter will focus on the root causes of water stress in Pakistan, ways to address those causes, and what actions the U.S. can take to improve the situation.

Main Water Challenges

Pakistan currently faces a water insecurity crisis.[8] The Pakistan Council of Research in Water Resources predicts that the country may face absolute water scarcity as soon as 2025.[9] According to the 2013 Asian Water Development Outlook for Asia-Pacific, Pakistan is the region's fifth most vulnerable country in terms of urban water security, with average per-capita water availability at 964 cubic meters per annum.[10] Unsustainable water use, poor water management, and ineffective governance practices are contributing to decreased water quantity and quality, which in turn lead to serious social and economic repercussions. Rapid industrialization and urbanization—37% of people in Pakistan currently live in cities—add to air, water, and land pollution. Inadequate or improper waste management compounds these problems. According to the UN's "World Population Prospects," Pakistan is one of the 10 most populous

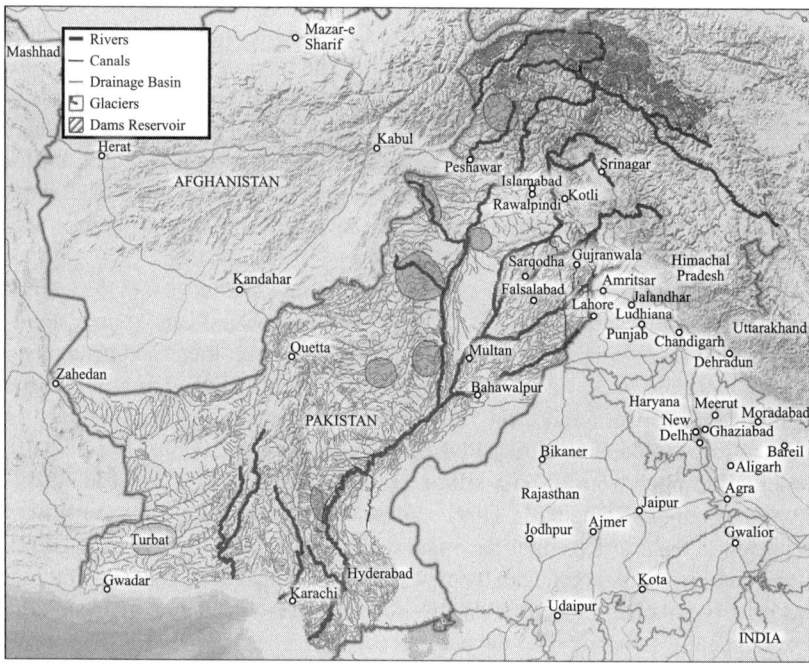

Figure 12.1 Map: Waters of Pakistan[11]

countries in the world, and by 2025 its projected population will reach 300 million, which will put extra pressure on water resources.[12] These issues, coupled with the high incidence of poverty, are serious threats to sustainable development and national security.

The current water scarcity scenario is further exacerbated by poor governance and water management, which suffer from the colonial legacy of Pakistan's water bureaucracy (hereafter referred to as the *hydrocracy*). The hydrocracy has historically focused on engineering dams and irrigation to support agriculture, to the detriment of other sectors. As a result, more than 90% of the country's surface water is allocated for agricultural needs, leaving little for other sectors. Its policies also tend to overlook underlying environmental and socioeconomic factors that contribute to water issues.[13]

Water insecurity in Pakistan can be attributed to three main causes, which will be examined in this chapter. These provide the basis for understanding the current water crisis, as well as potential avenues for positive change.

1   A lack of comprehensive groundwater management has contributed to overuse of existing groundwater resources, as well as insufficient sanitation. This has led to worsening water scarcity, to increased competition for water, and to public health crises.
2   Historically, Pakistan's water management has relied almost exclusively on dams. However, it is necessary to consider whether hard engineering solutions alone will be sufficient to address the new challenges presented by climate change, as weather patterns become more unpredictable and floods more severe.
3   Pakistan lacks control of its riparian freshwater sources, as they originate in India. The Indus Basin Water Treaty currently entitles Pakistan to water from three rivers—the Jhelum, Chenab, and Indus—but tensions between the two countries are growing due to growing populations in both Pakistan and India, as well as increased demand for water due to industrialization and expanding domestic water use.

Groundwater as an Orphan but Contested Resource

Groundwater and surface water resources in the Indus Basin are closely linked, both hydrologically and socioeconomically, yet no single department has the mandate to manage the groundwater. Responsibility for its management is divided on an ad hoc basis between the Irrigation, Power, and Sanitation departments, rendering groundwater an orphan when it comes to holistic management. In addition, Pakistan lacks comprehensive data on groundwater recharge, so it is difficult to gauge rates of abstraction and implications for current and future water availability. However, groundwater quality and quantity are already compromised by overabstraction of groundwater and pollution of surface water.[14]

Pakistan is the world's fourth-largest user of groundwater (after India, the U.S., and China), and is already experiencing the crunch of the overexploitation.

In the central Punjab region (interfluves of the Jhelum and Chenab rivers), the water table is declining by 0.91 meters per annum. In many areas of Balochistan, groundwater tables have fallen at a rate of 2 to 3 meters per annum. The province of Balochistan is currently facing its most severe water scarcity in terms of groundwater depletion. Salt affected soils caused by overabstraction of groundwater now affect 4.5 million hectares, amounting to more than 22% of Pakistan's irrigated lands and making groundwater an even more important resource.[15]

Surface water irrigation dominated Pakistan's agriculture sector until the beginning of the 1990s. However, there was a shift toward groundwater irrigation when scarcity and related administrative issues limited surface water availability. More than 50% of Pakistan's irrigated lands are now served by groundwater wells.[16] According to water expert Shahid Ahmed, annual groundwater pumpage from 2006 to 2007 was around 62 billion cubic meters, which is a major contribution to irrigation as compared with the available surface water. A 2016 LEAD (Leadership for Environment and Development) Pakistan policy brief states that there are around 1.2 million tube wells currently operating in Pakistan extracting about 52 billion cubic meters of water. Eight hundred thousand of these wells are installed in Punjab alone.[17]

Nationwide, about 70% of the private tube wells are located in areas where groundwater is used in conjunction with canal water, whereas the rest provide irrigation that relies on groundwater alone. This implies that groundwater use is probably more widespread than is known. Many farmers in irrigated areas use groundwater only when surface water systems are not functioning, but surface water can fulfill only 30% of the irrigation demand.[18] Although agricultural productivity has historically relied on both groundwater and surface water for irrigation, data suggest that easy access to groundwater has increased the total water use per hectare. Current unmetered use of groundwater is causing serious consequences for agriculture, including falling water tables and salt water intrusion in many areas of the Indus Basin. Excessive lowering of the water table is making water extraction more expensive for all users, and salinization associated with the use of poor-quality groundwater for irrigation has exacerbated the problem.

Of equal importance are competing sectors that rely upon the same groundwater supplies. Nearly 80% of domestic users in major cities of Pakistan rely on groundwater for drinking, bathing, and household chores.[19] The industrial and service sectors, including multinational companies, are also dependent on groundwater for their operations in the main cities. The scarcity generated by overuse of groundwater by the agriculture sector sets up competition between domestic consumers and the business sector. A 2014 study by WWF-Pakistan provides an illustrative case study of the water management of the city of Lahore.[20] It concluded that the water balance is asymmetrical, and on average the city's groundwater level is decreasing at a rate of 0.55 meters annually. Lahore, which has always been considered a water-rich city, is now facing scarcity in many areas, particularly in areas where less-privileged populations reside. Moreover, industrial discharge of wastewater in open channels is creating water-quality issues for domestic use, as the arsenic levels in the majority of Lahore's tested samples were higher than World Health Organization standards (10 parts per billion).

In other parts of the country, the situation is even more serious. In Islamabad, the federal capital, only 15% of the tested samples by the Pakistan Council for Research in Water Resources (PCRWR) were found fit to drink, whereas the rest were infested with bacterial contamination. In Faisalabad in central Punjab, the industrial hub of the country, only three of the tested 13 samples were fit to drink, while the rest were contaminated with bacteria and high levels of sulphates, sodium, potassium, and fluoride.[21] In Khyber Pakhtunkhawa Province, which used to have access to pristine stream and river water (from the Mangora, Mardan, and Peshawar rivers), anthropogenic interventions have led to high levels of iron and total dissolved solids. In the Sindh Province, samples were taken from the affluent cities of Karachi (the business hub of Pakistan) and Hyderabad. No sample taken was considered safe in Hyderabad and 90% of the samples in Karachi were contaminated with bacteria, iron, nitrates, and turbidity.[22] Finally, the strategic province of Balochistan, which is rich in mineral wealth, faces the country's greatest scarcity and quality challenges. The Kuchlagh basin, which has been one of the main reservoirs for the past 30 years, has completely dried up, and groundwater samples contain arsenic, fluoride, and nitrates.[23]

> *"Our livelihood is completely dependent on rain, but for the [p]ast five years we are experiencing either complete drought or erratic rains in Nushki and its adjacent districts. It hasn't rained here for the [p]ast seven months now. I and many other farmers couldn't sow any crop this year, which will cause food shortage for us in upcoming days."*
>
> *Mir Zahir, 43-year-old farmer in Nushki district*[24]
>
> According to a UN report, nearly a million hectares of agricultural land remained uncultivated in Balochistan and 1.75 million livestock died by 2001 as a result of droughts. Many families had to migrate to urban centers due to the 1997–2003 severe droughts in Balochistan.

Climate Change and the Floods Scenario

The Indus River is considered the lifeline of Pakistan. It is the 12th-largest river in the world, originating from the Tibetan Plateau, running the length of Pakistan, and terminating in the Arabian Sea. It supports the largest contiguous irrigation system in the world, ranked as the most important in the world in terms of human dependence because it supports around 215 million people directly and indirectly. Influenced by changing geopolitical, social, economic, and environmental conditions, the basin faces increasing stress and is losing its capacity to support the future water needs of the region.

Ninety percent of the water used for agriculture in Pakistan is supported by the river and its tributaries, and agriculture employs around half of the country's workforce. However, the agriculture sector is vulnerable to the effects of climate change, including increased temperatures, changing rainfall patterns, and more severe flooding. In the Hindu Kush Himalayan region, where glacial and snow melt are the main contributors to the flows of the Indus River, a temperature increase of 4–5 degrees Celsius is already being observed, and the current rainfall average is less than 240 millimeters per year. Global warming will also have repercussions in terms of imbalance of snow and variation in rainfall patterns. Models predict that this will lead to more stabilized flows till 2050, at which point shrinking glaciers will cause water scarcity for the Indus basin.[25]

In 2010, Pakistan suffered the most severe riverine floods in its history. About 20 million people, or one-tenth of Pakistan's population, were affected, with more than 1,980 deaths and nearly 2,946 injuries reported. About 1.6 million homes were destroyed, and thousands of acres of crops and agricultural lands were damaged, with major soil erosion in some areas. The total damage in the crops, livestock, and fisheries subsectors was estimated at about US$5 billion. The losses were largest for crops, with direct damage to 2.1 million hectares of standing crops—mainly cotton, rice, sugarcane, and vegetables. The environmental damages incurred by the floods amounted to US$11.67 million and comprised heightened environmental health risks and damages to forests, wetlands, and other natural systems. The floods also caused contamination of drinking water, proliferation of disease vectors caused by stagnant water ponds, and accumulation of solid waste. The floods caused a loss of US$49.67 million to the health sector, causing mild to moderate damage to the country's public health infrastructure, including basic health units and dispensaries, which suffered the most damage. The irrigation and flood protection damages incurred by floods were estimated at about US$277.6 million.[26]

> *"Now my only source of livelihood is goats and sheep. I am now landless. Yet I hope I will have enough money one day to recover my field and regrow crops on it."*
>
> **Ishaq Khan, Ghizer, Gilgit Baltistan**
>
> The above statement comes from a farmer who became a herdsman following the loss of his farmland after the 2010 floods. He suffered a loss of US$14,000 in just one hour. Ishaq Khan is one of the 20 million people who lost their livelihoods as a result of the 2010 floods in Pakistan.

The year 2012 also saw major flooding in southern Punjab, northern Sindh, and eastern Balochistan. At least 450 people lost their lives. More than 4.8 million people were affected, with an estimated 636,438 homes damaged or destroyed and an estimated 1,172,045 acres of crops lost. The floods left more than 850,000 people relying on food assistance and more than a million people needing assistance to resume farming activities.[27]

These incidents are flowing in sync with global trends of temperature increases and more-severe weather patterns. According to the Global Climate Risk Index 2015, developed by the German Watch, Pakistan was the country most affected by climate change in 2012, both in terms of human losses and financial losses.[28] The socioeconomic effects of climate change are being felt at the subnational level across several sectors. For instance, in northern Pakistan (Chitral), variations in the frequency and duration of rain are the main reason for prolonged droughts, which threaten the livelihoods of smallholder farmers. In the plains, small-scale farmers (cultivating an area of less than 5 hectares) make up 86% of the total farming community, with around 40% of the farming area. They, along with the rain-fed agriculture of Balochistan, will be most vulnerable to the changing climate patterns, as productivity of staple crops will decrease due to variation in temperature in Punjab. Decreased agricultural outputs, driven by climate change, will further jeopardize the 22% of Pakistan's GDP that is associated with agriculture, which makes up 60% of Pakistani export goods.[29]

Transboundary Issues

In the second half of the 20th century, challenges arose surrounding transboundary rights to the water supply of the Indus, which supplies water to China, India, Afghanistan, and Pakistan. The division of the Indian subcontinent into separate states was a result of British colonialism, and boundaries were drawn based on social rather than geographical realities. The then-British government intended for India and Pakistan to jointly manage irrigation waters between the two states. It was a naïve ask on the part of the British government, leaving Pakistan feeling insecure because all its water heads emanated in India, and it resulted in no compromise between the two nations until 1952, when the World Bank jumped in as a mediator. At that time the World Bank wanted to give loan money to both countries to develop hydropower resources, but funding was not granted due to a 1948 conflict during which India halted the flow of the Sutlej River into Pakistan. A consultative process between the two countries under the World Bank umbrella eventually succeeded in the form of the Indus Basin Water Treaty. Signed in 1960, it allocated three eastern rivers to India (the Ravi, Sutlej, and Beas) and three western rivers to Pakistan (the Jhelum, Chenab, and Indus). The U.S. was instrumental in the treaty's negotiation and in financing the canals and storage facilities that made equitable allocation of Indus water resources possible. It is considered a landmark water treaty between two countries that have historically been at odds politically.

Prevailing public sentiment in Pakistan holds that India is violating the treaty by building reservoirs and is depleting the common groundwater pool by unchecked abstraction in Indian Punjab. Overabstraction poses a high risk to the socioeconomic well-being of Pakistan by jeopardizing its agriculture and irrigation systems. There is also a great deal of criticism of the fact that geopolitical borders between India and Pakistan divide the Indus River Basin, making holistic basin management nearly impossible.

In the present context, the driving force behind Pakistan's mistrust of the treaty originates from the political division of watercourses that favors India geographically as being the upper riparian. The political rhetoric on both sides makes the issue emotionally charged, as India sees it as a territorial sovereignty issue, while Pakistan sees a threat to the security of its food, energy, and economy. Much of the competition between the two nations centers on the race to harness hydropower and enhance water storage capacity. India has surpassed Pakistan in both, thus increasing Pakistan's vulnerability. Estimates suggest that India has tapped around 80% of the hydropower potential on the eastern rivers, whereas Pakistan has been able to harness only 13% of the potential on its allocated western rivers. Pakistan has struggled in this context mainly due to lack of political will and provincial disharmony on the issue of large dams.[30]

Despite the political animosity between the two countries, the Indus Basin Water Treaty has withstood the test of time for half a century. In many ways, it has facilitated both India's territorial sovereignty and Pakistan's right to economic development, and no wars have been waged over water issues. However, the treaty needs to be revisited in light of new realities such as climate change, demographic transition, environmental flows, and aquifer recharge of the floodplains considering the eastern rivers.[31]

Recommendations for Pakistan

Groundwater Management

- In order to ensure a holistic approach to groundwater management at the basin level, a single governing body is needed. Under the 18th Amendment of Pakistan's constitution, which gives provinces more control of resource management, a single provincial authority can be made responsible for holistic groundwater management under the federal patronage.
- A comprehensive groundwater policy should be approved by the Council of Common Interests—made up of the prime minister and the chief ministers of each province—and should clearly differentiate roles relating to aquifer management and to service provision. This plan could then be tailored to meet the idiosyncratic needs of each province.
- Creating a data repository will be an important first step in providing the basis for improved groundwater management. Mapping of hydraulic parameters, recharge zones, confined and unconfined areas, and natural discharge zones will help responsible institutions prioritize their actions.
- Based on this data, the local nature of groundwater abstraction for different purposes (domestic, agriculture, industry, commercial use, etc.) can then be considered in relation to each economic sector's significance. Maximizing Pakistan's water use efficiency through better management practices

Challenges for a Water-Secure Pakistan 231

and sustainable consumption will minimize its dependency on India and lower the stakes for renegotiation of transboundary water treaties.
• In the agriculture sector specifically, the government should promote and support demand-based irrigation and cropping patterns that are suitable for high-delta crops.

Climate Change and Flood Management

> *"My petition aimed to compel the concerned departments and ministries to take action and consider climate change an important issue before it is too late."*
>
> *Asghar Leghari, Rahim Yar Khan*
>
> This farmer from southern Punjab was fed up with unpredictable weather that threatened his crops. He filed a petition against the Pakistani government for violating his rights by failing to take measures to mitigate climate change. In response, Justice Syed Mansoor Ali Shah formed a climate change commission to push forward policies to promote climate resilience.

Pakistan's historically techno-centric approach has largely relied on hard-infrastructure solutions (dams, dykes, etc.) to build flood resilience, as the country is among those with the lowest water storage capacity in the region.[32] With this in mind:

• In planning infrastructure projects, strategic environmental assessments (SEAs) have historically been carried out by the projects' proponents. In order to increase transparency and accountability, SEAs should instead be conducted by third-party evaluators.
• Soft engineering solutions (environmental reserves, buffer zones, etc.) should be considered in conjunction with hard infrastructure for maximum flood resilience.
• In the aftermath of severe floods in 2010, the Climate Change Ministry in Pakistan invited the RAMSAR Advisory Mission to advise the government on future flood-mitigation measures. In its recommendations, the mission asked for an integrated approach to flood management and stressed the use of soft engineering approaches. The ministry cited China as an example of successful implementation; soft engineering enabled the storage of floodwater equivalent to three times the capacity of Three Gorges Dam.[33]

Transboundary Agreements

The shaky peace between Pakistan and India is dependent upon both countries' adherence to the terms of the Indus Basin Water Treaty. Increased cooperation between all nations in the basin is the key to ensuring food security, stabilizing societies, and improving livelihoods in the region. Optimized efficiency is vital if limited water supplies are to withstand increasing demand across borders. In pursuit of this goal, Pakistan should:

- Enlist the help of organizations like the World Bank and U.S. Agency for International Development (USAID) to mediate negotiations between all stakeholders, in order to update existing transboundary agreements.
- Engage all sectors, as well as non-governmental organizations (NGOs) and academics, to ensure that multiple perspectives are considered in updates to existing agreements.[34]
- Promote contextual data sharing on all sides—both across sectors and across borders—to ensure that cooperation is based on current realities and the challenges of climate change.

Recommendations for the U.S. Government

Since Pakistan gained its independence in 1947, its relationship with the U.S. has been through many changes and remains today in a transitional phase. The U.S. has historically been involved in highly relevant changes to Pakistan's water resource management, both through its involvement in brokering the Indus Basin Water Treaty and through the Green Revolution of the 1960s. More generally, however, USAID has focused most of its water efforts on health and sanitation issues, with subsidiary attention paid to agricultural efficiency. Its approach has notably lacked a long-term strategic plan and larger basin-level management for the Indus in the past five years. With this in the mind, the U.S. could plausibly engage in water management in the following ways:

- First and foremost, the U.S. can provide technical support for instituting holistic groundwater management practices. In particular, support in data gathering and monitoring will equip new and existing departments to plan better and come up with a strategic policy on the issue. The Gravity Recovery and Climate Experiment imagery used by the U.S. to highlight the groundwater situation at the boundary of India and Pakistan has already become the basis of concern for the proper management of groundwater aquifers; this awareness could be expanded to provide an overview of the whole country's groundwater situation.
- The U.S. can deploy its experience and research capabilities to help set up a water-pricing and -metering mechanism for various sectors that would oblige water users to pay for the used volume in a market-based atmosphere to manage the operation and maintenance costs of service providers efficiently and to simultaneously ensure the sustainable use of limited water resources.
- While the hydrocracy of Pakistan has historically relied on hard engineering solutions like dams, the U.S. should support the additional

implementation of soft engineering solutions such as improving connectivity of rivers with wetlands and creating environmental reserves as a buffer to flood catastrophes that are increasing in severity as climate change occurs.
- The Kabul River from Afghanistan represents an integral part of the Indus River system. In light of Afghanistan's planned hydropower interventions on it, the U.S. has an opportunity to technically support the two neighbors on reaching an amicable agreement for future water use.
- The U.S. can encourage NGOs, academics, and private-sector actors in Pakistan to inform and support the government in broadening and improving water management. The idea of water stewardship is gaining momentum in Pakistan as a new paradigm to manage water resources by not only examining physical risk but also taking into account regulatory and reputational risks that may surface in the wake of irresponsible water management. Moreover, inclusive governance is a step toward creating a congenial environment and lessens the risk of conflict.

Conclusions: Preparing for Tradeoffs

Water management in the Indus River Basin is growing increasingly complicated as scarcity generates competition for limited water resources. The nation and region will face challenges in trying to meet a broad range of water needs moving forward, and Pakistan's leaders must prepare to make difficult tradeoffs. The nation's simplistic, agriculture-focused water policies can no longer account for the realities of modern water resource management, and compromise will be needed on many scales:

- India and Pakistan must find ways to fairly allocate transboundary fresh water, taking into account increased population and demand in both countries, as well as the ecological challenges posed by a changing climate.
- Within Pakistan, agriculture will increasingly vie with other sectors for limited water resources. More than 90% of Pakistan's surface water is currently allocated for the agriculture sector, which contributes 20%–25% of Pakistan's GDP. The growing industrial and service sectors will demand a greater share of the water.
- Domestic consumers also have an increasing claim to existing water supplies, as they enjoy preferential treatment under the UN mandate declaring domestic water availability to be a right of citizens by law.
- Current demographic trends suggest that 60% of Pakistan's population will be urbanized by 2030, which will increase pressure on already-insufficient water infrastructure and sanitation in urban centers.[35] It will also require a larger percentage of water to be diverted from rural areas.
- Falling water tables indicate that continuing groundwater access is not assured. Metering and regulating the country's groundwater is a question of balancing the water needs of today with those of tomorrow.

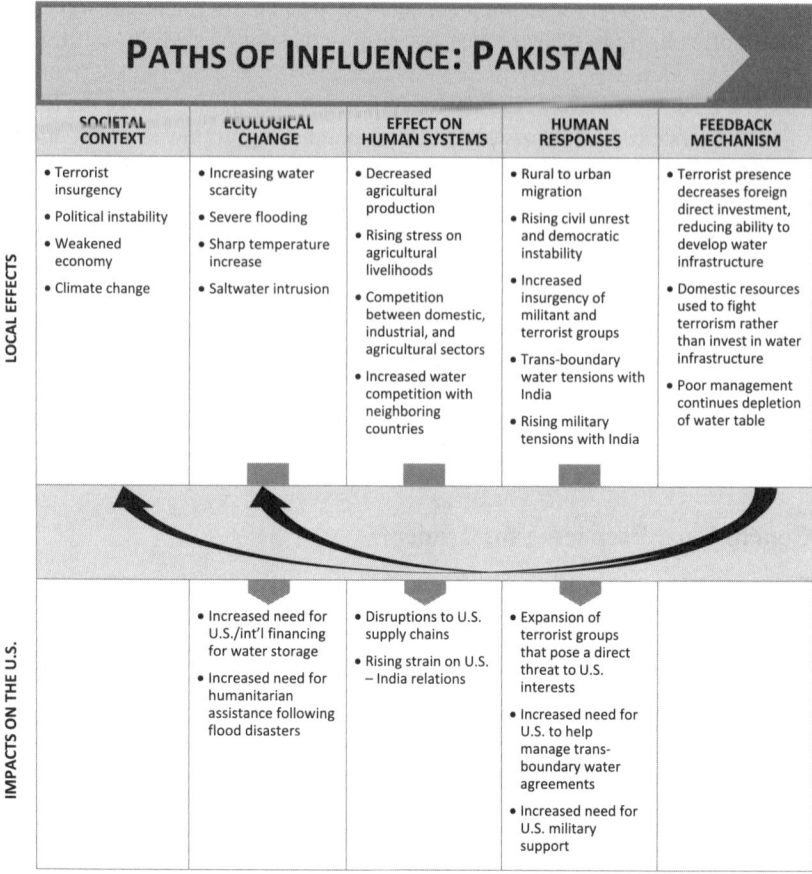

Figure 12.2 Paths of Influence: Pakistan

These tradeoffs may seem difficult initially, but their careful consideration is necessary to ensure that limited water resources are able to meet the diverse needs of all sectors in all the countries of the Indus River Basin for the long term, and to ensure the region's future stability and prosperity.

Notes

1 Kugelman, Michael. "Why the India-Pakistan War Over Water Is So Dangerous." *Foreign Policy*, 2016. http://foreignpolicy.com/2016/09/30/why-the-india-pakistan-war-over-water-is-so-dangerous-indus-waters-treaty/.
2 The Economist Intelligence Unit. 2016. "The Global Food Security Index." http://foodsecurityindex.eiu.com/.
3 International Food Policy Research Institute. 2016. "2016 Global Hunger Index." http://ghi.ifpri.org/countries/PAK/.

Challenges for a Water-Secure Pakistan 235

4 "Pakistan: A Country Snapshot" (The World Bank, 2015), documents.worldbank.org/curated/en/619971467987825539/Pakistan-Country-snapshot.
5 "Cost of War on Terror for Pakistan Economy" (Pakistan Department of Finance, 2011),www.finance.gov.pk/survey/chapter_11/Special%20Section_1.pdf.
6 "Task Force on Climate Change," 2011, http://mocc.gov.pk/gop/index.php.
7 Ali Arshad, "Economic Cost of Terrorism: A Case Study of Pakistan," accessed online on October 26, 2016, http://issi.org.pk/wp-content/uploads/2014/06/12995 69657_66503137.pdf.
8 "International Decade for Action 2005–2015" (UN Department of Economic and Social Affairs, 2015),www.un.org/waterforlifedecade/.
9 "Pakistan May Run Dry by 2025," *The Nation*, June 1, 2016, http://nation.com.pk/business/01-Jun-2016/pakistan-may-run-dry-by-2025.
10 Asia Pacific Water Forum, "Asian Water Development Outlook 2013: Measuring Water Security in Asia and the Pacific," 2013, www.adb.org/sites/default/files/publication/30190/asian-water-development-outlook-2013.pdf.
11 Map generated using tools and database provided publicly by the National Disaster Management Authority's official website, www.ndma.gov.pk/. The following link, provided on the website, was accessed online on September 22, 2016, to generate the map: http://203.124.39.68/webmaps1/. The base map used by NDMA is provided by the Environmental Systems Research Institute.
12 "World Population Prospects" (UN Population Division, 2015).
13 Daanish Mustafa, "Theory Versus Practice: The Bureaucratic Ethos of Water Resource Management and Administration in Pakistan," *Contemporary South Asia* 11, no. 1 (2002): 39–56.
14 "Pakistan: A Country Snapshot."
15 Rabia Aslam, "Pakistan's Water Vulnerability and the Risk of Inter-State Conflict in South Asia," *Forman Journal of Economic Studies* 9 (2013): 19–41.
16 Asad Sarwar Qureshi, Mushtaq A. Gill, and Asrar Sarwar, "Sustainable Groundwater Management in Pakistan: Challenges and Opportunities," *Irrigation and Drainage* 59, no. 2 (2010): 107–116.
17 LEAD Pakistan, Policy Brief, "Groundwater Challenges, Opportunities and Governance in Pakistan's Indus Basin," March 2016, www.lead.org.pk/lead/Publications/Groundwater%20challenges,%20opportunities%20and%20governance%20in%20Pakistan%E2%80%99s%20Indus%20basin.pdf
18 Ijaz Hussain, Z. Hussain, M. H. Sial, W. Akram, and M. F. Farhan, "Water Balance, Supply and Demand and Irrigation Efficiency of Indus Basin," *Pakistan Economic and Social Review* (2011): 13–38.
19 Munir A. Hanjra and M. Ejaz Qureshi, "Global Water Crisis and Future Food Security in an Era of Climate Change," *Food Policy* 35, no. 5 (2010): 365–377.
20 Dr. Asad Sarwar Qureshi and Ali Hasnain Sayed, "Situation Analysis of the Water Resources of Lahore": Establishing a Case for Water Stewardship (WWF Pakistan, 2014).
21 M. A. Kahlown, A. Majeed, and M. A. Tahir, "Water Quality Status in Pakistan" (PCRWR, Ministry of Science & Technology, Government of Pakistan, 2002).
22 Ibid.
23 Ibid.
24 Nadir Jailani, "Drought Threatening Livelihood in 29 Districts of Balochistan," http://thebalochistanpoint.com/drought-threatening-livelihood-in-29-districts-of-balochistan/
25 A. B. Shrestha, N. K. Agrawal, B. Alfthan, S. R. Bajracharya, J. Maréchal, and B. van Oort eds., "The Himalayan Climate and Water Atlas: Impact of Climate Change on Water Resources in Five of Asia's Major River Basins" (ICIMOD, GRID-Arendal and CICERO, 2015). The models are climate and precipitation time series analysis seen with Representative Concentration Pathways (RCPs).

26 "Pakistan Floods 2010: Preliminary Damage and Needs Assessment," Flood Emergency Reconstruction Project, 2010, www.adb.org/sites/default/files/linked-documents/44372–01-pak-oth-02.pdf.
27 Shaheen Chughtai, "Pakistan 2012 Monsoon Floods: A Prolonged Disaster," 2013.
28 S. Kreft, D. Eckstein, L. Junghans, C. Kerestan, and U. Hagen, "Briefing Paper: Who Suffers Most From Extreme Weather Events? Weather-Related Loss Events In 2013 and 1994 to 2013," *Global Climate Risk Index*, 2015.
29 "Task Force on Climate Change."
30 Asad Sarwar Qureshi, P. G. McCornick, A. Sarwar, and B. R. Sharma, "Challenges and Prospects of Sustainable Groundwater Management in the Indus Basin, Pakistan," *Water Resources Management* 24, no. 8 (2010): 1551–1569.
31 Gareth Price, R. Alam, S. Hasan, F. Humayun, M. H. Kabir, C. S. Karki, S. Mittra, T. Saad, M. Saleem, S. Saran, and S. Shakya, "Attitudes to Water in South Asia," *Chatham House Report*, 2014.
32 John Briscoe and Usman Qamar, *Pakistan's Water Economy: Running Dry* (Oxford University Press World Bank document, 2006).
33 "RAMSAR Advisory Mission Report on Pakistan Floods," 2010.
34 N. Mirumachi, *Transboundary Water Politics in the Developing World* (Abingdon: Routledge, 2015).
35 "Government of Pakistan's Agriculture and Water Policies with Respect to Climate Change: Policy Gap Analysis," 2009, https://cmsdata.iucn.org/downloads/policy_gap_analysis_report.pdf.

13 Water Scarcity and Regional Security in India

Cecilia Tortajada, Udisha Saklani, and Asit K. Biswas

Introduction

India's water management has been on an unsustainable path for centuries. In the 16th century, Akbar the Great, the celebrated Mughal emperor, decided to build a new capital for his vast empire. The best architects of the realm were invited to design a magnificent palace in Fatehpur Sikri (City of Victory) in the dry plains of northern India. The cream of the crop of Indian artisans worked for several years to complete the capital, and vast amounts of resources were spent realizing the emperor's dream.

In 1589, shortly after the completion of the new capital, Robert Fitch, a gentleman merchant and one of the earliest English travelers to India, noted that Agra and Fatehpur Sikri were "two great cities, either of them much greater than London and more populous."[1]

As any modern-day traveler to Fatehpur Sikri will attest, it is an excellent testimonial to the Indian architects and artisans. Not surprisingly, it is now a UNESCO World Heritage site. However, the history of the new capital was not so auspicious. Akbar used it for only 13 years, 1572–1585, and then abandoned it rather ignominiously to return to his old capital permanently. The main reason for the abandonment of the new capital was serious water scarcity.

Fatehpur Sikri is a magnificent monument to bad and unsustainable water management. Some 440 years later, water management in India has improved only marginally compared with rapidly increasing requirements, because of a rising population, rapid urbanization and industrialization, steadily increasing aspirations of the people for better quality of life, and information and communication revolutions in the country that have made its citizens aware of what may be possible. All these developments, along with poor water planning for centuries, have ensured that India would face a perfect storm as it tries to provide a reliable supply of water to a burgeoning population and satisfy its escalating domestic, agricultural, industrial, thermonuclear, and environmental needs.

Population and Water

As India's population has increased, so have its water requirements. Before India was partitioned in 1947 into two countries, its population was around

390 million: 330 million in India and another 60 million in what later became Pakistan. By 2016, the population of India had increased to 1.33 billion, with another 193 million in Pakistan and 163 million in Bangladesh (Pakistan was divided into Pakistan and Bangladesh in 1971). By 2050, India is expected to have some 1.7 billion people, Pakistan 344 million, and Bangladesh another 202 million.

Compared with the total population of 390 million in 1947 undivided India, by 2016, these three countries have a combined population of 1.686 billion. By 2050, the total population of these countries is estimated to reach 2.206 billion. Also by 2050, India is expected to be the most populous country in the world, overtaking China by around 2022. The three countries that were part of India in 1947 are expected to become three of the 10 most populous countries of the world by 2050, with India at number one, Pakistan at number six, and Bangladesh at number eight.

Increases in population, in the absence of serious and sustained efforts to improve water use efficiencies, have become a major driver of water requirements in the country. India's neighbors—Pakistan, Nepal, and Bangladesh—are facing similar problems. This is an important consideration because many of the major rivers of the subcontinent are shared by two or more countries.

The net result has been steady increases in water requirements all over India as well as in Pakistan, Bangladesh, Nepal, and Bhutan. Another neighbor, China, with the legacy of its strict one-child policy from 1979–2015, does not have similar population growth issues, but it is facing different types of water-related problems because of nearly 40 years of explosive economic growth and continued poor water quality management.

Population growth, in the absence of proper domestic and industrial wastewater treatment, has created a serious water quality problem. Improperly treated wastewater and indiscriminate discharges of untreated or partially treated wastewater have contaminated surface and groundwater bodies within and near Indian population centers with known and unknown pollutants from domestic, industrial, and agricultural sources. This has had major implications in terms of human and ecosystem health.

Urbanization, Economic Growth, and Water

In addition to population, two other factors—urbanization and economic growth—are also important drivers of increased water requirements.

With steady economic growth, higher literacy rates, and increasing skill levels, the number of middle-class families in India has been increasing steadily. In 1985, less than 10% of the population constituted the middle class. It is now estimated that by 2030 the median income of Indian households is likely to increase by 90% to reach over $10,000, in 2014 prices. The number of middle-class families will increase from about 70 million to more than 90 million. India's middle class will still be much smaller than that of China, which has witnessed phenomenal growth over the past four decades. Even by 2030, when

India will be the world's most populous country, its number of middle-class households will be less than China's in 2014. This is assuming India's growth rate will continue to be high, around 7%, in the foreseeable future.

The biggest impacts in terms of higher water requirements have been due to changes in the dietary habits of the increasing number of affluent and educated consumers who are consuming more animal protein through meat and fish or milk products. They are also consuming more oil and sugar. All these products take tremendous quantities of water to produce. Thus, as Indians become more affluent, their direct and indirect water requirements increase as well. Historically, determined attempts have not been made to improve water use efficiencies. Even now, with a major water crisis looming in the near future, there are no signs that real improvements in water use efficiencies are likely to happen.

India is facing another major problem: As the water requirements of the domestic and industrial sectors go up, the quantities of wastewater generated increase as well. As Indians manufacture more and more sophisticated products, the wastewater generated contains increasingly complex chemical substances that are expensive and difficult to treat. Domestic and industrial wastewaters are point sources of pollution. These can be treated cost effectively when there is political will, public demand, and good legal and regulatory systems that can be enforced. Unfortunately, extensive and pervasive corruption across the entire country has ensured that enforcement seldom takes place.

Nonpoint sources of pollution due to increasing agricultural and livestock activities means that water bodies are being further contaminated by increasing use of agrochemicals like fertilizers and pesticides, as well as by animal waste. For a developing country like India, there has been no real attempt to efficiently manage nonpoint sources of pollution.

India is facing a perfect storm in terms of managing water. First, water requirements are increasing steadily. This has meant that in years of moderately low rainfall there is not enough water in rivers and lakes to meet human demands. The river waters are already mostly overallocated.

Second, the problem is compounded by the fact that even in normal-rainfall years, because of past policies and politics groundwater use is increasing steadily. The situation has worsened since the Green Revolution. Farmers do not pay for the electricity needed to pump water. Consequently, India now uses more groundwater than the United States and China combined. The net result has been that groundwater extraction has reached critical and overexploited levels in many states such as Andhra Pradesh, Gujarat, Karnataka, Maharashtra, Madhya Pradesh, Punjab, Rajasthan, and Telangana. In many places, groundwater levels are falling by more than one meter each year. This is contributing to land subsidence and many other environmental hazards.

Third, as demands have increased, more and more water can no longer be used due to increasing pollution.

Fourth, problems are further magnified by India's federal structure. Under the Indian Constitution, water is primarily under the purview of the states. The Central Government has very limited power to regulate how water is managed in the states. Since all the major Indian rivers are interstate in nature, continuing conflicts on allocation of waters in most rivers have become a serious challenge to regional stability in the country.

Interstate River Conflicts and Regional Security

The latest series of interstate river conflicts has triggered numerous protests, violence, and property destruction in many Indian states over existing water allocation decisions. With consistently poor water policies and inadequate management practices in all sectors and provinces, continued improper governing practices, and climate change threatening to alter water access and distribution over the entire country, river water allocation disputes may prove to be one of the biggest political constraints to India's future economic growth and social cohesion. In light of such developments, several procedural as well as political obstacles preventing efficient, equitable, and timely dispute resolution need to be examined to gain a better understanding of the issues. More efficient, timely, and cost-effective processes that would be acceptable to all the states need to be formulated. This will prove to be a herculean task.

One of the most important challenges in confronting interstate river conflicts is the absence of permanent and efficient dispute resolution mechanisms. With the aim to facilitate speedy and decisive decision-making, the Inter-State Water Disputes (ISWD) Act was passed by Parliament in 1956. This allowed the setting up of ad hoc tribunals on a case-by-case basis whenever water conflicts between two or more states could not be solved by mutual discussion.

Initially, the idea of tribunals in India was to allow states an opportunity to discuss the conflicts and resolve them, before engaging in adjudication as provided under Article 262 and the ISWD Act. Tribunals have often contributed to long, drawn-out negotiation processes that have often led to the hardening of the positions of the individual states. Tribunals have also accentuated rivalries between states to try to receive higher allocations of river waters. Unfortunately, contrary to initial expectations, tribunals have failed to resolve river disputes despite several attempts over the years to give teeth to their functioning through successive amendments to the ISWD Act. Recently, Prime Minister Narendra Modi called tribunals a "barrier" to just allocation because they are often ambiguous and the process continues to remain opaque and often inconsistent.

There are several problems with the current tribunal system that prevent a long-term and sustainable solution. First, in the absence of a uniform, logical, and common process, considerable discretion has been left to the various tribunals in terms of the processes adopted and the underlying concepts on the basis of which the final awards were made. Sometimes the tribunals have acted contrary to international practices or earlier verdicts given by other interstate water tribunals on the same river and to the same parties. This has left the disputed parties dissatisfied and suspicious of the impartiality of the final awards and their rationale and fairness.

For example, in the case of the Ravi-Beas interstate conflict, Rajasthan was allocated an unjustifiably large share of water despite being a non-riparian party in the dispute. The rationale and justification for this decision were never properly explained. A similar principle was followed by the Narmada River Tribunal that was constituted prior to the Ravi-Beas Tribunal. Under this award, Rajasthan was granted some water from the Sardar Sarovar Dam because of "national interest" despite being a non-riparian party to the conflict.[2] In this case, Rajasthan's share was considerably smaller in comparison with the two major riparian parties, Gujarat and Madhya Pradesh. In sharp contrast, in the Ravi-Beas case, Rajasthan was granted 8 million acre feet (MAF)[3], which was even greater than the 7.2 MAF granted to an undivided Punjab. This later had to be shared with Haryana and New Delhi, resulting in further diminished water availability for Punjab, despite its being the most significant riparian party in the dispute. The logic behind this award was never properly explained.

The tribunal rejected Punjab's argument, which would have used the doctrine of riparian rights and the theory of ownership rights of a state in river waters to prevent Rajasthan, which is not within the basin, and Haryana, which lost its riparian character due to division from Punjab, from gaining access to Ravi-Beas waters by terming them "non-riparian" states.[4] Instead, the tribunal upheld the view that the "doctrine of riparian rights" and the "theory of proprietary rights of a State" apply to private parties and are not applicable in interstate water disputes in India.[5] This judgment goes against the conventional international practice of following riparian rights for water division, which originated in English common law. It is followed by many countries, such as Canada and Australia, and by eastern states in the U.S. Not surprisingly, Punjab has questioned the legality and fairness of the award and has remained hostile to the arrangement from the very beginning.

Second, arbitration results are nonbinding to the states involved. Tribunals are constituted on a case-by-case basis and are not backed by any statutory legitimacy. Consequently, and not surprisingly, state governments have often refused to comply with the final award. In the Ravi-Beas dispute, the Punjab state government passed legislation in 2004 aimed at neutralizing two Supreme Court judgments of 2002 and 2004 requiring Punjab to construct the Sutlej-Yamuna Link (SYL) canal on its territory to share water with other parties, in line with the tribunal order. This is an example of willful opposition to the spirit of federalism.

Third, there has been a clear reluctance on the part of the Central Government to establish institutions for the implementation of the tribunal awards. This has created an institutional vacuum, which has significantly contributed to the failure of the dispute settlement processes.

Fourth, there is no fixed stipulated time frame for negotiation and adjudication. The Cauvery Tribunal took 17 years to declare its final award. This delay was further aggravated by Karnataka's decision to file a Special Leave Petition at the Supreme Court to thwart the final order. By refusing to submit to the tribunal's decision to validate the 1924 agreement and by claiming historic injustice, Karnataka has gained additional time, which it used to finish new irrigation works that would ultimately diminish the water supply that could be made available to Tamil Nadu.[6]

Numerous suggestions have been made to provide for a permanent tribunal.[7] The first such recommendation was mooted in 2011 by the then-law minister M. Veerappa Moily, a decision that received support from the current ruling government. Giving statutory backing to a permanent committee instead of constituting ad hoc and temporary tribunals could help remove discrepancies, increase accountability and transparency, and maintain consistency in finding definitive and acceptable solutions to conflicts over river water allocations.

The 2012 National Water Policy draft included such a policy[8] as part of its recommendations, which were given to the United Progressive Alliance government in 2013. More recently, the Modi government has discussed the possibility of constituting a tribunal that would replace all other adjudication mechanisms in the country, including those related to interstate water conflicts, in line with its "minimum government, maximum governance" philosophy. It should, however, be noted that the mere constitution of a permanent tribunal is unlikely to create substantial changes and improvements if procedural aspects of dispute resolution are not appropriately modified. There is no reason to believe that state governments would abide by a permanent tribunal's decisions when there are no constitutional measures in place to ensure cooperation by the disputed parties through clear, transparent penalties in case of noncompliance.

Additionally, it is imperative for the Central Government to implement the tribunal awards promptly, in part by establishing the required infrastructure that would facilitate their implementation. Political calculations, especially in the face of upcoming elections at both the Central and state levels, often prevent a genuine attempt from the Central Government to mediate appropriate settlements between the states, as seen in the Ravi-Beas and Cauvery conflicts. Complex issues are left to the judiciary as political parties shrink from making difficult decisions that may displease one or more state governments involved in the disputes. Consequently, states continue to dominate water-sharing negotiations and use their political power to lobby for their individual interests. Such political calculations and/or bargaining have wide influence on the final awards given by the tribunals. This is because decisions on water allocation are based primarily on the arguments put forward by the states.

As noted earlier, the Indian Constitution has a provision for the formulation of temporary and ad hoc tribunals because adjudication was envisioned as the last-resort option in handling various river disputes. Discretionary measures involving mutually negotiated agreements were seen as a potential solution for resolving most conflicts pertaining to interstate water sharing.[9] States were expected to engage in constructive discussions, and only after they failed to resolve water conflicts through mutual discussions was the route of adjudication offered. An absence of formal institutional arrangements for facilitating such interstate negotiations has resulted in state governments resorting to the tribunal option as their first preference.

The Inter-State Council (ISC) came into being in 1990 as a constitutional body with the sole purpose of serving as a facilitating platform to discuss and resolve interstate disputes within reasonable time frames. The process has remained neglected and the constitutional body has fallen into disuse. Despite the current ruling government's emphasis on "cooperative federalism," the ISC met in 2016 for the first time since the new government came to power in May 2014. This was after a long gap of 10 years since the previous ISC session was held in December 2006. Such reluctance to strengthen a potential platform that can act as a binding mechanism between the Central Government, the states, and the union territories for sharing common policies and facilitating interstate cooperation and dispute resolution has resulted in a lost opportunity. This may prove to be a heavy burden on India's national integrity, as well as on its further social and economic development.

While weak and inefficient legal and institutional mechanisms have contributed to the rising number of interstate river disputes during the past decade, current policies of the present government may further exacerbate challenges in resource sharing through unintended consequences.

The philosophy of "competitive federalism" is aimed at achieving efficiency and inclusive growth in the country. Inspired by the U.S. model targeted at improving performance at the subnational level, the Modi government has promoted several good governance initiatives in a spirit of cooperative and competitive federalism. These keep national objectives in mind while allowing states to compete with each other for investment opportunities. For this reason, funding patterns have also changed, with states getting greater autonomy in designing their development programs.[10] While this has most certainly expanded the political space to plan and implement state development initiatives without any, or much, interference from the Central Government, it has also increased the accountability of the state governments. States find it more difficult to lay the blame on the union government for their own poor performance. Additionally, in centrally financed schemes such as Smart Cities or the Rashtriya Krishi Vikas Yojana (RKVY) program,[11] states are encouraged to compete with each other for funding by showcasing their strengths and enforcing rapid reforms aimed at reducing administrative and regulatory inefficiencies, increasing public investments in infrastructure, and controlling corruption to attract private investments.

The potential for such policies to further aggravate conflicts over fair and speedy allocation of interstate river waters needs to be assessed. As states become more aware of the implications of interstate competition for investments, they are bound to consider reliable availability of resources to ensure that their self-interest is properly guarded and that they do not fall behind in the race with other states. Most industries are likely to locate in regions where availability of infrastructure and resources is assured on a long-term, reliable basis. For instance, reliable availability of water, electricity, and land at reasonable prices, and flexible labor laws, may become significant factors that are essential for attracting new projects and thus generating employment and contributing to regional economic development. States are now in a rush to improve their competitiveness because they are being compared in terms of "ease of doing business" rankings calculated at the national level by the World Bank and the Department of Industrial Policy Promotion under the Ministry of Commerce and Industry.[12] Hence, the policy of competitive federalism has huge political and economic significance for state governments that are constantly eyeing the next election.

India has witnessed intersectoral competition for water for many decades, owing to its poor water management strategies in the face of rapidly rising water requirements. In the absence of significant water sector reforms at the Central and state levels, competition for the shares of river waters will increase significantly in the coming decades. The new concept of federalism promoted by the current government is built on values of cooperation as well as competition. An unbalanced competitive-cooperative dynamic, mired in historical animosity, linguistic rivalries, and unresolved issues between neighboring states, is bound to prevent voluntary compromises from taking place even in years of normal rainfall, let alone during droughts. This can be seen in the Cauvery River dispute, in which Tamil Nadu has steadfastly refused to sacrifice even 20 or 30 thousand million cubic meters (TMC)[13] to Karnataka during droughts, in the absence of an acceptable water-sharing formula during distress periods. Tamil Nadu is aware that this voluntary gesture would go a long way in exacerbating the crisis.[14]

In contrast to disputes with other countries, interstate water disputes have a special dimension that makes their resolution far more complex: state languages. Languages have played very important roles in shaping the political, social, and economic destiny of India. India was reorganized postindependence as groups demanded separate states based on language. Despite clear reservations by the Constituent Assembly of India regarding the possibility of such political division stoking regionalism and linguistic chauvinism, the demand was accepted.

Given the high relevance of linguistic ethnicity that still prevails, in the aftermath of violence in Bengaluru over the Cauvery dispute, it appears that social media has played pervasive and particularly harmful roles in encouraging aggression toward "the other" linguistic groups. Residents of Tamil Nadu and Karnataka were seen engaging in verbal spats and posting violent videos and hate speeches on Twitter and Facebook. Provincial newspapers seem to be far more chauvinistic than their national counterparts; they emphasize what divides them linguistically rather than promoting just, equitable, and acceptable solutions. Online communities are increasingly playing major roles in shaping conflict discourses, which is bound to add to the existing tensions[15] and also delay the acceptance of rational water allocation awards.

Conventional news platforms have further added to these woes. A video clip showing the beating of a Tamil boy in Bengaluru by a pro-Kannada fringe group for mocking protests by Kannada actors on the Cauvery issue was picked up by Tamil television channels and played multiple times. Almost immediately, Kannadigas whose families had lived in Chennai for several generations faced retaliatory attacks by Tamilians.[16] Media insensitivity to the impact of telecasts contributes to the dissemination of aggressive, violent, and sometimes even false content that inevitably fuels riots within highly charged, ill-informed, and aggrieved communities. Civil society initiatives to break interstate political logjams fueled by politicians and extremist groups are also conspicuous by their absence. Few scholars and intellectuals have come forward to engage in mediation processes or participate constructively in holding dialogues to prevent misinformation and false allegations from spreading across a region like wildfire, as has been the case recently. It seems as if the intelligentsia and civil society stand no chance to dampen the political frenzy and linguistically divisive propaganda promoted by vested institutions and individuals.[17]

One of the unintended consequences of several rounds of carving out new states from existing ones has been that there have been dramatic increases in river water conflicts. In 1956, 14 states and six union territories constituted the entire country. In less than seven decades, the number has more than doubled, with India now composed of 29 states and seven union territories, most of them divided on the basis of language. The nature of demand for statehood has changed considerably since the beginning of the 21st century. It is worth noting that the four most recent cases of statehood—granted to Uttarakhand, Chhattisgarh, Jharkhand, and Telangana—were based not on language or ethnic differences but on the political apathy of mother states over resource sharing,[18] water being the most contentious issue for Telangana.

While some political analysts consider the creation of new economically and administratively viable small states an inevitable future reality for India, the process itself is driven by desperation and anger, instead of by peaceful negotiations between and within states. This phenomenon threatens India's national integrity and exposes serious shortcomings in the existing dispute resolution mechanisms. These are being increasingly burdened by a large number of water-related conflicts between existing parties. They simply do not have the capacity to handle the additional complexities of new parties and their claims. For instance, if the proposed plan to divide the northern state of Uttar Pradesh into three or four smaller states becomes a reality, the challenges that future tribunals may have to face to allocate water from the rivers interspersing the Gangetic plains will multiply. Allocating water would be a major exercise involving multiple rounds of political bargaining and renegotiations of current water-sharing arrangements between existing riparian states.

Ongoing statehood agitations based on resource conflicts have a tendency to destabilize the country by invoking strong feelings of deprivation and anger among minority groups, further strengthening regional identities. This could conceivably contribute to the balkanization of the country. This issue is important in an Indian context where demands for the creation of new states have significant potential to further aggravate and complicate existing regional conflicts over the sharing of natural resources, particularly in interstate rivers. This is especially true for water, which is essential for human survival, economic development, and poverty alleviation, and has strong emotional bonds to human beings, more so than any other resource.

Another issue linking water disputes to state politics is the power of destructive state campaigns to distract voters from real issues of poor governance and lack of administrative skills and action. Water has assumed the role of a political weapon. This is particularly true in the case of the Ravi-Beas conflict, in which both Punjab and Haryana have been on the receiving end of public wrath on account of consistently poor administration and continued nonperformance. Punjab, which was once the symbol of India's progress because of the Green Revolution, now has a dwindling economy as a result of years of militancy (including the demand for a new state based on religion), corruption, and distorted economic incentives in water, energy, and fertilizer use. These have contributed significantly to a severe crisis of groundwater that is likely to affect agricultural production.

On one hand, Punjab's focus has largely remained confined to its fight for water resources with its Hindi-speaking neighboring state of Haryana, which is also its immediate political and economic competitor. Haryana, on the other hand, has grown, especially in terms of cities such as Gurugram, Faridabad, and Karnal, which have witnessed rapid development in recent years, even though the state as a whole has been witnessing a steady decline in economic growth rates.[19] Haryana has been a victim of major law-and-order situations, poor social indicators (especially the child sex ratio and crimes against women), financial mismanagement, and more recently the complete failure of the state administration to manage political crises fueled by the Jat reservation demand for fixed employment quotas.[20] Hence, fighting for water rights often serves as a successful way of diverting the growing frustration and desperation of agricultural communities in the two states, at least over the short and medium terms, to an external issue.

Regional political parties have used river water allocation conflicts to position themselves as the greatest guardians of state rights and interests. This holds great significance, especially for a democratic country like India, where the continuity of political regimes is uncertain. Hence, intrastate politics mostly reduce the chance of arriving at a mutually acceptable solution with neighboring states. When Andhra Pradesh lost out on its share of surplus water in the Krishna River due to the Krishna Water Disputes Tribunal (KWDT)-II award,[21] opposition parties in the state such as the Telugu Desam Party, the Communist Party of India, and others immediately laid the blame on the Indian National Congress, which ruled the state and Center at that time, for failing to bargain for a better deal for Andhra Pradesh.[22] Such posturing, opportunism, and short-termism shown by political groups push state governments to take an extreme stand to counter allegations that might cause the ruling political party to lose power in the next election. This is because water has been historically, and continues to be today, a very emotional subject in India.

In a similar case, Karnataka refused to allow any independent party to determine how much water it must "release" or "spare" to the lower riparian state, Tamil Nadu. This means, in reality, that it completely violated its upper riparian responsibility, whether examined under the Helsinki rules, which stress equitable sharing for beneficial use, or the 1997 UN Convention, which calls for equitable utilization of the waters.[23] This situation is further complicated by the tendency of the Central Government to take political stands depending on the party in power in either state. Such cases of political collusion are most visible in the Ravi-Beas conflict, in which the union government and the Haryana state government are led by the same political party. These dynamics have probably played a role in triggering a much more definitive and aggressive stand from the Central Government against Punjab's constitutional violations, in comparison with the other ongoing interstate river disputes, in which the Center has chosen to remain at arm's length. Hence, the motivations driving water conflict resolution in India are most often aimed at immediate electoral gains and petty politics, which compromises the long-term interest of the country as a whole.

A recent political strategy used by many states engaged in legal battles over water sharing is depicted in the Krishna dispute, which points to a larger issue at stake. In 2005, a few years before the KWDT-II tribunal was going to release its award revising its earlier verdict, Andhra Pradesh accelerated the development of its 16 irrigation projects with the objective to turn its temporary "liberty" to use the surplus waters of the Krishna River granted by the earlier tribunal into a permanent "right."[24] Project Jalayagnam was launched by the state government based on an election promise to double the irrigated area in the state. As part of the project, up to R46,000 crore (US$6.734 billion) was spent from the state exchequer, including expenditure on irrigation projects that relied on the surplus waters of the Krishna and had no clearances from the Central Government.[25] The Andhra Pradesh government went ahead with its plans despite being aware of the objection raised by Karnataka for more than a decade to restrain the upper riparian state from executing irrigation projects that were dependent on the utilization of surplus waters of the Krishna River. This was simply a political maneuver by the state to mount pressure on the second tribunal, hoping for a verdict that would permanently settle the surplus water issue in its favor.

When the KWDT-II award was declared, Andhra Pradesh's exclusive liberty of using surplus water was taken away and it was compelled to share the water with the other riparian parties. This created political turmoil, with the state government blaming the tribunal and the Central Government agencies for endangering its development plans and robbing its farmers of their irrigation facilities.

A similar yet much more controversial strategy was employed by the Punjab government when it passed the Punjab Termination of Waters Agreement legislation in 2004, and more recently in 2016, the Punjab SYL Canal (Rehabilitation and Re-vesting of Proprietary Rights) Bill in response to growing pressure from the Center and the judiciary to complete the SYL Canal construction. Both cases indicate the extent to which state governments are willing to fight for their share of water, resorting to unfair means including, but not limited to, the brazen violation of the constitution upon which India was established as a nation.

At present, with a number of state governments defying orders of the tribunals as well as of the Supreme Court, water has assumed the position of an important threat to India's federalism and its future economic and social development. As noted by a former prime minister of India, Manmohan Singh, in many ways it is easier for India today to negotiate water-sharing agreements with its international neighbors than to manage its domestic water disputes.[26] The conflicts have taken a violent shape in many parts of the country, and the Central and state governments have failed the constitutional and legal provisions that were put in place for their abatement and/or solution. Legal experts have strongly recommended that Parliament take up the jurisdiction and management of all major interstate rivers in the country, regardless of the state boundaries that they traverse.[27] Unfortunately, in the present political scenario, the Union Government is often highly dependent on the support of the regional political parties to remain in power. This has meant that the Center's dominance and its potential as an honest broker to resolve interstate water disputes have been eroded steadily and considerably over the past several decades. This has been replaced by the increasing role of state-level politics, which is likely to make future resolution of interstater river water allocations more complex and difficult to reach. In addition, the solutions reached are likely to be suboptimal to all the parties, and thus in all probability will prove to be temporary in nature.

Implications for U.S. Foreign Policy

During the past five decades, U.S. policies toward India have often been reactive and inconsistent, and have sometimes shifted dramatically. For example, in 1971, the U.S. sent the aircraft carrier *Kitty Hawk* to the Bay of Bengal, ostensibly to show its support of and friendship to Pakistan. A generation later, in 2007, the same aircraft carrier participated in joint military exercises with India's *INS Vibrant* as a show of good relations and friendship.

Similarly, when Narendra Modi was the chief minister of Gujarat, the U.S. government would not even give him a visa. As soon as it appeared that he was likely to be elected the prime minister of India, American policy changed dramatically. Modi, for his part, during his address to the U.S. Congress used the words "partner" or "partnership" 15 times. The official joint communique noted that India is a "major defense partner." Thus, within a single generation the U.S.-India relationship has changed from "cool" to "strategic."

This, of course, is to be expected. In politics, countries really do not have long-term "friends." Their policies are primarily based on short- to medium-term visions and interests. When countries' interests converge, they are mostly friendly to each other. The reverse is also true. Also, countries can be concurrently friends, competitors, and even enemies depending on specific interest areas and their importance to the countries concerned. The U.S.-India relationship over the past five decades has gone through many ups and downs. Our view is that the future will not be any different.

While the overall direction of relationships is likely to be more of the same, it will probably be more complex and uncertain in the coming decades. This is because the pace of change in all spheres will accelerate, and because of new uncertainties due to political, economic, environmental, and technological changes. If India's gross domestic product grows at around 6% per annum over the next 10–15 years, there is no question that the country will be more assertive in the future. This will be reflected in its relationship with its neighbors and with other major powers like the U.S. or China.

For India, at least over the medium term, power is likely to become more diffused within the states, with non-state actors garnering harder and softer power. This is likely to make the Central and state governments increasingly unable, or unwilling, to manage or control internal debates and conflicts, especially for highly emotionally charged issues like allocation of interstate waters. The power of regional, language-based political parties, in all likelihood, would create a policy vacuum and paralysis in terms of action.

In the absence of a logical and long-term framework for water governance, and ever-increasing water demands on states facing accelerating water scarcities, our expectations are that these factors will lead to more hesitancy by the Central and state governments to make essential but hard political decisions. The issues are going to be further complicated by the increasing number of parties with vested interests. This could lead to delays in making decisions or even result in no action for a prolonged period of time.

There is also an increasing belief that India is for the Indians and Asia is for the Asians. This will probably lead to reduced chances for external governments like that of the U.S. to play any constructive and meaningful role in India. Based on current and future trends, the problems are likely to be defined within India, and the process to solve and manage them will become exclusively Indian because of social and political pressures. Thus, it is unlikely that the U.S. can play a meaningful role in issues that are under the jurisdiction of the Indian states, such as water, except in general terms, such as capacity building.

Concluding Remarks

Water management in India is now at a crossroads. The country can continue with its centuries-old strategy of meeting its burgeoning water requirements simply by increasing supplies. However, expanding supplies is increasingly becoming expensive, and the projects' environmental and social costs may not be acceptable. This is because most of the good and economical sites have already been developed.

With a large population that is expected to become even larger in the coming decades, the country has no choice but to seriously look at demand management through economic instruments such as water pricing and water rights, education, public awareness, and increasing water use efficiency practices by all available means. Water conservation and continually increasing efficiencies of water uses in all sectors must now become important items on the country's agenda.

Nowhere is demand management more essential than in water allocation to the different states on all the interstate rivers. Already such rivers do not have enough water even in moderate-drought years. Yet, on not even one interstate river has demand management been high up on the political agenda of the states. Unless demand management is considered very seriously in the immediate future, India will face serious water crises that no previous generation has witnessed. This will lead to economic hardships that could precipitate political crises, social unrest, and regional instability in the coming years.

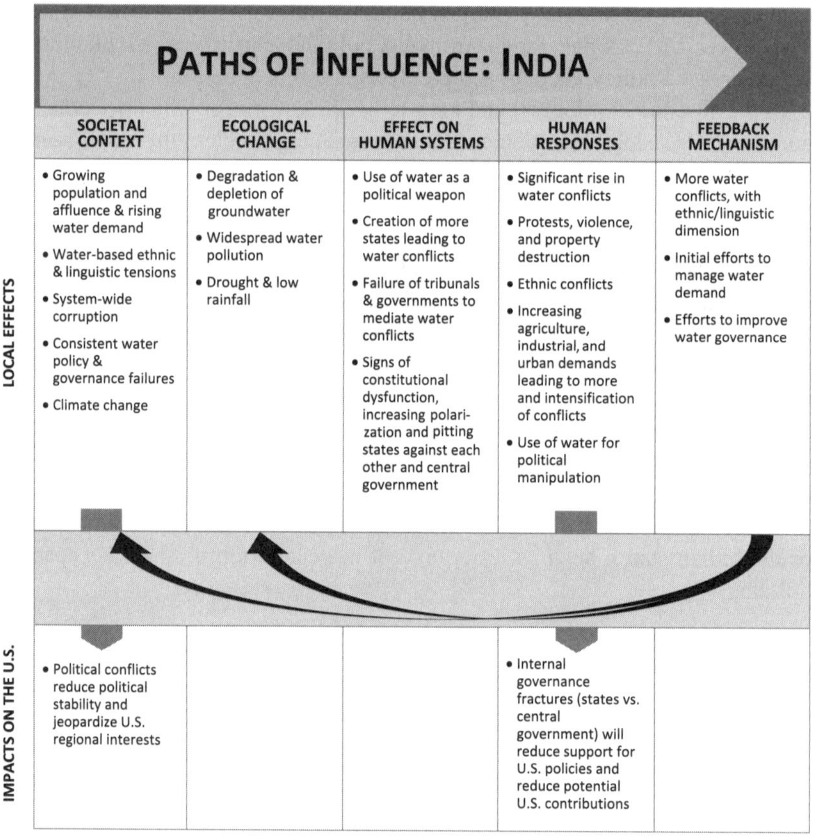

PATHS OF INFLUENCE: INDIA

	SOCIETAL CONTEXT	ECOLOGICAL CHANGE	EFFECT ON HUMAN SYSTEMS	HUMAN RESPONSES	FEEDBACK MECHANISM
LOCAL EFFECTS	• Growing population and affluence & rising water demand • Water-based ethnic & linguistic tensions • System-wide corruption • Consistent water policy & governance failures • Climate change	• Degradation & depletion of groundwater • Widespread water pollution • Drought & low rainfall	• Use of water as a political weapon • Creation of more states leading to water conflicts • Failure of tribunals & governments to mediate water conflicts • Signs of constitutional dysfunction, increasing polarization and pitting states against each other and central government	• Significant rise in water conflicts • Protests, violence, and property destruction • Ethnic conflicts • Increasing agriculture, industrial, and urban demands leading to more and intensification of conflicts • Use of water for political manipulation	• More water conflicts, with ethnic/linguistic dimension • Initial efforts to manage water demand • Efforts to improve water governance
IMPACTS ON THE U.S.	• Political conflicts reduce political stability and jeopardize U.S. regional interests			• Internal governance fractures (states vs. central government) will reduce support for U.S. policies and reduce potential U.S. contributions	

Figure 13.1 Paths of Influence: India

Central and state water institutions continue to have inconsistent, inefficient, substandard, and overlapping policies. Furthermore, water-quality management has been consistently ignored. This has resulted in steady deterioration of water quality in rivers and aquifers, making their water difficult to use without extensive and costly treatments.

It is a paradox that even though interstate disputes in rivers like Cauvery have been going on for over a century, data on water availability, use, and quality leave much to be desired. Without reliable data over a reasonable period of time, it is not possible to develop a long-term, rational, and equitable water management plan.

In the absence of functional and efficient water institutions at both the Central and state levels, and with a lack of the political will it would require to make difficult decisions that have not been made for the past several decades, the water allocation problems in interstate rivers can only become increasingly difficult to resolve. This does not include any consideration of complexities and uncertainties that may be imposed by future changes in climate.

Without serious demand management practices, interstate water allocations will be reduced to a zero-sum game for which the states and/or the Central Government will not be able to find any acceptable and equitable solutions. This will prove Mark Twain's wisdom: "Whisky is for drinking; water is for fighting over."

Notes

1 William Foster (ed.), *Early Travels in India, 1583–1619* (London: Oxford University Press, 1921), 122–187.
2 F. S. Nariman, "Inter-State Water Disputes in India: A Nightmare!," in Ramaswamy R. Iyer (ed.), *Water and the Laws in India* (New Delhi: SAGE Publications, 2009), 43.
3 Million acre feet, or MAF, is the volume of water that would cover 1 million acres to a depth of one foot. One acre-foot is equivalent to 325,851.427 gallons.
4 B. Chauhan, "Punjab-Haryana-Rajasthan Dispute," in B. Chauhan (ed.), *Settlement of International and Inter-State Water Disputes in India* (New Delhi: Indian Law Institute, 1992), 279–301.
5 B. Chauhan, "Inter-State Water Disputes in India: Appraisal of the Problems," B. Chauhan (ed.), in *Settlement of International and Inter-State Water Disputes in India* (New Delhi: Indian Law Institute, 1992), 316–328.
6 A. Richards and N. Singh, "Inter State Water Disputes in India: Institutions and Policies," *International Journal of Water Resources Development* 18, no. 4 (2002): 611–625.
7 S. Chokkakula, "The Water Tribunal Trap," *The Hindu*, June 3, 2005.
8 Government of India, "Setting Up Inter-State River Disputes Tribunal" (New Delhi: Government of India, February 10, 2014).
9 Chokkakula, "The Water Tribunal Trap."
10 The 14th Finance Commission report recommended increasing states' share in taxes from 32% to 42% from the Central funds, giving much greater fiscal autonomy to state governments.
11 RKVY or the National Agriculture Development Scheme is a state scheme in which the amount of central assistance provided to a state is contingent upon the state

maintaining or increasing the percentage of its expenditure on agriculture and allied sectors with respect to the total state plan expenditure.
12 "A.P., Telangana Top in Ease of Doing Business," *The Hindu*, November 1, 2016.
13 One TMC, or tmcft, refers to the volume of water in a reservoir or river flow. It is equivalent to 1 billion cubic feet (28,000,000 m^3).
14 R. R. Iyer, "Cauvery Dispute," *Economic and Political Weekly* 48, no. 13 (March 30, 2013).
15 K. Singh, "Bengaluru Violence and the Social Media: It's Time We Clicked With Responsibility," *The Indian Express*, September 16, 2016.
16 M. Maramkal and S. Ravishankar, "Cauvery Dispute: Is the Media Responsible for the Violence in Karnataka and Tamil Nadu?" *Scroll.in*, September 13, 2016.
17 Ibid.
18 K. S. Shrivastava, "Fight Over Regional Resources Drive Demand for New States," *Down to Earth*, July 15, 2013.
19 Aditi Phadnis, "India Facing Its Worst Water Crisis Ever: Himanshu Thakkar," *Business Standard*, May 14, 2016,www.business-standard.com/article/opinion/india-facing-its-worst-water-crisis-ever-himanshu-thakkar-116051400704_1.html.
20 Demand for reservation from the Jat community, a relatively well-off agrarian community, fueled violent protests in Haryana in early 2016, owing to exclusion from the Other Backward Classes (OBCs) list of the National Commission for Backward Classes. As part of India's affirmative action, communities that feature on the OBC list are given preferential treatment through reservations in educational institutions and government jobs. The Jat community in nine states of India has been demanding reservations beyond the 27% reservation granted to OBCs to secure jobs and admission in higher education institutions.
21 The Krishna Water Dispute Tribunal-II, or KWDT-II, award was declared on 31 December 2010 for allocation of the Krishna River water between three riparian states of Maharashtra, Karnataka, and Andhra Pradesh. More recently, the newly constituted state of Telangana, bifurcated from Andhra Pradesh, has become the fourth riparian party in the dispute; however, the second tribunal has refused to reallocate the water to accommodate a fresh petition filed by Telangana, which will be receiving a share of water from Andhra Pradesh. The next review of the verdict is scheduled for 2050. For more information, please refer to the Ministry of Water Resources website: http://wrmin.nic.in/forms/list.aspx?lid=371.
22 K. Venkateshwarlu, "Water Conflict," *Frontline*, Vol. 28, Issue 2, January 15, 2010.
23 Iyer, "Cauvery Dispute."
24 W. Chandrakanth, "Andhra Pradesh's Case," *Frontline*, Vol. 22, Issue 9, May 2005.
25 Ibid.
26 "Water Rows, Disparities Challenge Federalism: PM," *Business Standard*, October 6, 2007.
27 Nariman, "Inter-State Water Disputes in India: A Nightmare!"

14 Water-Energy Nexus in the Himalayas

Keith Schneider

Himanshu Thakkar directs the South Asia Network on Dams, Rivers, and People, his 18-year-old public-interest group, from an office on the roof one floor above the north Delhi apartment building where he lives.

The view from a desk corralled by file cabinets filled with technical documents and shelves crowded with science texts encompasses a capital city that mirrors the room. Delhi bulges. A daily tide of rural emigrants—pushed out of their villages by droughts, floods, crop failures, or immense hydropower and coal-fired power station construction—surges into the city. Delhi's street population is swollen with beggars, especially children dispatched by their mothers to plea for rupees at the intersections of major avenues. Shelters of blue plastic sheeting and brown cardboard rise from medians and grassy lots.

Thakkar's goal for starting the dams and rivers group is as straightforward as it is worthy. Trained at the Indian Institute of Technology, Thakkar was convinced he would make a difference, and a living, by using his expertise in water and energy supply to persuade India to turn aside what he considered a reckless and damaging industrial development strategy. Thakkar's research, and his wide network of like-minded authorities, made a powerful case that India's devotion to big dams, big mines, big power stations, and big water transport projects as a means for achieving well-being and prosperity was not operating very well.

India churns with economic disruptions and religious hostility that are not helped at all by frequent blackouts and confrontations over scarce water. Its governance often seems shaky. State and national agencies are burdened by corruption that impedes enforcement of pollution control laws or a fair permitting process for new electrical generation or hydropower plants. In villages across India—especially in the coal-rich states of the east and in communities built on the shorelines of the plunging rivers of the Himalayas—civic stability and security falter as rural people organize to block mammoth new coal-fired power plants and hydroelectric dams.

India is in need of a new way to pursue the future, Thakkar and his allies argue. The country's social stability faces so many challenges in large measure because its life-giving resources are in trouble.

India is not enduring its distress alone. Similar conditions of severe stress caused by erratic markets and meteorological disorder exist in India's nearest neighbors: Pakistan, Bangladesh, Nepal, and Myanmar. This stretch of south Asia lying at the base of the Himalayas is among the most ecologically and socially unstable regions on Earth. Pakistan and India rank fifth and tenth, respectively, in the latest annual climate risk index prepared by Germanwatch, a Berlin-based environmental and economics research group.[1]

India, Myanmar, and Bangladesh rank, in order, fourth, sixth, and eighth among nations with high levels of civil unrest, according to Verisk Maplecroft, a respected British risk consultancy.[2]

In the more than two years since Indian Prime Minister Narenda Modi took office, he has set in motion an aggressive economic reform agenda to respond to some of his nation's insecurities. The agenda includes measures to join the world in addressing climate change, accelerate renewable-energy development, curtail an expensive program coal for building huge coal-fired power plants, and clean up the country's polluted rivers.

The United States, for its part, is displaying uncommon interest in India's stability and in recruiting much stronger alliances with its charismatic prime minister. It was not terribly long ago that the U.S. viewed Modi warily—when he was the powerful chief minister of Gujarat—going so far as to deny him a visa in 2005. Since Modi was elected in 2014, though, the U.S. has courted and fully embraced the prime minister.

At the top of the list of shared interests are climate, energy, and agriculture. Modi visited the U.S. four times during his first two years in office, and told a joint session of Congress in June 2016: "In every sector of India's forward march, I see the U.S. as an indispensable partner. Many of you also believe that a stronger and prosperous India is in America's strategic interest."[3]

The U.S. shares with India a common heritage of democratic government and counts on India to be the most stable partner in an economically, ecologically, and socially volatile region. Dwindling water supplies in the Indus River basin, for instance, are a factor in the escalating tension between India and Pakistan.[4] Terrorist activity is an omnipresent risk. Climate change is causing seas to rise, inundating portions of Bangladesh, which increases the threat of a refugee crisis on the India-Bangladesh border.

Just as important, the U.S. views India as an economic engine capable of driving a portion of the global growth that benefits India and the American economy. In 2015, the two nations agreed to a U.S.-India Joint Strategic Vision for the Asia-Pacific and Indian Ocean Region to "support sustainable, inclusive development, and increased regional connectivity by collaborating with other interested partners to address poverty and support broad-based prosperity."[5] India's GDP growth, after sinking to under 5% in 2008, rebounded to around 7% in 2016.

Yet diplomatic negotiations do not obscure commanding evidence that ecological pressures are injuring India's economic performance, a trend that is worrisome for India and for the U.S.[6] Thakkar asserts that "when we conduct water-intensive activities in water-starved regions, that is an invitation to an inequitable, unsustainable, conflict-generating situation."[7]

This chapter considers carefully the confrontation between ecological disruption, economic performance, and civil unrest in India. It focuses on three objectives. The first is to thoroughly explore India's economic industrialization policies, which are designed for a less crowded, more ecologically stable century that has passed. The second is to understand the consequences to India's environment, economy, and social stability. The third is to report on how Prime Minister Modi and his administration have recognized that the world's second-largest nation must make fresh and courageous leaps in economic and environmental logic to assure its security.

Problems Anticipated

Environmental specialists and academics saw India's impasse coming almost a decade ago. In 2008, the International Panel on Climate Change, in a technical paper, projected that India would be on the brink of a water crisis by 2025.[8]

In two prescient reports in 2009, the U.S. National Intelligence Council described the considerable risks to India from high heat and water stress.[9,10] It predicted that India's growing population, coupled with harmful climate impacts, would increasingly hinder socioeconomic development. Sea level rise, unpredictable and severe storms and flooding, more frequent drought, and exacerbated water stress were projected to destroy crops, property, and infrastructure, and to impact human health. These impacts were expected to cause mass migrations both internally—from rural to urban areas—and across the borders of neighboring countries, ultimately leading to social disruptions.

Those projections have come to pass. At the end of 2015, and again in the summer of 2016, a record-setting flood drowned Chennai, a city of more than 8 million people along India's southeast coast. More record flooding in northeast India wrecked tens of thousands of homes, inundated hydropower reservoirs, and killed roughly 1,000 people.[11]

Between the winter and summer floods, the deepest drought in India's 69-year history as a democracy dried up water wells in Punjab, prompted riots in central India, and dried rivers across northern India, the most densely irrigated region on Earth. The same dry spell closed coal-fired power plants in at least six states that were unable to secure sufficient water for cooling. The drought also touched off ferocious street fighting over water supplies in southern India between farmers in Karnataka and residents of Tamil Nadu served by the receding waters of the Cauvery River.

The drought joined the hottest week ever recorded in India. Temperatures soared past 51 degrees Celsius (124 degrees Fahrenheit) in May in Rajasthan, a desert state southwest of Delhi. Almost 200 people died.[12] In 2015, similarly extreme heat killed 2,000 people.

Millions of farmers and rural residents packed into India's teeming cities, adding to joblessness, putting pressure on municipal services, and causing rising social instability. Thousands of farmers committed suicide.[13] "This year has been the worst," Ghana Devi, a 55-year-old grower from Bundelkhand told a reporter. She said her village, with 2,000 residents, was deserted as farmers and their families left to find work, most of them as laborers living on the streets in Delhi.[14]

Drought Consequences to Energy

In the energy sector, the drought aggravated India's chronic shortages of electricity. In Karnataka, a highland state in southeast India, the 1,720-megawatt Raichur Thermal Power Station operated far below its capacity due to low water levels in the Krishna River. One or more of the plant's eight generating units shut down almost continuously due to lack of water for much of the year. Shankar Sharma, a power policy analyst from Karnataka, told me

> there have been massive impacts of drought on the lives of the ordinary people in a country largely dependent on rainwater. Since coal power plants depend on massive quantities of fresh water, many power plants were shut either fully or partially.[15]

Electrical generation losses also occurred in India's hydropower sector as a result of the drought, according to the Central Electric Authority. Total hydropower generation during the 2015–2016 period was 6% below the 2014–2015 period, even as the operating hydropower capacity in 2015–2016 was 1,151 megawatts higher than in the previous year.[16] "This difference is even more pronounced in March 2016 and April 2016, when generation from hydropower projects in these months was 19% and 17% lower compared with the same months the previous year," Thakkar told me.[17]

Government Response to Worsening Water Scarcity

The most recent findings by the National Meteorological Department and India's universities indicate that the severe drought that first appeared in 2015 and returned in 2016 is part of a developing climate pattern. Over the past three decades, the frequency of drought years has been increasing. The period between 1950 and 1989 had 10 drought years. Since 2000, there have been five droughts, and the frequency is projected to increase between 2020 and 2049.[18]

According to Rajendra Singh, winner of the 2015 Stockholm Water Prize and one of India's premier water policy experts, "The government does not have a strategy to deal with the permanent water shortages that we've been seeing for a decade."[19]

India, in short, is snared in what Thakkar calls the "worst hydrological distress ever." He told the *Business Standard*:

> What we are seeing this year is unprecedented in many respects: major perennial rivers like the Ganga, Godavari, Krishna, and Netravati have

dried up at several locations, which was unheard of earlier. Groundwater levels are at a record low. In many places, hand pumps have dried up completely. The number of people impacted and the intensity of the impact are huge. This is only the fourth time in a century that there has been a back-to-back drought, but on all previous occasions groundwater, an insurance in times of drought, had provided relief. That is no longer an available option in several places. Our rivers are in a much worse situation today than ever in the past, due to all the ill treatment we have meted out to them, including multiple and often unnecessary, unjustified damming. All this makes the situation this year much worse.[20]

A Unifying Thesis of Risk

Thakkar is not alone in arguing that the growing confrontation between rising demand for energy and declining freshwater supplies is a force multiplier affecting India's economy and rising social disorder. He is joined by a number of other analysts, including Raj Pandit, a professor of ecology at the University of Delhi; Peter Bosshard of Berkeley, California–based International Rivers; and ThirdPole.net, a newsroom that covers the Himalayas. Their work forms a powerful and perceptive critique of a development strategy in desperate need of replacement. In essence, India insists on pursuing a Western-style, 20th-century concept of industrial development and prosperity that does not fit 21st-century conditions. Stripped to its basics, the thesis is fairly simple to explain.

Arguably the central economic principle of the 20th century was achieving greater efficiency through mass and scale. Big dams. Big power plants. Big roads. Big mines. Big farms. The pursuit of the "bigger is better" economic construct shaped the West's prosperity because it fit 20th-century economic and ecological conditions. Resources and land were plentiful and accessible. Markets were steadily growing. Government and personal wealth increased. Weather systems were more predictable. Energy was cheap. The population was much smaller. Popular resistance, especially in rural areas targeted for big construction projects, was disorganized and isolated, and lacked influence.

Well into the second decade of the 21st century, all of those conditions have flipped. A new state of ecological and economic menace has formed. Resources are more scarce and less accessible. Markets are shrinking and erratic. Weather systems are less predictable and much more dangerous. Energy is expensive. The world's population, almost 7.5 billion, is two times larger than it was 45 years ago and growing by 70 million people annually. There are more people in more places, better connected, better financed, and more prepared to resist immense new industrial projects.

Connect the dots, and examples of the futility of continuing to build huge energy projects in the face of such economic and ecological instability are easy to find in India: big coal-fired power plant projects on the Mahanadi River in eastern India near the Indian Ocean shutting down because of process water shortages or fierce protest over water supplies; proliferation in Punjab in northwest India of ever-deeper, ever-more-powerful electric groundwater pumps that drain freshwater supplies and electricity and contribute to blackouts that hamper every business in the region; hydropower reservoirs drained of water by droughts; dams washed away in floods; and the Cauvery River unrest.

Consequences to Energy

India's ecological havoc is hindering its energy production. Even after 35 years of trying to match China's transition from a rural economy to an urban manufacturing economy, India's total electrical generating capacity reached only 306 gigawatts in September 2016, according to the Central Electric Authority.[21]

A gigawatt is 1,000 megawatts, or about the level of generating capacity that a large coal-fired power plant has. Since 2000, when India's total electrical generating capacity was 116 gigawatts, utilities and industrial electricity producers have increased India's generating capacity an average of 11.8 gigawatts annually.[22] In the past five years, annual generating capacity has more than doubled to nearly 26 gigawatts.

It is not nearly enough electricity, though, for a nation of 1.3 billion people, where 300 million to 400 million people have no access to power, and where blackouts and brownouts are endemic and a chronic impediment to business development.

At the current pace of increases in generating capacity—even at 26 gigawatts a year—it will take India 40 years to produce the levels of electricity now generated in the U.S. By that time climate change will have dried up more of India's rivers and groundwater, melted Himalayan glaciers, and put big portions of its coastal cities under water. With India's population projected to reach 1.7 billion people, it will require 2,000 gigawatts of generating capacity by mid-century to serve the power demands required of a resource-consuming, Western-style industrial economy.[23]

Achieving that much power using the current tools that India relies on—particularly coal-fired power plants and hydropower, which together account for 77% of generating capacity—appears improbable, if not impossible. Where should the country turn to liberate itself from power shortages?

Energy Paths

Prime Minister Modi is the latest Indian leader to pursue the answer to this question, though he is doing so along competing paths. During his campaign, and in his first two years in office, Modi pledged to bring power to the thousands of villages that do not have electricity. His government wants to more than double domestic coal production to 1.5 billion metric tons annually, from just over 600 million metric tons in 2015, to fuel new coal-fired power plants proposed across the country. In theory, if India follows through on its coal-fired strategy, the nation's carbon emissions, currently the third highest on the planet, could triple.[24]

But Modi joined U.S. President Barack Obama and Chinese President Xi Jinping to help lead international efforts to limit climate change. India sees hydropower development as a low-carbon energy source and has proposed to add 17 more gigawatts to domestic hydropower production by 2022, as well as assist its neighboring Himalayan countries, particularly Nepal and Bhutan, to build new dams for electricity for purchase by India. India wants to increase the national share of electricity generated by hydropower to 30% by mid-century, from 13% in 2016.[25] That means generating 150 gigawatts of hydropower. India's current hydropower-generating capacity is 43 gigawatts and is growing slowly—increasing an average of less than 1 gigawatt annually since 2012.[26]

The Modi administration's climate change and electricity access policies also include increasing India's renewable-energy capacity to 175 gigawatts, including 100 gigawatts of solar energy, by 2022. That is just over 130 more gigawatts of renewable energy than India produces today.[27]

The various pieces of the Modi program are almost certain to be adjusted to meet contemporary conditions. That is because Prime Minister Modi and his advisors propose very expensive power production strategies that 1) test the limits of India's treasury and technical capacity, 2) defy treacherous environmental conditions, and 3) require significant support from wary citizens.

Diving into three electrical production sectors, the picture about why and how the plan is likely to change gets much clearer.

Coal

Arguably there is no better region for exploring the confrontation between India's resource-driven electrical generating ambitions and its unstable ecology and civic support than the energy-rich eastern state of Chhattisgarh.

Chhattisgarh is perennially neck and neck in the race with neighboring Jharkhand for the title of largest coal-mining state. Last year, Chhattisgarh alone produced 20% of the total national coal production.[28] The state also is home to four big rivers, among them the Mahanadi, the state's primary source of fresh water. Long and wide, the Mahanadi River runs north and then east 550 miles through central Chhattisgarh and neighboring Odisha to the Bay of Bengal. The river drains a 25,000-square-mile region.

The proximity of prodigious coal reserves and apparently ample access to fresh water for producing steam and cooling coal-fired generating stations is attractive to India's big electricity supply companies. Two of India's 10 largest coal-fired power stations are in Chhattisgarh.[29] Of the state's existing coal-fired power plants, which produce more than 19 gigawatts of generating capacity, 73% rely on the Mahanadi for process and cooling water.[30]

A typical one-gigawatt coal-fired plant needs 24 million gallons of water an hour for steam generation and cooling. About 1 million gallons an hour evaporate. The quandary along the Mahanadi is that over the past 50 years the uses of water from the river have increased sevenfold, as both Chhattisgarh and Odisha states industrialized and food production increased.

Some 40 million people—most of them farmers—in Chhattisgarh and Odisha depend on the river. As of 2012, Chhattisgarh has planned 58,000 megawatts of coal-fired generation using mostly Mahanadi water; Odisha has planned for 75,000 megawatts.[31] "There should be joint movement from both the states to fight rapid industrialization and its impact on the Mahanadi," said Prafulla Samantara, an environmental leader.[32]

It is not certain whether the Mahanadi has enough water to serve the power sector and all of the other communities and businesses that depend on it. During the summer monsoon, it certainly does. Chhattisgarh's water resources department maintains gauges that show the river carries a flood of fresh water—51,000 cubic meters (13.5 million gallons) per second—in the summer. Much of this can be stored downstream in Odisha, behind the 16-mile-long Hirakud Dam, opened in the mid-1950s, which is the longest earthen dam in the world and which also contains one of the largest man-made reservoirs in Asia.[33]

But during the dry season, the Mahanadi's flows typically shrink by 70%. Even with modern water-conserving cooling systems, new coal-fired plants that are planned for the next decades will require Chhattisgarh's government to issue permits that annually allocate millions of cubic meters of Mahanadi water for electrical generation, both in the wet season and the dry.

There is also the question of how India expects to build that many new power stations along the river. Although India supplies about 70% of its electricity from burning coal, the colliding trend lines in water supply and energy demand that have developed along the banks of the Mahanadi are consistent with India's national struggle to electrify the country.

Opposition to new power plants along the Mahanadi is deep, and often persuasive. In August 2012, for instance, 3,000 villagers concerned about disruptions to their water supply protested construction of a 1,320-megawatt coal-fired plant planned along the river near Pitamahul, in Odisha. The project's developer, KU Projects Ltd., has since suspended its work on the power station.[34]

One of the largest demonstrations of concern about energy and water supply occurred in December 2012 when representatives of 32 communities, farm groups, and nonprofit organizations from Chhattisgarh and Odisha held a two-day conference on the tightening contest for the Mahanadi's water. The meeting was held in Sambalpur, a downriver city of 200,000 residents located near the foot of the big Hirakud Dam.

"The river is a common resource of people. They should get priority in its water," Gautam Bandhopadhyay told reporters. Bandhopadhyay leads Nadi Ghati Morcha, a Raipur-based environmental group that cosponsored the meeting.[35] Five months after the Sambalpur conference and demonstration, the National Thermal Power Corporation—the largest state-owned utility in India—announced that it had shifted the site of a 1,600-megawatt coal-fired plant that was to be cooled by the Mahanadi to a new site in Madhya Pradesh, a state to the west. The company said that its decision was due to "delays in environmental clearances and land acquisition" in Odisha.[36]

Water-Energy Nexus in the Himalayas 261

In June 2016, the Indian Ministry of Power scrapped plans to build a 4,000-megawatt coal-fired power plant in Chhattisgarh, another in Odisha, and two more in Karnataka and Maharashtra. It was the first time India had walked back its expensive and technically flawed 11-year-old program to add 40 gigawatts of generating capacity by pursuing its 20th-century-style Ultra Mega Power Plants (UMPPs) construction program.[37]

India proposed to build 10 UMPPs, each four gigawatts, which would be among the largest coal-fired stations in the world. Two of the UMPPs were completed, though they each cost over $4 billion, twice as much as initially proposed. The two plants have proved difficult to operate efficiently and stirred active local citizen opposition campaigns principally focused on air and water pollution. Developers have expressed little interest in building more. Indian news sources have declared that "the UMPP experiment . . . has failed."[38]

Hydropower

It has been no simpler with hydropower development.

On May 24, 2003, as part of a national plan to generate more electricity from sources other than coal, former prime minister Atal Bihari Vajpayee directed India to pursue one of the most daring energy production campaigns in history. Vajpayee called on his nation to break through corruption, bureaucracy, and its own doubts to build 162 big hydroelectric power projects by 2025. The dams and power stations would be capable of generating the equivalent energy of 50 big coal or nuclear-fired power plants. At the time, India's utility sector had the capacity to generate 108 gigawatts—one-tenth as much as the electrical sector in the U.S. Some 27 gigawatts, or a quarter of India's total energy production, came from hydropower.

Thirteen years later, Prime Minister Modi expresses similar allegiance to the kinetic energy of running water as a cure for the country's endemic electricity shortages—specifically, the swift currents that pour from the Himalaya range. In the three months after his election in May 2014, Modi personally dedicated three new hydropower plants and a 233-mile-long transmission line in Jammu and Kashmir, a Himalayan state in northwest India bordered by Pakistan, Afghanistan, and China. Modi visited Bhutan, which shares a border with several northern Indian states, to lay foundation stones for the 60-megawatt Kurichhu hydroelectric project and the 600-megawatt Kholongchu hydropower project.

In August 2014, Modi was the first Indian prime minister to visit Nepal since 1997. While in Kathmandu he negotiated an agreement to finance the construction of two transmission lines and to set up a project office for the 5,600-megawatt Pancheshwar hydropower station on the Mahakali River.

Modi's run-of-the-river diplomacy culminated in February 2015 when an advisory committee overruled a previous government decision and approved construction of the 3,000-megawatt Dibang power project in the steep Himalayan peaks in Arunachal Pradesh, a state in northeastern India bordered by Bhutan, China, and Myanmar.

Yet India's ambitious program for diversifying its electrical power supply has been stymied by a powerful convergence of ecological and economic trends. Rampaging floods, landslides, and earthquakes have wrecked hydropower dams in India and Nepal. Powerful civic opposition has developed in India's Himalayan states, dramatically slowing dam development. In the past 13 years, India has added only 16 gigawatts of hydropower-generating capacity.[39]

Two examples of distress in Himalayan hydropower development are especially apt.

In Assam, a northeast Himalayan state, the 2,000-megawatt Lower Subansiri hydropower project has been abandoned. The silent construction site has become an expensive stage for a national drama over Himalayan hydropower development. It pits public demands for lower financial and safety risks, more public involvement, and more reasonable scale—all 21st-century values—against 20th-century government ideas about the efficiency and economy of building mega hydropower projects in the world's tallest and most treacherous mountains.

When construction started in 2005, the $1.6 billion Lower Subansiri project was viewed in Delhi as a paramount example of Indian engineering and a jewel of India's hydropower campaign. The project would be India's largest when it started generating power by 2010. The dam would also be the first of three on the Subansiri, which include a 1,600-megawatt project in the river's middle reach and another 2,000-megawatt power dam near the headwaters.[40] The Subansiri dams were viewed as showcase installations for gaining access to the 63 gigawatts of hydropower capacity—43% of the country's hydropower generation—that government studies said were ready to be tapped in Arunachal Pradesh and seven neighboring states in northeastern India.[41]

But farmers of the Subansiri River valley objected to being displaced by the 29-mile-long lake formed by the dam. Residents downriver along the banks of the Brahmaputra—which starts in China, tumbles through 7,000-meter-deep valleys of the Himalayas, and transports more fresh water than any river but the Amazon and the Congo—were alarmed by what they viewed as an authentic threat of a big dam in their midst.[42] Citizens' groups organized protests in neighboring Assam and established blockades on roads leading to the Subansiri dam construction site.[43] In December 2011, construction shut down. Every summer since the shutdown, National Hydropower Corporation (NHPC) and Delhi hydropower administrators have announced a new start date that slips by.[44]

Two years after the Lower Subansiri dam was halted, an even more powerful example emerged in Uttarakhand, a Himalayan state north of Delhi. In mid-June 2013, during the Hindu pilgrimage to a temple high in the Himalayas, a convergence of hydrological events unleashed the worst flood ever recorded in the mountains.

One exacerbating factor was the early arrival of the annual monsoon, causing accelerated snowmelt and higher-than-normal rainfall. Next, a cloudburst dumped at least 12 inches of rain on the Himalayan ridges that feed the

Alaknanda and Bhagirathi river basins. As a direct result of the cloudburst, the banks holding back the waters of Chorabari Lake—located at 13,000 feet—collapsed. Chorabari Lake released all of its water in 10 minutes.[45]

Floodwaters damaged everything in their path. Villages were buried, and more than 100 landslides brought down or damaged over 600 miles of highway and destroyed an unknown number of hotels, homes, shops, and government buildings. The estimated death toll ranged from 6,000 (Government of Uttarakhand) to 30,000 (residents and the Wadia Institute of Himalayan Geology). The torrent produced consequences that no engineer anticipated and no Uttarakhand resident had ever seen.[46] Studies estimate that damage to public infrastructure—roads, water transport, buildings—amounted to nearly $700 million. There has been no formal estimate of the financial damage to the state's hydropower projects.[47]

The June flood also drowned India's long campaign to diversify its energy production with big Himalayan hydropower projects. The flood seriously damaged at least 10 large projects in operation and under construction in Uttarakhand. One dam under construction washed away. Another 19 small hydropower projects that generated less than 25 megawatts were destroyed. A study overseen by the Indian Supreme Court blamed dam construction practices, including the storing of excavation debris along riverbanks, for amplifying the flood's destruction.[48]

The consequences of the flood in Uttarakhand continue to influence hydro development projects across the Himalayan states. The government proposal to build the 3,000-megawatt Dibang Dam in Arunachal Pradesh has been met with powerful civic opposition and public protests. Two people were killed and 10 were injured in May 2016 when security police fired on a public demonstration.[49]

"There are close to 200 big hydropower projects planned for the Himalayas in northeast India. Most of them are yet to be approved," Thakkar told me in 2015. He continued:

> Almost all have generated significant protest from people in the region, and from local government leaders. The big projects are difficult to build and dangerous to manage in mountains that are on highly silt-laden rivers, in a region rich in biodiversity and prone to earthquakes and flooding. The lives and livelihoods of so many millions are dependent on these resources. Most of the dams will never be built.[50]

Renewable Energy Makes Sense

The stage has already been set for India to shift much of its energy production to solar and wind technologies. On Sunday, October 2, 2016, India ratified the Paris global climate control agreement. Among the agreement's central provisions was India's pledge to lower its greenhouse gas emissions by generating at least 40% of its electricity from non–fossil fuels by 2030.

India's commitment was simultaneously a concession and a diplomatic breakthrough. For more than a decade, India and the U.S. engaged in regular negotiations to decide shared strategic and economic objectives under the U.S.–India Strategic Dialogue. Near the top of the achievements cited by leaders of both countries are 1) the cooperation to limit climate change and 2) the work under the Partnership to Advance Clean Energy that the U.S. and India started in 2009 to dramatically expand India's renewable-energy sector.

Prime Minister Modi sees the global effort to cool the planet and heat up India's clean energy economy as essentially the same task. Most of India's climate-changing emissions are generated by burning coal for electricity. As Gujarat's chief minister, Modi led investment in that state's solar and wind energy construction. As prime minister, he put aside India's previous resistance to diminish its climate emissions and joined Obama and Xi Jinping—leaders of the two biggest carbon polluters—in negotiating an accord in Paris in 2015 that is meant to keep global temperatures from rising to more ruinous levels.

In the days before the Paris climate treaty was approved, Modi also worked with France and 119 other nations to establish the International Solar Alliance. The group, started with a $30 million five-year grant from India, has a single overarching mission: to establish solar power industries in developing nations. "Solar technology is evolving, costs are coming down and grid connectivity is improving," Modi said when introducing the Solar Alliance in Paris. "The dream of universal access to clean energy is becoming more real. This will be the foundation of the new economy of the new century."[51]

Modi's confidence in starting the solar alliance was influenced by the U.S. The U.S.–India clean energy partnership had already committed the U.S. to a $125 million joint clean energy research center, a $1.7 billion account to finance clean energy projects, access to American technology, and a pledge to finance 5.4 gigawatts of solar energy in India.[52]

Modi needs every bit of that U.S. assistance to achieve the goal of generating 40% of India's electricity from non-fossil generating sources by 2030. At current annual rates of increase in generating capacity, India requires roughly 170 more gigawatts of non-fossil capacity to meet its Paris Agreement commitment.

In December 2015, Modi installed a new energy plan that the government called the "largest renewable capacity expansion program in the world."[53] The new plan pushes India's renewable generating capacity to 175 gigawatts by 2022, or 131 more gigawatts than India's renewable-energy plants currently produce. The plan also embraces the new reality that India can really count on just two technologies to achieve most of the 40% non-fossil goal: wind and solar power. The prime minister and his aides anticipate having 100 gigawatts of solar-generating capacity by 2022 and 60 gigawatts of wind-energy capacity.

Then, in December 2016, Modi upped the clean energy ante. India released a draft energy plan that set a target of generating nearly 60% of the country's electricity, around 275 gigawatts, from wind, solar, biomass, and small hydropower plants by 2027. India currently relies on coal for about 70% of its electricity. Modi pledged to "achieve energy security for India based on clean fuels."[54]

Indeed, the brightest outlook in India's deeply troubled energy sector—arguably the only truly encouraging prospect India holds for actually achieving its ambitious energy production goals—comes from drawing upon water-conserving wind and solar resources and developing much larger domestic renewable-energy industries to tap them. The 2016 edition of India's national energy statistics estimates that the potential generating capacity of the renewable sector is 896.6 gigawatts. Of that total, 850 gigawatts are due to the potential for 102 gigawatts of wind-generating capacity and 748 gigawatts are from the sun.[55]

Before Modi took office, India had already begun to recognize non–fossil fuel sources of electricity. In 2010, India established the Jawaharlal Nehru National Solar Mission to focus on solar generation. Its ambition was modest. At the time, India believed it could connect 22 gigawatts of solar-powered energy to India's transmission grid by 2022.

Similarly, India counted on wind energy to diversify its sources of electricity. The Indian Wind Power Association counts 17 wind manufacturing companies that are operating in India, which national leaders assert is the second-largest wind sector in the world.[56] As of July 2016, India's wind-generating capacity reached 26.7 gigawatts.[57] India has now built the world's fourth-highest number of wind installations, and it is the fifth-largest wind energy producer as of 2016.[58]

Among the states that India counts on to drive the nation's renewable-power industry, Rajasthan sits at the top. A desert state nearly as large as Germany, and with 75 million residents, Rajasthan's primary 21st-century resource is sunshine. The sun heats Rajasthan's water-scarce and rural landscape 300 to 330 days a year.

Outside of Jodhpur, a city of 1 million residents 205 miles from the state capital of Jaipur, photovoltaic solar energy plants direct broad arrays of blue panels to the sun. This stretch of desert, lightly populated and ripe with solar potential, represents almost precisely what Rajasthan and Delhi envision for a 21st-century renewable-energy sector. Land is ample for energy development. Of the more than 8 gigawatts of solar energy capacity that India now produces, 1.26 gigawatts are generated in Rajasthan.[59]

Much more is coming. The state government installed a solar policy two years ago to recruit, finance, and provide land and technical assistance to renewable developers in order to install 25 gigawatts of solar-generating capacity.[60] The Adani Group, one of India's largest energy developers, signed a memorandum of understanding with the state in February 2015 to build two solar parks to generate 10,000 megawatts of capacity. Two other companies signed agreements for plants of similar size. The state budget includes provisions to provide public funds to connect villages with "local solar grid, stand-alone solar systems and smart grid systems."[61]

One more essential ingredient in Rajasthan's renewable-energy development formula: diving costs for solar energy.[62] Jai Shanker Verma, the manager of an AES solar plant near Jodhpur, told me that it cost $20 million to build the plant in 2011. AES, an American energy supplier based in Arlington, Virginia, chose First Solar, also a U.S.-based manufacturer, to supply the 65,000 photovoltaic panels that cover the plant's 74 acres of ground.

If the same facility had been built three years later, it would have cost only $11.5 million due to an abundance of undeveloped land and the shrinking cost of solar technology.

In order to reach the Modi government's 175-gigawatt goal, India will need to add 22 gigawatts of renewable generating capacity annually over the next six years, which will mean quadrupling the current growth rate.

What's not in doubt is the amount of energy potential from India's wind and sun resources, and the influence a change in electrical generation strategy would have on India, and the world. For one, it will provide a chance for India to leapfrog the resource-consuming, polluting, centralized electrical production and distribution networks that were the norm of the past half-century, instead moving toward energy sources that are clean, abundant, and increasingly affordable.

Recommendations

By encouraging smaller renewable-energy plants and transmission systems, India recognizes the distributed, decentralized energy production trends of the 21st century. Wind and solar plants typically take two or three years to design, build, and start. They can be located close to villages. Rooftop solar panels or ground solar arrays make it possible to link multiple homes and businesses in a local grid. Wind and solar require scant water, produce no pollution, and can end the era in India when government looked to enormous engineering projects to remake the nation for public benefit.

Such a shift amounts to one of the most challenging public interest pursuits that India—or any other nation, for that matter—will undertake this century. Unyielding trends in pollution, hydrology, meteorology, and technology essentially mean that India has no choice. Even though India's coal-mining industry employs at least 350,000 people, and thousands more work in coal-fired utilities, construction companies, engineering firms, and the railroad and trucking industry, coal is the dirtiest, and quickly becoming the most expensive, power-generating option.[63]

Updating India's electricity sector means that India embraces a new model for building its economy, reducing threats to the environment, and sustaining a better quality of life for more of its people. Indian renewable-energy companies already employ more than 400,000 people.[64] The sector rests on a foundation of new operating practices fit for the times: drastically reducing climate emissions, increasing energy efficiency, requiring net zero energy use in new buildings, conserving water, electrifying transportation, preventing pollution, and pursuing cleaner energy production.

Too much of the economic, ecological, and civic disturbance unfolding in India, and around the world, is a slow collision between the operating systems of the resource-wasting, vertically managed, centralized systems of the 20th century and the much more ecologically volatile, economically erratic, and decentralized conditions of the 21st. The old order, it's clear, is undergoing a severe stress test in India. The power of nature to unleash its fury and subdue India's surprisingly unstable transactional systems—reservoirs and dams, power plants, roads, transmission lines, food production and distribution networks—becomes clearer with every passing year.

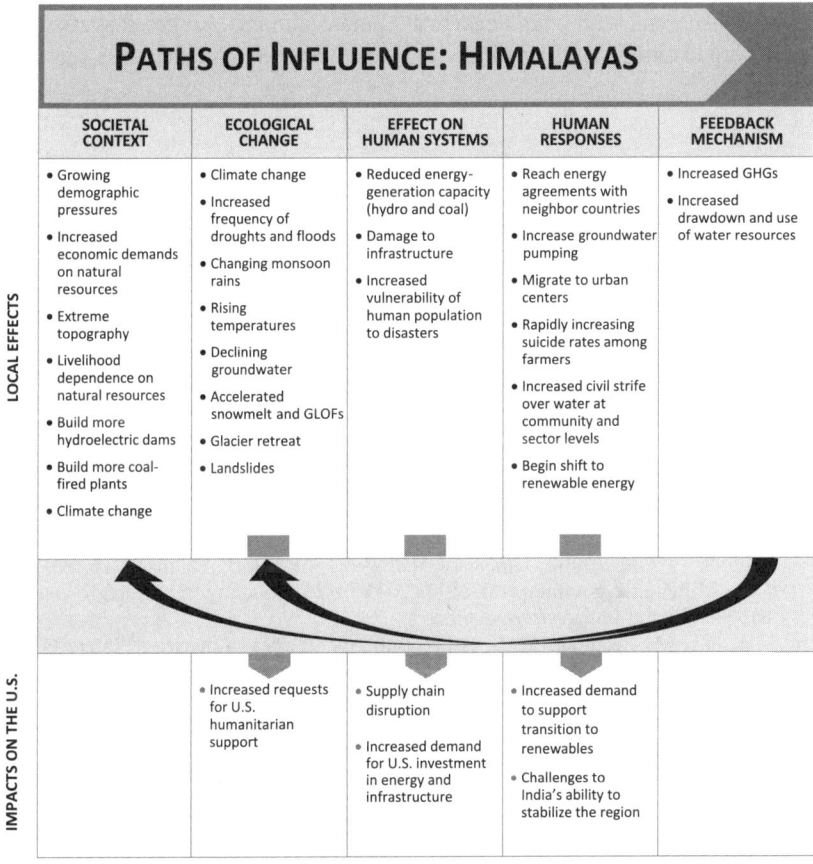

Figure 14.1 Powerful and Menacing Ecological and Economic Trends Are Pushing India to Consider Abandoning Its 20th-Century Approach to Development and Embracing a Different Path to Prosperity and Well-Being in This Century

When world leaders gathered in Paris at the end of 2015 for the United Nations climate talks, they committed to reducing climate-disrupting greenhouse gas emissions. Climate change is now producing deeper droughts, deadlier floods, and more-extreme heat waves. It is a force multiplier that weakens energy production, diminishes food production, and causes destabilizing human dislocation. All of that is now apparent in India. In order to adapt to 21st-century conditions, India will need continued support from the U.S. It is in both countries' best interest to ensure that India is able to become energy secure by achieving its 2030 goal of generating 40% of its electricity from power sources that are less dependent on dwindling resources—primarily solar and wind.

The actual work of realigning economies and industrial practices to reflect changing ecological realities, and directing nations to saner development strategies, doesn't end with world leaders at climate summits. Rather, it starts with leadership in cities, states, and individual countries. In India, that work has just begun.

Notes

1 Sönke Kreft, David Eckstein, Lukas Dorsch, and Livia Fischer, "Global Climate Risk Index 2016," *Germanwatch* (Berlin), https://germanwatch.org/fr/download/13503.pdf.
2 Will Martin, "The 16 Countries With the Most Civil Unrest," *Business Insider* (August 3, 2016), http://read.bi/2dshi9F.
3 "Text of the Prime Minister's Address to the Joint Session of U.S. Congress," *The Hindu* (June 8, 2016), www.thehindu.com/news/resources/text-of-the-prime-minister-narendra-modis-address-to-the-joint-session-of-us-congress/article8706251.ece.
4 Michael Kugelman, "Why the India-Pakistan War Over Water Is So Dangerous," *Foreign Policy* (September 30, 2016), http://foreignpolicy.com/2016/09/30/why-the-india-pakistan-war-over-water-is-so-dangerous-indus-waters-treaty/.
5 "U.S.-India Joint Strategic Vision for the Asia-Pacific and Indian Ocean Region," *White House Office of the Press Secretary* (January 25, 2015), www.whitehouse.gov/the-press-office/2015/01/25/us-india-joint-strategic-vision-asia-pacific-and-indian-ocean-region.
6 Marshall M. Bouton, "U.S.-India Initiative Series America's Interests in India," *Center for New American Security* (October 2010).www.files.ethz.ch/isn/122831/2010_10_USInterestsinIndia_Bouton.pdf.
7 Aditi Phadnis, "India Facing Its Worst Water Crisis Ever: Himanshu Thakkar," *Business Standard* (May 14, 2016), www.business-standard.com/article/opinion/india-facing-its-worst-water-crisis-ever-himanshu-thakkar-116051400704_1.html.
8 "Intergovernmental Panel on Climate Change," *Climate Change and Water*, IPCC Technical Paper VI (June 2008), www.ipcc.ch/pdf/technical-papers/climate-change-water-en.pdf.
9 National Intelligence Council (NIC), "India: The Impact of Climate Change to 2030, a Commissioned Research Report" (NIC, Washington, D.C. 2009): 7.
10 National Intelligence Council (NIC), "India: The Impact of Climate Change to 2030, Geopolitical Implications" (NIC, Washington, D.C. 2009): 7.
11 "Floodlist," *European Union*, http://floodlist.com.

12 Ada Carr and Sean Breslin, "330 Million Impacted by India Heat Wave That Has Killed at Least 160, Officials Say," *The Weather Channel* (April 26, 2016), https://weather.com/safety/heat/news/deadly-southern-eastern-india-heat-wave-2016.
13 Deeptiman Tiwary, "Farmer Suicides Up 40 Per Cent in a Year, Karnataka Shows Sharpest Spike," *The Indian Express* (August 19, 2016), http://indianexpress.com/article/india/india-news-india/farmer-suicide-case-in-india-crop-failure-drought-dry-zones-indian-monsoon-2984125/.
14 Shriya Mohan and Vikas Kumar, "Bundelkhand Refugees Find Shelter and Wait for Hope Under a Delhi Flyover," *Catchnews.com* (June 5, 2016), www.catchnews.com/india-news/drought-drives-lakhs-away-from-bundelkhand-to-delhi-many-find-shelter-under-sarai-kale-khan-flyover-1465129250.html.
15 Interview with author, June 15, 2016.
16 Central Electric Authority, "Power Supply Position Report," www.cea.nic.in/monthlypowersupply.html.
17 Interview with author, June 16, 2016.
18 Nisha Nambiar, "Drought Years May Become More Frequent in India, Says Study," *The Indian Express* (May 31, 2016), http://indianexpress.com/article/india/india-news-india/maharashtra-gujarat-drought-waterless-monsoon-crisis-years-may-become-more-frequent-in-india-says-study-2826500/.
19 Nirmala George, "As India Reels From Drought, Gov't Slammed for Poor Policies," *Associated Press* (April 14, 2016), http://bigstory.ap.org/article/f8cce868a35d49dd8be793360cdbcace/india-reels-drought-govt-slammed-poor-policies.
20 Aditi Phadnis, "India Facing Its Worst Water Crisis Ever: Himanshu Thakkar," *Business Standard* (May 14, 2016), www.business-standard.com/article/opinion/india-facing-its-worst-water-crisis-ever-himanshu-thakkar-116051400704_1.html.
21 Central Electric Authority, "Installed Capacity Report" (September 2016), www.cea.nic.in/reports/monthly/installedcapacity/2016/installed_capacity-09.pdf.
22 Central Statistics Office, "Energy Statistics 2012" (March 2012), http://mospi.nic.in/mospi_new/upload/energy_statistics_2012_28mar.pdf
23 Department of Economic and Social Affairs, "World Population Prospects Volume II," *United Nations Population Division* (2015), https://esa.un.org/unpd/wpp/Publications/Files/WPP2015_Volume-II-Demographic-Profiles.pdf.
24 "Summary of CO_2 Emissions From Fuel Combustion," *IEA Statistics 2015 Edition*, www.iea.org/publications/freepublications/publication/CO2EmissionsFromFuelCombustionHighlights2015.pdf.
25 M. M. Madam, "The Hydro Power Development in India-Challenges and Way Forward," *Jindal Power Limited Abstract* (July 4, 2015), http://inae.in/wp-content/themes/inae-theme/pdf/hydro_power_development.pdf.
26 Central Statistics Office, "Energy Statistics 2016" (April 2016), http://mospi.nic.in/mospi_new/upload/Energy_statistics_2016.pdf.
27 Central Electric Authority, "Installed Capacity Report."
28 "Top Ten Coal Producing States in India," *Listz*, http://listz.in/top-10-largest-coal-producing-states-in-india.html.
29 "The Top 10 Biggest Thermal Power Plants in India," *power-technology.com* (October 10, 2013), www.power-technology.com/features/feature-the-top-10-biggest-thermal-power-plants-in-india/.
30 "24X7 Power for All Chhattisgarh," *Joint Initiative Government of India and Government of Chhattisgarh* (2014), http://powermin.nic.in/sites/default/files/uploads/joint_initiative_of_govt_of_india_and_chhattisgarh.pdf.
31 Ranjan Panda, "Treat Mahanadi as a River, Not a Commodity," *Scribd* (December 24, 2012), www.scribd.com/document/119610777/First-Report-of-Mahanadi-Dialogue-Dec-2012.

32 Kishore Dash, "Groups From Odisha and Chhattisgarh Demand People Management of Mahanadi," *Orissa Diary* (December 24, 2012), www.orissadiary.com/CurrentNews.asp?id=38462.
33 "Chhattisgarh Water Resources Department," www.cgwrd.in/water-resource/wr.html.
34 "Pitamahul Power Station," *Sourcewatch*, www.sourcewatch.org/index.php/Pitamahul_power_station
35 "Holistic Plan For Mahanadi Water Management Sought," *The Hindu* (December 24, 2012), www.thehindu.com/todays-paper/tp-national/tp-otherstates/holistic-plan-for-mahanadi-water-management-sought/article4233960.ece
36 "NTPC to Shift Odisha Project to Gadarwara in Madhya Pradesh," *The Economic Times* (April 4, 2013), economictimes.indiatimes.com/industry/energy/power/ntpc-to-shift-odisha-project-to-gadarwara-in-madhya-pradesh/articleshow/19378138.cms.
37 Jai Sharda and Tim Buckley, "India's Questionable Ultra Mega Power Plans," *IEEFA* (August 3, 2016), http://ieefa.org/wp-content/uploads/2016/08/India's-Questionable-Ultra-Mega-Power-Plans-Viability-Issues-Continue-to-Complicate-New-Coal-Fired-Projects-August-2016.pdf.
38 Jyoti Mikul, "India's Mega Power Projects Grind to a Slow Halt," *Business Standard* (August 3, 2016), www.business-standard.com/article/companies/indias-mega-power-projects-grind-to-a-slow-halt-115090301248_1.html.
39 Central Electric Authority, "Installed Capacity Report."
40 Biswajit Das, "Planning and Building the Subansiri Lower Dam and Hydro Project," *HydroWorld* (January 1, 2012), www.hydroworld.com/articles/2012/01/planning-and-building.html.
41 Neeraj Vagholikar and Partha J. Das, "Damming Northeast India," *Aaranyak* (November 2010), https://chimalaya.files.wordpress.com/2010/12/damming-northeast-india-final.pdf.
42 "Lower Subansiri Project In Trouble Yet Again," *The Times of India* (October 13, 2015), http://timesofindia.indiatimes.com/city/guwahati/Lower-Subansiri-project-in-trouble-yet-again/articleshow/49330817.cms.
43 Tanmoy Sharma, "Mega Dams: Campaigning Against The Plans of the Indian Government," *openDemocracy* (January 20, 2012), https://archive.nyu.edu/bitstream/2451/33618/2/Mega_dams_campaigning_against_the_plans_of_the_Indian_government.pdf.
44 Teresa Rehma, "Experts Call for Major Redesign of India's Largest Dam," *thethirdpole.net* (November 7, 2013), www.thethirdpole.net/2013/11/07/experts-call-for-major-redesign-of-indias-largest-dam/.
45 Deepti Singh, Daniel E. Horton, Michael Tsiang, Matz Haugen, Moetasim Ashfaq, Rui Mei, Deeksha Rastogi, Nathaniel C. Johnson, Allison Charland, Bala Rajaratnam, and Noah S. Diffenbaugh, "Severe Precipitation in Northern India in June 2013: Causes, Historical Context, and Changes in Probability," *Bulletin of the American Meteorological Society* (September 2014), www.researchgate.net/profile/Daniel_Horton2/publication/266554791_Severe_precipitation_in_northern_India_in_June_2013_Causes_historical_context_and_changes_in_probability/links/5433fb270cf2dc341daf29e2.pdf.
46 Saurabh Dani and Anil Motwani, "India—Uttarakand Disaster 2013: Joint Rapid Damage and Needs Assessment Report," *World Bank, Asian Development Bank, Government of Uttarakhand* (August 2013), http://documents.worldbank.org/curated/en/322951468041658053/pdf/826430WP0P146600Box379864B00PUBLIC0.pdf.
47 "Rapidly Assessing Flood Damage in Uttarakhand, India," *World Bank and Asian Development Bank* (July 29, 2014), www.worldbank.org/en/results/2014/07/29/rapidly-assessing-flood-damage-uttarakhand-india.

48 "Report of Expert Committee on Uttarakhand Flood Disaster & Role of HEPs: Welcome Recommendations," *South Asia Network on Dams and Rivers* (April 29, 2014), https://sandrp.wordpress.com/2014/04/29/report-of-expert-committee-on-uttarakhand-flood-disaster-role-of-heps-welcome-recommendations/.
49 Gupta Kashyap, "Arunachal Anti-Dam Protests: Buddhist 'Monk' Among Two Killed in Tawang Police Firing," *The Indian Express* (May 3, 2016), http://indianexpress.com/article/india/india-news-india/monk-among-two-killed-in-tawang-police-firing-2781542/.
50 Interview with the author, June 2015.
51 Arthur Nelson, "India Unveils Global Solar Alliance of 120 countries at Paris Climate Summit," *The Guardian* (November 30, 2015).
52 "FACT SHEET: The United States and India—Moving Forward Together on Climate Change, Clean Energy, Energy Security, and the Environment," *The White House* (June 7, 2016), www.whitehouse.gov/the-press-office/2016/06/07/fact-sheet-united-states-and-india-moving-forward-together-climate.
53 "Report of the Expert Group on 175 GW RE By 2022," *National Institute for Transforming India* (December 31, 2015), www.theguardian.com/environment/2015/nov/30/india-set-to-unveil-global-solar-alliance-of-120-countries-at-paris-climate-summithttp://niti.gov.in/writereaddata/files/writereaddata/files/document_publication/report-175-GW-RE.pdf.
54 Ibid.
55 Central Statistics Office, "Energy Statistics 2016" (April 2016).
56 "'Revised List of Models and Manufacturers of Wind Turbines—ADDENDUM—I List' to 'Main List dated 28.09.2015,'" *Indian Wind Association* (March 2, 2016), www.indianwindpower.com/pdf/RLMM_Addendum-I_List_03.02.2016.pdf.
57 "State Wise % of Wind Potential Utilized," *Ministry of New and Renewable Energy* (March 31, 2016), http://mnre.gov.in/file-manager/UserFiles/State-wise-wind-power-potential-utilized.pdf.
58 "Global Wind Report 2016," *Global Wind Energy Council* (2016), www.gwec.net/wp-content/uploads/vip/GWEC-Global-Wind-2015-Report_April-2016_19_04.pdf.
59 Yogendra Singh and Armin Rosencranz, "Rajasthan Leads India's Solar Power Ambitions," *The Statesman* (June 1, 2016),www.thestatesman.com/news/opinion/rajasthan-leads-india-s-solar-power-ambitions/145743.html.
60 "Renewable Energy Policies," *Rajasthan Renewable Energy Corporation*,www.rrecl.com/State.aspx.
61 "Renewable Energy Policies," *Rajasthan Renewable Energy Corporation*.
62 Tom Randall, "Wind and Solar Are Crushing Fossil Fuels," *Bloomberg* (April 6, 2016), www.bloomberg.com/news/articles/2016-04-06/wind-and-solar-are-crushing-fossil-fuels.
63 Swarup Santra and Nidhi Bagaria, "Labour Productivity in Coal Mining Sector in India: With Special to Major Coal Mining States," *Journal of Human Resources* (January 2014), http://papers.ssrn.com/sol3/papers.cfm?abstract_id=2390524.
64 "Renewable Energy and Jobs," *International Renewable Energy Agency Annual Review 2016*, www.se4all.org/sites/default/files/IRENA_RE_Jobs_Annual_Review_2016.pdf.

15 A Perfect Storm in the Greater Mekong Subregion
Climate Change Impacts on Food, Water, and Energy

Arjun Thapan

Strategic Significance

The Greater Mekong subregion (GMS) comprises six countries, and the Mekong River runs through it. The countries are China (Yunnan and Guangxi Zhuang provinces), Cambodia, Laos, Myanmar, Thailand, and Vietnam. The GMS countries (with the exception of China) are also members of the Association of Southeast Asian Nations (ASEAN). The subregion has exhibited strong economic growth since 2001. Total gross domestic product (GDP) (in current U.S. dollars) grew from $893 billion in 2001 to $2,449 billion in 2013. GDP per capita also grew, from an average of $2,973 in 2001 to $7,335 in 2013. Intra-regional trade boomed and grew from 4% in 2000 to 7.4% in 2013.[1] Cross-border connectivity has improved significantly, thanks to several infrastructure initiatives in the transport, energy, and communications sectors.[2]

The Mekong River originates in Tibet and flows about 4,350 kilometers through the GMS countries before discharging into the South China Sea. Its basin is estimated at 810,000 square kilometers; only the Yangtze and Ganges river basins are larger.[3] A population of 325 million people, of whom between 38% and 74% work in agriculture, depend on the river for their economic livelihood and food sustenance. In Cambodia, Laos, Thailand, and Vietnam, up to 83% of people in the economically active population are engaged in a water-related activity as their primary occupation.[4] The basin's inland fisheries production is the highest in the world, estimated at up to $17 billion annually, and provides between 47% and 80% of the animal protein consumed.[5] Rice is a staple carbohydrate and accounts for 76% of the average daily caloric value per person in the lower Mekong basin (Thailand, Laos, Cambodia, and Vietnam). Hydropower and thermal power are crucially dependent on the basin's water resources. Energy demand is growing rapidly.[6] At the same time, the GMS is witnessing increased industrial and urban expansion, a substantial contribution by agriculture to economic growth, and the debilitating impacts of climate change on the subregion's water resources.

The river's strategic significance derives mainly from a combination of elemental factors rooted in the food-water-energy nexus and exacerbated by climate change. But it is also a shared, and now increasingly contested, resource among the six riparian countries, which have a mixed record of cooperation over the past 30 years. An added element is the location of the subregion between the two large continental economies of China and India, a position that can potentially be either disruptive or beneficial in terms of a larger pan-Asian economic integration. The polities and governance systems of the countries are diverse, as are their perspectives of threats to national security. Although they are members of ASEAN, and ostensibly working toward an ASEAN Economic Community, there is a significant historical overhang of mutual suspicion among the six countries that often constrains the development of sustainable economic partnerships. The fact that Cambodia, China, and Vietnam are also directly embroiled in the contentious South China Sea dispute makes an already dry tinderbox a potential flash point for conflict.

It is against this background that this chapter will examine some key aspects. These relate to the impact of declining water resources on food security in the GMS, the energy-food nexus; the potential for conflict; and the politics of Mekong river management. It will also examine the security implications of the impacts for economic and political stability in the subregion and the role the United States could play through a new set of policy and institutional instruments.

Declining Water Resources—Impact on Food Security

The Mekong's basic hydrological regime has not changed over the past 50 years. The interannual variability of water in the basin is low to medium, and is seen mostly in Yunnan Province in China, Thailand's northeast, and Cambodia, but seasonal variations are very high across the basin.[7] However, changes have occurred as a result of storages built in the upper reaches of the river in China and in the lower Mekong basin that alter wet- and dry-season flows based on the needs of hydropower generation and flood control. Table 15.1 shows country flows as a share of total basin area.

Although overall water resources in the GMS have remained relatively stable over time, annual per-capita water resources have been declining for decades. They have more than halved in the approximately 50 years since 1962. The rate of decline—between 1.7% and 2.2% per annum—has been higher

Table 15.1 Hydrological Flows as a Percent of Total Basin Area

	China	Myanmar	Laos	Thailand	Cambodia	Vietnam	Total
Area in Basin (Sq. Km)	165,000	24,000	202,000	184,000	155,000	65,000	795,000
Catchment as % of Basin	21	3	25	23	20	8	100
Flow as % of Basin	16	2	35	18	18	11	100

than the average rate of efficiency gains in water use between 1990 and 2010, which is less than 1% per annum.[8] While the population has grown, so has the volume of water drawn by consuming sectors.

Agriculture alone withdraws between 64% and 98%[9] of accessible fresh water. This translates to 41.8 billion cubic meters (bcm) in the lower Mekong basin. More than half (26.3 bcm) is used in Vietnam in the delta region, followed by Thailand (95 bcm), Laos (3 bcm), and Cambodia (2.7 bcm). The area under irrigated rice (and other crops) in the lower Mekong basin is about 6.8 million hectares, of which almost 4 million are in the Mekong Delta in Vietnam.[10] During 2001–2009, average yields varied from 2.45 tons per hectare in Cambodia to 4.60 tons in Vietnam.[11] Rice is a staple in GMS diets; it is eaten at all meals.

Withdrawals by other sectors are considerably lower: the industrial sector withdraws between 1.5% and 24% of the total,[12] and the domestic sector between 1.47% and 4.77%.[13] However, both sectors are growing and will demand more water. The 10 largest cities in the GMS, for instance, are projected to grow from a combined population of 33 million in 2010 to 45 million in 2025. They are forecast to require 3.2 bcm of water in 2025, up from 1.76 bcm in 2010.[14] Projections of urban water needs in the GMS as a whole increase from 4.82 bcm in 2010 to 10.82 bcm in 2030.[15] This is hardly surprising given that 203 million people are forecast to live in towns and cities in 2050, as compared with 76.8 million in 2000.[16]

Energy demand is also forecast to grow as a consequence of industrial and urban growth. Recent estimates project a doubling of energy demand from 243,439 kilotons of oil equivalent in 2010 to 499,241 in 2025.[17] Projections of installed capacity show that the 2012 capacity of 118,967 megawatts will increase to 329,232 megawatts by 2025. This is an increase of 177%. From the perspective of the water footprint of this increase, we should note that an additional 42 coal- and lignite-fired plants will be set up (from the current 41). Gas-fired stations will increase from 39 to 54. Most significant, large hydropower plants are expected to increase from 116 to 254.[18] Clearly, the water footprint of the energy sector in the GMS is likely to double by 2025. There is little evidence of the GMS moving up the technology chain to reduce energy's dependence on water. Also, the numbers may be understated, as there is no evidence that consumptive water withdrawals in hydropower plants have been taken into account. As pointed out by Mekonnen and Hoekstra,[19] the water footprint of hydropower is significant: a global study of 35 hydropower projects (including Nam Ngum in Laos) in 2011 concluded that their annual water evaporation footprint was 90 cubic gigameters, equivalent to 10% of the total blue water needs of global crop production in 2000. Therefore, factoring in hydropower's water footprint in an era when the GMS is significantly expanding its hydropower assets is necessary to inject greater realism into our understanding of the subregion's water balance.

If we stay with the 2025 forecasts, we conclude that water demand by the domestic and industrial (including energy) sectors will, at a minimum, increase by 100%. In that event, water for agriculture will need to decline unless the irrigation sector is able to significantly increase its efficiency of water use. Creating new water through storages is already difficult because of large-scale urbanization and possible sites located in protected areas. In the Mekong Delta, the principal rice-growing region of the GMS, saline ingress is already playing havoc with rice production; bringing new areas under cultivation is improbable. The precipitation balance is also likely to alter with climate change, and extended dry periods with consequential evapotranspiration will put further pressure on water resources. An economic water scarcity scenario for the GMS by 2025 is entirely possible.

In its State of the Basin Report for 2010[20] (the 2015 report is not yet available), the Mekong River Commission (MRC) predicted climate change parameters for 2030 based on the Intergovernmental Panel on Climate Change's Scenario A1B using the general circulation models that best simulated past climate conditions in the Mekong basin (11 out of 24 available model simulations were selected). The study's forecasts of the average future climate effects for the Mekong basin can be summarized as follows:

- Temperature will increase by 0.79 degrees Celsius; there will be higher increases for colder catchments in the north of the basin (ranges from 0.68 to 0.81 degrees Celsius).
- Annual rainfall will increase by 200 millimeters, equivalent to 13.5%, mainly on account of increased wet season precipitation, going as high as 360 millimeters.
- During the dry season (February to July), an increase in rainfall in the northern catchments and a decrease in most of the lower Mekong basin can be expected.
- The total annual runoff is likely to increase by 21%. This will maintain or improve annual water availability in all catchments, but with pockets of high levels of water stress remaining during the dry season in some areas, such as northeast Thailand and the Tonle Sap Great Lake.
- Increased flooding will occur in all parts of the basin, with the greatest impact in downstream catchments on the mainstream of the Mekong River.

These changes will exacerbate pressures on accessible fresh water throughout the basin. It is difficult to say by how much, but the markers are already there by way of droughts, increased interseasonal flooding, and extreme weather events. It seems wise for the governments to understand these better through a series of sub-catchment–based vulnerability assessments.

The projected growth in the volume of water required by the industrial, energy, and municipal sectors notwithstanding, agriculture will continue to dominate access to water. It is estimated that agricultural labor in Cambodia, Laos, and Vietnam ranges from 65% to 85% of the workforce. Even in Thailand, where agriculture accounts for less than 10% of GDP, almost 70% of the workforce in northeastern Thailand is engaged in agriculture.[21] Because of Thailand's dependence on rice farming, or fishing, its vulnerability to changes in the basin's hydrology is high. A decline in the availability of water for agriculture will have an adverse impact on the labor force in terms of return to poverty or near-poverty conditions.

In the GMS context, no discussion of water and food can be divorced from that of fisheries. About 40 million people are engaged in fishing and related businesses, mainly in the lower Mekong basin. The current annual value of fisheries is estimated at $17 billion, equivalent to 3% of the combined GDP of the four countries and to 13% of the international trade value of fish as measured by the export forecasts of $130 billion for 2015 by the UN Food and Agriculture Organization. In Cambodia and Laos, fish production is worth an estimated 18% and 12.8% of GDP, respectively. Of the total production of 4.4 million tons, capture fisheries produced about 2.3 million tons and aquaculture yielded 2.1 million tons.[22] Much like rice, fish is a dietary staple in the region and accounts for more than half the animal protein consumed, ranging from almost 50% in Laos and Thailand to 60% in Vietnam and 80% in Cambodia. There can be no gainsaying the fact that fisheries, like rice, are crucially central to both economy and society in the GMS. But how vulnerable are the Mekong's fisheries, and to what? A quick analysis of the expanding hydropower sector can help us understand this better.

The Energy-Food Nexus

The increase in the number of large hydropower plants in the GMS from 116 in 2010 to 254 in 2025 is the principal cause of worry for the fisheries economy. Almost 40% of the total capture fisheries production comprises long-distance migratory fish; the dams would bar their migration paths. The impact of mainstream dam development (12 dams are being planned, of which one, Mengsong—the last in the Chinese cascade—has reportedly been postponed) is the estimated loss of 550,000–880,000 tons of migrant fish per year. In the case of Cambodia, fish consumption is likely to decline to 44 kilograms per capita per year from 2030, compared with 63 kilograms in 2011 (mainly on account of population growth and a small reduction in inland fish yields). However, in the case of mainstream dam development, the decline in fish consumption is estimated to range from 29 kilograms to kilograms kg (best- and worst-case scenarios).[23] In Laos and Cambodia, up to 30% of the national protein supply would be at risk if all 11 mainstream dams were built to plan. An impact at this scale on public health and nutrition, especially among some of the most vulnerable communities in the region, would be unacceptable. Collateral damage would include $3.53 billion every year in lost value of fish yields,[24] and losses suffered by ancillary and processing industries. Add to this the loss of primary and secondary livelihoods for several million people, and the GMS will likely witness a gargantuan socioeconomic disaster with serious ramifications for social stability and economic security.

Are there mitigating circumstances? The received wisdom suggests that reservoir fisheries cannot compensate for the loss of capture fisheries except marginally. In the long term, the reduction by 50%–75% of the nutrient and sediment outflow by 2030 is likely to have a negative impact on coastal fisheries, especially in Vietnam. Increased aquaculture also cannot replace capture fisheries; it can only complement them. Also, fish passes are not an assured way of mitigating risks. The Mekong has more than 130 migrant species, huge densities during migration peaks, and several migration pulses per year. Globally, fish ladders operate efficiently only when they are designed for a few species that migrate only once a year in limited numbers. Only three of the 11 mainstream dams have incorporated explicit fish ladder designs, and none have identified the specific species, or the final site locations.[25]

There will be impacts on agriculture too. About 135,000 hectares are likely to be inundated by the 11 projects. Land acquired for transmission lines and access roads will also be taken from agricultural and forest areas. Some 150,000 hectares of riverbank gardens, agricultural lands, and irrigation schemes will probably be directly affected by the reservoirs created by the 11 projects between Chiang Saen and Kratie. Twenty percent of affected agricultural lands are likely to be permanently lost through inundation or clearing, while the use and productivity of the remaining 80% under irrigation schemes may experience increased complications in management and system performance.

There are also likely to be impacts on the integrity and productivity of the Mekong aquatic system through (1) a permanent inundation of the majority of the river's aquatic habitats, (2) elimination at the local level of the seasonal distinctions of the river hydrology, and (3) cutting the transport of sediment and nutrients between the upland areas and the floodplains. Based on loss of habitat alone, the mainstream dams are projected to induce a 12%–27% reduction in the primary productivity of the aquatic systems (e.g., vegetal productivity), with implications for the overall productivity of the river and in the reservoirs themselves. Considering the estimated 75% reduction in nutrient loading as a cumulative impact of the mainstream dams, primary productivity could reduce to a small fraction of present values, with severe implications for the aquatic food chain, fish habitat, and fisheries. The mainstream projects are conservatively expected to be responsible for one-third of the reduction in nutrient and sediment loads of the Mekong. The Yunnan cascade of dams and other tributary developments expected by 2030 would be responsible for the other two-thirds of this reduction.

Finally, the mainstream dams are likely to have a negative impact on ecosystems of international importance. Many globally endangered species may become extinct. Habitat loss would encourage the proliferation of generalist species that do not migrate over long distances. If all mainstream dams proceed, 55% of the Mekong between Chiang Saen and Kratie would be converted into reservoir, shifting the environment from riverine to lacustrine. At least 41 riverine fish species found only in the mainstream, upstream of Vientiane, would be threatened. The loss in biodiversity would be permanent and irreplaceable—a global loss that cannot be compensated for. Of course, this relates to the mainstream dams only; impacts of the tributary dams are not factored in here. Cumulatively, they are likely to be substantial.[26]

The thermal-power expansion plans in the GMS also need to be incorporated in the water-energy-food balance. An accretion of 51,000 megawatts between 2012 and 2025 will pose significant demands on water. Coal and lignite plants take up to 17,000 gallons of water per megawatt-hour. Most of this water is released at high temperature into the river or reservoir and puts aquatic and marine life at risk. Also, pollution issues concerning coal and ash handling become manifest in soil and water regimes. This expansion will occur at a time when climate change will mean significant temperature increases and extended dry seasons. Water for thermal power stations cannot be taken for granted. As experience in India and China shows, thermal power stations are frequently shut down for extended periods for want of water, with consequential significant economic losses.

The Potential for Conflict

The analysis so far suggests that the socioeconomic future of the GMS is imperiled. Are communities concerned? They are, and conflict situations are increasing as livelihoods are adversely affected, and as local stresses find their way to country capitals for cross-border articulation. About 140 nongovernmental organizations (NGOs) called for MRC intervention to prevent dam construction on the mainstream dams in 2013.[27] In March 2014, communities from Stung Treng, Kratie, and Kampong Cham provinces in Cambodia went on a three-day march against the Don Sahong and Xayaburi dam projects that threaten to seriously impact their fisheries' livelihoods.[28] Communities in Thailand are now engaged in legal proceedings against Thai state agencies for "illegally" signing an agreement to purchase power for the 1,285 megawatt Xayaburi hydropower project in Laos.

The anti-dam sentiment has burgeoned since 2003 when the Pak Mun Dam first raised community hackles in the subregion. Originally a local issue, as in the Mekong Delta where saline ingress began destroying rice fields,[29] it gathered strength thanks to a mushrooming of civil society organizations that have played a key role in shaping and mobilizing public opinion.[30] Over the past five to seven years, the sentiment has acquired a political significance that national leaders are finding difficult to pat away. The Vietnamese president said in September 2012 that tensions over water resources are not only threatening economic growth but also presenting a source of conflict.[31] In April 2014, at the second MRC Summit in Ho Chi Minh City, the Vietnamese Prime Minister said that Laos was in violation of the 1995 Mekong Agreement under which any mainstream project required prior consultation. To set an example, Vietnam reportedly cancelled about 400 dam projects in 2014.[32]

Of course, it would be naïve to imagine that the anti-dam movement is either static or confined to affected local communities. It is a dynamic process with links to trade-offs. The Myanmar government had stayed work on the Myitsone Dam on the Irrawaddy River in 2011, but the project (designed to supply 6,000 megawatts to Yunnan Province) is now expected to be back in play with Aung San Suu Kyi's current visit to China.[33] In the case of the Xayaburi and Don Sahong projects that were once suspended, and on which work has recommenced, the Lao government would have taken a calculated decision in the context of the broader political economy of the subregion notwithstanding protests from Thailand, Cambodia, and Vietnam.[34]

Do these stresses and strains have the potential to develop into wider and more serious conflicts? They probably do. Cambodia and Thailand skirmished over the Preah Vihar border temple only eight years ago. Water is a far more serious business with a lower flash point. The temperature has been increased recently with sides being taken in the South China Sea dispute. Vietnam and China have a history of conflict and the construction of the Yunnan cascade of hydropower projects is already affecting Vietnam's delta region. Conflict may not be imminent, but the potential grows. The disregard shown by Laos in flouting protocols established by the MRC is bound to be more seriously challenged sooner than later. And if the several million rice and fish farmers in the lower Mekong basin are going to see diminished or lost livelihoods, social unrest is a foregone conclusion. From there, conflict is usually a very short step.

Giant Catfish and Dolphins: The Human Impact

About 10 years ago, the giant catfish, one of the world's largest freshwater fish species, was an icon of the Mekong River. It is one of the approximately 850 species that live in the Mekong, and among the 135 that are migratory. Pornsawan Boontun, a community senior in the northern Thai village of Huay Luk in Chiang Mai Province, remembered how villagers had caught a catfish that weighed 615 pounds. He confesses to this being a distant memory—the giant catfish is now hard to find throughout the Mekong basin.[35]

Has its death knell been sounded? It would seem so. The seasonal fluctuations in the river are extreme, with the volume of water in the dry season being up to 30 times less than the wet season. This manner of change in the habitat promotes mobility and migration among fish populations. And that is what the Xayaburi Dam, adjacent to a spawning area, is unlikely to allow. As the first dam in the lower Mekong basin, specialists say that its impact is likely to be severe.

So, Pornsawan and his fellow villagers today work with the Thai-based "Living River Siam Association" in what seems like a feeble attempt to ward off the inevitable. They study riverbank erosion and document changes to the river's habitats. Fish catches are in decline in stretches of the river adjacent to their communities, riverside gardens are affected, and harvests of *kai*, or Mekong water weed that nourishes human and fish populations, are much reduced.

Not far from Huay Luk, near the site of the Don Sahong Dam in Champassack Province, Yin Vuth, a 53-year-old Cambodian national, joined hundreds of others in protesting against the project. Like Pornsawan, he has a livelihood issue. He runs a tourism business related to dolphin spotting. Recent assessments show only about 80 freshwater dolphins left in the wild, down from an estimated 200 in 1997.[36]

The dolphins are native to Anlong Cherr Teal in Cambodia's Preah Rumkel commune. Dam Chan, a 55-year-old resident who has farmed and sold food all her life, is desperately worried as she watches the decline in the dolphin population. As a child, she used to see 20–30 dolphins playing every day, as compared with two to three today. She finds it deeply ironic that just as the government constructed roads and other infrastructure that saw an increase in the number of tourist arrivals, the dam construction commenced and the dolphin migration patterns changed. Sightings are rare now, and tourist traffic has perceptibly declined. She knows that the trajectory of her life will change after the dam is constructed. Today's impacts are mainly construction related—use of dynamite and heavy machinery, and water diversion. But future impacts arising from a change in the river's hydrology will be "dolphin ending." Her thoughts are echoed by Bun Thorn, a 44-year-old dolphin-spotting tour operator. He sees no prospects of a recovery from the downturn his business has taken in recent times.[37]

All three individuals see a blighted future. Will they accept their altered condition with equanimity? Or will they agitate against it? Much depends on how their leaders see the light of day and what steps they take to pull their futures back from the brink.

The Politics of Mekong River Management

The MRC is the principal river basin management organization. Its members are Cambodia, Laos, Thailand, and Vietnam. China and Myanmar have declined membership—they have "Dialogue Partner" status. MRC's mandate requires its members to "cooperate in all fields of sustainable development, utilization, management and conservation of the water and related resources of the Mekong river basin" and "ensure reasonable and equitable use" of the Mekong River system. It is largely dependent on donor funding. Development partners include 12 countries in Europe and North America, the European Union, and the World Bank. Partner organizations include the Asian Development Bank (ADB), ASEAN, the UN Development Programme, and the UN Economic and Social Commission for Asia and the Pacific.[38]

Considerable analysis is available in the public domain on the effectiveness of MRC, the impact of China and Myanmar not being full members, the commitments of the four members, and prospects for effective river management.[39] A few issues stand out. As a Dialogue Partner, China has agreed to supply wet-season river data to the MRC but has declined to provide dry-season data. Its cooperation has been minimal, and it has made no bones about it. It is easy to speculate that China's lack of full membership makes matters difficult. Although membership would impose discipline and responsibility, the reality is that China's strategic interests are never constrained by the niceties of formal organizational memberships. That said, the Mekong is an international river, and any arrangements for shared management can work only if all parties join hands. For China to become a full member of the MRC would be in line with its professions of a "peaceful rise." It is a moot point whether this can, or will, happen.

Next is the perceived limited value of the MRC's work. Although its mandate is broader, it has mainly managed to gain cooperation from its member states on apolitical issues. Several river-monitoring programs have gathered significant data but with limited effectiveness in analyzing trade-offs to inform decision-makers and creating a platform for discussion.[40]

Some analysis also shows a considerable disconnect between the MRC's program objectives and those of the individual members' national Mekong committees or water resources organizations.[41] This seems so because of the inherent structural dichotomy in the MRC: while its priorities are mainly driven by its development partners and are focused on principles of good water-resources management, the member countries are more driven by national priorities. Also, the MRC lacks the authority to drive transboundary water governance. This is a design limitation and not a functional weakness.

Flouting the MRC's agreed-upon working procedures does not help. The Lao and Thai governments were in dispute with MRC over processes in late 2011 on the Xayaburi project. Construction on the dam commenced, was suspended subsequently because of strong protests from Cambodia and Vietnam, and recommenced thereafter. This also happened with the Don Sahong project in Laos, where ground-breaking was reported on August 16, 2016.

Given the restriction of MRC's consultation processes to mainstream dam projects, activities on the Mekong's tributaries acquire particular significance. The combined effects of the proposed tributary dams in Laos are estimated to have a greater detrimental impact than the proposed mainstream dams.[42] Similarly, the Thai Water Grid project that aims to divert water from the Mekong to northeastern Thailand, but which has not been executed so far, is also outside the purview of the MRC. If implemented, it would exacerbate river management issues. Pornsawan, Yin Vuth, Dan Cham, and Bun Thorn would all be affected. But so would millions of others, and not merely by their immediate environment and loss of livelihoods; they would have lost any trust that remained in their leaders and governments to care for their welfare.

Under these circumstances, it may be tempting to dismiss MRC as a toothless intergovernmental bureaucracy. This would be hasty. The MRC has, by all accounts, provided value as a platform for discussion of mainstream dam issues, flood management, and civil society engagement. It has also served as a hub for academics and the media to engage with member governments.

Perhaps the MRC's singular weakness has been a lack of ownership by its members. Their perception that it is a donor-driven organization has brought MRC to a virtual standstill. Against a need of $115 million over the 2011–2015 period, only $53 million has been committed for the 2016–2020 budget. Up to half of the 150 staff are proposed to be relocated to member governments in a so-called decentralization exercise, over which there is considerable skepticism.[43] Donors are questioning MRC's value and have not made further financial commitments, pending a full review of the organization's plans.

With MRC at the crossroads, what can we expect next? While the commission's demise is far from certain, China has shrewdly observed the MRC's diminished ability to handle substantive river management issues. In April 2015 it seized the initiative and proposed the Lancang-Mekong River Community of Common Destiny (LMCCD).[44] All six GMS countries are members of this initiative, and meetings at the senior-official level have commenced to develop a cooperative framework. In March 2016, the Chinese Foreign Minister said that his government saw this initiative as supporting the China-ASEAN relationship.[45]

Clearly, the latent asymmetry suggested by China's geopolitical influence, and its control over the Mekong in its territory, are now explicitly manifest. The LMCCD is an unvarnished attempt to impose a China-led cooperative exercise on an existing patchwork quilt of subregional economic cooperation programs. That China has deep pockets and will be willing to use them with significant collateral benefits for its 21st Century Maritime Silk Road project is obvious. How it will play out in the next decade is more difficult to predict.

A Role for the U.S.?

As this analysis has demonstrated, the GMS is potentially worrisome from a geostrategic security perspective. Its crucial dependence on the Mekong River for food, energy, and economic security; the inability of countries to design and implement effective, cooperative river basin management arrangements; the overlapping and competing interests in the South China Sea dispute; and the prospective erosion of current economic baselines given climate change forecasts make it a toxic cocktail of potential instability.

The U.S. has been involved in the development agenda of the GMS for decades. Its most recent program is the Lower Mekong Initiative (LMI) of 2009, which is a broad program encompassing agriculture and food security, energy security, environment and water, gender, health, and other crosscutting issues. In 2012, $50 million was committed over three years to expand the program. Most components address capacity-building issues, and overall budgets are small when compared with amounts provided bilaterally by China or Japan, or multilaterally by ADB or the World Bank. The ADB alone had committed $6.65 billion from 1994 to 2015 over 76 projects that brought in contributions of $4.65 billion from the member governments, and $6.45 billion from cofinanciers, including the private sector. The reality is that while programs such as the LMI are useful in building capacities in key development areas, substantive leverage is essentially in the hands of the big investors.

How should the U.S. seek to develop a more effective role in helping the GMS countries attain global good practice levels of effective river basin management? With China having taken the initiative through the LMCCD, there is not much wiggle room left for the U.S. to directly influence infrastructure-related development. However, a practical, near-term role for the U.S. would be to influence the broader policy and technical environment with substantially higher levels of engagement, and with greater use of technological and financial resources. Employing a two-track approach would be a useful starting point, with carefully designed synergies and crossover points between the two.

One track would suggest a complete overhaul of the MRC. The MRC would need to be renewed with the full membership of China and Myanmar, and with full responsibility for transboundary river management and dispute resolution. Since the LMCCD is essentially focused on infrastructure development within the basin, a renewed MRC could be the substantive technical and regulatory platform on which the infrastructure investments are based. The timing is also right given that MRC's future is at an inflection point. If the U.S. were to take leadership, in alliance with current MRC partners, this would be seen as an open and substantive commitment to helping the LMCCD attain its grand design. Redesigning the MRC, however, must not be seen as an opportunity to wipe the slate clean. The good work done hitherto should be built upon.

The need for a redesigned MRC stems from several factors. A paramount need is to strengthen the information and knowledge base. Building a transboundary decision support system that integrates water, energy, and food poses huge challenges for international river basins. MRC's focus on hydrology, sediment transport, and water-quality parameters is not enough. The Basin Development Strategy (2016–2020) is a good beginning but more needs to be known about trade-offs at the basin scale to underpin the development of strategies and plans. Perhaps a "State of the Basin Report and Road Map" that moves beyond MRC's traditional reports and integrates sector assessments, and the latest climate change data and scenarios, is needed to help the LMCCD make informed investment decisions. This would, ideally, be based on a multi-sector assessment that integrates biophysical and socioeconomic aspects in the basin but also assesses the water-resources development options and benefits of cooperation based on analysis of all water-related sectors including wetlands, tourism, and navigation. Various development options could then be scenario-analyzed and used as the basis of a 15-year road map that would assist the LMCCD at a regional level and the countries at national level, and be a model regional protocol.

The other track is for the U.S. to play a more effective role in shaping the policies and programs of the multilateral development banks (MDBs) operating in the region. Both the World Bank and the ADB have large programs. The latter's GMS Economic Cooperation Program supplements the country-level programs that both banks are engaged in. In ADB's case, it has stayed away from transboundary water resources programs because of its reluctance to be involved in current or prospective disputes. But its inability to see the big picture in the basin has meant that its country-level water resources programs are driven by country priorities that do not necessarily accord with what might have been regional priorities. Its country water projects are typically not set in a basin context. This is also true of the World Bank, even though it has a smaller development footprint in the GMS when compared with ADB.

If the U.S. intervened more directly and thoughtfully, together with its development partners, in the design of the GMS strategy and program, it might result in better coherence between its efforts and the individual countries' strategies and programs. These are overseen by boards of directors, and there is no reason for the U.S. not to exercise a more active voice in their design and execution. At a higher level, it is not incongruent for the MDBs to be playing a role in supporting the LMCCD or the MRC. After all, it was the World Bank that brokered the Indus Waters Treaty, which has stood the test of time, between Pakistan and India in 1960. It was also instrumental in promoting the Nile Basin Initiative of 2010, which involved nine countries. For the Mekong, the design of a similar initiative, promoted by the MDBs and with support from the U.S., would not be a leap too far.

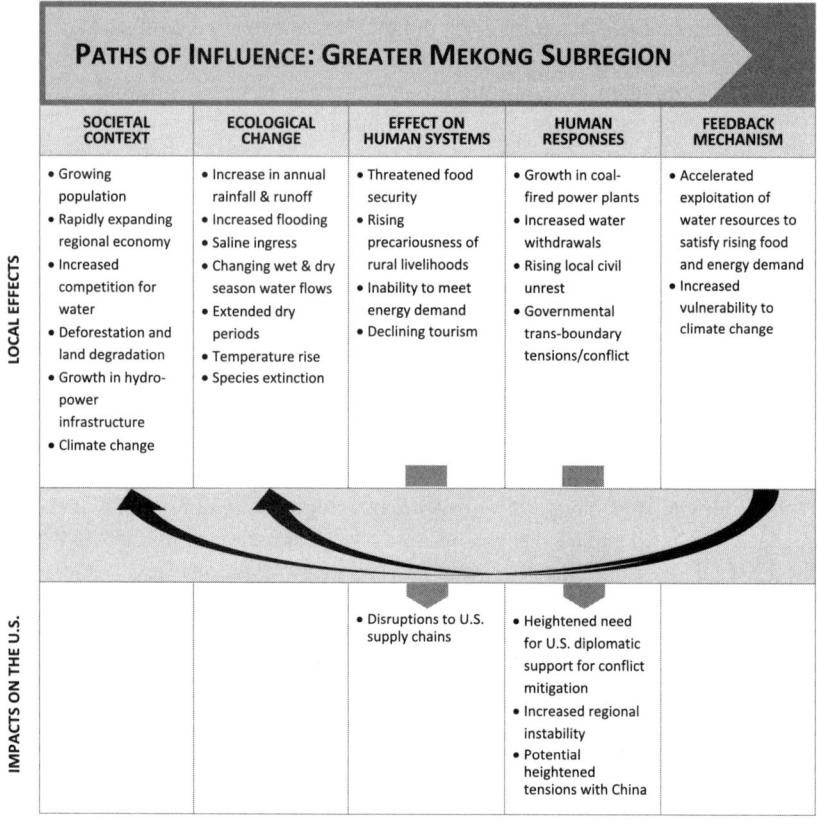

Figure 15.1 Paths of Influence: Greater Mekong Subregion

Finally, it seems necessary for the U.S. to take on its leadership role in the GMS with a single voice. It does not help that the LMI, designed and managed by the U.S. State Department, and the programs run by USAID are not unified. Perhaps it is time for the U.S. to ensure that future assistance is integral not only to its own objectives but, more important, to those of the GMS and in partnership with others'. A good example is the USAID-supported Mekong Adaptation and Resilience to Climate Change Project that has piloted various adaptation options at the community level in collaboration with ADB's GMS Core Environment Program. Scaling up and replication should also be undertaken in partnership with larger investment programs that are on the design anvil.

As they stand on the banks of the Mekong, Pornsawan in Thailand, and Yin Vuth and Dam Chan in Cambodia speculate endlessly about their future. The irony that they are now part of the "Lancang-Mekong Community of Common Destiny" is perhaps lost on them. Their destiny is really in the hands of those who have so far shown little evidence of the ability to efficiently manage the water resources on which their lives, and their children's lives, depend. History is replete with examples of communities being driven to the wall and forging their own destinies, often in violent ways. There is time yet, but not much, to gently shape a secure economic and social future for the GMS.

Notes

1 "GMS Statistics on Growth, Connectivity, and Sustainable Development" (Asian Development Bank, September 2015).
2 The Asian Development Bank–supported Greater Mekong Subregion Economic Cooperation Program, initiated in 1992, remains the principal external impetus to build a peninsular economic community.
3 Gilbert F. White, "The Mekong River, Southeast Asia," May 2014.
4 Scott William David Pearse-Smith, "The Impact of Continued Mekong Basin Hydropower Development on Local Livelihoods," *Consilience: The Journal of Sustainable Development* 7, no. 1 (2012): 73–86.
5 "Investing in Natural Capital for a Sustainable Future in the GMS" (Asian Development Bank, September 2015).
6 "GMS Energy Sector Assessment, Strategy, and Road Map" (Asian Development Bank, June 2016).
7 F. Gassert et al., "Mekong River Basin Study" (Washington, DC: World Resources Institute, 2012).
8 2030 Water Resources Group. *Charting our Water Future. 2010*. This is a global figure and not specific to the GMS.
9 The 64% figure relates to China; however, no figures are available for Yunnan and Guangxhi Zhuang provinces. The range for the other five countries is from 90.3% to 94.78%.
10 Mekong River Commission. Crop Production for Food Security and Rural Poverty. Baseline and Pilot Modeling. 2014.
11 Mekong River Commission. Working Paper 2011–2015.
12 The 24% figure relates to China; however, no figures are available for Yunnan and Guangzhi Zhuang provinces. The range for the other five countries is from 1% to 4.86%.

13 World Bank. *World Development Indicators (Mekong)*. 2015.
14 Rogers, Peter. *Water-Energy Nexus: Sustainable Urbanization in the GMS*. 2012.
15 "International Conference on GMS 2020: Balancing Economic Growth and Environmental Sustainability Focusing on Food–Water–Energy Nexus" (Asian Development Bank, 2012).
16 Asian Development Bank. *Urban Development in the GMS—Trends and Forecasts*. 2016.
17 "International Conference on GMS 2020: Balancing Economic Growth and Environmental Sustainability Focusing on Food–Water–Energy Nexus."
18 Asian Development Bank. *GMS Energy Sector Assessment Strategy and Roadmap*. 2016.
19 Mekonnen, M. M., and Hoekstra, A. Y. *The Water Footprint of Electricity from Hydropower*. 2011. UNESCO-IHE Institute of Water Education.
20 Mekong River Commission. *State of the Basin Report*. 2010.
21 Mekong River Commission. *State of the Basin Report*. 2010. Page 48.
22 Mekong River Commission. *Catch and Culture—Fisheries Research and Development in the Mekong Region* 21, no. 3 (December 2015).
23 Baran, E., Cheng, P., and Nao, T. *Food Security in the Mekong: The Contribution of Fisheries*. 2013. SIWI/UNDP/IWMI Workshop on Food Security in the Mekong. Chiang Rai, Thailand.
24 Estimated at 40% of the value of capture fisheries (a total of 2.3 million tons valued at $8.84 billion in 2015). However, the 2010 *State of the Basin Report* estimated a combined fisheries and agriculture annual loss of $500 million, partially offset by a $30 million annual gain from reservoir fisheries and new irrigation potential.
25 International Center for Environmental Management. *Strategic Environmental Assessment of Hydropower on the Mekong Mainstream*. Final Report. 2010.
26 International Center for Environmental Management. *Strategic Environmental Assessment of Hydropower on the Mekong Mainstream*. Final Report. 2010.
27 Amelia Woodside, "Halt Building of Dams on Lower Mekong: NGOs," *Phnom Penh Post* (August 6, 2013), www.phnompenhpost.com/national/halt-building-dams-lower-mekong-ngos.
28 Laignee Barron, "Dam Critics Unswayed," Text, *Phnom Penh Post* (March 14, 2014), www.phnompenhpost.com/national/dam-critics-unswayed.
29 According to the Ministry of Planning and Investment in Vietnam, up to 45% of the Mekong Delta will be affected by saline intrusion by 2030 if the proposed hydropower projects obstruct water flows. See VN Express International, *Vietnamese Farmers Indignant as Mekong Delta Prays for Floodwaters to Arrive*. August 15, 2016.
30 Ron Corben, "Communities Band Together to Oppose Mekong Dams," *VOA*, September 18, 2014, www.voanews.com/a/communities-band-together-to-oppose-mekong-dams/2454113.html.
31 Parameswaran Ponnudurai, "Water Wars Feared Over Mekong," *Radio Free Asia*, September 30, 2012, www.rfa.org/english/commentaries/east-asia-beat/mekong-09302012160353.html.
32 Kraisak Choonhavan, June 10, 2014, "Vietnam Demands Halt to Mekong Dams," *The Third Pole*, June 10, 2014, www.thethirdpole.net/2014/06/10/vietnam-demands-halt-to-mekong-dams/.
33 Htet Naing Zaw, "Peace Process, Myitsone Dam on Agenda for Suu Kyi's China Trip," *The Irrawaddy*, August 15, 2016, www.irrawaddy.com/news/burma/peace-process-myitsone-dam-on-agenda-for-suu-kyis-china-trip.html.
34 Kate Ross, "Don Sahong Dam Casts Wide Shadow Over Mekong," *The Bangkok Post, International Rivers*, July 2016, www.internationalrivers.org/resources/don-sahong-dam-casts-wide-shadow-over-mekong-the-bangkok-post-11519.

35 Joshua Zaffos, "Life on Mekong Faces Threats as Major Dams Begin to Rise," *Yale Environment 360*, February 20, 2014, http://e360.yale.edu/feature/life_on_mekong_faces_threats_as_major_dams_begin_to_rise/2741/.
36 Phak Seangly, "Hundreds Protest Laos Dam," Text, *Phnom Penh Post* (March 31, 2014), www.phnompenhpost.com/national/hundreds-protest-laos-dam.
37 May Titthara, "Don Sahong vs Dolphins: How the Dam Is Affecting Local Residents," *Khmer Times | News Portal Cambodia*, June 13, 2016, www.khmertimeskh.com/news/26037/don-sahong-vs-dolphins--how-the-dam-is-affecting-local-residents/.
38 The Mekong River Commission, www.mrcmekong.org/about-mrc/development-partners-and-partner-organisations/.
39 See, for instance, Nathaniel Matthews' "Water Grabbing in the Mekong Basin: An Analysis of the Winners and Losers in Thailand's Hydropower Development in Lao PDR," 2012, and "Paper Tiger meets White Elephant—An Analysis of the Effectiveness of the Mekong River Regime" by Ellen B. Backer, Fridtjof Nansen Institute, 2006.
40 John Dore and Kate Lazarus, "De-Marginalizing the Mekong River Commission," *Contested Waterscapes in the Mekong Region: Hydropower, Livelihoods and Governance* 2 (2009): 357.
41 Suhardiman, D. et al., 2012. As reported in Matthews, "Water Grabbing in the Mekong Basin."
42 Ziv et al., 2012. As reported in Matthews, ibid.
43 Igor Kossov and Lay Samean, "Donors Slash Funding for MRC," Text, *Phnom Penh Post* (January 14, 2016), www.phnompenhpost.com/national/donors-slash-funding-mrc.
44 *China Daily–USA*. April 7, 2015.
45 "China Wants Closer Community of Common Destiny with ASEAN: FM—Xinhua," English.news.cn, *Xinhua*, March 8, 2016, http://news.xinhuanet.com/english/2016-03/08/c_135166991.htm.

16 Building Resilience for Peace
Water, Security, and Strategic Interests in Mindanao, Philippines

Roger-Mark De Souza

Introduction

Security dimensions in Mindanao, the second-largest and most southern island group in the Philippines, are multifaceted. Contemporary conditions pose multiple threats to the Philippines' defensive capabilities and political stability. These threats also endanger the well-being and self-determination of its peoples, as well as the viability of strategic natural resources and economic security. Ultimately, international security is also at risk as a result of terrorist activities in the region.[1] Control over and access to water resources, availability in rural and urban areas, and connections to other key sectors of the economy underlie these security concerns. Additionally, climate change is bringing unstable weather patterns and larger storms to the region, putting increased stress on Mindanao's political and economic stability. Storm damage and outbreaks of violence have displaced millions of people and placed a burden on the central government.

Mindanao denotes one of the three island groups that, together with Luzon and Visayas, make up the Philippines. It consists of the island of Mindanao and smaller outlying islands. These include Maguindanao, Basilan, Lanao del Sur, Sulu, and Tawi Tawi, the five predominantly Muslim provinces that make up the Autonomous Region in Muslim Mindanao (ARMM). The three largest cities in Mindanao are Davao, Zamboagna, and Cagayan de Oro, all located on the island of Mindanao.[2]

Mindanao's history is rife with conflict, and the use of force and violence has emerged as a legitimate vehicle for both the state and aggrieved groups to settle grievances. Together, these factors represent a nexus of military, political, societal, environmental, and economic dynamics. These dynamics intertwine and enmesh in the narratives and acts of discontent—an improvised explosive device waiting to detonate and fling its shards far and wide, across the island, the country, and beyond. This chapter will explore the position of Mindanao in this security framework and discuss challenges and opportunities in this geographically strategic and complicated area in terms of U.S. military, diplomacy, and development interests.

Figure 16.1 Map of the Philippines

Source: The World Factbook 2013–14. Washington, DC: Central Intelligence Agency, 2013. www.cia.gov/library/publications/the-world-factbook/index.html

Current and Emerging Water Conflict Dynamics

In the Philippines, 85% of the country's water demand is for agriculture, with the remaining 15% going to the industry, commercial, and domestic sectors.[3] Of the country's five principal river basins, one is located in the Mindanao River.[4] Mindanao hosts 262 watersheds. The two largest are the Agusan and Pulangi watersheds. The Pantaron Mountain Range is central to these important water sources and affects more than half of the Mindanao mainland. It covers approximately 12,600 square kilometers and covers about 12.4% of Mindanao.[5] There are also dams in the region. The National Power Cooperation operates and manages three dams in Mindanao, with a capacity of approximately 27.7 million cubic meters.[6] In Mindanao, as in the country overall, the households that get limited water include those whose supply comes from unprotected or poor-quality sources, those whose needs extend beyond local water resources, and those unable to access sufficient potable water due to poor governance (particularly in conflict areas).[7] These water-poor populations are on the brink of instability.

In those cases where water is not directly connected to the proximate causes of conflict, water dynamics fit into the larger context. Specific events—the exploitation of an important marsh such as the Liguasan Marsh or severe droughts—can spur people's decision to join a protest. Disputes over access or usage rights can turn violent if there are no strong, legitimate governing institutions that cover resource rights, including water rights. Water may also be used as a tactic against local communities, such as by depriving access.

Some key water conflict dynamics include the following.

Drought and urban water security: In April 2016, drought drove some 6,000 hungry farmers and their families in Kidapawan City, the capital of the Cotabato Province, to demand food assistance, with thousands protesting in the streets and two people being killed. In 2016, some speculate that a food and water crisis was developing in Mindanao in the wake of what had been anticipated to be one of the most severe El Niño episodes ever recorded. Local reports indicated that farmers and indigenous groups would have been in a better position to manage such impacts if their watersheds and water systems had not been compromised by the activity and growth of agribusiness plantations, coal-fired power plants, and mining operations.[8]

Zamboanga City has also been struggling with a three-year drought and has had to ration its water. With support from the Water Security for Resilient Economic Growth and Stability Project ("Be Secure") of the U.S. Agency for International Development (USAID), the city created an urban water demand management plan anchored in five key components: expanding local water treatment capacity, improving staff members' ability to determine and repair leaks, promoting rainwater harvesting, performing residential and commercial water audits, and upgrading the sewage system to limit contamination of water supplies.[9] This is part of an effort that is housed under the Cities Development Initiative, working to advance secondary cities as drivers of economic growth. As the World Bank specifies, however, economic growth in Mindanao is tied to managing its conflict dynamics—this offers an important entry point for water management, peacebuilding, and economic development.[10]

Energy production, water insecurity, and land tenure: Persistent and lingering insecurity around water—in terms of energy, public access, and livelihoods—has contributed to a sense of disenfranchisement and economic isolation among rebel groups. At the heart of their grievances is a proposed dam on a tributary of the largest river in Mindanao. For the Moro Islamic Liberation Front (MILF), a militant group seeking independence from the central government, the dam was seen as an attempt by the government to extract oil and gas from the Liguasan Marsh, as the dam would drain the marsh and make the area drillable. Additionally, the dam could flood several thousand hectares of farmland, threatening the livelihood of several Muslim communities.[11] The Liguasan Marsh is also a very contentious area of potential development and exploration. It covers 220,000 hectares of the provinces of North Cotabato and Maguindanao, which ethnic Muslims (the Moro people) have claimed as part of their ancestral domain.[12] Under a 2001 expanded version of earlier legislation that granted autonomy to the ARMM, the regional ARMM government lost control over the development of its natural resources such as the Liguasan Marsh, which were ceded to central government ownership.[13]

As highlighted in the case of ancestral domain claims, economic development in Mindanao is intimately linked to land tenure, indigenous rights, and conflict. Development of resource-rich areas has been curtailed because some of these areas are under MILF control and the central government is not able to convince investors of their safety, even when there are temporary ceasefires. Questions of ancestral domain and land claims need to be resolved before any government projects can be implemented.[14]

Impacts from other sectors on water resources: One lingering concern regarding water security is the impact of other sectors on water resources. In the T'boli municipality, for example, a 1980s gold rush brought many fortune seekers, many of whom ended up disappointed and turned to illegal mining and logging instead. As rivers have dried up, remaining water resources have been poisoned by mercury and cyanide, which were used to separate gold from mined ores.[15] With a lack of clarity about who controls natural resources, major industries have been granted permits by the central government to exploit these areas with little regard for environmental standards. The impacts have been harmful throughout Mindanao. Such actions have contributed to heavy siltation in four major river tributaries of Lake Lanao, the largest lake in Mindanao, located in the Lanao del Sur Province. Subsequently, water flows into the lake have been restricted, limiting electricity generation that could help curb recurring incidences of blackouts and brownouts in the region.[16]

Displacement and water scarcity: An associated issue is the provision of water services for those who are displaced by instability, droughts, and storms. After more than 250,000 people were displaced due to prolonged fighting between MILF and government forces in central Maguindanao Province in 2009, reports indicated dire situations in evacuation centers. The centers struggled to provide access to potable water, with acute shortages of latrines, water points, and sanitation. Those camps in higher elevations had very low groundwater tables and others were marshy or flooded.[17]

Since 2000, more than 4 million people have had to relocate from Mindanao due to violence.[18] Another 4 million were forced to relocate because of Typhoon Yolanda in 2013 and additional climate change impacts.[19] This has created a food crisis in displacement camps. During an "all-out war" against MILF secessionists in 2000, 30%–40% of preschool children were malnourished, according to the Internal Displacement-Monitoring Centre.[20]

Historical and Geophysical Roots of Instability

These environmental triggers are situated in deep-seated historical and geophysical dynamics. Conflict has long been part of Mindanao's history. The conflict landscape is multidimensional and dependent upon the interplay of a variety of stressors that include socioeconomic, institutional, external, and geographic issues.

The island has experienced a communist insurgency as well as armed Muslim separatist movements, both of which are partially reflected in its current administration. The ARMM was established in 1989 by the Organic Act for the Autonomous Region in Muslim Mindanao, a legislative act that provided for ARMM's own government. In a 2014 peace deal, the national government signed the Comprehensive Agreement on the Bangsamoro (CAB) with the MILF, which proposed that the autonomy be redesigned to focus on power sharing and the ARMM become known as the Bangsamoro Autonomous Region.[21] The Bangsamoro Basic Law, which would have enforced the propositions of CAB, did not pass when it was deliberated upon during the 16th Congress of the Philippines in June 2016 due to questions of constitutionality.

The Indigenous Peoples, or Lumad, are the original inhabitants of Mindanao and are composed of 35 tribes. The Moro population, composed of mostly Muslim communities, is made up of 13 ethnic groups. The most recent settlers are Christians from Luzon and Visayas. Although exact estimates vary, the settlers and their descendants make up the largest proportion of the population, followed by the Moros and the Lumad.[22] Contentions in Mindanao include conflict between these groups, the government, and the MILF, and between the central government of the Philippines and the Islamic Moro people of Mindanao.

Two persistent factors feed the violence, according to analysts. First is the absence of the rule of law. Political and social institutions are ineffective and the judicial system is both ineffective and corrupt, which spurs violence among citizens as they seek to resolve their own grievances. This has exacerbated so-called *rido*, or clan violence, which may persist over generations. Second are the lingering effects of the dictatorship of Ferdinand Marcos (1965–1986), who brought state-sanctioned, politicized violence against those sectors of society that were opposed to government policies. These acts of violence led to more support for the MILF and a subsequent splinter group called Abu Sayyaf.[23] These factors have been exacerbated by simultaneous socioeconomic problems including religious differences and widespread poverty.

Socioeconomic factors include population settlement policies, displacement, and ethnic tensions. Sustained underdevelopment of the region is also a key driving factor. The interplay of these factors can be observed in the municipality of Maluso on the island province of Basilan. In the 1900s, investors from Zamboanga, Negros, and Luzon occupied the island for economic opportunities. There is some evidence of initial peaceful coexistence of Christian, Muslim, and indigenous groups, but displacement and economic isolation played on tensions.

Muslim tribes of Tausugs and Samals were concentrated in coastal towns, some indigenous tribes were driven far inland, and the Christians settled in plantations and cities. The Christians owned most of the arable land and businesses, while indigenous peoples' ancestral claims were not fully recognized and many Muslim groups had no title to land. With no formal or informal dispute resolution mechanisms, the result was uncertainty and insecurity.[24] The central government's resettlement program ultimately increased the ratio of Christians to Muslims in the region, adding fuel to the fire given the relatively underdeveloped economic status of the Muslim population. Local rebel groups attribute this underdevelopment to the government's inability to effectively integrate their ethno-religious groups.[25] Inequitable access to and ownership of natural resources have pushed many in the Moro population to separatism.

Mindanao has been described as being "trapped in a vicious circle of conflict and underdevelopment," which leads to constrained economic integration and development.[26] But conflict has not affected all economic development. The World Bank notes that the economy in conflict zone areas of Mindanao with high economic density and integration has grown at a higher rate than in non-conflict zones in areas that are less integrated.[27] These integrated areas demonstrate high "geographic compactness of economic activity in one place."[28] The key is to figure out how to harness that compactness with peacebuilding efforts.

Finally, a number of geophysical changes are anticipated in the region and are already being observed, particularly around climate change. Rising temperatures due to climate change and heightened El Niño–Southern Oscillation effects are causing extreme drought in the southern islands of Mindanao.²⁹ The El Niño and La Niña climate patterns are growing stronger and oscillating on a three-year interval, as opposed to the prior five-year interval.³⁰ Global climate change is causing the waters to warm, the seas to rise, and the storms to grow stronger. Since 2006, the Philippines has recorded five of the 10 strongest storms in its history.³¹

These changes and the political instability are having an effect on the economy. Mindanao is the farming and agricultural core of the Philippine archipelago.³² More than 48% of jobs in the region are in the agriculture sector, and overall Mindanao accounts for 14% of the nation's gross domestic product (GDP) (2014).³³ Without sufficient water supply and storm-recovery mechanisms, agriculture may no longer be economically profitable and many people will face displacement. Droughts brought on by the effects of El Niño and heavy monsoons caused by La Niña are already making some agricultural land unusable.³⁴ Environmental degradation of watersheds due to droughts and flash floods also creates an inconstant source of water.³⁵ By 2050, the average rainfall is projected to decrease by 11%.³⁶ This reduction in water supply will hinder the ability to generate hydropower and stunt crop production.³⁷ This could potentially create a food security crisis for the area and the nation as a whole. Indeed, the national government has lifted all export tariffs and minimized importing quotas in an to attempt to stabilize the agricultural sector and promote crop diversification.³⁸ Food instability threatens rural livelihoods and the national economy. The challenge is to determine the interplay of these factors, including water, with resource conflict dynamics (see Figure 16.2).

U.S. Strategic Interests and Changing Political Winds

In 2012, the U.S. intelligence community warned that countries important to U.S. interests would experience water problems in coming years, which could lead to instability and greater tension, and distract the countries from U.S. policy objectives.³⁹ The potential problems identified include shortages, poor water quality, and floods, which when combined with poverty, social tensions, environmental degradation, ineffectual leadership, and weak political institutions can result in state failure. These factors are all very salient in Mindanao.

The Global Climate Risk Index 2015 listed the Philippines as the country most affected by climate change according to recent data.⁴⁰ Mindanao is home to 10 of the 16 poorest provinces in the Philippines.⁴¹ Ongoing violence makes the region highly susceptible to the negative ramifications of climate change.

Displacement has created space for terror operations to assimilate into rural Mindanao.⁴² Some extremist groups in the area have been deemed "foreign terrorist organizations" by the United States and have carried out deadly strikes on western targets.⁴³ Recently, the Abu Sayyaf militia, formerly linked to al-Qaeda, was appointed leader for the Islamic State of Iraq and Syria (ISIS) in the Philippines.⁴⁴,⁴⁵ There are also reports that ISIS training camps have been established in Mindanao, and there is fear that water may become a weapon of war.[46] As in Iraq and Syria, extremist groups could take control of water sources during a drought, force agriculture workers to flee, and damage infrastructure.

Managing these implications is particularly important given U.S. geostrategic interests in the region. Mindanao's proximity to the South China Sea and disputed Spratly Islands—which may hold significant oil and gas deposits and where a number of overlapping claims from various states, including China, have led to increased tensions—makes it important for the United States and its allies.

Negative posturing from the newly elected Duterte government toward the United States lends urgency to this relationship. In 2014, the United States and the Philippines signed the Enhanced Defense Cooperation Agreement allowing the United States to rotate troops into the Philippines for extended stays and facilitating the building and operating of facilities in the Philippines. Such agreements, it has been suggested by the U.S. Pacific Command, are necessary for maritime security, counterterrorism, and humanitarian and disaster relief operations.[47] They are also likely an important part of U.S.-Philippine collaboration to counter al-Qaeda–linked terrorist groups in the southern Philippines.

Since 2001, Mindanao has been part of a "second front in the war on terror" with the U.S. Joint Special Operations Task Force–Philippines combating the Islamist groups Jemaah Islamiyah and Abu Sayyaf.[48] In 2013, it was reported that the United States was attempting to establish a drone base on Mindanao. The current Philippine president, Rodrigo Duterte, was mayor of Davao City at the time and disclosed that he had rejected a request from the U.S. government to use the city's old airport as a base for launching unmanned aerial vehicles.[49] In response, the U.S. embassy noted that the U.S. military would deploy drones only at the request of the Philippine government and only to help in aerial surveillance for humanitarian responses, such as those warranted by Typhoon Pablo, which struck Mindanao in December 2015 and killed more than 1,100 people.

The 2014 defense cooperation agreement is contentious and is now under attack by Mindanaoan groups and President Duterte for putting U.S. interests before the Philippines'.[50] Some have called it a "negotiated surrender of national sovereignty to U.S. imperialism," criticizing it for legitimizing existing U.S. military bases and paving the way for additional facilities.[51]

Indeed, the conflict and tensions in Mindanao are now manifesting at the national and international political levels. Even though the Moro population may be disillusioned by the Philippine electoral process, Mindanao experiences high voter turnout in most elections. One researcher noted that in the 2004 general election the province of Maguindanao witnessed the highest voter turnout (89%) in the entire country, and its overwhelming support for incumbent President Gloria Macapagal Arroyo was a determining factor in her reelection.[52] Mindanao's troubles, described as a "spiral of crisis, contestation, and conflict,"[53] were punctuated by water and electricity shortages. Such shortages fueled discontent as aquifer and other water sources dried up because of a lingering drought, particularly in Lake Lanao and the Pulangui River, a major tributary of the Rio Grande in Mindanao. Discontent fed into already-accelerating levels of conflict marked by spikes in violence in the region, higher than in the previous three years, and led to the election of Duterte in 2016 as the first Mindanaoan president in the country's history.

Duterte promised to combat "imperial Manila" and "imperialist foreigners" on a platform of sustainable peace, constitutional reform, and development of the countryside. But he is also allegedly involved in extrajudicial killing of criminals and drug lords in Davao and is reputed to be close to the family of former dictator Ferdinand Marcos.[54] Despite his bellicose approach to other issues, he has espoused a more conciliatory approach to China over lingering territorial disputes in the South China Sea.[55] In late 2016, he repeatedly called for the removal of U.S. troops from the Philippines, presenting this in the context of seeking a closer relationship with China and a critique of U.S. foreign policy. While meeting with Japanese leaders he stated, "We will survive without the assistance of America—maybe a lesser quality of life, but as I said, we will survive."[56]

Duterte's ascent to power elevates the importance of Mindanao and of the Philippines overall and complicates a previously reliable partnership. One commentator summarizes it as such:

> It should be clear that there's more to the Philippines than Manila, more to its politics and society than upper-class Catholicism, and more to its security concerns than partnering with the United States to push back against the PRC [People's Republic of China] in the South China Sea. There's Mindanao, there's Moros, there's separatism, there's issues of justice that have been papered over by the Manila establishment to present a neat neo-liberal narrative that complements the U.S. pivot to Asia. And there's Duterte.[57]

Future Directions

The Philippines and the United States have a long and not always positive history, but also genuine shared strategic interests and a mutual desire for stability in the region. There are strong trade and cultural ties, with some 4 million Filipinos and Filipino-Americans in the United States and more than 150,000 Americans in the Philippines.[58]

As we disentangle the nexus of water, security, and national interests in Mindanao, there are some directions to consider moving forward. These considerations revolve around three key areas:

1. Using water programs as points of entry to invest in peacebuilding and climate resilience
2. Building trust among the key stakeholders including local actors, the national government, and international aid groups

3   Linking defense, diplomacy, and development efforts tied to water management programs

Using Water Programs to Invest in Peacebuilding and Climate Resilience

The U.S. government should invest in peacebuilding and resilience in Mindanao, harnessing water and natural resources in an underlying, integrated approach to achieve both. This recognizes three key intractable realities in Mindanao: the persistence of conflict, the saliency of natural resources, and the state of underdevelopment. USAID, together with the Wilson Center, developed a water and conflict toolkit in 2014.[59] The toolkit is meant to address the challenges of promoting, engaging, and designing cross-sectoral collaboration among development practitioners and to get to a peaceful resolution. It helps ensure that water resources can meet community needs for food, economic development, and health services provision, as well as helps develop longer-term responses that deal with climate change impacts.

The water and conflict toolkit offers some specific examples of programmatic interventions, including advice on how to achieve peace dividends and how to improve public relations with police and security forces, and provides a rapid appraisal guide to identify and evaluate the conflict risk and peacebuilding potential of water programs.

Some guiding principles that can be aptly applied to the situation in Mindanao include:

- Recognize that all water issues are part of a complex and dynamic system.
- Enhance information management, public awareness, and public engagement, including by ensuring women are involved in the process.
- Build formal and informal institutional capacity for collaborative governance.
- Strengthen equitable and affordable water access.
- Coordinate water-related aid and investment.
- Ensure conflict-sensitive design and capitalize on peacebuilding opportunities.

USAID's Be Secure Project is working toward strengthening the enabling environment for sustainable water and wastewater treatment service delivery as well as strengthening the analysis, communication, and use of water resource and climate data, ensuring that end users can base their decisions on that information.[60] As USAID and other humanitarian and aid actors conduct water activities in the region, they should incorporate conflict mitigation and peacebuilding dimensions into their programming.

Building Trust Among Key Stakeholders Through Water Peacebuilding

Increasing levels of trust among the various interest groups in Mindanao is paramount to building peace. In the long run, trust must be reestablished between the ethnic groups of Mindanao, the Philippine Army, and the U.S. military to protect human security and allow space for development efforts that could address inequities in water security and other development objectives.

One such example is the case of Mindanao's Pantaron Mountains, wherein lie the remaining 1.8 million hectares of virgin forest that supply water to the major rivers of the island. Local groups fear that the biodiversity in Pantaron Range might disappear along with the indigenous knowledge of the Talaingod Manobos, or "mountain people," who have protected the forests over the past decades but face pressure from troop deployments and military operations.[61]

Today's leaders might look to the 1996 peace agreement signed between the Moro National Liberation Front and the government of the Philippines that included providing basic economic and social infrastructure in the most conflict-ridden and impoverished areas of Mindanao. The World Bank help set up a social fund under its Special Zone for Peace and Development Project for quick financing and water supply and sanitation in these areas. This mechanism was helping rebuild trust in government, but was abandoned in 2014 as the government transitioned to a new mechanism, the Bangsamoro Transition Authority.[62] At the heart of such peace-friendly development in the region there must be a recognition of the importance of providing basic services to the populace, which could go a long way toward building trust. The government must ensure that basic social services, including water, are provided for all social groups, regardless of ethnic origin, gender, or status.

Using Water to Link Defense, Diplomatic, and Development Initiatives

U.S. development and diplomatic efforts in Mindanao should align with "do no harm" principles in the conflict arena. These principles recognize the importance of examining and contextualizing local fragilities and drivers of conflict dynamics.[63] They emphasize that donor interventions should not undermine broader efforts at peacebuilding. These efforts should also implement post-conflict strategies that are cognizant of the stresses around water, with a focus on mechanisms that increase participation and improve governance. Conflict management and mitigation should be incorporated into all programs working in the region. More attention should also be paid to resolving legal frameworks for resource management, especially of mining, which has a very harmful effect on water resources. Questions around ancestral domain need to be resolved to provide a clear framework for equitable resource extraction and management.

These principles could also be applied to humanitarian and climate assistance. More destructive storms as a result of climate change are leading to displacement problems and threatening the productivity of the dominant commercial sector, agriculture. The Philippine government is working to adapt to climate change impacts, including by implementing early warning systems for large storms and droughts to help farmers plan ahead, as well as establishing metrological stations to track and measure drought severity.[64] The government has promoted crop diversification methods for farmers to grow corn, coconuts, legumes, and other crops to supplement the dependence on rice.[65] It is hoped that these crops will be more drought resistant and can be planted in rotation with rice. Oxfam and other non-governmental organizations (NGOs) are teaching "resilient farming strategies." There is also a movement for reforestation in Mindanao, which especially around riparian corridors could help restore ecosystems, limit flash floods, and bring a consistent flow of water back to the area.[66]

At the same time, raids by the military and anti-terror operations are contributing to the displacement of people. NGOs like the International Committee of the Red Cross and the International Organization for Migration are donating funds and assisting in displacement camps.[67] Water and food allocation is a challenge in these camps because there is often not enough to go around.[68] The displacement camps are also a major liability during any droughts or typhoons in the future, potentially creating a feedback loop of vulnerability. The quick resettlement of the displaced must be a priority of the Philippine government and its partners.

It is in the best economic interest of the Philippines to ensure the viability of the agricultural sector and improve stability, and it is in the best interest of the United States to keep terrorist groups confined to a small region, if not to eliminate them, and to keep them from controlling critical resources. Water scarcity is a clear factor in these objectives.

There are many avenues of overlap for the two governments that encourage coordinated water security strategies. The United States can provide assistance to the Philippine government, the military, and other development partners to monitor and allocate water resources for agriculture and urban uses. Military protection of dams is needed as well to ensure terrorist groups are not tempted to hijack vital resources or power sources during times of drought. The U.S. government and military must consider the long-term implications of climate and water changes in the region and work collaboratively, internally and externally, to facilitate a climate-resilient future for Mindanao.

— —

While Mindanao's conflict dynamics are largely defined by struggles at the local level for political self-determination, natural resources and water dynamics are important dimensions of these grievances. Historic instability, limited resolution mechanisms, and underlying mistrust have allowed ideologically driven Islamic terrorist movements to gain a foothold. But this nexus of military, political, societal, environmental, economic, and conflict dynamics is not intractable. Investing in resilience, building trust, and integrating defense, development, and diplomatic efforts will support an

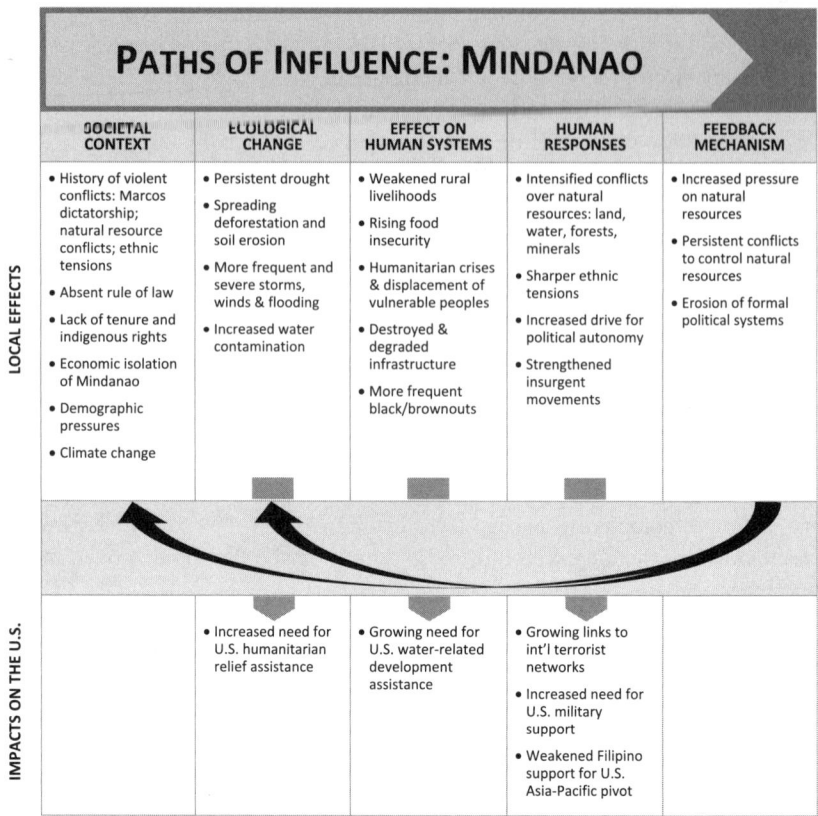

Figure 16.2 Paths of Influence: Mindanao

enabling environment to position the peacebuilding dividend of water and natural resource initiatives.

Acknowledgements

Thank you to Margaret B. Bowman, Lynae Bresser, Victoria Johnson, Lauren Herzer Risi, and Schuyler Null for their review and assistance with this chapter.

Notes

1 Ewing, John Jackson. *Environmental Security in Mindanao: A Case for Comprehensive Methods*, PhD, ePublications@bond, Faculty of Humanities and Social Sciences, 2011. Accessed online on October 1, 2016, at http://epublications.bond.edu.au/theses/44/.
2 Wikipedia. "Mindanao." Accessed online on August 20, 2016, at https://en.wikipedia.org/wiki/Mindanao.

Mindanao: Water and Insurgency 301

3 Asian Development Bank. "Philippines: Water Supply and Sanitation Sector Assessment, Strategy, and RoadMap." January 2013.
4 Food and Agriculture Organization of the United Nations (FAO). 2016. AQUASTAT website. Accessed online on November 8, 2016, at www.fao.org/nr/water/aquastat/countries_regions/phl/index.stm.
5 Mindanao Development Authority. "Mindanao 2020: Peace and Development Framework Plan (2011–2030)." Accessed online on November 8, 2016, at http://minda.gov.ph/resources/Publications/Mindanao_2020/m2020_full_doc_for_web.pdf.
6 Ibid.
7 The Water Dialogues. "The Philippine Water Situation." Accessed online on November 8, 2016, at www.waterdialogues.org/documents/PhilippinesCountryContext.pdf.
8 Repollo, Zephanie. "The Kidapawan Standoff—The Real Face of the Global Climate Crisis." *Common Dreams*, April 8, 2016. Accessed online on August 27, 2016, at http://commondreams.org/views/2016/04/08/kidapawan-standoff-real-face-global-climate-crisis.
9 Sticklor, Russel. "Changing Climate, Changing Minds: How One Philippine City Is Preparing for a Water-Scarce Future." *Global Waters* 7, no. 2. Accessed online on October 20, 2016, at https://medium.com/usaid-global-waters/changing-climate-changing-minds-how-one-philippine-city-is-preparing-for-a-water-scarce-future-29327b5c5bfa#.3sfmp67wv.
10 The International Bank for Reconstruction and Development/The World Bank. "Behind the Veil of Conflict Moving Toward Economic Integration for Sustained Development and Peace in Mindanao." Washington, DC, May 2010.
11 Slack, Alyson. "Separatism in Mindanao, Philippines." *ICE Case Studies*, Number 118, May 2003. Accessed online on August 25, 2016, at www1.american.edu/ted/ice/mindanao.htm.
12 Savchuk, Mariya. "Natural Resources of Mindanao: Fueling the Conflict?" in *Mindanao: Understanding Conflict*. SAIS, Johns Hopkins. Accessed online on August 27, 2016, at www.sais-jhu.edu/sites/default/files/Mindanao-Report_Complete_Report%20April%205_0.pdf.
13 Ibid.
14 Ibid.
15 The International Bank for Reconstruction and Development/The World Bank. "Behind the Veil of Conflict Moving Toward Economic Integration for Sustained Development and Peace in Mindanao." Washington, DC, May 2010. p.23.
16 Savchuk, Mariya. "Natural Resources of Mindanao: Fueling the Conflict?" in *Mindanao: Understanding Conflict*. SAIS, Johns Hopkins. Accessed online on August 27, 2016, at www.sais-jhu.edu/sites/default/files/Mindanao-Report_Complete_Report%20April%205_0.pdf.
17 Swanson, David. "Water and Sanitation Situation in Mindanao 'Critical.'" *IRIN*, October 6, 2009. Accessed online on October 10, 2016, at www.irinnews.org/feature/2009/10/06/water-and-sanitation-situation-mindanao-critical.
18 The Internal Displacement Monitoring Centre. "Philippines IDP Figures Analysis." IDMC, 2013. Accessed online on June 17, 2016, at www.internal-displacement.org/south-and-south-east-asia/philippines/figures-analysis.
19 Sherwood, Angela, Bradley, Megan, Rossi, Lorenza, Guiam, Rufa, and Mellicker, Bradley. "Post-Disaster Resource: Insights from the Philippines after Typhoon Haiyan (Yolanda)." *Brookings IOM*: 1. SAS, June 15, 2015. Accessed online on June 16, 2016, at www.brookings.edu/~/media/research/files/reports/2015/06/15-philippines-typhoon-haiyan-displacement-solutions/resolving-postdisaster-displacementinsights-from-the-philippines-after-typhoon-haiyan-june-2015.pdf.
20 Gavalin, Jodesz. "How Conflict Can Lead to Food Insecurity and Hunger." *Rappler*. February 7, 2016. Accessed online on June 16, 2016, at www.rappler.com/move-ph/issues/hunger/83166-conflict-food-insecurity-hunger.

21 Wikipedia. "Autonomous Region in Muslim Mindanao." Accessed online on August 20, 2016, at https://en.wikipedia.org/wiki/Autonomous_Region_in_Muslim_Mindanao.
22 According to the 2000 census, the Lumad constitute about 9% of Mindanao's 18.1 million population. The Moros account for 18.5%, and the settlers and their descendants make up 72.5% and are the leaders in economics and politics. The International Bank for Reconstruction and Development/The World Bank. "Behind the Veil of Conflict Moving Toward Economic Integration for Sustained Development and Peace in Mindanao." Washington, DC, May 2010. p.9
23 Martin, Eugene G. "Extrajudicial Killings in the Philippines: Strategies to End the Violence." United States Institute of Peace, March 2007. Accessed online on August 25, 2016, at www.usip.org/publications/2007/03/14/extrajudicial-killings-in-the-philippines-strategies-end-the-violence.
24 The International Bank for Reconstruction and Development/The World Bank. "Behind the Veil of Conflict Moving Toward Economic Integration for Sustained Development and Peace in Mindanao." Washington, DC, May 2010. p.21
25 Savchuk, Mariya. "Natural Resources of Mindanao: Fueling the Conflict?" in *Mindanao: Understanding Conflict*. SAIS, Johns Hopkins. Accessed online on August 27, 2016, at www.sais-jhu.edu/sites/default/files/Mindanao-Report_Complete_Report%20April%205_0.pdf.
26 The International Bank for Reconstruction and Development/The World Bank. "Behind the Veil of Conflict Moving Toward Economic Integration for Sustained Development and Peace in Mindanao." Washington, DC, May 2010. p. 1.
27 Ibid. p. 3.
28 The International Bank for Reconstruction and Development/The World Bank. "Behind the Veil of Conflict Moving Toward Economic Integration for Sustained Development and Peace in Mindanao." Washington, DC, May 2010, p. 1.
29 Ranada, Pia. "100% Crop Damage Reported in Parts of Mindanao Due to Drought." *Rappler*. May 6, 2015. Accessed online on August 20, 2016, at www.rappler.com/nation/92234-mindanao-crop-failure-drought-oxfam.
30 Tejada, Dr. Silvina Q., Tuddao, Dr. Vincente B. Jr., Juanillo, Edna, Ms., and Engr. Brampio, Ernesto. "Drought Conditions and Management Strategies in the Philippines." pp. 2–4. Accessed online on June 16, 2016, at www.ais.unwater.org/ais/pluginfile.php/597/mod_page/content/79/Philippines.pdf.
31 *Ecowatch*. "How Is Climate Change Affecting the Philippines?" *EcoWatch*. January 22, 2016. Accessed online on June 15, 2016, at http://ecowatch.com/2016/01/22/climate-change-affecting-the-philippines/.
32 Gavalin, Jodesz. "How Conflict Can Lead to Food Insecurity and Hunger." *Rappler*. February 7, 2016. Accessed online on June 16, 2016, at www.rappler.com/move-ph/issues/hunger/83166-conflict-food-insecurity-hunger; Fairfood International. "Mindanao Economy: Comes with Benefit, Come to the Benefit of All." *The Land of Promise*. May 23, 2013. Accessed online on June 17, 2016, at http://landofpromise.fairfood.org/mindanao-economy-comes-benefit/; "Mindanao Share of Philippine GDP up at 14.4%." *Business World Online*. August 4, 2015. Accessed online on June 17, 2016, at http%3A%2F%2Fwww.bworldonline.com%2Fcontent.php%3Fsection%3DEconomy%26title%3Dmindanao-share-of-philippine-gdp-up-at-14.4%25%26id%3D112684.
33 "Mindanao Share of Philippine GDP up at 14.4%." *Business World Online*. August 4, 2015. Accessed online on June 17, 2016, at http%3A%2F%2Fwww.bworldonline.com%2Fcontent.php%3Fsection%3DEconomy%26title%3Dmindanao-share-of-philippine-gdp-up-at-14.4%25%26id%3D112684.
34 United Nations Office for the Coordination of Humanitarian Affairs (OCHA). "Humanitarian Needs." *Humanitarian Bulletin Philippines* 51, no. 7 (2014): OCHA, July 31, 2015. Accessed online on June 16, 2016, at www.humanitarian

response.info/en/system/files/documents/files/ochaphilippines_humanitarian_bulletin_no7_july_2015_final.pdf.
35 Magdalena, Federico V. "Population Growth and the Changing Ecosystem in Mindanao." *JSTOR*. Institute of Southeast Asian Studies (ISEAS), April 1, 1996. Accessed online on June 22, 2016, at www.jstor.org/stable/pdf/41056929.pdf?_=1466529829132.
36 World Bank. "Vulnerability, Risk Reduction, and Adaptation to Climate Change in the Philippines." (2011): 4. *Climate Risk and Adaptation Country Profile*. April 2011. Accessed online on June 16, 2016, at http://sdwebx.worldbank.org/climateportalb/doc/GFDRRCountryProfiles/wb_gfdrr_climate_change_country_profile_for_PHL.pdf.
37 Magdalena, Federico V. "Population Growth and the Changing Ecosystem in Mindanao." *JSTOR*. Institute of Southeast Asian Studies (ISEAS), April 1, 1996. Accessed online on June 22, 2016, at www.jstor.org/stable/pdf/41056929.pdf?_=1466529829132.
38 Espino, Rene Rafael C., and Atienza, Cenon S. "Crop Diversification in the Philippines." *FAO Corporate Document Repository*, 2000. Accessed online on June 17, 2016, at www.fao.org/docrep/003/x6906e/x6906e0a.htm.
39 "Intelligence Community Assessment." *Global Water Security*, February 2012. Accessed online on October 20, 2016, at www.dni.gov/files/documents/Special%20Report_ICA%20Global%20Water%20Security.pdf.
40 *EcoWatch*. "How Is Climate Change Affecting the Philippines?" *EcoWatch*. January 22, 2016. Accessed online on June 15, 2016, at http://ecowatch.com/2016/01/22/climate-change-affecting-the-philippines/.
41 Gavalin, Jodesz. "How Conflict Can Lead to Food Insecurity and Hunger." *Rappler*. February 7, 2016. Accessed online on June 16, 2016, at www.rappler.com/move-ph/issues/hunger/83166-conflict-food-insecurity-hunger.
42 "Philippines-Mindanao Conflict." *Thomson Reuters Foundation News*. June 3, 2014. Accessed online on June 15, 2016, at http://news.trust.org//spotlight/Philippines-Mindanao-conflict/?tab=briefing.
43 Ibid.
44 Ibid.
45 Slack, Alyson. "Separatism in Mindanao, Philippines." *ICE Case Studies*, Number 118, May 2003. Accessed online on August 25, 2016, at www1.american.edu/ted/ice/mindanao.htm.
46 Stratchan, Anna Louise. "Conflict Analysis of Muslim Mindanao *Online*. December, 2015. Accessed online on June 17, 2016, at https://assets.publishing.service.gov.uk/media/57a0898e40f0b64974000140/ConflictAnalysisARMM.pdf.
47 Letman, Jon. "As the US 'Enhances' Military Cooperation in the Philippines, a Complicated Relationship is Challenged." *Truthout*. December 29, 2014. Accessed online on August 30, 2016, at www.truth-out.org/news/item/28245-as-the-us-enhances-military-cooperation-in-the-philippines-a-complicated-relationship-is-challenged.
48 Ibid.
49 Manlupig, Carlos. "Duterte Rejects US Request to use Davao as Drones Base." August 16, 2013, Tarra Quismundo/@inquirerdotnet. Accessed online on September 15, 2016, at http://globalnation.inquirer.net/83435/duterte-rejects-us-request-to-use-davao-as-drones-base.
50 Avendaño, Christine O. "Duterte threatens to stop EDCA." *Philippine Daily Inquirer*. October 2, 2016. Accessed online on November 1, 2016, at http://globalnation.inquirer.net/145962/duterte-threatens-to-stop-edca.
51 Salamat, Marya. "Mindanao Natural Resources, Protectors Under Siege." May 25, 2014. Accessed online on August 20, 2016, at http://bulatlat.com/main/2014/05/25/mindanao-natural-resources-protectors-under-siege/.

52 Donahoe, Kirk. "The Philippines Democratic Deficit as an Obstacle to Peace" in *Mindanao: Understanding Conflict*. SAIS, Johns Hopkins. Accessed online on August 27, 2016, at www.sais-jhu.edu/sites/default/files/Mindanao-Report_Complete_Report%20April%205_0.pdf.
53 de la Rosa, Nikki. "Mindanao's summer of discontent." @inquirerdotnet. April 18, 2016. Accessed online on August 15, 2016 at http://opinion.inquirer.net/94338/mindanaos-summer-discontent
54 Iyengar, Rishi. "The Killing Time: Inside Philippine President Rodrigo Duterte's War on Drugs." *Time*. August 25, 2016. Accessed online on October 15, 2016, at http://time.com/4462352/rodrigo-duterte-drug-war-drugs-philippines-killing/.
55 Arugay, Aries A. "The 2016 Philippine Elections: Democracy's Discontents and Aspirations." *Canada Asia Agency*. June 24, 2016. Accessed online on August 25, 2016, at www.asiapacific.ca/canada-asia-agenda/2016-philippine-elections-democracys-discontents-and.
56 Calonzo, Andreo. "Duterte wants foreign troops out of Philippines in two years, says 'we will survive' without Americans." *Bloomberg News*. October 26, 2016. Accessed online on October 27, 2016, at http://news.nationalpost.com/news/world/duterte-wants-foreign-troops-out-of-philippines-in-two-years-says-we-will-survive-without-americans.
57 ChinaHand blog post, "Mindanao, Duterte, and the Real History of the Philippines." May 23, 2016. Accessed online on August 25, 2016, at http://chinamatters.blogspot.com/2016/05/mindanao-duterte-and-real-history-of.html.
58 Campbell, Kurt M., assistant secretary, Bureau of East Asian and Pacific Affairs, U.S. Department of State. "The U.S. Philippines Alliance: Deepening the Security and Trade Partnership." February 7, 2012. Accessed online on August 24, 2016, at www.state.gov/p/eap/rls/rm/2012/02/183494.htm.
59 Moses, Jackman. "USAID Launches New Water, Conflict, and Peacebuilding Toolkit." *New Security Beat*. April 2014. Accessed online on October 20, 2016, at www.newsecuritybeat.org/2014/04/usaid-launches-water-conflict-peacebuilding-toolkit/.
60 AECOM International Development. "Water Security for Resilience Economic Growth and Stability (Be SECURE) Quarterly Report" (3rd Quarter). Accessed online on October 20, 2016, at http://pdf.usaid.gov/pdf_docs/PA00JV6P.pdf.
61 Salamat, Marya. "Mindanao Natural Resources, Protectors Under Siege." May 25, 2014. Accessed online on August 20, 2016, at http://bulatlat.com/main/2014/05/25/mindanao-natural-resources-protectors-under-siege/.
62 USAID Office of Conflict Management and Mitigation, "Water and Conflict." p. 30. Accessed online on August 20, 2016, at www.usaid.gov/sites/default/files/documents/1866/WaterConflictToolkit.pdf. Kawabata, Yasuhiro. "Ex-Post Evaluation of Japanese ODA Loan Project ARMM Social Fund for Peace and Development Project." Accessed online on November 4, 2016, at www2.jica.go.jp/en/evaluation/pdf/2014_PH-P235_4.pdf. Arquillas, Carolyn O. "MNLF, ARMM legislators will be in Bangsamoro Transition Authority." *Minda News*. May 24, 2015. Accessed online on November 4, 2016, at www.mindanews.com/peace-process/2015/05/mnlf-armm-legislators-will-be-in-bangsamoro-transition-authority/.
63 USAID Office of Conflict Management and Mitigation. "Water and Conflict." Accessed online on August 20, 2016, at www.usaid.gov/sites/default/files/documents/1866/WaterConflictToolkit.pdf.
64 Tejada, Dr. Silvina Q., Tuddao, Dr. Vincente B. Jr., Juanillo, Edna, Ms., and Engr. Brampio, Ernesto. "Drought Conditions and Management Strategies in the Philippines." pp. 2–4. Accessed online on June 16, 2016, at www.ais.unwater.org/ais/pluginfile.php/597/mod_page/content/79/Philippines.pdf.
65 Ibid.
66 Oxfam. *To Climate Change Case Study: Climate Resiliency Field Schools Help Farmers in Mindanao, Philippines, Prepare for an Uncertain Future*. July 2015.

Accessed online on June 15, 2016, at www.oxfam.org.au/wp-content/uploads/2015/07/2015-12-CC-case-study_phil_FA.pdf

67 ReliefWeb. "IOM Aids over 100 Families Displaced by Conflict in Mindanao." *ReliefWeb*. International Organization for Migration. April 15, 2011. Accessed online on June 16, 2016, at http://reliefweb.int/report/philippines/iom-aids-over-100-families-displaced-conflict-mindanao.

68 Dominguez, Gabriel. "Mindanao Conflict Uproots Tens of Thousands." March 20, 2015. Accessed online on June 15, 2016, at www.dw.com/en/mindanao-conflict-uproots-tens-of-thousands/a-18330640.

Part III
Financing Water Infrastructure

Framing Note: Persistent Challenges

E. Patrick Coady

Water infrastructure is critical in a functioning society to manage water flows and ensure timely delivery of water resources to agriculture, business, and citizens. A key element in making sure that infrastructure is in place is mobilizing the necessary financing. The terms established by the financing agencies—be they public, private, or self-financed—frequently dictate the very characteristics of infrastructure systems, influencing their sustainability, their integration into national water management regimes, and often the actual beneficiaries.

The financing of water infrastructure and municipal water systems has experienced considerable change over the past 50 years, notably in the developing world. Following independence during the decolonization period, many recently formed governments relied exclusively on parastatal companies to provide fresh water to their respective populations. As a rule, major water infrastructure works, whether designed to expand agricultural production, to provide energy to mining companies and industries, or to supply growing urban populations, were managed through specific line ministries. Because those projects were aligned with and managed to respond to needs in different economic sectors such as agriculture, mining, and energy, a frequent result was the fragmentation of the nation's water management regime, with little prospect of developing integrated water management systems. A principal source of financing for those projects up to the 1990s—whether for municipal systems or large infrastructure—was the multilateral development banks, which channeled financial resources through the corresponding ministerial offices. Investment from private companies was forthcoming to the degree that multilateral bank guarantees were firmly in place.

The debt crises of the late 1980s challenged those financing arrangements. The inability of scores of developing countries to service debt obligations to public and private lenders alike necessitated massive cash infusions from the multilateral financial system to stabilize economies. A byproduct of the debt crisis was a comprehensive macroeconomic policy response from developed societies, notably the United States and Great Britain, which sought to replace state-driven economies with market-based ones. Those economic reforms of the early 1990s, framed by the principles of the Washington Consensus, focused first and foremost on macroeconomic reforms designed to restore balance between expenditures and revenues and to encourage foreign investment in virtually all sectors of the national economy.

While macroeconomic policy across the developing world went through this Teutonic shift, emerging protest movements demanded changes to the prevailing blueprint for building large infrastructure projects and managing municipal water systems. In the late 1980s through the 1990s, public protests emerged in the developing world in opposition to the proliferation of large-scale infrastructure projects, notably large dams. Some of the world's largest dams, often financed by the World Bank and other multilateral institutions, were being built in Brazil, India, South Africa, China, Indonesia, Nepal, and scores of other countries. Over the course of the decade, local protests coalesced into a global anti-dam movement that culminated in the convocation of the World Commission on Dams (WCD) and the 2000 release of the commission's report. The report was pivotal in establishing a new code of standards and procedures for building large water infrastructure projects. In addition, the World Bank imposed a de facto moratorium on financing for large dams that continued for the better part of a decade.

It was also at the turn of the millennium that the focus of macroeconomic reforms sponsored by multilateral institutions shifted to reforming public utilities, notably water and energy parastatals. The stated intent of this focused privatization effort was to modernize management of the public utilities and, in parallel, attract private capital as a major funding source to respond to the growing needs of developing countries. It was not difficult to justify the push away from state-owned and state-managed water utilities in light of their endemic financial losses, highly inefficient water management systems, and repeated failure to extend access to broad segments of the population, particularly in urban areas.

Privatization initiatives in the early 2000s seeking to divest public utilities of control and give it to the private sector ran into broad opposition. One major cause of the protests was efforts to price water, including those social sectors least able to pay the transitional costs of the management change. Promises of future benefits did little to assuage the ire of vulnerable populations straddled with higher water bills. Hundreds of municipalities faced hostile resistance from popular movements, often forcing local governments to end or delay reforms for many years.

But public demand for safe, steady water provision has continued to rise. Today, public policymakers and planners are being forced to respond to the increased variability of rainfall and the uncertainties of water flows for energy generation and industrial uses.

Today, there is growing interest in combining public and private financial resources to address the rising demand for safe, clean water across the developing world. Public finance, even when combining resources from both developed and developing countries, cannot possibly cover the full range of urgent demands to expand clean water supply in developing countries, particularly as urban populations continue to explode. Parallel challenges apply to financing new infrastructure projects. Yet, while private investors are cautiously exploring ways of increasing their water portfolios in the developing world, the risks facing private investors remain high and, as weather variability and climate uncertainties intensify, such risks will only increase.

Adequate water infrastructure is a necessary component of a functioning society for food security, economic activity, and a healthy population. This infrastructure requires a substantial amount of capital. The first two chapters in this section look at the current situation with respect to financing from both public and private sources. The third chapter addresses the unique issues related to large river basin projects with impacts that are huge and complex.

Assessing the role of financing is itself a complex task because water infrastructure projects vary by size, political jurisdiction, region, and the institutional capacity of each country. The authors make clear that there is a dilemma of massive dimensions, noting the regular failure of water infrastructure projects to obtain financing. If financed, the entity is often not sustainably managed. Water pricing, cost recovery, and structuring of subsidies are almost always inadequate. Proper consideration of side effects such as environmental impacts and fair distribution of benefits add to the problem. Rising population and pending climate change turn a daunting challenge into an overwhelming one.

The authors note that there is no lack of capital—public, private, and multilateral—but it can be accessed only if the use of capital meets the typical standards for a businesslike operation. Because water infrastructure is so critical, it has been considered the highest priority by the international community for many, many decades, but with success that falls short of the need. The potential for major conflict at the local and regional levels of the competition for water resources multiplies the problem, reaching the threshold of a global security concern.

17 Helping Weak Water Utilities Climb the Financial Ladder

Aldo Baietti

I. Introduction

For many decades the global water problem and the related financing challenge have been central to the development agenda of developing economies, donor agencies, development partners, and multilateral and regional financial institutions. Much has been written on the subject, and many financial instruments and interventions have been developed and implemented in the hope of finding solutions that can be fully replicated throughout the developing world in order to get a foothold on the growing financing gap. Unfortunately, we are not there yet.

Water is essential for life and for our well-being. It is the most difficult of infrastructure services to substitute, and as such, its absence or deficiency creates a dramatic impact on the poor.[1] Despite this, the global water problem has grown over the years, and water entities in many countries and especially in the developing world continue to underperform their counterparts in other sectors. The preponderance of water entities in developing countries is failing, and scarce water resources are not properly managed. Based on recent estimates, more than 1 billion people in the world lack access to safe drinking water and another several billion lack access to adequate sanitation. This translates into a serious financing challenge that is becoming even more pronounced due to the effects of climate change.

This chapter does not purport to have the definitive answers in this regard. The sector is complex and multidimensional, and requires appropriately sequenced interventions that go beyond the field of finance. However, this note does dissect the challenges that are being confronted in financing water investments, explains why some financial interventions have or have not worked, and offers a series of policy recommendations for U.S. foreign assistance for continued engagement, particularly for improving the status of financially weak and unsustainable water entities.

II. The Daunting Water Finance Challenge

There has been much debate on how to address the financing challenge in the water sector. Following the failures of entirely public solutions, the

1980s–1990s witnessed a paradigm shift toward replacing inefficient publicly managed and financed water utilities with privately managed and financed ones. It was hoped that the private sector could step up to the challenge, mobilize significantly more investment capital, foster transparency in governance, improve operational and management efficiency, and greatly expand service to unserved areas. Unfortunately, public-private partnerships (PPPs) did not live up to the expectations.

By the year 2000, the formulation of the Millennium Development Goals (MDGs) for water[2] and sanitation significantly increased the stakes, as this brought home the daunting finance challenge of achieving the targets established for 2015. According to the Commission on Water,[3] the investment gap in water for flood and draught mitigation, hydropower, environmental improvement, irrigation, and water and sanitation amounted to more than $105 billion annually.[4] Water supply and sanitation comprised about one third of all investment and would need to double from $15 billion to $30 billion annually.

The MDGs were replaced with the Post-2015 Development Agenda and the Sustainable Development Goals (SDGs). While progress has been made since the MDGs,[5] the current state of affairs shows exactly how difficult it is to resolve this funding gap. Close to 800 million people still lack access to safe drinking water, and 2.4 billion lack sanitation. But more important, estimates as to what would be needed just for water supply, sanitation, and wastewater treatment are now closer to $100 billion annually.[6] Even more concerning is the added investment burden that will arise due to the growing effects of climate change.

In reviewing the experiences over the past several decades, it is clear that there have been countless initiatives—revolving and special-purpose funds, guarantee facilities, greater public finance support through the international financial institutions (IFIs), other innovative programs such as output-based aid (OBA), different PPP approaches, technical assistance (TA) for regulatory and utility reform, management improvement, commercialization, creditworthiness, and many other programs to prop up domestic capital markets. So why the lack of sustainable progress?

III. Two Sides of the Problem: The Ladder of Financial Sustainability

In order to fully understand the financing challenge in the sector, two sides to the problem need to be well understood: on one hand, the availability of financially viable and creditworthy water entities, and the other hand, the availability of predictable or reliable sources of affordable finance.

Private finance is generally far more abundant than public finance, but it is also substantially more risk averse and wants to be compensated for its risk. It will not be channeled if the risk-return criteria of private investors and lenders is not met. However, it is also an indisputable fact that public finance alone cannot close the funding gap in order to achieve the SDGs. Private finance needs to play a significant role. So the key to making further progress in this area is how to strategically utilize the limited sources of public finance in order to crowd-in private participation.

The Ladder of Financial Sustainability (Figure 17.1)[7] depicts this particular dilemma for water entities. As utilities become more and more financially sustainable through adherence to cost recovery principles, the opportunities for financing expand dramatically from the extremely limited sources of public grant funding at the lowest level of the ladder to more sophisticated private financing structures such as bonds, securitization, and pooled facilities in more mature financial markets at the top.

In the middle of the ladder lies the opportunity for direct financing and deployment of various guarantees by international donors and IFIs. This is where these institutions can play an effective role by lowering the cost of finance or by mitigating some of the risk elements in order to crowd-in private money, albeit in modest amounts.

However, the dilemma is such that most utilities are kept at the lower end to the middle of the ladder, where they have difficulty recovering their operating and maintenance costs. Many of these entities have minimal or no capacity to service debt and, as such, are predominantly dependent on public grants to fund their financial losses and/or other needs to expand or diversify services.

The important point here is that the water entities that fall in the lower half of the ladder confront a "financial viability gap" that private finance cannot close using its own financial instruments or with IFI guarantees. Unfortunately, this

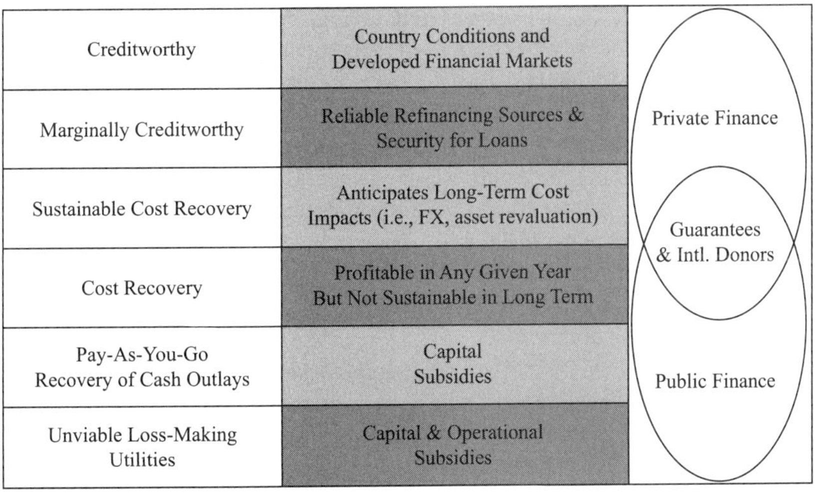

Creditworthy	Country Conditions and Developed Financial Markets	Private Finance
Marginally Creditworthy	Reliable Refinancing Sources & Security for Loans	
Sustainable Cost Recovery	Anticipates Long-Term Cost Impacts (i.e., FX, asset revaluation)	Guarantees & Intl. Donors
Cost Recovery	Profitable in Any Given Year But Not Sustainable in Long Term	
Pay-As-You-Go Recovery of Cash Outlays	Capital Subsidies	Public Finance
Unviable Loss-Making Utilities	Capital & Operational Subsidies	

Figure 17.1 Ladder of Financial Sustainability

Source: Adapted from A. Baietti & P. Raymond, Financing Water and Sanitation Investments: Utilizing Risk Instruments, World Bank

has often been misunderstood, and many prescriptive programs have focused on solutions without addressing this fundamental reality.

IV. At the Heart of the Finance Problem

A. Why Water Is Different

Water, like other infrastructure sectors, confronts particular financing challenges. Power, transport, and water are all capital intensive, with high up-front cost and long payback periods. This necessitates financing in long-term maturities and with sufficient grace periods to accommodate fairly long construction schedules. For most infrastructure sectors in the developing world, the financing should be in local currency. In power and water especially, utilities do not earn foreign exchange (FX), and a revenue-cost mismatch can prove devastating if not prudently managed.[8]

Most infrastructure investments in developing countries also confront a high level of political oversight, but the regulatory and governance frameworks are often not transparent or "time tested" in applying consistent and equitable treatment to private parties. Such frameworks are also typically associated with weak sector institutions that allow governments to change policies on a whim and force operators into difficult financial circumstances.

Certainly, these twin challenges have impeded the potential for private investment and financing in all infrastructure sectors. But the effects on water investments have been even more pronounced. Since 1991, operational PPPs in all infrastructure sectors have totaled approximately 11,429.[9] Of this amount, only 600 (5.2 percent) were in water. By comparison, power had almost 5,000 PPP transactions during the same period, or close to half of all PPPs.

So why the difference? Why has it been so difficult to crowd-in private investment and financing in water?

Indeed, water confronts unique problems that complicate its financing opportunities. For example, most water assets are underground, consisting of civil works, and for practical purposes are unrecoverable as collateral. As such, lenders have less recourse for recovering outstanding loans in the event of default. Also, many water utilities are decentralized, and are owned and overseen by local governments, making it more difficult for central governments to institute uniform policies for investment requirements, cost-recovery tariffs, and financial sustainability.

These factors may in part explain the difference in private investment, but water's most fundamental finance-related problem arises from its social dimensions. Water is commonly regarded as a public good and a necessity for life, prompting greater interference from policy makers and a "politicization" of tariffs. Politicians have been hesitant to authorize cost recovery tariffs, especially to low-income communities, and in developing countries, this is by far the largest consumer segment.

Helping Weak Water Utilities 315

The difference may also stem from the fact that the water service can be gradually starved of resources without causing a total collapse of the operation. Water supply and sanitation (WSS) services often deteriorate gradually over time and without changes being noticeable by consumers. By comparison, withholding resources to power and other sectors can suddenly disrupt service, causing serious consequences to the productive sectors, international competitiveness, and productivity.[10]

These factors allow water policy makers and oversight agencies to defer confronting a deteriorating situation, and many poorly performing utilities are often relegated to a minimal standard of operation and financial sustainability in an effort for keep tariffs low and affordable for household consumers. The consequence is an inordinate number of water supply and sanitation entities being locked in a vicious cycle of poor performance, insufficient funding, and a deterioration of assets. Water customers lose out but often lack the voice to force meaningful change.

B. Few Financially Viable Investments

Of a representative sample of almost 700 utilities within the International Benchmarking Network for Water and Sanitation Utilities (IBNET) database maintained by the World Bank of financial and operational data of almost 5,000 utilities in 186 countries and territories, less than 60 percent were able to fully cover their operations and maintenance expenses, and far fewer had either the profitability or the capacity to fully service debt obligations of any significant amount. For these reasons, the clear examples of financially viable water investments in the developing world are few. While wealthy governments can perpetuate a public finance model, experience has shown that most developing economies cannot achieve the SDGs through public funding alone; as such, water utilities must move up the financial sustainability ladder and become creditworthy so that they may access the more abundant sources of private capital.

Cost recovery must be implemented with fairness to consumers. Indeed, PPPs have offered an expedient way to achieve this by transferring management and financing tasks to the private sector. Theoretically viable financial transactions would be structured on the basis of increasing tariffs, financed by international loans, and secured by back-to-back agreements and sovereign guarantees. However, when adding up all the costs, including management fees and the mitigation of FX risks for the international loans, the resulting tariff in many cases became fundamentally prohibitive.

Noteworthy is also the fact that private equity is the most expensive form of finance and can raise total financing cost measurably in a traditional private financing structure consisting of at least 30–35 percent equity contribution. Such structures have created pressure points and have been highly vulnerable to public scrutiny, macro volatility, and global crises.

There have been countless cases where bulk treatment plants were proposed on a build, operate, and transfer (BOT)[11] basis. On the surface, many of these structures appeared robust, but if the utility was under financial hardship, it could not honor its financial obligations to the PPP sponsor. Equally, there are numerous instances of failures in water concessions such as Aguas Argentina, Dar es Salaam, Manila West Side, Cochabamba, and others, which have fundamentally changed the risk perception of investors. Because of the continuing difficulties with PPP concessions, performance contracts and other forms for PPPs that do not necessarily source private capital are now preferred and, as such, negate the original intent.

Several years ago a major study[12] documented the profitability of infrastructure concessions[13] in Latin America. The study found that the water concessions particularly stood out, with an average return significantly lower than that of the other infrastructure sectors: not only were the returns of the water sector the lowest, but also the returns were the most volatile and very risky. It is no wonder that investors have a low appetite for entering the private water market.

The lackluster performance of PPPs in stepping up to resolve the finance problem again helped reshape the policy for interventions in the sector. The objective was replaced to focus more on "improving performance" rather than continue the debate of "public versus private." This shift gives priority to fixing the core business model, resting on the principle that "you cannot just price your way out of the viability gap"—not in water and not if the utility is experiencing a severe financial viability gap!

Furthermore, the gradual deterioration of many utilities over the years has in many cases shifted the performance burden onto consumers through higher tariffs. In such cases, consumers are not only paying for the cost of poor service, but also subsidizing a myriad of inefficiencies and bad policies that have impaired performance.

C. Underdeveloped Financial Sectors and Capital Markets

The lack of depth and breadth of financial markets in developing countries is indeed a constraint. Financial institutions in many of these countries do not have the capacity to provide long-term financing in local currency with grace periods and at affordable rates. Moreover, the capital markets in these same countries are largely underdeveloped and cannot offer more sophisticated equity and bonds instruments, pension funding, asset securitization, guaranteed or pooled facilities, etc. But more fundamentally, most commercial banks in these countries are also not familiar with the risk-return profiles of water investments and have few business relationships with water entities.

Given these constraints, foreign loans are the only long-term and affordable debts that can be raised, but these carry substantial FX risks. However, in the wake of the Asian, Russian, Argentinean, and (more recently) global crises, currency devaluations continue to be one of the foremost concerns of private lenders in developing countries.

Effective FX cover would create a significant bridge in the mobilization of long-term financing. IFIs have focused on this particular problem by seeking ways to expand local currency financing to their clients. However, such measures do not increase the flow of private finance. There has also been discussion about creating a special-purpose facility in order reduce FX risks. Yet with the exception of one such mechanism for power, the success here has also been disappointing. As such, after more than a decade of work on this issue, there is still no meaningful FX cover for water infrastructure in developing countries.

Given the lack of effective FX cover, central governments can step up to assume currency risk for their sectoral institutions. This practice, which has been applied in a significant number of on-lending arrangements with IFIs, consists of the central government taking on the FX exposure on foreign currency loans and charging an interest rate premium to the ultimate borrower. The interest rate premium charged to the borrower defrays the costs to the government in the event of a devaluation, and with proper management, central governments could create their own FX facility for their infrastructure projects, including PPPs.

V. Financing Solutions That Bridge the Private Financing Gap

A. *Financing Instruments and Lending Supported by Donors and IFIs*

Over the past decades, official development aid (ODA) entities and IFIs have supported governments by increasing the flow of public financing and by creating numerous financing instruments to reduce the inherent risk of private lenders and sponsors and to bridge the financing gap. Such initiatives include, among others:

- **Direct Grants or Concessional Lending to the Utility.** This is the most common approach to funding the sector through IFIs and ODA, but is often oriented to financing systems for expanding coverage and access to the poor.
- **On-lending with National Development and Commercial Banks**. IFI wholesale lending addresses local banks' lack of capacity to offer long-term loans with grace periods at affordable rates.
- **Revolving Funds, Municipal Development Funds, and Other Special-Purpose Funding Vehicles**. These provide long-term financing for water sector projects, and are specifically targeted to increase finance at the local government level and for small utilities where transactions costs are high.
- **Partial Credit and Partial Risk Guaratees**. These are offered for both specific investment projects and in the form of facilities; for extending maturities and rates of private lenders and bond investors; and for reducing policy, regulatory, and contractual risks associated with PPPs.

318 *Aldo Baietti*

- **Political Risk Insurance.** This is offered to support investments against expropriation, war and disturbances, and currency convertability and transferability.
- **OBA.** This is offered for extending service to poor communities by providing subsidy support on a performance basis.
- **Pooled Finance Mechanisms.** Pooling different water investments can diversify the risk of any given utility.

With the exception of direct grants, many of these instruments and programs have worked best when the utility has already surpassed its major deficiencies in cost recovery or is deemed capable of satisfactorily fulfilling its debt service obligations. They do not work when utilities are in the lower half of the financial sustainability ladder, which again points to their relative lack of historical success in crowding-in private finance for water. But even IFI on-lending arrangements require borrowers to clearly demostrate their debt service capacity, albeit with a significant amount of leeway placed on the borrower.

Moreover, many ODA arrangements are viable only if a donor agency underwrites the associated risks. As demonstration projects, they can certainly show successes and lessons learned, but if these initiatives cannot be replicated by private capital to a wider national, regional, or global market, they serve a limited purpose in addressing the profound financing challenge.

Perhaps too much focus has also been placed on identifying that "innovative" financing mechanism rather than on addressing the pervasive viability gap or the governance deficiencies in the sector, and unfortunately, there is no "magic solution." The fundamentals of finance are the same for water as they are for the other infrastructure sectors. Either consumers or taxpayers have to pay![14] (See figure 17.2.) While there is ample room for innovation, many programs, instruments, and financing mechanisms exist in the market already and, if channeled correctly, these can be much more effective in helping close the funding gap.

ODA has also supported financing efforts by providing special technical assistance for worthwhile causes to deal with, including (among others) (i)

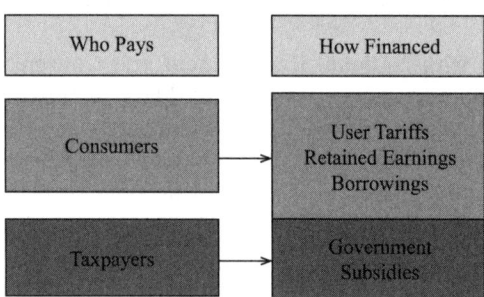

Figure 17.2 There Is No Magic Solution—Someone Has to Pay!
Source: Author's Elaboration

information gaps through credit ratings and creditworthiness indices; (ii) project development funds for better designing investment projects; (iii) commercialization and management improvement initiatives; (iv) tariff payment programs; and (v) strengthening policy, governance, and regulation. These should continue but, again, merit closer scrutiny of their appropriateness.

So essentially, we have a significant contradiction in the water financing market: (i) a substantial number of financing programs and instruments that are more effective at the middle to the top tier of the financial sustainability ladder, but (ii) a preponderance of water entities that operate at the lower middle to bottom level.

B. Strategically Blending Financing for Weak Utilities

Solutions for the contradictions described do exist, but these have to be purposefully tailored to address the specific problem. For poorly performing utilities, the problem can be resolved by using public sources of finance, including ODA, much more efficiently in order to improve performance and to strengthen the utility's financial standing. Public and private sources of finance have different advantages if structured correctly, and limited public funds can be used sparingly to lengthen maturities, reduce the overall cost of funds, and de-risk and leverage private finance in order to come up with financially sustainable solutions. But the approach also has to be staged.

In such cases, the solution lies, first, in creating first-stage financing arrangements that can minimize the immediate increase on tariffs. This can be achieved through "viability gap financing," which aims to convert a projected negative cash flow stream into a positive one by reducing the overall cost of funds and by lengthening maturities. At this stage, the focus should be less on the overall rate of return of the intervention and more on achieving positive cash flows for the utility as a whole.

Second, at the outset an attempt should be made to crowd-in private finance in a blended structure, even if in small amounts. This can reduce information gaps among private lenders and can give private lenders more comfort in second-stage financing and in future lending to the sector. As indicated, private lenders in developing countries still have extremely limited exposure to the water sector or its entities. So bringing them into the process early on, with the backstop of the IFIs and donors, can potentially change their perception of risk.

Blending in private finance can be achieved by taking advantage of the flexible public financing instruments to insert less flexible forms of private finance into a given structure, and by de-risking the exposure of the private lender.

For illustration purposes, consider a case offering two alternative financing options as shown in Table 17.1:[15] (i) a PPP financing structure as "base case," and (ii) a "blended finance" option that combines public with private finance to lower the cost of financing and extend maturities. Under the base case option, the financing arrangement is not able to yield positive cash flows, rendering the entire investment not "bankable" for the private parties. Typically commercial

Table 17.1 Effects of Gap and Blended Financing on Leveraging Private Finance

	Base Case		Blended Option	
Investment Program				
CAPEX	30.0	98%	30.0	99%
Interest During Construction	0.7	2%	0.2	1%
Total Financing Required	30.7	100%	30.2	100%
Financing Sources				
Equity	10.9	35%		
Commercial Financing	19.8	65%	4.2	14%
Up-Front Subsidies	–	–	8.0	27%
Concessional Loans	–	–	18.0	60%
	30.7	100%	30.2	100%
Financial Structuring Results				
Financing WACC		11.25%		1.73%
Consolidated Utility NPV (MM)		–15.9		25.9
Consolidated Utility IRR (%)		4.76%		7.33%
Projected Cash Flow Positive		no		yes
Lowest Annual DSCR (Total)		.28x		1.22x
Lowest Annual DSCR (Senior Debt Only)		.28x		1.32x

The calculations are based on the following: (i) 19% return on private equity; (ii) 7% interest on commercial debt with maturity of 5 years and no grace period; (iii) concessions I debt at .05% with 20-year maturity and five-year grace period on principal repayments; and (iv) government CAPEX subsidy at zero percent. (Amounts in millions except where otherwise noted.)

Source: Author's elaboration

lenders will require, at a bare minimum, a debt service coverage ratio (DSCR) greater than 1.1 times. The result in this case is only 0.28 times. This PPP financing structure would require more than a threefold increase in tariffs in order to work, and still would yield only a DSCR of 1.06 times.

As an alternative, consider viability gap financing under a blended finance option. In this case, grants and concessional financing would be combined with commercial financing, as shown in Table 17.1. The commercial lender would be given seniority on the debt repayments, and the concessional loans would be subordinate to the senior lender. Alternatively, a guarantee could be offered to reduce the commercial lender's exposure. The result is remarkably different in that it: (i) reduces the overall cost of capital to 1.75 percent; (ii) generates a positive negative net present value (NPV) of $25.9 million; (iii) achieves a debt service coverage ratio of 1.22 times overall, and 1.32 times for the senior commercial debt; and (iv) achieves a financially sustainable structure with no immediate tariff adjustment.

This illustration underscores three important points:

- First, it shows the difficulty in channeling private finance or implementing PPP options when utilities are struggling financially, even with enhancements. However, the concept of viability gap financing for PPPs is also gaining more acceptance among government policy makers and donors.

- Second, while the blended option shown here requires a significant contribution of public financing as first-stage financing, the illustration is not totally different from the reality of the current market. Governments and donors must recognize this and step up to the challenge if they are to squarely address the funding gap.
- Third, viability gap financing creates a robust platform for second-stage financing where private lenders can assume a greater financing role. Further, if these proposed investments are oriented to performance improvements, substantial operational savings could also be realized, which could then be applied to second-stage expansion of services—and potentially without the need for additional public financing support.

VI. Dealing With Climate Change Risks

There is now indisputable evidence that climate change will have a profound impact on the water sector through increased hydrologic variability. According to the Intergovernmental Panel on Climate Change, climate change will contribute to an increased frequency of floods, droughts, and other extreme weather events, with an overwhelming effect on most regions and countries.[16]

Undoubtedly, investments for both delivering water services and managing them will be affected.[17] However, some water infrastructure—including high capital cost projects with long useful asset lives, multipurpose infrastructure systems, irreversible projects, and urban water supply—will be affected more than others. Changes in river flows impact hydropower generation, and soil erosion from increased rainfall intensity can lead to sedimentation in reservoirs.[18] Droughts will have similar consequences, particularly on the availability of water for drinking and for agriculture, and will increase requirements for water storage capacity and irrigation systems. All this means additional investments to a sector that is already strapped by a severe funding gap.

Adapting to climate change impacts requires better water infrastructure planning and financial risk assessments. Water professionals have always incorporated some level of uncertainty in sizing specific water infrastructure. However, in the past, water infrastructure was planned and designed under more stationary hydrologic patterns that are no longer valid given the wide degree of climate-based volatility. This means that planning must now be based on risk assessments under uncertain conditions, where probabilities and the cost of climatic events are specifically considered.

A. "No Regrets" and "Climate Justified" Strategies

The Intergovernmental Panel on Climate Change categorized adaptation strategies into two types. First the "no regrets" strategy assesses specific cost/benefit of actions irrespective of a climate-based event occurring. The alternative is the "climate justified" strategy, which assesses the benefit of actions only if the climate impacts do occur.

Uncertain climatic events under no-regrets strategies are largely immaterial, as the cost/benefit analysis would be the same as under the more conventional economic, financial, social, or environmental analyses. However, such strategies also do serve the purpose of mitigating the effects of climate change and disaster-type events. They focus on water conservation measures, reduction of operational inefficiencies that contribute to water losses, increased financial sustainability for increasing financing capacity for protecting watersheds, expanding networks, increasing treatment capacity, and recycling wastewater. In comparison, climate-justified strategies may include retrofitting existing investments and constructing new investments such as dams, storage reservoirs, water transfer systems, and irrigation.

No-regrets strategies reinforce objectives for financial sustainability and build resilient water infrastructure entities to deal with different types of hazards. Moreover, they can significantly postpone new infrastructure investments while also creating more financially resilient institutions given the eventual need for climate adaptation investments.

B. Climate Finance and Risk Instruments

Because climatic events can be catastrophic, a number of financial instruments have been developed in order to substantially reduce the fiscal impacts on governments when events do occur, and to increase possible sources of disaster-related finance for poorer nations, beyond the more traditional ex-post humanitarian aid. Disaster risk instruments can be divided into two broad categories: (i) loss financing instruments, and (ii) catastrophic risk (CAT) insurance.

Sovereign funding for extreme events is the more common of the various funding sources. These are typically set-asides from budgetary allocations in order to provide immediate emergency funding as well as long-term funding for reconstruction and development. This funding would also finance contingent liabilities for government-supported agricultural insurance schemes and social safety nets following a disaster.

Given the increasing incidence of natural disasters, in 2008 the World Bank approved the first contingent loan, called the Catastrophe Drawdown Option. Such a credit line allows governments to secure funds ahead of a disaster, which can be drawn down quickly in cases of emergency.

CAT insurance transfers the loss to another party in exchange for a premium. Like any insurance, the premium amortizes potential losses over time instead of requiring a government to finance a single sizable loss. These policies can be either indemnity or parametric. Indemnity insurance will cover the actual loss but requires an estimation of actual damages, while parametric insurance predefines the payment based on a specific event occurring and is administratively much simpler to implement, especially in developing countries.

Some of the more interesting products include CAT insurance for property and agriculture, catastrophic bonds, and multi-country pooled facilities. In 2007, the Caribbean nations introduced the first multi-country facility, called the Caribbean Catastrophe Risk Insurance Facility (CCRIF), on the basis of parametric policies to provide short-term liquidity to the participating countries in the event of climatic disasters.

Catastrophic risk bonds work just like other parametric insurance contracts (also indemnity insurance contracts) except that the insurers are investors who receive a bond yield rather than a premium. The bond proceeds transfer to the insured if the event occurs, in which case the investor loses the entire principal value of the bond. CAT bonds typically have a short-term maturity of no more than three to four years.

VII. Concluding Remarks and Recommendations

The funding gap is undeniably daunting, and achieving the SDGs requires a realistic approach to understanding the problems of the sector, as well as a much better track record in tailoring solutions and deploying financing instruments.

Private finance is significantly more abundant than public sources, but the fundamentals of finance must be adhered to in order to effectively harness it. There is no magic solution, and investors and private lenders seek investments that offer a reasonable risk-adjusted return that many water entities simply cannot provide.

The other infrastructure sectors have done far better in mobilizing private finance. The financial instruments and facilities are the same as they are for water, and yet the success rate of these sectors is markedly better. The reason is that the majority of water entities in the developing world do not offer any return whatsoever, and the risk factors are complicated by a politicization of water tariffs by government leaders.

Moreover, an overwhelming number of water utilities confront performance inefficiencies and a "financial viability gap" that need to be corrected before expansion strategies are adopted. No private finance instrument or facility can effectively deal with this. If the pipeline of viable projects does not exist, it is senseless to continue to introduce new instruments or additional privately financed facilities.

A. Policy Considerations for U.S. Government

From a U.S. foreign engagement perspective, it is imperative that substantive reforms in the sector-complement capital expenditures (CAPEX) investments for expanding access. In this regard, there is a compelling reason to squarely deal with the core business model of weak water entities and make them financially viable and creditworthy. There is also a continued need for TA to plug information gaps among private lenders and to utilize public sources of finance in order to draw them into a financing structure, even with minimal initial exposure.

Investment financing can be made available to improve performance, but water tariffs must remain equitable and affordable. Water has far lower average cost recovery levels than power does, and pricing your way to a viable project will not work. Instead, project analysts must create suitable financing structures using a mix of grants and concessional loans in order to maintain affordable tariffs. This is not to say that tariffs cannot be increased, but rather that affordability and willingness to pay are particularly critical in the decision to do so

Donors need to fully support the notion that the financing problem must be anchored in a strategic continuum with a starting point and an end goal, and a system that shepherds faltering utilities through various and distinct stages of reform and financing. Unfortunately, this does not point to quick-fix or innovative solutions, but rather to more mundane and rigorous interventions that will, however, pay off over time. Moreover, during the first-stage financing, the public-to-private leveraging factor may be extremely low and difficult for U.S. policy makers to rationalize.

While there is scope for private finance facilities where the pipeline of viable projects exists, the need is greater for donor funding of country or regional viability gap facilities consisting of grants and concessional loans that can be blended with commercial debt. Such funding should be made available on a competitive basis in order to drive proper policy reforms in the sector. This would incentivize government policy makers to institute proper reforms within their sectors. Donors and IFIs must stand firm on this point in order to drive real and substantive change.

While this is often said, it is critical that donors work cooperatively on these initiatives and avoid negating each other's efforts. U.S. policy should be oriented toward creating such cooperation. Sectorwide programs with other donors and IFIs are effective since they also take into consideration the comparative advantages of each as part of a collective country effort.

Innovation has an important role to play, but donors must ensure that new mechanisms or approaches they support can stand on their own without the continuing support of donors in taking up risks that should be ultimately borne by the market in order to be sustained. Alternatively, a strategy would need to be established a priori in order to ensure replication.

Climate change and water security will force a need for additional funding in the sector, but will also create the opportunity for donors to reinforce financially sustainable policies and operational efficiencies among water entities. Building resilience into water infrastructure is entirely compatible with efficient water allocation, performance improvements, and financial sustainability, and must be a clear priority for U.S. policy support.

Notes

1 Baietti, A. and Raymond, P. 2005. "Financing Water Supply and Sanitation Investments: Utilizing Risk Mitigation Instruments to Bridge the Financing Gap," Water Supply and Sanitation Sector Board Discussion Paper Series No. 4, World Bank, Washington, D.C.
2 The MDGs sought to halve the proportion of the population without improved drinking water and sanitation between 1990 and 2015. According to the UNICEF and The World Health Organization in their report titled "25 Years Progress on Safe

Drinking Water and Sanitation—2015 Update and MDG Assessment," the target for drinking water was met in 2010 when 2.6 billion people gained access, bringing to 58 percent the global population with safe drinking water. Although 2.1 billion people gained access to improved sanitation between 1990 and the present, and the global proportion of people practicing open defecation has been nearly halved, sanitation remains a significant global challenge.
3 The World Water Commission for the 21st Century was established by the World Water Council in December 1998 to guide long-term thinking on water and environment policy into the next century.
4 Report of the World Panel on Financing Water Infrastructure, "Financing Water for All," p. 1.
5 Ibid., UNICEF and World Health Organization report on MDG progress.
6 "Investing in Water Infrastructure, Capital, Operations and Maintenance," Water Partnership Program, November 2012, p. 8.
7 For more discussion on the Ladder of Financial Sustainability, see Baietti, A. and Raymond, P. 2005. "Financing Water Supply and Sanitation Investments: Utilizing Risk Mitigation Instruments to Bridge the Financing Gap," Water Supply and Sanitation Sector Board Discussion Paper Series No. 4. World Bank, Washington, D.C. https://openknowledge.worldbank.org/bitstream/handle/10986/17235/320280WSS1Investments.pdf;sequence=1.
8 Baietti, A. 2001. "Private Infrastructure in East Asia: Lessons Learned in the Aftermath of the Crisis." World Bank Technical Paper No. 501. World Bank, Washington, D.C.
9 World Bank–PPIAF PPI database.
10 Baietti, A. et al. May 2006. "Characteristics of Well-Performing Public Water Utilities." Water Working Notes No. 9. Water Supply and Sanitation Sector Board. World Bank. Washington, D.C.
11 BOTs are a form of PPP where the private sponsor assumes the responsibility for building and financing a specific infrastructure asset for rendering a service over a long-term period, typically 15 to 20 years. The asset is returned to the government at the expiration of the agreement. Given the risks associated with BOTs, the private parties will typically request guarantees from the government or sectoral agencies.
12 Guasch, J. and Foster, V. January 2005. "How Profitable Are Infrastructure Concessions in Latin America? Empirical Evidence and Regulatory Implications." PPIAF.
13 The study reviewed profitability of water, power, transport, and telecom concessions.
14 Taxpayers include funding from Official Development AID and borrowings include IFI lending.
15 With all other things being equal, consider a total financing program of $30.7 million with a private finance structure of 35 percent equity and 65 percent debt. The commercial debt terms are 7 percent interest, at five years maturity with no grace period, while the cost of equity is estimated at 19 percent. Under such terms, the investment yields a NPV of −$15.9 million, which simply means that the project is earning substantially less than its weighted average cost of capital (WACC) of 11.25 percent.
16 "Water and Climate Change: Understanding and Making Climate-Smart Investment Decisions." November 2009. The World Bank, Washington, D.C., p. 1.
17 Water systems for delivering water services include irrigation; urban water, sanitation, and drainage; rural water and sanitation; and ports and navigation. By comparison, systems for managing water resources include those for delivery of bulk irrigation, watersheds, and water resources broadly, as well as multipurpose systems (including hydropower) and flood control.
18 "Water and Climate Change," p. xix.

18 Financing Water and Sewer Infrastructure in the Developing World

William Streeter

Summary

> *Water is distressingly underfinanced compared to other types of infrastructure. . . . Water issues are important, but their management is difficult and often fragmented or ignored.*[1]
>
> —United Nations Secretary-General's Advisory Board on Water and Sanitation, November 2015

In developing countries, public- and private-sector financial resources for essential water and sewer infrastructure have traditionally lagged behind other essential infrastructure sectors, such as electricity and roads. For the latter sectors, much progress has been made in some emerging-market countries to provide clearer lines of service responsibility; greater certainty of regulatory oversight; greater confidence in the reliability of revenue streams to pay for the maintenance, operation, and debt service of infrastructure projects; and improved contractual certainty for lenders and investors. Where these characteristics have developed in concurrence with growth in local capital markets, financing has become available (in varying degrees) for infrastructure projects, mostly through bank loans.

These characteristics are less developed for water and sewer services, where public-sector decentralization to state and local governments has progressed but often remains fragmented and incomplete. This means that the local government sector, including municipal authorities or districts that provide water and sewer services, are less likely to be creditworthy entities from the perspective of a local institutional lender or investor. This also means that the evolution of private-sector financing for local water and sewer projects, whether through municipal debt or through public-private partnership (PPP) special purpose vehicles (SPVs), will most likely require innovative ways to lower the default risk of these obligors, while also strengthening their internal capability to plan, design, finance, operate, maintain, and regulate water and sewer projects.

The first track, of lowering default risk through innovative financial strategies, is necessary in order to create or supplement the local debt markets for water and sewer projects, where the placement of project debt (either bank loans or bonds) would be difficult due to creditworthiness concerns. The second track, of strengthening local capabilities, should be viewed as a condition for innovative financial support (i.e., reforms in exchange for support) in order to maximize the chance that newly found access to capital is sustainable over the longer term. The public policy, technical assistance, and funding support of these two tracks by the United States government and its agencies will improve the capability of water projects to support their financial obligations, and the receptivity of institutional investors to water projects as investible and attractive assets.

Large-scale examples of these dual strategies in developing countries are nonexistent, which—together with the nascent stage of many of their capital markets—explains the scarcity of private-sector finance for the sector. Nevertheless, proven smaller-scale initiatives in a number of developing countries could be adapted and replicated across the developing world. While these innovations cannot resolve the financing gap for essential water and sewer services, they could greatly supplement and even leverage existing national funding efforts as well as the assistance of multinational and development bank institutions. They would also provide a necessary means to broaden and deepen the domestic capital markets of emerging-market countries.

This chapter describes some emerging-market water and sewer-funding initiatives, what makes them important financial role models for other developing countries, and lessons learned about what should be either encouraged or avoided. Examples include commercial bank and investor comfort in the regulatory and administrative framework for a municipal water authority in Culiacan, Mexico; the layers of protection and enhanced access to capital afforded by pooling municipal default risk, structuring reserves, and providing intercepts to intergovernmental transfers for the debt of a water and sanitation facility in Tamil Nadu, India (as well as in a newer but national-level program being developed for Kenya); the successes and failures of an urban funding and institution building exercise for the 60 largest cities in India (which had a heavy focus on water and sanitation services, as well as on municipal transparency and accountability); and the ability of both a private water project SPV and a strengthened municipal corporation to adapt to fluctuations in the demand and supply of water services for growing urban and politically charged environments.

The Underfunding of Water Projects in Developing Countries

Financing for water and sewer projects in developing countries has lagged that of other essential infrastructure sectors, such as electricity and roads. The author believes that there are three predominant reasons for this: (1) differences between water and other utility sectors in how the services are provided and regulated, which result in a generally weaker institutional framework to plan, operate, and finance water services; (2) differences in public perception of how water should be priced, in comparison with other utility sectors; and (3) the nascent stage of development of financial markets in many emerging-market countries, which affects all infrastructure finance (not just water).

How Water Projects Differ from Other Utilities in the Provision and Regulation of Services

Whereas the provision and regulation of electricity in many emerging-market countries are managed centrally in cooperation with state governmental entities (and increasingly with private-sector project companies or privatized utilities), the provision and regulation of water have become increasingly decentralized. In the Philippines, for example, the Department of Energy maintains a medium-term Philippine Energy Plan, which provides a road map for energy demand and supply nationwide.[2] In 1990, the country enacted a build, operate, and transfer (BOT) law, largely to facilitate private-sector participation in power generation. In 2001, the Philippines enacted the Electric Power Industry Reform Act (EPIRA), which privatized the National Power Corporation's generation assets; created a wholesale electricity spot market, as well as the Energy Regulatory Commission (ERC) as an independent regulator; and formed the Joint Congressional Power Commission as a "champion" to oversee the implementation of the new law.[3] By contrast, the Philippine Institute for Developmental Studies in a recent report stated that the water sector did not have a national government department with the responsibility of "translating government's policies, strategies and goals into a comprehensive water supply program."[4] It went on to state that there were not enough changes in central government agency programs to develop the capabilities of local government units (LGUs) to establish, operate, and finance water utilities. It also mentioned that the country's water master plans were outdated.

In addition to differences in service provision and regulation, electricity and telecom lend themselves to competition, whereas water is a natural monopoly. Airports are also a monopoly, but their activities are commercially oriented for those who can afford to fly, and therefore there is no public resistance to fees for service. Water, as an essential element for life, carries a perception of being an entitlement, even in the developed world. The World Commission on the Ethics of Scientific Knowledge and Technology (COMEST) identified some fundamental principles about water, including a principle for human dignity that stated that "there is no life without water, and those to whom it is denied are denied life."[5] This is why water charges are only for extraction, treatment, and distribution, and not for the commodity itself. Contrast this with the electricity and telecom sectors, where pricing takes into consideration the availability and usage of the service commodity, with prices for the latter often varying by time-of-day fluctuations in demand.

The Devolution of Water Services

Water services in many countries have devolved to the local government level. In many emerging-market countries, however, this level of government generally has weaker administrative, planning, and operational capabilities. James Winpenny, in the World Water Council report of 2003, stated that local governments and water authorities have the responsibility for water services, but "not all of them have the necessary skills, efficiency and financial powers."[6] A study of the local water and sewer authorities (*organismos operativas de agua y saneamiento*) by the government-owned Instituto Mexicano para la Competividad (Mexican Institute for Competitiveness) cited five main shortcomings of these local authorities, which this author believes could easily apply to water authorities or departments in many emerging-market countries.[7] The report's findings are that the local Mexican water authorities generally lack the following:[8]

1. Autonomy to determine tariffs, a sense of professional direction, and mechanisms for both public participation and remaining free of influence by outside interest groups
2. Sufficient scale of operations to reduce operating costs and to remain economically viable
3. A mandate for financial self-sufficiency, since the water authorities are not obligated to operate as commercial enterprises, covering their costs with user fees, but instead are often dependent on subsidies from higher levels of government or transfers from their parent municipality
4. Standardized and transparent financial information, as well as disclosure about contractual processes and tenders
5. Technical, human resource, and financial capabilities, as well as long-term planning processes (often tied to shorter-term election cycles), the lack of which limits the water authorities commercial efficiency and ability to mediate service shortcomings

The confluence of these factors contributes to diminished lender and investor confidence in the reliability of water and sewer revenue streams to pay for the maintenance, operation, and debt servicing of water projects. It also contributes to a diminished sense of contractual certainty. For these reasons, the water sector sees proportionately less financing for its projects than do other utility sectors.

Underfunding of Water Infrastructure Assets in Comparison to Other Infrastructure Assets

The World Bank maintains a database on infrastructure projects that reached financial closure in 101 low- and middle-income countries. For the East Asia and Pacific region, for instance, project data is included for financed projects in

China, the Philippines, and Indonesia, but is not included for projects financed in Japan, South Korea, Taiwan, and Singapore.[9] Project financings for public authorities, state-owned enterprises, and PPPs are included. The author selected data from 2008 through 2015 in order to show the levels of infrastructure finance activity during and since the global financial crisis (see Table 18.1 for a summary of project financial closings by region, number of projects, value of projects, and asset class).

The database shows that since the global financial crisis, developing countries have financed 2,483 infrastructure projects in all sectors, for investment totaling the equivalent of USD 1.2 trillion. This record contradicts two commonly held notions: that the emerging markets are unable to finance the pipeline of infrastructure projects, or alternatively that because of the absence of a project pipeline, they have been unable to finance infrastructure. Instead, the facts support different arguments: that some emerging-market countries have been doing better than others in financing infrastructure, and that a legitimate discussion can be had as to whether the pace of actual investments is adequately accommodating the pace of economic growth and urbanization.

The data also show the predominance of the electricity sector, with 61 percent of funded projects, in comparison with the water sector, with only 12 percent of funded projects.[10] The reasons for this disparity were covered in the last section. Funded water projects also tended to concentrate in two regions: East Asia and the Pacific (the data show that this is primarily in China, where the government-controlled banking sector is active), and Latin America (primarily Mexico and Brazil, which have active national development banks and, to a lesser extent, viable capital markets).[11] The data also show that most projects were financed with local currency bank loans, although some projects also received financing from multilateral and export credit agency loans (this information is not depicted in Table 18.1).[12] Very few projects were financed by bonds, for reasons discussed next.

The Nascent Stage of Developing Country Debt Markets

The potential for debt financing of water projects in emerging markets is a factor of the nascent stage of debt capital market development in these countries. A generally disappointing record of debt financing for infrastructure projects is likely to coincide with a very limited bond market for any type of corporate debt issuance, irrespective of credit quality. In addition, it is likely that the credit quality of stand-alone, nonrecourse projects, even if rated investment grade on the national scale, is still below the minimum investment thresholds of the country's domestic insurance and pension funds. The State Bank of India guidelines for pension fund investments, for instance, allow for investment in infrastructure debt, but require a minimum rating threshold of "AA" for such investments.[13] The author participated in many project ratings in India from 2007 to 2011, and during that time, the distribution of infrastructure project ratings was primarily between "BB" and "A" on the national scale, which

Table 18.1 Infrastructure Projects Reaching Financial Closure by Region, 2008–2015 (Q2)

Region	Number of Countries	Number of Projects Since 2008	Value of Projects Since 2008 (USD millions)	Number of Projects by Sector							
				Airports	Electricity	Natural Gas	Railroads	Roads	Seaports	Telecom	Water & Sewerage
Latin America and Caribbean	17	716	477,850	25	452	7	9	110	35	7	71
Europe and Central Asia	18	292	227,831	9	231	15	6	9	12	7	3
East Asia and Pacific	15	616	131,529	5	337	21	8	17	25	14	189
South Asia	7	640	266,123	2	360	2	5	236	20	9	6
Middle East and North Africa	11	56	49,267	1	27	–	–	–	7	7	14
Sub-Saharan Africa	33	163	92,532	4	105	–	–	3	11	36	4
Totals	101	2,483	1,245,132	46	1,512	45	28	375	110	80	287
% of Total Projects				2%	61%	2%	1%	15%	4%	3%	12%

Source: Private Participation in Infrastructure Database, Regional Snapshots, www.worldbank.org/snapshots/region

meant that none of the pension funds would have been allowed to invest in this project debt.

The following is a case study on characteristics of the Indonesian corporate bond market, which the author believes are typical across many emerging-market countries. These characteristics demonstrate the challenges facing stand-alone, nonrecourse, and low-investment-grade rated infrastructure projects in these countries. The Indonesian bond market is growing, but it is small and highly concentrated. Total bonds outstanding equal just 15 percent of GDP, and 85 percent of this amount represents government bonds.[14] Corporate bond issuance is dominated by a relatively small number of highly rated government-owned or finance-related corporations, which typically issue bonds with one- to five-year maturities, and with ratings in the "AAA" and "AA" categories comprising 87 percent of debt issuance.[15] The lack of long-term corporate debt in the Indonesian market is a constraint to local governments and infrastructure project companies that might want to enter this market. In the opinion of the Indonesian Stock Exchange, the insurance and pension fund investors (which are the country's primary institutional investors) concentrate their interests on the "AAA"- and "AA"-rated corporates, while the mutual funds, which are much smaller, like the yield pickup provided by the "A"-rated corporates.[16]

Finally, Indonesia's corporate bond market exhibits a high degree of observed randomness or inefficiency in the pricing of corporate bonds, which acts as a hindrance to greater corporate bond market activity. This seems to be exacerbated by too few government benchmark bonds, significant foreign ownership of government bonds,[17] and adverse treatment for infrequent corporate issuers. During the 2014–2015 study period, the spread of monthly 10-year government bond rates was 186 basis points (bps),[18] which is extreme for such a short period of time. In the author's opinion, this volatility was caused by foreign investor concerns over potential increases in U.S. interest rates, the economic fallout in Indonesia of a slowing Chinese economy (due to Indonesia's dependence on resource exports), and the ability of the new government to manage these macroeconomic challenges while also instituting regulatory reforms. The median spread in corporate bond rates over the commensurate 10-year government bond rates during the same period was 217 bps, which is also high, since 87 percent of the corporate issuance volume was rated either "AAA" or "AA."[19]

The author believes that in order to improve the potential for water project financing, innovative strategies that lower project default risk and strengthen local planning, administrative, and operational capabilities of water service providers will improve the environment for such financings. While large-scale examples of these dual strategies in developing countries are nonexistent, there are proven smaller-scale initiatives in a number of developing countries, which could be adapted and replicated across the developing world. These innovations cannot resolve the financing gap for essential water and sewer services, but they could greatly supplement and even leverage existing national funding efforts, as well as the assistance efforts of multinational and development bank institutions. They would also provide a necessary means to broaden and deepen the domestic capital markets of emerging-market countries.

Creating a Regulatory and Administrative Framework for Water Service Debt

In areas where the authority for the provision of water and sewer services has devolved to the local government level, the legislative authority must exist for these governments to issue debt and to raise revenue that can be pledged as security for their debt. Institutional capabilities of local governments need to be augmented in advance of introducing local borrowing, including the abilities to assess and collect taxes and user fees; conduct operating budgetary exercises; prepare capital plans; successfully tender projects with contractors and service providers; solicit and process constituency opinions in community forums; and prepare and disseminate accurate, transparent, and timely financial statements. These considerations are important not only for government regulators who want to monitor changes in the financial performance and indebtedness of local units of government, but also for the promotion and maintenance of investor confidence in local government bonds.

Perhaps the most elegantly designed but imperfectly implemented local government institution building and infrastructure funding exercise in an emerging market was the Jawaharlal Nehru National Urban Renewal Mission (JNNURM), which was launched by the government of India's Ministry of Urban Development (MoUD) in 2005 and which ran for seven years through 2012.[20] The JNNURM provided conditional infrastructure investment funding for India's largest municipal corporations. Eligible project areas included urban renewal, water, sewerage, solid waste, environmental improvement, street lighting, roads, urban transport, and civic amenities. Funding of capital projects was conditional upon the municipal corporations and their parent states submitting to a series of mandatory reforms. Municipal reforms included the following:[21]

- Adoption of an accruals-based, double-entry accounting system
- Introduction of various e-governance and technological applications
- Reform of the property tax system
- Implementation of systems to recover utility costs through user fees
- Budgetary provisions for basic urban services for the poor
- Preparation and approval of a city development plan (CDP), presenting the current status future direction of the city's development
- Institution of a city planning function
- Acquisition of a municipal debt rating (issuer level)
- Encouragement of PPPs
- Provision of a local funding share of 25 percent for approved projects, to come from available municipal revenues, municipal debt, or solicited

PPPs; the remainder was to be provided by the government of India (50 percent) and the parent state (25 percent)

While this program achieved a great number of technical initiatives that had never been produced before at a local government level in India, it failed to change behavior patterns between levels of government. The final report by the Ministry of Urban Development on the JNNURM program in 2011 noted that there was no provision for constituting a city-level committee to monitor the JNNURM guidelines.[22] This forced technical support for the exercises to be performed by parastatal agencies and private consultants, minimizing local government participation and ownership in the preparation of community development plans and the implementation of municipal reforms.[23] The author, who was involved in a number of local government ratings under the program, also observed an inherent level of mistrust between levels of government, which hampered implementation of the program.

Securing Project Cash Flows

Infrastructure finance relies on securing project cash flows, since the actual project assets continue to be either owned or heavily regulated by the host government. The need for pledging a physical asset as security for debt is one common misconception of infrastructure finance, as it is often portrayed in studies that focus on emerging markets. In the case of a PPP, for example, the concession is contractually a "conditional grant" to project cash flows, in exchange for the provision of certain project improvements and service provision to the public (such as lane miles of road with a predetermined drivability standard to the driving public, or square footage of a hospital building designed and maintained so that the government can provide specified health care services). The same is true for a municipal revenue bond, secured by user fees from the city's revenue-generating assets. In the case of an off-take agreement between a privately owned project and a public utility, there is a contractual conditionality based on the availability and delivery of an agreed-upon service output (such as million cubic feet of natural gas or kilowatt hours of electricity). A corporate-style mortgage lien on these assets (a typical form of debt security for commercial debt) is meaningless if the asset is government owned, and it is dubious for a privately owned project if the asset is strategic to a government-regulated network.

Finally, in the case of a local government bond, the security for that debt (whether pledged revenues come from government taxes, user fees, or transfers to the local government from a higher level of government) is predicated on the establishment of a trust estate and escrow agreement created for the benefit of lenders or investors.

A second common misconception of emerging-market infrastructure finance studies is that water and sewer debt needs to be fully self-supporting from project user fees. Governments can have the public policy goal of making utility services self-supporting from user fees. The benefits of such a policy include broadening a municipality's revenue stream and allowing a redistribution of limited tax revenues from self-supporting utilities to other important services. Such a goal may even be a condition of financial support from a multilateral agency. These are valid and useful objectives, but they do not need to be a prerequisite for financing water projects.

Where there is confidence in the adequacy of water user fees, they can be used as security for water project debt. Where confidence in the adequacy of water fees is missing, but where such confidence exists for local government taxes, the taxes can also be pledged as a security for water project debt or as a backstop for deficiencies in user fees until such time that the fees become more predictable and robust. Where there is a lack of confidence in both a local government's user fees and taxes, government transfer payments (to the extent that they are available) can be used as a form of debt security. This broader utilization of available revenues for water project debt does not prevent the promulgation of public policies or multilateral agency conditions to financing, whereby a commitment is required to improve the financial adequacy of water user fees over time.

Mexican Structured Certificates

Mexico borrowed from structured finance techniques to provide an early and solid form of cash flow security for project debt. In 2001, the Mexico Securities Market Law (Ley del Mercado de Valores) was amended in order to allow states, state enterprises, municipalities, municipal enterprises, and project SPVs to enter into fiduciary trusts that could issue fiduciary stock exchange certificates (*certificados bursatiles*).[24] The certificates provide investors with a beneficial interest in pledged revenues of the trust, by ring-fencing pledged cash flows. They also allow credit institutions and exchanges to act as fiduciary trustees.[25]

Subnational governments were allowed to pledge only a fixed percentage of their revenue streams to the trust, which minimized the potential for overleveraging their limited resources.[26] In addition, three other innovations made this structure particularly effective as a means of securing cash flow:[27]

1 The creation of a separate trust for debt service payments outside of the subnational government's treasury
2 The advance segregation of revenues as they are collected or received by the trustee, prior to a debt service payment date
3 The overcollateralization of the debt service requirement by pledged revenues, with the release of excess revenue back to the subnational government only after the debt service payment was made

The securities are listed on the Mexican Stock Exchange (Bolsa Mexicana de Valores). Mexican law allows the trust to be the obligor, so bondholders have no recourse against the underlying state or municipality. This feature, which also borrows from structured finance, is useful for a commercial asset where there can be a "true sale," but in the author's opinion, the securitization of municipal revenues is at best a quasi-securitization and would not meet the true sale test. The main risks for Mexican certificate holders are that revenue collections may fall short of expectations or that there may be delays in transfer payments from another government (if such transfer payments are used as security).[28] In the opinion of the author, these risks are largely mitigated with debt service reserves, by carefully managing the degree to which these revenue streams are leveraged by debt, and by spacing out principal payments on debt significantly after the time when pledged revenues are expected to be collected or received.

The fiduciary trust structure has allowed states, larger municipalities, and autonomous enterprises (including state and municipal water authorities, ports, and even a few universities) to enter the debt market through the issuance of structured certificates.[29] A pattern emerged in 2012–2013 of municipal defaults on short-term lines of credit from banks, but there have been no recorded events of default on long-term debt secured under a fiduciary trust structure.[30]

Enhancing Credit Quality by Pooling Project Risk

In many cases, as was discussed earlier in the section on the nascent stage of capital markets in developing countries, the credit quality of a municipal borrower or a stand-alone water project would not meet the country's investment guidelines. Smaller towns and villages, with weak credit quality and small projects, would have no access at all. Therefore, to become viable as investments, these credits require pooling in order to attain scale of investment (most institutional investors require an investment of a minimum size), and credit enhancement in order to attain the ratings threshold required by domestic institutional investors. The concepts of pooling and credit enhancing a basket of municipal loans for water and sewer projects began in the U.S. in 1987 with changes to the Clean Water Act, which created the state revolving funds.[31] Emerging-market applications of this model were replicated with mixed success in India, with a pooled finance scheme at the state level, and are soon to be applied at the national level in Kenya.

Water and Sanitation Pooled Fund (WSPF)

WSPF was the first rated pooled finance issue in India, representing a major breakthrough in Indian public finance.[32] From 2002 to 2013, WSPF issued five separate bonds to the domestic bond market, for a total of INR 222.3 crores (roughly USD 44 million), with 10- to 15-year maturities.[33] The bonds funded projects in a total of 45 urban local bodies (ULBs). Three of the bonds had reserves sized at 22 percent of proceeds, while the remaining two issues had reserves sized at 37 percent of proceeds.[34] All of the bonds were rated "AA(SO)" on the Indian national scale by Fitch Ratings ("SO" stands for "structured obligation").[35] There have been no defaults on any of the bonds. The MoUD, under its Pooled Finance Scheme, established guidelines for other states to create pooled finance vehicles for water and sanitation projects, as well as budgetary authority to set up reserve funds for these vehicles.[36] One other state, Karnataka, set up a pooled finance vehicle and successfully issued bonds in the domestic debt market. In the author's opinion, the lack of broader utilization by states of this very beneficial program can be explained by two now familiar factors: the absence of a champion within the MoUD to promote the program for a sustained period of years, and the inherent lack of trust between levels of government in India, whereby new central government schemes are accommodated by lower levels of government but rarely embraced.

The basic premise of the pooled finance structure is that it reduces the risk of a single borrower loan default causing a bond default. The greater the size and diversity of a pool—even with the inclusion of small and less creditworthy borrowers—and the more the concentration of the largest participants is minimized, the more the default risk is spread, improving the creditworthiness of the pool and lowering the cost of funds. The benefits of this form of internal credit enhancement are as follows:[37]

- Each individual borrower has access to the capital markets at a much lower interest rate than it would otherwise obtain if it borrowed on its own. This provides an attractive means of financing for small and medium-sized ULBs, some of which are too small or too weak from a credit perspective to enter the markets on their own, to access the capital markets.
- Transaction costs are spread among the participants, providing further efficiency.
- Resources once used to fund grants can instead be used to make subsidized loans, spreading the resources to a larger group of beneficiaries.
- Bonds used to finance loans can receive higher ratings than those assigned to the underlying borrowers owing to the diversity of the pool and other structural credit enhancements.

The second form of internal credit enhancement under the WSPF program is a fully funded debt service reserve fund.[38] If the fund is drawn, it must be replenished using the same state aid intercept mechanism as that used for the ULB loans. The initial WSPF bonds had a USAID guarantee in an amount equal to 50 percent of the principal amount of the bond issuance. The final two issues by WSPF had reserves funded by the MoUD under its Pooled Finance Scheme.[39]

In aggregate, the loan agreements with the ULBs mirror the repayment obligation on WSPF's bonds. Monthly loan repayments are transferred to the WSPF trustee-held account in advance of WSPF's bond payment date.[40] In the event of any shortfall in the ULB loan contributions, the financing agreement contains provisions to intercept future state transfer payments to the ULBs.

This state aid intercept feature is a major external credit enhancement for the bonds. In practice, some ULBs make their loan payments using their own budgetary revenues, while others rely on the intercept mechanism to make their loan payments.[41]

The Kenya Pooled Water Fund (KPWF)

A new pooled finance vehicle for water projects is being established in Kenya by the Netherlands Water Partnership and the World Bank.[42] Whereas WSPF in Tamil Nadu operates at the state level, the KPWF will exist at the national level. This should allow it to attract a greater variety and number of projects. The KPWF hopes to issue its first bond next year, in the Kenyan shillings equivalent of USD 25–40 million.

The World Bank and Kenya's Water Services Regulatory Board (WASREB) found that 13 of Kenya's 117 water utilities (or just 11 percent of operating utilities) have the potential for an investment-grade, national-scale rating.[43] If the financial health of more water utilities improves, KPWF hopes to be able to issue one bond per year. The Kenya National Water Master Plan of 2014 estimates water project funding requirements of the Kenyan shilling equivalent of USD 976 million through 2030.[44] Existing levels of public funding are expected to meet only 60 percent of this requirement. The hope is that the KPWF will bridge the funding gap.

In Kenya, it is not possible to utilize a state aid intercept mechanism (the legal authority of a project or utility to intercept intergovernmental transfers should there be a deficiency in the ability of water charges to meet debt service requirements). Nevertheless, it is hoped that the segregation of utility loan installment payments in advance of a bond debt service payment will strengthen the program and allow utility payment difficulties to be spotted early.[45] A fully funded debt service reserve and a potential source of external credit guarantee are being explored. It is also hoped that these structural features would allow for the program to have a higher rating, a longer tenure, and a lower interest rate than would be possible for each individual water utility.

Instilling a Focus on Quality Service and Adaptation

Urban areas in developing countries are systems of people, services, and institutions that are subject to many changes—some expectedly, from the pace of growth, and others less expectedly, from the interactions of that growth with natural disasters, health concerns, or social tension. Whether water and sewer services are provided by a local government or through a PPP, it is important for their providers to maintain a focus on quality services, and to adapt to the deficiencies in the infrastructure and institutions governing them. This need to adapt, and to assess and plan for risks, is described by an organization called Shack/Slum Dwellers International as the "need for bottom-up development."[46] While this chapter has discussed a number of shortcomings in the planning, financing, and implementing of water and sewer service improvements in developing countries, these adaptations are taking place in both private- and public-sector examples, such as the Manila Water Company concession in Manila, Philippines, and the municipal water department in the Pune Municipal Corporation in India.

The Manila Water Company

The Manila Water Company holds a long-term concession agreement with the Metropolitan Waterworks and Sewerage System for the East Zone of the Manila metropolitan area, which encompasses parts of Manila and 22 other cities and municipalities.[47] Prior to the concession, the region suffered from water contamination, inconsistent service, severe leakages, and a high incidence of water-borne disease.[48] Since the concession started in 1997, 24-hour potable water has increased from 26 percent of users to 99 percent even as the number of customers grew rapidly; furthermore, water losses decreased from 63 percent to 11 percent, and water quality now meets the clean drinking water standards set by the Philippines Department of Health.[49] The company focused on providing "care for the customer," which included customer care specialists to the area's different communities,[50] and on retraining the absorbed public-sector employees from a seniority-driven political culture to a meritocracy-driven corporate culture. While the company's debt grew through FY 2013 as it financed water and sewer system improvements, its revenue base and EBITDA have also grown, allowing it to retire almost one-quarter of outstanding debt by FY 2015.[51] Importantly, the concession has a tariff regime that allows for cost recovery plus a "market-based real rate of return," which is reset every five years.[52] These provisions reduce the discretion of local regulators and politicians.

Early on, Manila Water defined a vision to "become a leader in the provision of water, wastewater, and other environmental services which will empower people, protect the environment, and enhance sustainable development."[53] Along the way, it has recognized key challenges to its ability to meet that vision and has developed plans for adapting to these challenges. These include urbanization and population growth, regulatory risk (a regulatory office monitors the concession), climate change, and developing new talent within the water sector.[54] The company's orientation toward public service, metrics-driven performance indicators, continual investment, positive financial performance, development of expertise, and key risk-based planning is exemplary.

The Pune Municipal Corporation

The Pune Municipal Corporation, a young and growing municipality in the state of Maharashtra, is embarking upon a water services expansion project that is likely to be funded both by existing tax revenues and by municipal bonds. The municipality currently has water losses of 45 percent and, because of its hilly terrain, has uneven water availability in different parts of the city.⁵⁵ Its capital plan is focused on expanding the availability of water services not only for its population today, but also for its expected population level over the next 30 years. Capital expenditures are concentrated on fixing existing leaks, installing metering for all water customers, implementing a new water user fee (that will eventually replace the existing water tax on property value), and increasing pressurized storage facilities to provide 24-hour service throughout the city.⁵⁶ While the plan calls for some increase in water supply, most of its activities are directed at more efficient usage of its existing supply. The municipality has held public consultations on its water capital plan, and has attained both intergovernmental approvals for the project and initial council approvals for a schedule of water fees over the next 30 years. It is planning its first municipal bond issuance in order to finance this citywide improvement program.⁵⁷

Policy Recommendations for Water Project Financing

U.S. public policy, technical assistance, and funding support can be instrumental in supporting the dual tracks of lowering project default risk and strengthening local capabilities to plan, design, build, and operate water projects. This will improve both the capability of water projects to meet their financial obligations and will increase the receptivity of institutional investors to water projects as investible and attractive assets. While the initiatives highlighted in this chapter are small in scale, U.S. support for such initiatives will increase the scale of financing for water projects. While it will not close the recognized financing gap for water projects, it can reinforce and amplify the impact of institutional, governance, and financial-sector reforms that are taking place in developing countries, as they pertain to water projects. Key recommendations for U.S. public policy, technical assistance, and financial support include the following:

1   Promote improvements to the architecture of financing documents for both loans and bonds in developing countries, which migrate away from traditional forms of unsecured corporate lending to secured project lending and enhance the ability of project debt to survive normal variances in construction, economic, and regulatory cycles. These structural mitigants already exist, but they require a national recognition for the benefits of rules-based financing. Improvements in debt security, such as a trust estate around project revenues, appropriate operating and debt service reserves, investor recourse, and flexible amortization, will be beneficial to both domestic and international investors.

2   Develop secure revenue streams for project debt. Where there is confidence in the adequacy of water user fees, these fees can be used as security for water project debt. Where this confidence is lacking, local government taxes can also be pledged as a debt security, and where this confidence is also lacking, government transfer payments can be used. While the ultimate goal may be to promote stand-alone water projects, sum-sufficient

in their financial needs solely from water tariffs, this can be repositioned as a secondary-stage goal. Water projects are underfinanced, and to some extent, various forms of debt security are available but underutilized.

3 Strengthen the local planning, administrative, and operational capabilities of water service providers, whether the authority for the provision of water and sewer services has devolved to the local government level, to PPPs, or to both. These include local institutional capabilities to pursue domestic borrowing; collect taxes or user fees; conduct operating budgetary exercises; prepare capital plans; successfully tender projects with contractors and service providers; solicit and process constituency opinions in community forums; and prepare and disseminate accurate, transparent, and timely financial statements.

4 Encourage more predictable and more supportive regulatory regimes, including the ability to predictably raise rates or other revenues when needed, and to issue and service debt, while also meeting other important environmental, social, and labor concerns of the national government. This also requires establishing appropriate rules for national and local as well as public and private service providers to interface in a collaborative manner, around the implementation of national plans and with greater trust between counterparties.

5 Promote much-needed national and regional planning and performance standards for water services, as has been effectively done in the electricity and telecom sectors. These base goals and standards for developing countries provide metrics for public- and private-sector involvement, for creative local solutions, and for project performance monitoring. They also keep the focus not only on today's service delivery deficit, but also on where the country wants to be in the future.

6 To the extent that these recommendations are insufficient or will take time to implement, seek ways to develop pooling of credit risk and the provision of external credit enhancement. Project pooling disperses credit default risk and provides larger-sized debt tranches to the domestic market; in other words, liquidity matters. External credit enhancement further mitigates default risk and, depending on its structure, can elevate the credit quality to meet the investment criteria of the most discerning domestic institutional investor. Institutional investors will not bend their investment criteria to meet the higher default risk of individual projects or even project pools. Through credit enhancement, however, the credit quality of the projects and project pools can bend upward to meet the investment criteria of institutional investors.

Notes

1 "The UNSGAB Journey" (United Nations Secretary-General's Advisory Board on Water and Sanitation [UNSGAB], November 18, 2015), https://sustainabledevel opment.un.org/content/documents/8701unsgab-journey-web.pdf.

2 "The Energy Report Philippines: Growth and Opportunities in the Philippine Electric Power Sector, 2013–2014 Edition" (KPMG, 2013), www.aseanconnections.com/pdf/Advisory-ENR-The-Energy-Report-Philippines.pdf.
3 Ibid.
4 Gilberto M. Llanto, "Water Financing Programs in the Philippines. Are We Making Progress?," PIDS Discussion Paper Series No. 2013-34 (Philippine Institute for Development Studies [PIDS], May 2013), http://dirp3.pids.gov.ph/ris/dps/pids dps1334.pdf.
5 "Best Ethical Practice in Water Use" (UNESCO: Paris: World Commission on the Ethics of Scientific Knowledge and Technology [COMEST], 2004), http://unesdoc.unesco.org/images/0013/001344/134430e.pdf.
6 "Financing Water for All: Report of the World Panel on Financing Water Infrastructure" (World Water Council, March 2003), http://www.worldwatercouncil.org.
7 "Guia Para La Creacion de Organismos Metropolitanos de Agua Potable Y Saneamiento En Mexico" (Instituto Mexicano para la Competividad A.C. [IMCO] and The British Embassy of Mexico, February 2014), http://imco.org.mx/wp-content/uploads/2014/03/AguaPotable.pdf.
8 Ibid.
9 "East Asia and Pacific Regional Snapshots—Private Participation in Infrastructure (PPI)—World Bank Group," n.d., https://ppi.worldbank.org/snapshots/region/east-asia-and-pacific.
10 Ibid.
11 Ibid.
12 Ibid.
13 "Investment Policy for Government Sector," Investment Policy, Government Sector (SBI Pension Funds P. Ltd., June 2016), http://sbipensionfunds.com/docs/Investment%20Policy%20Govt.pdf.
14 "Capital Markets Overview" (Jakarta: World Bank, June 25, 2015).
15 "Indonesia Bond Market Directory 2014–2015, PT Bursa Efek Indonesia" (Jakarta: World Bank, n.d.), www.idx.co.id/.
16 Poltak Hotradero, Head of Research Division at PT Indonesia Stock Exchange, October 2015.
17 "Indonesia Bond Market Directory 2014–2015."
18 Indonesia 10-Year Bond Yield, Investing.com/rates-bonds/Indonesia-10-year-bond-yield-historical-data, with further calculations by William Streeter.
19 "Indonesia Bond Market Directory 2014–2015."
20 N. Raju, S. Nandakumar, and W. Streeter, "India's Public Finance Outlook: Toward an Emerging Market Model," *Fitch Ratings*, March 20, 2006, www.fitchratings.com.
21 Ibid.
22 " Appraisal of Jawaharlal Nehru National Urban Renewal Mission (JnNURM)," Final Report—Volume I (Grant Thornton, March 2011), http://jnnurm.nic.in/wp-content/uploads/2012/06/Appraisal-of-JnNURM-Final-Report-Volume-I-.pdf.
23 Ibid.
24 "Financing of Mexican States, Municipalities, and Agencies: Alternatives and Strategies," *Fitch Ratings*, January 31, 2002, www.fitchratings.com.
25 "Ley Del Mercado de Valores, Articulo 62 & 63," http://mexico.justia.com/federales/leyes/ley-del-mercado-de-valores/titulo-iii/capitulo-i/.
26 "Financing of Mexican States, Municipalities, and Agencies."
27 Ibid.
28 "Securitizations of Tax Revenues in Mexico," *Project Finance Newswire* (Chadbourne & Parke LLP, June 2008), www.chadbourne.com/SecuritizationsofTax Revenues_Jun08_projectfinance.
29 "Listado de Calificaciones," Finanzas Publicas, *Fitch Ratings*, October 31, 2013, www.fitchmexico.com.

30 "Casos de Incumplimiento Recientes En Mexico," Fitch Ratings, n.d., www.fitch ratings.com.
31 OW U.S. EPA, "Learn about the Clean Water State Revolving Fund (CWSRF)," Overviews and Factsheets, www.epa.gov/cwsrf/learn-about-clean-water-state-revolving-fund-cwsrf.
32 Raju et al., "India's Public Finance Outlook."
33 "Tamil Nadu Urban Infrastructure Financial Services, Ltd. (TNUIFSL)," n.d., www.tnuifsl.com/wspf.asp.
34 Ibid.
35 Raju et al., "India's Public Finance Outlook."
36 Ibid.
37 Ibid.
38 Ibid.
39 "Tamil Nadu Urban Infrastructure Financial Services, Ltd."
40 Raju et al., "India's Public Finance Outlook."
41 Ibid.
42 "Pooling Resources to Bridge Kenya's Funding Gap," *Global Water Intelligence* 17, no. 5 (May 2016), www.globalwaterintel.com/global-water-intelligence-magazine/17/5/general/pooling-resources-to-bridge-kenya-s-funding-gap.
43 Ibid.
44 Ibid.
45 Ibid.
46 Caroline Walker, SDI Secretariat, "Well-Run Cities Are Resilient: The Importance of Responsive Relationships Between Local Governments and Slum Communities," *Shack/Slum Dwellers International*, February 17, 2014, http://old.sdinet.org/blog/2014/02/17/well-run-cities-are-resilient-importance-responsiv/.
47 "Business Profile," *Manila Water,* n.d., www.manilawater.com/Pages/OurCompany-BusinessProfile.aspx.
48 Ibid.
49 Ibid.
50 Ibid.
51 Ibid.
52 Virgilio C. Rivera, Jr., "The Case Study of Manila Water Company: An Exercise in Successful Utility Reform in Urban Water Sector," in *Tap Secrets: The Manila Water Story* (Mandaluyong City, Philippines: Asian Development Bank & Manila Water Company, 2014), www.adb.org/sites/default/files/publication/42755/tap-secrets.pdf.
53 Ibid.
54 Ibid.
55 "Water for Pune's Future, 24X7 Water Supply Project," Pune Municipal Corporation, n.d., www.punecorporation.org.
56 Ibid.
57 Ibid.

19 A New Chapter in Developing Water Infrastructure

Marc Jeuland

This chapter aims to offer a critical perspective on a set of questions related to changes in water infrastructure planning, and their potential implications for the prosperity and security of the diverse interests that depend on water resources. Following a brief historical introduction, the discussion first considers how new risks and costs, which are emerging due to the tightening connections between human and water systems in the Anthropocene, are affecting investment decisions. Investment decisions are also evolving in response to changes in the nature of international financing. I thus discuss how the shifting financial landscape for water infrastructure may be altering riparians' willingness and ability to develop their water resources, and changing the types and designs of these development projects. I then explore the potential consequences of these combined changes, for the stability of relationships between countries, and for social and environmental security more generally. This leads to a number of reflections on how policy makers in agencies seeking to promote sustainable water resources management and in the U.S. foreign policy establishment might attempt to realign financial incentives with global norms.

Brief Historical Background on the Development of Large Water Infrastructure

The installation of water infrastructure has for millennia facilitated economic development by allowing societies to exert control over highly unpredictable surface water resources, or tapping new supplies. Water infrastructure is inherently disruptive, however. It alters natural river hydrology, with sometimes serious implications for the downstream riparians and ecosystems that depend on pre-existing flow dynamics.[1] As river basin infrastructure is increasingly built up, the potential for distributional harm may also rise, and established historical rights may become threatened.[2,3] Asymmetries that stem from riparians' relative positions and development actions have often led to disputes and diplomatic controversies.[4,5] Though the majority of transnational interactions involving water globally tend to be cooperative, a large portion of conflictive interactions involve water quantity or infrastructure.[6,7,8] At the subnational level also, it has been argued that infrastructure plays a role in economic marginalization of specific groups. For example, following dam construction in the Senegal Basin, elites in Mauritania rewrote legislation governing land ownership in order to facilitate their acquisition of suddenly valuable irrigable land near the riverbanks, at the expense of black Africans who lived there.[9]

By the end of the 20th century, amid a number of controversies over river basin infrastructure and a growing protest movement that culminated in the Curitiba Declaration of 1997,[10] many of their most important funders had come to believe that the planning process for such investments was flawed. Planners seemed to routinely overstate benefits and understate costs, and did not show sufficient appreciation for the technical, distributional, and long-term challenges posed by large water projects. This new awareness was perhaps most clearly demonstrated in the work of the World Commission on Dams (WCD), a global multi-stakeholder body initiated in 1997 by the World Bank and the International Union for Conservation of Nature and Natural Resources (IUCN). The work of the WCD led to crystallization of a new set of international norms and safeguards related to dams and other similar projects.[11] It is difficult to quantify the influence of these norms on practice, especially in light of the mixed reactions to the final report by governments in the global South. These governments mainly opposed the imposition of new project conditionalities that the developed countries had not faced during their own development process.[12] Nonetheless, the WCD has had a profound influence on the way multilateral and bilateral aid organizations decide whether and how to invest in water resource infrastructure projects.

The evolution toward a normative consensus over best practices for constructing dams and other similar projects has pushed many traditional planners and financiers to proceed much more slowly than they did previously.[13] This deliberative process has empowered technocrats who concern themselves with integrated water resources management, but it has also increased costs. These additional costs, which many consider essential to ensure proper management of water, arise both from the need for new studies and stakeholder engagement processes, and from the delay in production of services generated by large water resources infrastructure (e.g., enhanced water storage, power production, and flood control), which are often in short supply. Furthermore, the newly recommended deliberative processes and greater scrutiny of projects have heightened perceptions of the already sizable investment risks for large projects, since projects' problematic aspects now often garner more, and global, attention.[14] It has thus become more difficult for more scrupulous actors in this sector to achieve acceptable financial returns over a sufficiently short time horizon, and this reality has reduced their activity.

Some may see this reduced activity by traditional development banks as precisely what was needed in the sector, given the failings apparent in the history of preparation and implementation of such projects. Yet one of its consequences has been to reduce the relative cost of alternative sources of capital for projects that are not fully compliant with the new norms. This change, coupled with the generally increasing use of transnational capital flows that are not part of official flows of development assistance (Figure 19.1), has profoundly affected water resources infrastructure financing.[15,16,17] Resources for investments can now be obtained from private sources, or from new competitors (e.g., emerging bilateral sources, the Asian Infrastructure Investment Bank [AIIB]); some argue, albeit with limited evidence, that these may be less exacting in their application of global best practices than the more traditional funding agencies,[18,19] and may have their own strategic interests in mind.[20]

Thus, the question remains open as to what the broader effects of this shift in financing will be on the types of projects that are selected, the extent and nature of their downstream impacts, and the consequences for overall social well-being of the societies they directly affect. Furthermore, the immediate local consequences of this shift in capital flows may spill over to affect the economic prosperity and social well-being of people well beyond the borders

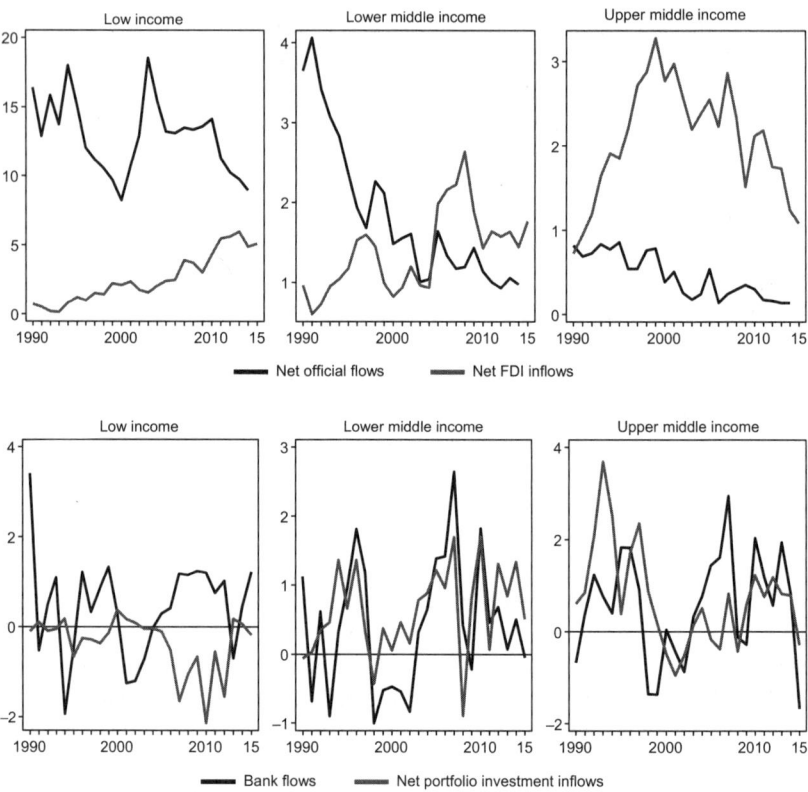

Figure 19.1 (**Top panel**) Capital Inflows, Official vs. FDI (**Bottom panel**) Portfolio and Banking Flows

Note: All in percentage of GDP. Net official flows include net ODA and net other official flows.

Sources: OECD, BIS Locational Banking Statistics, IMF Balance of Payments Statistics, World Bank World Development Indicators

of the countries involved, depending on the interdependencies in particular water resources systems.[21,22]

How Are New Planning Realities Being Considered in Today's Planning Process?

Water resources infrastructure has high up-front costs, and then generates a time series of costs and benefits that extends over decades or even centuries. Such long planning horizons create tremendous uncertainty about these investments' long-term impacts. Well aware of this fact, water systems engineers have long been adept at using tools and methods for planning under uncertainty.[23] Until very recently, such methods were widely used to make projections of infrastructure benefits across scenarios of potential future demographic and economic change, assuming that the physical behavior of the hydrological system would remain stable.[24,25] And though human societies have been aware of their capacity for affecting the natural environment,[26] only very recently have scientists begun to speak of a new era of dominance in the nature of this influence.[27,28] With respect to water resources, growing recognition of two-way coupling between human and natural systems has led to calls for more sophisticated socio-hydrological perspective on the long-term co-evolution of these systems.[29]

Unfortunately, this new awareness of the importance of feedbacks between human and water systems does not offer easy answers for planners of water resources infrastructure projects. Scientific understanding of the true dynamics of socio-hydrological trajectories remains rudimentary and highly stylized.[30] This threatens both generalization from historical examples and practical application of the temporal and spatial resolution that is required for effective planning. Yet the tightening interplay of population, economic, and hydrologic factors heightens risks; renders seeming impossibly rare ("black swan") events possible;[31] and raises the possibility of future regrets (as witnessed in the strength of the dam removal movement in many developed countries).

Climate change offers perhaps the starkest example of the complexity induced by human-nature system interconnections; this phenomenon is expected to have varied impacts on water resources around the globe, the distribution of which is highly uncertain.[32] These effects will include spatial and temporal disturbance of patterns of precipitation and local water availability, increased evaporation, and impacts on land use and the demand for services produced by water infrastructure (e.g., renewable energy, flood control, drought mitigation).[33] The sheer complexity of the suite of potential changes, and the particular exposure of the water sector to them, have led researchers to propose a range of new methods that focus on identifying infrastructure solutions that are robust across widely divergent scenarios of change.[34,35] The recommendations that emerge from these analyses emphasize the need to invest in ways that maintain adaptive flexibility.[36] This adds another layer to an already difficult planning process.

There is only limited evidence, however, that planning approaches aimed at ensuring robustness are being mainstreamed into the planning of new water infrastructure investments today. To be sure, the use of integrated "hydro-economic" or other decision models that are favored by academic water resources economists has never been widespread among engineering firms and institutions responsible for overseeing project feasibility studies, which largely determine how infrastructures should be designed.[37] Perhaps for reasons of scope, such agents often prefer to take a project-by-project approach that focuses on optimizing the direct outputs (e.g., hydropower or water supply reliability) and financial returns of a particular project, without fully considering the incremental, system-wide perturbations that the project will induce throughout an inter-connected river basin. Meanwhile, simple financial analyses of projects tend to downplay projects' distributional consequences, and full accounting of nonmarket impacts also remains relatively rare. Exceptions that take a more holistic planning perspective include historical examples of coordinated public investment, e.g., the Tennessee Valley Authority in the United States; the occasional river basin master plans or strategic basin assessments that include economic analyses such as those carried out with sponsorship from the World Bank in recent years,[38,39] and policy models developed to evaluate difficult water management tradeoffs in systems already facing acute scarcity in the developed world, such as in California.[40]

Related institutions or reforms that aim to improve holistic water resources planning and management include integrated water resource management (IWRM) and environmental impact assessment (EIA), both of which were given greater support by the WCD.[41] While these have been mainstreamed in many legislative and policy documents worldwide, and are generally supported by traditional financiers of large water projects, the extent of their application and influence in decision-making remains inconsistent. IWRM, for example, has been criticized for being conceptually appealing but impractical, not to mention ambiguous.[42] National governments appear to have failed in many cases to establish and sustain truly integrative programs,[43] and the benefits of integration have not been empirically demonstrated.[44] EIAs, meanwhile, typically discuss downstream impacts, but often only in qualitative or semi-quantitative terms. The quality, comprehensiveness, and influence of these assessments also remain inconsistent, reflecting the highly variable capacity of regulators in different settings.[45] Especially in low- and middle-income settings, such institutions are typically weak and community participation mechanisms underdeveloped. Thus, the processes that lead to IWRM and EIAs often remain aspirational at best;[46] at worst, they may be viewed as paternalistic impositions of the international community's preferences on the global poor.

The Shift in International Financing Away from Official Sources and Its Link to Changes in Project Planning and Implementation

Until the early 1980s, financing for large water investments was mostly provided by donor-funded multilateral organizations like the World Bank and regional development banks, which could provide low-risk capital at relatively low cost.[47] This pattern changed dramatically as a result of three forces that began to strengthen in the 1980s. The first was the increasing prominence of critiques of the projects that were being supported with such funds. Opponents of the existing development paradigm found ample reasons to complain about the lack of transparency in selection of such projects, on environmental, social, and even financial grounds.[48,49] Following construction, a majority of projects failed to meet expectations, suggesting that there was systematic bias in the appraisals for large infrastructure projects.[50,51] Faced with growing political backlash, development banks pulled back from financing large dams in particular, and the WCD was created to evaluate and suggest ways to improve the planning process. This enhanced scrutiny ultimately raised the cost of the traditional funding sources. The second force was the increasing liberalization of global financial markets, which reduced the cost of alternative sources of capital. The third force was the strong economic growth in many emerging economies, which simultaneously heightened the demand for water and related capital requirements. These three forces coincided with an important increase in the use of private and bilateral capital financing for infrastructure projects. The consequences of this shift have largely been positive, as emerging economies now have enhanced possibilities to pursue new and more diverse types of investments.

It is nonetheless important to consider whether the nature of the water resources projects that are being implemented has been altered by this shift. Answering this question definitively is difficult because planning by the traditional financiers has itself changed. As discussed earlier, the shift among these traditional actors has largely followed increasing pressure for holistic, socially, and environmentally sensitive planning (the first force described earlier), but it also stems from the increasing demand for capital in emerging economies (the third force). The literature, meanwhile, contains suggestive evidence that project selection, design, and operations may evolve differently when financing comes from non-traditional sources (several important considerations are listed in Table 19.1). In particular, some of the most politically controversial recent projects—Turkey's Great Anatolia Project (GAP), the Three Gorges Dam in China, the Grand Ethiopian Renaissance Dam (GERD), and several hydropower projects on the Upper Mekong—were or are being financed using private or internal funding sources.[52,53,54] The GERD in particular would not have been approved by multilateral development banks, given the lack of prior agreements with downstream riparians, who continue to challenge the legitimacy of the project. In addition, it appears to have been designed to be far too large, perhaps for political reasons, and removes a number of options that would enhance flexibility.[55] Other projects, such as the Sardar Sarovar Project in India's Narmada Basin or the West Seti Project in Nepal's Karnali Basin, were initially under consideration by various development banks, but ultimately were pursued with other funds when these multilaterals pulled out due to social and environmental concerns. Finally, a number of scholars have

Table 19.1 Pros and Cons of the Shift to Increased Private Financing of Water Infrastructure

Pros	Cons
Focus on short-term returns favors selection of financially viable projects	Reduced focus on long-term risks and sustainability
More context-appropriate processes (i.e., participation and environmental protection according to local norms)	Lower adherence to global norms for public participation
Greater incentive to complete construction on time and efficiently	In the absence of regulation, reduced emphasis on environmental and social safeguards
Lower transaction costs	Less room for international diplomacy to resolve potential conflicts
Fewer conditionalities (which may raise project costs)	Non-preferential interest rates, unless packaged with other deals or incentives

argued that sources of bilateral finance (e.g., from China) are cheaper precisely because they tend to overlook environmental, social, and transparency risks,[56,57] perhaps in exchange for preferential bilateral trade deals.[58] It remains to be seen what role the new AIIB will play, but some observers fear it will back the types of projects that multilateral financiers have avoided. For example, several Mekong projects have been deemed to impose excessive social and environmental costs, and there is concern that the AIIB will extend the Chinese hydropower development agenda in the Upper Mekong into capital-starved downstream countries in the basin.[59]

There may be other, more subtle effects of the increasing substitution into private finance. First, it is a tautology to say that investors focus on achieving short-term financial returns, particularly in riskier developing economy contexts, but this objective may be poorly aligned with broader economic development objectives. Two aspects that are particularly problematic from an economic perspective are long-term uncertainties (e.g., climate change, the dynamics of socio-hydrological trajectories) and nonmarket spillovers (e.g., on fisheries and the environment, downstream water users, etc.). Other than avoiding reputational risks, there is limited incentive for private investors to consider such impacts, especially long-term risks (such as gradual shifts in societal preferences for environmental quality) that will typically outlast specific investors and firms. Second, empirical evidence has highlighted the importance of government regulation in initially establishing compliance with environmental laws and safeguards during construction, prior to local norms becoming established.[60] If compliance can be established, however, the government's role may become subsidiary to demand-side pressures from market consumers and communities, especially for larger firms that benefit from economies of scale and typically face larger reputational risks. In this regard, the trend toward smaller projects (e.g., run-of-the-river hydropower projects) and/or PPPs that are perhaps easier and less risky to finance with private capital may be problematic.[61]

Third, government regulation remains weak in many developing country contexts.[62] This raises the possibility of a "race to the bottom" in the market for capital and management of water infrastructure projects. For example, even in Bhutan, where the environment is afforded fairly high priority at the official level, critics of Indian-financed hydropower projects have argued that project environmental assessments are shrouded in secrecy, and that a weak civil society is unwilling to ask questions about top-level decisions.[63] Finally, use of private finance raises the incentive to achieve full cost recovery and effective service delivery from investments, in order to pay back loans. At first glance, this seems unambiguously positive,[64] given the poor record of cost recovery in this sector. Low prices also fail to provide socially efficient incentives for conservation. Yet water infrastructure investments worldwide have traditionally been supported with subsidies in order to keep user prices down for equity reasons. When the prices of water are perceived by beneficiaries to rise too quickly, and especially when these price hikes coincide with institutional changes, a destructive political backlash may ensue, as occurred in Bolivia in 1999,[65] and even after an objectively successful privatization experience such as that in Argentina.[66]

Thus, there are many theoretical and practical reasons to be concerned about the impacts of the shift in financing for water infrastructure. With increased use of private capital, incentives to minimize delays and achieve short-term financial returns are heightened, which may lead to greater long-term risks if important economic costs—both direct and nonfinancial, and indirect—are neglected. Of course, it is also possible that the new AIIB and bilateral lenders will become subject to the same pressures that led to reforms in the project appraisal process used by the World Bank, the International Finance Corporation (IFC), and other regional development banks, as environmental awareness in China and elsewhere continues to increase.[67]

What Are Some Possible Consequences of These Changing Dynamics for Political, Socioeconomic, and Environmental Security?

There is ample documentation in the literature and the popular press of disagreements and tensions over infrastructure. At the transnational level, for example, India and Pakistan have regularly clashed over infrastructure, though arbitration processes that are institutionalized in the Indus Basin Treaty have generally proven effective and resilient at resolving conflicts.[68] Turkey's GAP invoked concerns about downstream water scarcity and spurred disputes between the Turkish, Syrian and Iraqi governments, including threats to mobilize military action.[69] Similar dynamics can be found along the Nile, in response to the GERD project.[70] Even among otherwise friendly neighbors (e.g., Namibia and Botswana in the Okavango), differences of opinion over operating rules have engendered disputes.[71] Subnationally also, numerous projects—such as the Manantali Dam on the Senegal River in West Africa, and several irrigation dams along the Cauvery in India—have been implicated in tensions or violence between stakeholders.[72,73,74]

Against this background of often contentious infrastructure development and the inequality of its impacts, it is notable that the traditional financiers are now playing a much diminished role. These are the global champions of more integrated and complete weighing of costs and benefits, of the need for just compensation to those adversely affected, and of agreements by all riparian countries to proceed with infrastructure projects. As discussed earlier, powerful interests seeking to take unilateral action on infrastructure, or desiring to avoid careful project vetting, may deliberately avoid seeking assistance from multilateral sources and instead opt for other alternatives—as seems to have occurred in the case of many of the projects discussed so far. Alternatively, it may simply be cheaper to avoid traditional sources because of the fear of a long negotiation process.[75,76] Equally problematic, many controversial projects end up being implemented even when financiers such as the World Bank have decided to abandon them, which only confirms the increasingly marginal role of the multilateral financiers.

Even if decision-makers' intentions are not nefarious, choices to opt for alternative financing mechanisms may have unintended consequences. For one, given that many countries have failed to operationalize a truly integrated perspective even at the national level, the government may be unable to detect potential problems with projects that are being planned. Water management is typically handled by a variety of overlapping governmental institutions that cover divergent objectives—irrigation, energy, water supply development, and environmental protection—separately. The WCD processes adopted by the multilateral banks require consultations and attention to these divergent interests, but other financiers are not similarly exacting (especially as concerns environmental objectives). As argued elsewhere: "Controversy often arises when management decisions are formulated without sufficient participation, thus failing to take into account local rights and practices. Protests are especially likely when the public suspects that water allocations are diverting public resources for private gain."[77] Lower-level disputes may impose significant costs on basin riparians.[78] For example, interstate disputes could threaten cross-border trade and investment opportunities, limit possibilities for mutually beneficial water resources management, chill relations and lead to diversion of resources into military buildup, and raise other types of political costs. An additional concern is that spillovers from unilaterally planned projects could threaten the economic and environmental well-being of vulnerable populations, thereby exacerbating existing insecurities and inequalities, and leading these populations to move to undesirable locations.[79] Such threats may be particularly dangerous in countries with relatively weak economies and institutions or governance systems.[80] For example, infrastructure and environmental shocks have been linked to increased local poverty,[81] increased disease,[82] and subnational conflict and

Developing Water Infrastructure 353

rebellion.[83] The negative consequences of such changes have the potential to spill over to harm global security via a range of complex pathways.

How Can the Gap Between Financial Concerns and Global Norms be Narrowed?

It would be unreasonable to expect that the basic trajectory of development finance—moving toward more diverse investors and especially increased private flows of capital—will be changed anytime soon. Nor would this be desirable; this evolution has played a critical role in fueling economic growth in many countries that previously had only limited access to the benefits provided by the global financial system. Still, developed countries (such as the United States) and multilateral organizations should perhaps work to identify and implement policies that will close the gap between current investment in the sector and global norms for best practice. Many existing problems could be solved by improving institutions, but such capacity-building is difficult and often unrewarding. Institutional shortcomings are particularly crucial in light of the rising long-term risks in socio-hydrological systems today, and the harm they could do to socioeconomic well-being and government stability.

How precisely can this gap be closed? First, the international community and the U.S. foreign policy establishment should encourage creation of inclusive cooperative frameworks in international basins that do not have such agreements already, and should work to strengthen existing ones. There is ample evidence that river basin treaties are remarkable in their resilience, even in some of the world's most contentious locations, providing effective mechanisms to resolve disputes (e.g., the Indus, the Nile).[84] Third parties, such as the World Bank, have been key to empowering such institutions in a range of locations. Yet treaties often exclude key riparians; are insufficiently adaptable to changing conditions (e.g., in specifying water allocations that remain fixed over time); do not leave sufficient space to seize new opportunities for mutual gains (e.g., by focusing on sharing of water rather than benefits); or ignore key issues that may be difficult to negotiate (e.g., establishment of clear processes and institutions for developing new infrastructures, or monitoring and enforcement mechanisms).[85,86] The focus in the near term should be on targeting basins that are undergoing rapid changes, whether these are economic, hydrological, or geopolitical in nature. Providing economic or other incentives to countries that invest in improved water governance and multilateral negotiations could help encourage progress, and the international community could provide support by selecting a third party that would most objectively aid negotiations. Closely related to this is the provision of objective arbitration of disputes, as the World Bank has done in the Indus and as the International Court of Justice has done in other difficult water disputes.[87,88]

Second, the global community should continue to push for a consistent set of standards (based on globally discussed norms) to be applied by countries—for example, as expressed in the UN Convention on the Law of the Non-Navigational Uses of International Watercourses.[89] Though the convention has been criticized for being overly vague and compromising in key respects, it is undeniable that its language and discussion have heavily influenced a number of river basin agreements.[90] In most cases, because each shared water resource is unique, the latter can afford to be, and should be, much more specific than a global document such as the UN Convention. However, a consistent set of principles can be promulgated: e.g., clarification of just and equitable ways for specifying existing and future rights; description of exemplary mechanisms for information sharing, notification, and dispute resolution; statements of obligations to the natural environment; and guidance on dealing with emergencies. Meanwhile, violations of accepted global norms can perhaps be punished through connections to the global trade agenda; much has been learned, for example, in the past decade about how trade and environmental compliance instruments should and should not be linked.[91]

Third, sound water management does not stop with the establishment of clear international norms and rules in treaties; integrative institutions are also required, particularly because basin-level solutions are often more appropriate than international ones.[92] In recent years, the World Bank and others have invested heavily in creating and equipping river basin organizations (RBOs), whose numbers have doubled, with the personnel and tools needed to analyze complex water problems.[93,94] Despite their challenges, such efforts should surely continue in order to safeguard global public goods and avoid depletion of common property resources. Yet the multilateral development banks often have a conflict of interest in pursuing development projects or agendas alongside such institutional development. Bilateral aid may in *some* instances be able to support those institutional capacity-building efforts with fewer perceived conflicts.[95] It could also support more rigorous evaluations of RBO effectiveness.

Fourth, the United States and its international partners (including multilaterals like the World Bank) should continue to invest in the production of global knowledge about shared resources like water, providing support for the development of new data collection and information tools. It remains an unfortunate fact that domestic interests often trump attempts to improve global well-being, and obstacles to data sharing are a key feature of such strategies. Yet in this new age of remote sensing and information technology, treating hydrological and water use data as state secrets is becoming increasingly difficult, and new products promise to continue to erode information asymmetries, so long as the capacity to use new information is sufficiently developed and equitable. The global community should thus invest not only in tracking technology, but also in the capacity to interpret data obtained from that technology. This will facilitate enforcement of existing agreements, improve trust and arbitration processes, and help point to new opportunities for cooperation at the expense of entrenched interests. As with RBOs, however, evidence on the effectiveness of information investments is scant, and research is warranted to identify how to make them more cost effective.[96]

Finally, at a lower institutional level, the international community should continue to work to provide technical assistance aimed at improving coordination between what are typically fragmented water institutions and interests *within* countries. Such work would focus on improving the capacity of national and regional governments, establishing consultative processes with civil society to identify and tackle water governance challenges, and especially facilitating interactions during infrastructure planning, as recommended by the WCD. It could also focus on the need for policy instruments (e.g., scarcity pricing) that arrest overextraction of water and heighten the domestic salience of water challenges; lack of importance perhaps poses even greater challenges for cooperative negotiations than do fierce competition and rivalry.[97] A more robust domestic discourse could then feed into negotiations and interactions with the integrative RBOs discussed earlier. In transboundary water systems, there is often underappreciation of the complexity that is produced by the two-level game that affects both domestic and international discourse, perhaps because of the difficulty in discerning a clear signal from the typically fragmented domestic discourse around water challenges.[98]

Conclusion

Water resources infrastructure is expensive, capital intensive, and long-lived, and has important effects on human and ecological well-being. This chapter has explored the implications of the changing character of financing for such infrastructure, in an era of increasing socio-hydrological connectedness. As discussed, there is much suggestive evidence that new financing alternatives, mainly from private or bilateral sources, are increasingly being used to support unilateral and suboptimal project designs in low- and middle-income countries. It is unclear at present whether the negative implications of such choices will spill over to affect regional stability and security; indeed, the majority of disputes over water in the historical period have been and remain at the subnational level. Still, regionalization of conflicts remains possible and is highly disruptive. Given that multilateral and other development aid will likely continue to diminish over the coming years, the international community needs to be proactive in establishing and improving water governance and institutions that are capable of better regulating the influence of private finance for river basin infrastructure.

What are some possible interventions that U.S. agencies and international institutions could pursue to meet evolving water resources challenges? Many of the solutions proposed in the closing section of this chapter focused on enhancing river basin institutions and water governance. Resilient and flexible river basin treaties and organizations are essential for clearly establishing property rights over shared resources; providing capacity for analysis; and creating mechanisms for benefit-sharing, negotiation, and conflict resolution. The international community should continue to work to facilitate the development of such institutions, especially emphasizing the importance of adherence to global norms and best practices. This facilitation could happen via third-party arbitration, capacity building, or the use of trade and other incentives. Continued support for data sharing and use of technology to reduce information asymmetries is also key; such efforts will limit the gains that can be obtained from strategic behavior that damages overall social welfare. Finally, expanded investment in institutional capacity to better manage water at the national level in Southern countries will benefit both local populations and the global community by improving development outcomes and limiting risks to regional stability. Institutional solutions will not, and should not, arrest the growing importance of private capital in global water resources development, given the positive benefits these flows provide. Nonetheless, better governance will facilitate better harnessing of these benefits, to advance environmental and social well-being worldwide.

Acknowledgements

Acknowledgements are due especially to Lizzie Devitt, who provided invaluable research assistance in preparation of this chapter, and to Naotaka Sugawara at the World Bank, who helped provide data on financial flows in low and middle countries. Patrick Coady provided a valuable review of a prior draft of the chapter, and David Reed, Ryan Bartlett, Sarah Freeman, and others at the World Wildlife Fund provided helpful critical comments as it evolved from a concept note to a more comprehensive document.

Notes

1 The term *riparian* is used here to denote any party that derives value from the use of a specific shared water resource.
2 Esther Duflo and Rohini Pande, "Dams," *The Quarterly Journal of Economics* 122, no. 2 (2007).
3 Sheila M. Olmstead and Hilary Sigman, "Damming the Commons: An Empirical Analysis of International Cooperation and Conflict in Dam Location," *Journal of the Association of Environmental and Resource Economists* 2, no. 4 (2015).
4 Ben Crow, Alan Lindquist, and David Wilson, *Sharing the Ganges: The Politics and Technology of River Development* (New Delhi: Sage Publications Pvt. Ltd, 1995).
5 Dale Whittington, John Waterbury, and Marc Jeuland, "The Grand Renaissance Dam and Prospects for Cooperation on the Eastern Nile," *Water Policy* 16, no. 4 (2014).
6 Specifically, Wolf et al. find that 67.1 percent and 27.7 percent of all interactions related to transboundary water are cooperative or neutral, respectively (5.2 percent are neutral). Using a somewhat different methodology, Kalbhenn and Bernauer find that 18.3 percent of events are conflictive. In addition to this, a large majority of the conflictive interactions in both datasets are mild, involving only verbal hostility.
7 Anna Kalbhenn and Thomas Bernauer, "International Water Cooperation and Conflict: A New Event Dataset" (available at SSRN 21766092012).
8 Aaron T. Wolf, Shira B. Yoffe, and Mark Giordano, "International Waters: Identifying Basins at Risk," *Water Policy* 5, no. 1 (2003).
9 Thomas F. Homer-Dixon, "Environmental Scarcities and Violent Conflict: Evidence from Cases," *International Security* 19, no. 1 (1994).

10 Navroz K. Dubash et al., *A Watershed in Global Governance? An Independent Assessment of the World Commission on Dams* (Washington, DC: World Resources Institute, 2001).
11 World Commission on Dams, *Dams and Development: A New Framework for Decision-Making: The Report of the World Commission on Dams* (London: Earthscan, 2000).
12 Dubash et al.
13 Judith P. Braeckman and Peter Guthrie, "Loss of Value: Effects of Delay on Hydropower Stakeholders" (paper presented at the Proceedings of the Institution of Civil Engineers-Engineering Sustainability, 2015).
14 Indeed, even projects that are now considered to be successful cases of deliberative and careful planning, e.g., Nam Theun 2 in Lao PDR, invoke considerable controversy (Shannon Lawrence, "The Nam Theun 2 Controversy and Its Lessons for Laos," in *Contested Waterscapes in the Mekong Region: Hydropower, Livelihoods and Governance*, ed. F. Molle, T. Foran and M. Kakonen, 81–114 [London: Earthscan, 2009].
15 Generally speaking, as a percentage of GDP, foreign direct investment (FDI) has been increasing over time while official flows have been declining. The other two major categories of transnational financial flows—commercial loans and net portfolio investment—are highly variable over the recent period.
16 John Briscoe, "The Changing Face of Water Infrastructure Financing in Developing Countries," *International Journal of Water Resources Development* 15, no. 3 (1999).
17 Jonathan P. Doh and Ravi Ramamurti, "Reassessing Risk in Developing Country Infrastructure," *Long Range Planning* 36, no. 4 (2003).
18 Peter Bosshard, "China Dams the World," *World Policy Journal* 26, no. 4 (2009).
19 Raphael Kaplinsky and Mike Morris, "Chinese FDI in Sub-Saharan Africa: Engaging With Large Dragons," *The European Journal of Development Research* 21, no. 4 (2009).
20 Jane Perlez, "China Competes with West in Aid to Its Neighbors," *The New York Times*.
21 Olmstead and Sigman.
22 Xun Wu et al., "Interdependence in Water Resource Development in the Ganges: An Economic Analysis," *Water Policy* 15, no. S1 (2013).
23 Arthur Maass et al., *Design of Water-Resource Systems* (Cambridge: Harvard University Press, 1962).
24 Marc Jeuland and Dale Whittington, "Water Resources Planning under Climate Change: Assessing the Robustness of Real Options for the Blue Nile," *Water Resources Research* 50, no. 3 (2014).
25 P. C. D. Milly, J. Betancourt, M. Falkenmark, R. M. Hirsch, Z. W. Kundzewicz, D. P. Lettenmaier, and R. J. Stouffer, "Stationarity Is Dead: Whither Water Management?" *Science* 319, no. 5863 (2008).
26 George P. Marsh, *Man and Nature, Physical Geography as Modified by Human Action* (New York: Charles Scribner, 1865).
27 Paul J. Crutzen, "Geology of Mankind," *Nature* 415, no. 6867 (2002).
28 Jan Zalasiewicz Mark Williams, Will Steffen, and Paul Crutzen, "The New World of the Anthropocene," *Environmental Science & Technology* 44, no. 7 (2010).
29 Murugesu Sivapalan, Hubert H. G. Savenije, and Günter Blöschl, "Socio Hydrology: A New Science of People and Water," *Hydrological Processes* 26, no. 8 (2012).
30 Tara J. Troy, Mitchell Pavao Zuckerman, and Tom P. Evans, "Debates—Perspectives on Socio Hydrology: Socio Hydrologic Modeling: Tradeoffs, Hypothesis Testing, and Validation," *Water Resources Research* 51, no. 6 (2015).
31 Nassim N. Taleb, *The Black Swan: The Impact of the Highly Improbable* (New York: Random House, 2007).

32 J. Settele et al., "Terrestrial and Inland Water Systems," in *Climate Change 2014: Impacts, Adaptation, and Vulnerability. Part A: Global and Sectoral Aspects. Contribution of Working Group Ii to the Fifth Assessment Report of the Intergovernmental Panel on Climate Change*, ed. C. B. Field, V. R. Barros, D. J. Dokken, K. J. Mach, M. D. Mastrandrea, T. E. Bilir, M. Chatterjee, K. I. Ebi, Y. O. Estrada, R. C. Genova, B. Girma, E. S. Kissel, A. N. Levy, S. MacCracken, P. R. Mastrandrea, and L. L.White (Cambridge, UK and New York: Cambridge University Press, 2014).
33 Marc Jeuland, "Economic Implications of Climate Change for Infrastructure Planning in Transboundary Water Systems: An Example from the Blue Nile," *Water Resources Research* 46, no. 11 (2010).
34 David G. Groves and Robert J. Lempert, "A New Analytic Method for Finding Policy-Relevant Scenarios," *Global Environmental Change* 17, no. 1 (2007).
35 Jim W. Hall et al., "Robust Climate Policies under Uncertainty: A Comparison of Robust Decision Making and Info Gap Methods," *Risk Analysis* 32, no. 10 (2012).
36 Jeuland and Whittington.
37 Peter P. Rogers and Myron B. Fiering, "Use of Systems Analysis in Water Management," *Water Resources Research* 22, no. 9S (1986).
38 World Bank, "Ganges Strategic Basin Assessment: A Discussion of Regional Opportunities and Risks. Report No. 67668-Sas." (Washington, D.C.: South Asia Water Initiative, 2014).
39 BCEOM, BRGM, and ISL, "Main Report," in *Abbay River Basin Integrated Development Master Plan Project* (Addis Ababa, Ethiopia: Ministry of Water Resources, Federal Democratic Republic of Ethiopia, 1999).
40 Andrew J. Draper et al., "Economic-Engineering Optimization for California Water Management," *Journal of Water Resources Planning and Management* 129, no. 3 (2003).
41 IWRM has been defined by the Global Water Partnership as "a process which promotes coordinated development and management of water, land and related resources, in order to maximize the resultant economic and social welfare in an equitable manner without compromising the sustainability of vital ecosystems." See http://www.gwp.org/en/Press-Room/A-Water-Secure-World/.
42 Paul Jeffrey and Mary Gearey, "Integrated Water Resources Management: Lost on the Road from Ambition to Realisation?" *Water Science and Technology* 53, no. 1 (2006).
43 IWA UNEP, *Industry as a Partner for Sustainable Development: Water Management* (London: IWA UNEP, 2002).
44 Jeffrey and Gearey.
45 Stephen Jay, Carys Jones, Paul Slinn, and Christopher Wood, "Environmental Impact Assessment: Retrospect and Prospect," *Environmental Impact Assessment Review* 27, no. 4 (2007).
46 Jonathan Lautze et al., "Putting the Cart Before the Horse: Water Governance and IWRM" (paper presented at the Natural Resources Forum, 2011).
47 Briscoe.
48 Some of the greater controversies at that time occurred around the Sardar Sarovar Dam in India, the Ruzizi II hydropower project on the border between Rwanda and then Zaire, and the Arun III project in Nepal. Several more recent examples include the now complete Bujagali Dam in Uganda (community groups have protested failure to protect sensitive ecological areas), the Democratic Republic of Congo's planned Inga 3 Dam (the World Bank has pulled out, but the AfDB remains involved), and the Kandadji Dam that is under construction in Niger (which has significant resettlement and livelihood concerns).
49 John Briscoe, "The Financing of Hydropower, Irrigation and Water Supply Infrastructure in Developing Countries," *International Journal of Water Resources Development* 15, no. 4 (1999).

50 World Commission on Dams.
51 Atif Ansar et al., "Should We Build More Large Dams? The Actual Costs of Hydropower Megaproject Development," *Energy Policy* 69 (2014).
52 Vincent Merme, Rhodante Ahlers, and Joyeeta Gupta, "Private Equity, Public Affair: Hydropower Financing in the Mekong Basin," *Global Environmental Change* 24 (2014).
53 Whittington, Waterbury, and Jeuland.
54 Mehmet Tomanbay, "Turkey's Water Potential and the Southeast Anatolia Project," *Water Balances in the Eastern Mediterranean* (2000).
55 Jeuland and Whittington.
56 Perlez.
57 R. Edward Grumbine and Jianchu Xu, "Mekong Hydropower Development," *Science* 332, no. 6026 (2011).
58 Kaplinsky and Morris.
59 Hidetaka Yoshimatsu, "The United States, China, and Geopolitics in the Mekong Region," *Asian Affairs: An American Review* 42, no. 4 (2015).
60 Bing Zhang, Jun Bi, Zengwei Yuan, Junjie Ge, Beibei Liu, and Maoliang Bu, "Why Do Firms Engage in Environmental Management? An Empirical Study in China," *Journal of Cleaner Production* 16, no. 10 (2008).
61 Ritva Reinikka and Jakob Svensson, *How Inadequate Provision of Public Infrastructure and Services Affects Private Investment* (Washington, D.C.: World Bank, 1999).
62 Colin Kirkpatrick, David Parker, and Y-F. Zhang, "Foreign Direct Investment in Infrastructure in Developing Countries: Does Regulation Make a Difference?" *Transnational Corporations* 15, no. 1 (2006).
63 Vishal Arora and Chencho Dema, "Bhutan Should Come Clean on Hydropower Megaplan," *The Diplomat*, February 16, 2016.
64 Sebastian Galiani, Paul Gertler, and Ernesto Schargrodsky, "Water for Life: The Impact of the Privatization of Water Services on Child Mortality," *Journal of political economy* 113, no. 1 (2005).
65 Willem Assies, "David Versus Goliath in Cochabamba: Water Rights, Neoliberalism, and the Revival of Social Protest in Bolivia," *Latin American Perspectives* 30, no. 3 (2003).
66 Ariel A. Casarin, Jose A. Delfino, and Maria Eugenia Delfino, "Failures in Water Reform: Lessons from the Buenos Aires Concession," *Utilities Policy* 15, no. 4 (2007).
67 Oliver Hensengerth, "Interaction of Chinese Institutions with Host Governments in Dam Construction: The Bui Dam in Ghana," in *Evolution of Dam Policies*, ed. Waltina Scheumann and Oliver Hensengerth, 229–272 (Heidelberg, Springer, 2014).
68 Lydia Polgreen and Sabrina Tavernise, "Water Dispute Increases India-Pakistan Tension," *The New York Times*.
69 Leila M. Harris, "Water and Conflict Geographies of the Southeastern Anatolia Project," *Society &Natural Resources* 15, no. 8 (2002).
70 Whittington, Waterbury, and Jeuland.
71 Aaron T. Wolf, "Shared Waters: Conflict and Cooperation," *Annual Review Environment Resource* 32 (2007).
72 Homer-Dixon.
73 Amita Baviskar, *In the Belly of the River: Tribal Conflicts over Development in the Narmada Valley* (Oxford, UK: Oxford University Press, 1999).
74 Meredith Giordano, Mark Giordano, and Aaron Wolf, "The Geography of Water Conflict and Cooperation: Internal Pressures and International Manifestations," *The Geographical Journal* 168, no. 4 (2002).
75 Braeckman and Guthrie.
76 Wolf.

77 Ibid.
78 Claudia W. Sadoff and David Grey, "Beyond the River: The Benefits of Cooperation on International Rivers," *Water Policy* 4, no. 5 (2002).
79 Homer-Dixon.
80 Erwin H. Bulte, Richard Damania, and Robert T. Deacon, "Resource Intensity, Institutions, and Development," *World Development* 33, no. 7 (2005).
81 Duflo and Pande.
82 Jennifer Keiser et al., "Effect of Irrigation and Large Dams on the Burden of Malaria on a Global and Regional Scale," *American Journal of Tropical Medicine and Hygiene* 72, no. 4 (2005).
83 Edward Miguel, Shanker Satyanath, and Ernest Sergenti, "Economic Shocks and Civil Conflict: An Instrumental Variables Approach," *Journal of Political Economy* 112, no. 4 (2004).
84 Mary Miner, Gauri Patankar, Shama Gamkhar, and David J. Eaton, "Water Sharing Between India and Pakistan: A Critical Evaluation of the Indus Water Treaty," *Water International* 34, no. 2 (2009).
85 Jennifer Song and Dale Whittington, "Why Have Some Countries on International Rivers Been Successful Negotiating Treaties? A Global Perspective," *Water Resources Research* 40, no. 5 (2004).
86 Shlomi Dinar, David Katz, Lucia De Stefano, and Brian Blankespoor, "Climate Change and Water Variability: Do Water Treaties Contribute to River Basin Resilience?" *World Bank Policy Research Working Paper No. 7855* (Washington, D.C.: World Bank, 2016).
87 G. T. Pitman, "The Role of the World Bank in Enhancing Cooperation and Resolving Conflict on International Watercourses: The Case of the Indus Basin," *World Bank Technical Paper No. 414* (Washington, D.C.: World Bank, 1998).
88 Stephen Deets, "Constituting Interests and Identities in a Two Level Game: Understanding the Gabcikovo Nagymaros Dam Conflict1," *Foreign Policy Analysis* 5, no. 1 (2009).
89 United Nations, *Un Convention on the Law of the Non-Navigational Uses of International Watercourses* (New York: United Nations, 1994).
90 Stephen McCaffrey, "The Contribution of the Un Convention on the Law of the Non-Navigational Uses of International Watercourses," *International Journal of Global Environmental Issues* 1, no. 3–4 (2001).
91 Sadeq Z. Bigdeli, "Clash of Rationalities: Revisiting the Trade and Environment Debate in Light of WTO Disputes over Green Industrial Policy," *Trade, Law & Development* 6 (2014).
92 Andrea K. Gerlak and Susanne Schmeier, "Cooperation for the Sustainable Governance of International Watercourses: The Role of River Basin Organisations," *Global Dialogue (Online)* 15, no. 2 (2013).
93 Maurice Schiff and L. Alan Winters, "Regional Cooperation, and the Role of International Organizations and Regional Integration," *World Bank Policy Research Working Paper*, no. 2872 (2002).
94 Susanne Schmeier, "The Institutional Design of River Basin Organizations— Empirical Findings from around the World," *International Journal of River Basin Management* 13, no. 1 (2015).
95 Of course, there may still be political or economic conflicts of interest in these cases, and determining the proper support structure in politically sensitive regions is likely to remain challenging.
96 K. Hansen, H. Doherty, L. Eastman, M. Tchamkina, and M. Jeuland, *A Systematic Review of Studies Related to the Benefits of Water Information Systems* (Durham: Duke University, 2016).
97 Frederick W. Frey, "The Political Context of Conflict and Cooperation over International River Basins," *Water International* 18, no. 1 (1993).

98 Eran Feitelson, "Implications of Shifts in the Israeli Water Discourse for Israeli-Palestinian Water Negotiations," *Political Geography* 21, no. 3 (2002).

References

Ansar, A., B. Flyvbjerg, A. Budzier, and D. Lunn. "Should We Build More Large Dams? The Actual Costs of Hydropower Megaproject Development." *Energy Policy* 69 (2014): 43–56.
Arora, V., and C. Dema. "Bhutan Should Come Clean on Hydropower Megaplan." *The Diplomat*, February 16, 2016.
Assies, W. "David Versus Goliath in Cochabamba: Water Rights, Neoliberalism, and the Revival of Social Protest in Bolivia." *Latin American Perspectives* 30, no. 3 (2003): 14–36.
Baviskar, Amita. *In the Belly of the River: Tribal Conflicts Over Development in the Narmada Valley*. Oxford, UK: Oxford University Press, 1999.
BCEOM, BRGM, and ISL. "Main Report." In *Abbay River Basin Integrated Development Master Plan Project*. Addis Ababa, Ethiopia: Ministry of Water Resources, Federal Democratic Republic of Ethiopia, 1999.
Bigdeli, S. Z. "Clash of Rationalities: Revisiting the Trade and Environment Debate in Light of WTO Disputes Over Green Industrial Policy." *Trade, Law & Development* 6 (2014): 177.
Bosshard, Peter. "China Dams the World." *World Policy Journal* 26, no. 4 (2009): 43–51.
Braeckman, J. P., and P. Guthrie. "Loss of Value: Effects of Delay on Hydropower Stakeholders." Paper presented at the Proceedings of the Institution of Civil Engineers-Engineering Sustainability, 2015.
Briscoe, J. "The Changing Face of Water Infrastructure Financing in Developing Countries." *International Journal of Water Resources Development* 15, no. 3 (1999): 301–8.
———. "The Financing of Hydropower, Irrigation and Water Supply Infrastructure in Developing Countries." *International Journal of Water Resources Development* 15, no. 4 (1999): 459–91.
Bulte, E. H., R. Damania, and R. T. Deacon. "Resource Intensity, Institutions, and Development." *World Development* 33, no. 7 (2005): 1029–44.
Casarin, A. A., J. A. Delfino, and M. E. Delfino. "Failures in Water Reform: Lessons From the Buenos Aires Concession." *Utilities Policy* 15, no. 4 (2007): 234–47.
Crow, B., A. Lindquist, and D. Wilson. *Sharing the Ganges: The Politics and Technology of River Development*. New Delhi: Sage Publications Pvt. Ltd, 1995.
Crutzen, P. J. "Geology of Mankind." *Nature* 415, no. 6867 (2002): 23.
Deets, Stephen. "Constituting Interests and Identities in a Two-Level Game: Understanding the Gabcikovo-Nagymaros Dam Conflict." *Foreign Policy Analysis* 5, no. 1 (2009): 37–56.
Dinar, S., D. Katz, L. De Stefano, and B. Blankespoor. "Climate Change and Water Variability: Do Water Treaties Contribute to River Basin Resilience?" In *World Bank Policy Research Working Paper No. 7855*. Washington, DC, 2016.
Doh, J. P., and R. Ramamurti. "Reassessing Risk in Developing Country Infrastructure." *Long Range Planning* 36, no. 4 (2003): 337–53.
Draper, A. J, M. W. Jenkins, K. W. Kirby, J. R. Lund, and R. E. Howitt. "Economic-Engineering Optimization for California Water Management." *Journal of Water Resources Planning and Management* 129, no. 3 (2003): 155–64.

Dubash, N. K., M. Dupar, S. Kothari, and T. Lissu. *A Watershed in Global Governance? An Independent Assessment of the World Commission on Dams*: World Resources Institute, 2001.
Duflo, E., and R, Pande. "Dams." *The Quarterly Journal of Economics* 122, no. 2 (2007): 601–46.
Feitelson, E. "Implications of Shifts in the Israeli Water Discourse for Israeli-Palestinian Water Negotiations." *Political Geography* 21, no. 3 (2002): 293–318.
Frey, F. W. "The Political Context of Conflict and Cooperation Over International River Basins." *Water International* 18, no. 1 (1993): 54–68.
Galiani, S., P. Gertler, and E. Schargrodsky. "Water for Life: The Impact of the Privatization of Water Services on Child Mortality." *Journal of Political Economy* 113, no. 1 (2005): 83–120.
Gerlak, A. K., and S. Schmeier. "Cooperation for the Sustainable Governance of International Watercourses: The Role of River Basin Organisations." *Global Dialogue (Online)* 15, no. 2 (2013): 54.
Giordano, M., M. Giordano, and A. Wolf. "The Geography of Water Conflict and Cooperation: Internal Pressures and International Manifestations." *The Geographical Journal* 168, no. 4 (2002): 293–312.
Groves, D. G, and R. J. Lempert. "A New Analytic Method for Finding Policy-Relevant Scenarios." *Global Environmental Change* 17, no. 1 (2007): 73–85.
Grumbine, R. E., and J. Xu. "Mekong Hydropower Development." *Science* 332, no. 6026 (2011): 178–9.
Hall, J. W., R. J. Lempert, K. Keller, A. Hackbarth, C. Mijere, and D. J. McInerney. "Robust Climate Policies under Uncertainty: A Comparison of Robust Decision Making and Info-Gap Methods." *Risk Analysis* 32, no. 10 (2012): 1657–72.
Hansen, K., H. Doherty, L. Eastman, M. Tchamkina, and M. Jeuland. *A Systematic Review of Studies Related to the Benefits of Water Information Systems*. Durham: Duke University, 2016.
Harris, L. M. "Water and Conflict Geographies of the Southeastern Anatolia Project." *Society & Natural Resources* 15, no. 8 (2002): 743–59.
Hensengerth, O. "Interaction of Chinese Institutions With Host Governments in Dam Construction: The Bui Dam in Ghana." In *Evolution of Dam Policies*, 229–71. New York: Springer, 2014.
Homer-Dixon, T. F. "Environmental Scarcities and Violent Conflict: Evidence From Cases." *International Security* 19, no. 1 (1994): 5–40.
IWA UNEP. *Industry as a Partner for Sustainable Development: Water Management*. London: Author, 2002.
Jay, S., C. Jones, P. Slinn, and C. Wood. "Environmental Impact Assessment: Retrospect and Prospect." *Environmental Impact Assessment Review* 27, no. 4 (2007): 287–300.
Jeffrey, P., and M. Gearey. "Integrated Water Resources Management: Lost on the Road from Ambition to Realisation?" *Water Science and Technology* 53, no. 1 (2006): 1–8.
Jeuland, M. "Economic Implications of Climate Change for Infrastructure Planning in Transboundary Water Systems: An Example From the Blue Nile." *Water Resources Research* 46, no. 11 (2010).
Jeuland, M., and D. Whittington. "Water Resources Planning Under Climate Change: Assessing the Robustness of Real Options for the Blue Nile." *Water Resources Research* 50, no. 3 (2014): 2086–107.
Kalbhenn, A., and T. Bernauer. *International Water Cooperation and Conflict: A New Event Dataset*. Available at SSRN 2176609, 2012.

Kaplinsky, R., and M. Morris. "Chinese Fdi in Sub-Saharan Africa: Engaging With Large Dragons." *The European Journal of Development Research* 21, no. 4 (2009): 551–69.

Keiser, J., M. C. de Castro, M. F. Maltese, R. Bos, M. Tanner, B. H. Singer, and J. Utzinger. "Effect of Irrigation and Large Dams on the Burden of Malaria on a Global and Regional Scale." *American Journal of Tropical Medicine and Hygiene* 72, no. 4 (2005): 392–406.

Kirkpatrick, C., D. Parker, and Y-F. Zhang. "Foreign Direct Investment in Infrastructure in Developing Countries: Does Regulation Make a Difference?" *Transnational Corporations* 15, no. 1 (2006): 143.

Lautze, J., S. De Silva, M. Giordano, and L. Sanford. "Putting the Cart before the Horse: Water Governance and Iwrm." Paper presented at the Natural Resources Forum, 2011.

Maass, A., M. M. Hufschmidt, R. Dorfman, H. A. Thomas, S. A. Marglin, and G. M. Fair. *Design of Water-Resource Systems*. Cambridge: Harvard University Press, 1962.

Marsh, G. P. *Man and Nature, Physical Geography as Modified by Human Action*. New York: Charles Scribner, 1865.

McCaffrey, S. "The Contribution of the Un Convention on the Law of the Non-Navigational Uses of International Watercourses." *International Journal of Global Environmental Issues* 1, no. 3–4 (2001): 250–63.

Merme, V., R. Ahlers, and J. Gupta. "Private Equity, Public Affair: Hydropower Financing in the Mekong Basin." *Global Environmental Change* 24 (2014): 20–9.

Miguel, E., S. Satyanath, and E. Sergenti. "Economic Shocks and Civil Conflict: An Instrumental Variables Approach." *Journal of Political Economy* 112, no. 4 (2004): 725–53.

Milly, P. C. D., J. Betancourt, M. Falkenmark, R. M. Hirsch, Z. W. Kundzewicz, D. P. Lettenmaier, and R. J. Stouffer. "Stationarity Is Dead: Whither Water Management?" *Science* 319, no. 5863 (2008): 573–4.

Miner, M., G. Patankar, S. Gamkhar, and D. J. Eaton. "Water Sharing Between India and Pakistan: A Critical Evaluation of the Indus Water Treaty." *Water International* 34, no. 2 (2009): 204–16.

Olmstead, S. M., and H. Sigman. "Damming the Commons: An Empirical Analysis of International Cooperation and Conflict in Dam Location." *Journal of the Association of Environmental and Resource Economists* 2, no. 4 (2015): 497–526.

Perlez, Jane. "China Competes with West in Aid to Its Neighbors." *The New York Times* (September 18, 2006).

Pitman, G. T. "The Role of the World Bank in Enhancing Cooperation and Resolving Conflict on International Watercourses: The Case of the Indus Basin." In *World Bank Technical Paper No. 414*. Washington, 1998.

Polgreen, L., and S. Tavernise. "Water Dispute Increases India-Pakistan Tension." *The New York Times* (July 20, 2010).

Reinikka, R., and J. Svensson. *How Inadequate Provision of Public Infrastructure and Services Affects Private Investment*. Washington, DC: World Bank, 1999.

Rogers, P. P., and M. B. Fiering. "Use of Systems Analysis in Water Management." *Water Resources Research* 22, no. 9S: 146S–158S (1986).

Sadoff, C. W, and D. Grey. "Beyond the River: The Benefits of Cooperation on International Rivers." *Water Policy* 4, no. 5 (2002): 389–403.

Schiff, M., and L. A. Winters. "Regional Cooperation, and the Role of International Organizations and Regional Integration." *World Bank Policy Research Working Paper*, no. 2872 (2002).

Schmeier, S. "The Institutional Design of River Basin Organizations—Empirical Findings From Around the World." *International Journal of River Basin Management* 13, no. 1 (2015): 51–72.

Settele, J., R. Scholes, R. Betts, S. Bunn, P. Leadley, D. Nepstad, J, T. Overpeck, and M. A. Taboada. "Terrestrial and Inland Water Systems." In *Climate Change 2014: Impacts, Adaptation, and Vulnerability. Part A: Global and Sectoral Aspects. Contribution of Working Group Ii to the Fifth Assessment Report of the Intergovernmental Panel on Climate Change*, edited by C. B. Field, V. R. Barros, D. J. Dokken, K. J. Mach, M. D. Mastrandrea, T. E. Bilir, M. Chatterjee, K. L. Ebi, Y. O. Estrada, R. C. Genova, B. Girma, E. S. Kissel, A. N. Levy, S. MacCracken, P. R. Mastrandrea, and L. L.White, 271–359. Cambridge, UK and New York: Cambridge University Press, 2014.

Sivapalan, M., H. H. G. Savenije, and G. Blöschl. "Socio-Hydrology: A New Science of People and Water." *Hydrological Processes* 26, no. 8 (2012): 1270–6.

Song, J., and D. Whittington. "Why Have Some Countries on International Rivers Been Successful Negotiating Treaties? A Global Perspective." *Water Resources Research* 40, W05S06 (2004).

Taleb, N. N. *The Black Swan: The Impact of the Highly Improbable*. New York: Random House, 2007.

Tomanbay, M. "Turkey's Water Potential and the Southeast Anatolia Project." *In: Water Balances in the Eastern Mediterranean, David B. Brooks and Ozay Mehmet (eds.)* International Development Research Centre (Ottawa 2000): 95–112.

Troy, T. J., M. Pavao-Zuckerman, and T. P. Evans. "Debates—Perspectives on Socio-Hydrology: Socio-Hydrologic Modeling: Tradeoffs, Hypothesis Testing, and Validation." *Water Resources Research* 51, no. 6 (2015): 4806–14.

United Nations. *Un Convention on the Law of the Non-Navigational Uses of International Watercourses*. New York: United Nations, 1994.

Whittington, D., J. Waterbury, and M. Jeuland. "The Grand Renaissance Dam and Prospects for Cooperation on the Eastern Nile." *Water Policy* 16, no. 4 (2014): 595–608.

Wolf, A. T. "Shared Waters: Conflict and Cooperation." *Annual Review of Environmental Resources* 32 (2007): 241–69.

Wolf, A. T., S. B. Yoffe, and M. Giordano. "International Waters: Identifying Basins at Risk." *Water Policy* 5, no. 1 (2003): 29–60.

World Bank. *Ganges Strategic Basin Assessment: A Discussion of Regional Opportunities and Risks. Report No. 67668-Sas*. Washington, DC: South Asia Water Initiative, 2014.

World Commission on Dams. *Dams and Development: A New Framework for Decision-Making: The Report of the World Commission on Dams*. London: Earthscan, 2000.

Wu, X., M. Jeuland, C. Sadoff, and D. Whittington. "Interdependence in Water Resource Development in the Ganges: An Economic Analysis." *Water Policy* 15, no. S1 (2013): 89–108.

Yoshimatsu, H. "The United States, China, and Geopolitics in the Mekong Region." *Asian Affairs: An American Review* 42, no. 4 (2015): 173–94.

Zalasiewicz, J., M. Williams, W. Steffen, and P. Crutzen. "The New World of the Anthropocene." *Environmental Science & Technology* 44, no. 7 (2010): 2228–31.

Zhang, B., J. Bi, Z. Yuan, J. Ge, B. Liu, and M. Bu. "Why Do Firms Engage in Environmental Management? An Empirical Study in China." *Journal of Cleaner Production* 16, no. 10 (2008): 1036–45.

Part IV
New Challenges, New Directions

20 Paths of Influence

David Reed

The foregoing studies reflect the analysis of an extraordinary array of experts who have tried to answer the question of how water-related events create social disruptions capable of posing threats to U.S. interests around the world. The analysts include former ministers, negotiators of international water treaties, hydrologists, professors, managers of multilateral development institutions, advisors to presidents, architects of national water strategies and researchers with decades of on-the-ground research experience.

The stories these experts tell are equally extraordinary in both their complexity and their diversity, if not in their differences. The diversity in their analyses arises from three sources. First, WWF did not establish a unified methodological approach that sought to enhance comparability, or perhaps even uniformity, in their conclusions. We did not hold the analysts to a false expectation that it would be possible to construct a model of water-related social disruption with clear causalities and straight-line outcomes. Nor did we try to create a large-scale statistical research program or try to create controlled case study comparisons that would allow us to identify key variables and the relative weight of those variables in shaping social outcomes. Rather, we asked the analysts to reconstruct the trajectories of their respective countries and regions through their direct experiences, their own primary research and the rich analyses provided by others who seemed most capable of explaining the driving forces that shaped their countries' social experience of water. Our methodological choice was based on our view that the changes in water's presence and availability to society are extremely complex, multidimensional, multidirectional and highly sensitive to the issue of time—that is, shaped by the duration of water-related events, the timing of events in social cycles and the intersection of events with the stage of development of a society and the maturation of its governance systems. Consequently, an assessment of the contributions of this work should recognize the methodological option that we have chosen, with its associated strengths and limitations.

Second, the research approach we adopted reflects the increasingly complex reality that social disruptions are not shaped just by dynamics internal to a country. Increasingly, regional and global pressures influence social disruptions. Obviously, the changing climate affects all societies without exception, albeit in unique, differentiated ways. In equal measure, pressures and influences exerted by the increasingly integrated global economy weigh heavily on domestic markets and policy responses, exerting influence on relative prices, development strategies, political elites, local communities and ultimately on natural resource policies, including water allocation. Further, rising tensions and competition for water resources among countries in shared watercourses inject a regional dimension into domestic policies and behavior.

Third, perhaps most important, the diversity of the analyses the experts have provided reflects the very nature of the issue we are dealing with. Water is the foundation of life and social organization and water permeates every aspect of society and its many organized functions. Water scarcity and extreme weather events, because of water's foundational quality, can and do disrupt societies at multiple points of vulnerability, resulting in cascading consequences that defy simple chains of causation. Our point of observation is to understand when and how water-driven events intersect with the broader social dynamics to result in social dislocations that create destabilizing impacts to partner countries and to U.S. interests.

Let the reader understand: I have not presented formulaic straight-line interpretations of how water, in extreme absence or abundance, can expose social vulnerabilities or create new social fissures and jeopardize U.S. interests. What I do offer is an organized summary of experiences that illustrates how water-related ecological change moves along a diverse set of pathways that result in various expressions of social disruption. For the sake of clarity, we have used the term "social disruption" to signify a demonstrable negative change in social interactions that places in jeopardy the well-being of identified social groups, economic processes and relations of governance. Ultimately, ecological change translates down to the community and household levels, where, individually or collectively, members of society assess changing conditions and their options for responding to that change. A multitude of unpredictable factors can come into play at each level. At each one of those steps along the pathway of ecological change, adjusting social systems and human behavior can have considerable consequences for U.S. interests. Impacts on U.S. interests can be quite specific to a subnational geography or economic sector. By the same token, ecological change can significantly disrupt the very foundations of U.S. engagements with key allies in various regions of the world.

One further word of caution: Every chapter made reference to the actual or projected impacts of climate change; and, without exception, each expert asserted that the negative impacts of climate change will intensify and broaden in coming years. The reader, therefore, is encouraged to read this summary chapter not as an established statement of climate change's impacts but rather as a summary of initial directional indicators of what future shifts and adjustments each society will likely experience. This perspective also underscores the importance of examining the distinct dynamics in each country, being careful to avoid mechanistic interpretations of how changes in one part of a chain of influence will ultimately alter the society and U.S. interests.

Paths of Influence 369

The distillation offered in this chapter begins with a succinct statement of the different ecological changes experienced in the 14 geographies under consideration. Those ecological changes include ecological collapse, drought, groundwater depletion, severe storms and flooding, salinization, land degradation and desertification. I offer a brief summary of the experts' analysis of the societal forces and pressures that, one in combination with others, contributed significantly to the ecological change. Those pressures include governance failures, rising demand and climate change. In a second moment, we connect the ecological changes to impacts on the principal human systems on which society depends. Among those impacts, we include weakened rural livelihoods, destroyed and disrupted infrastructure, burdened health systems and humanitarian crises of displaced people.

The third step in tracing how ecological change leads to social disruption is through a summary of human responses to the combined effects of ecological and social change. In this summary we highlight rural-to-urban (and beyond) migration, intensified sub-national and ethnic conflicts, transboundary conflicts, the rise of insurgencies and expansion of the narcotics trade. I will also highlight the major obstacles, notably lack of robust domestic capital markets and lack of commercially viable water utilities, that forestall commercial investment in water projects in many developing countries. The final step in understanding the paths of influence of ecological and social change is to explore how current U.S. interests are negatively influenced. These issues include rising migration pressures; growth of insurgencies and criminal networks; regional instability; supply chain pressures; increased need for U.S. humanitarian, development and military support; and increased demand for financial support for water infrastructure.

I. Ecological Change

For those familiar with the ways that changes in water availability affect broader ecosystems, the enumeration of such impacts will not be surprising. What does strike home is the frequency and severity that change in water access and intensity have across the geographies we examined, as reflected below.

1 **Ecological collapse:** The draining of the Aral Sea in Central Asia is widely regarded as among the most catastrophic ecological events in past decades, as it is now reduced to a salinized lake unable to support robust marine life and agricultural activities. Lake Chad, which once straddled four countries (Chad, Cameroon, Nigeria, Niger), is now reduced to a meager presence only in Chad, at 5% of its original size, and its collapse has been accompanied by the displacement of 2.6 million people. The virtual desiccation of Lake Urmia, Lake Bakhetan and the Hamouns in Iran mark the collapse of significant ecosystems and agricultural systems with little prospect of recovery.
2 **Drought:** Of the 14 case studies, only the Greater Mekong Subregion does not report a significantly sharp rise in the severity, geographic extent

and duration of acute drought. The four-year El Niño drought that beset El Salvador, Guatemala, Mexico and Panama was deeper and longer than any in history, increasing vulnerabilities and migration across the region. Syria's historic drought became a precipitant of mass migration and political revolt against the al-Assad regime. Declining rainfall in Afghanistan, the countries of Central Asia, Pakistan and India has devastated rural livelihoods in regions of those countries, causing failing communities and region-wide migratory trends. Extended drought in Mindanao (Philippines) has dried up aquifers, rivers and lakes, exacerbating long-standing conflicts between local communities on one side and the central government and extractive industries on the other, as government and industries hope to tap oil and gas and mineral deposits.

3   **Groundwater depletion:** Falling groundwater levels and depletion of aquifers often accompanied droughts in many countries, resulting in crop failure and deeper drilling that further lowered the water table. Groundwater storage levels in six of Iran's major river basins are falling up to 11 millimeters per year. With 80% of Pakistan's domestic water consumption originating from groundwater, Lahore's groundwater level is falling more than half a meter a year. Groundwater extraction in India has reached critical levels in some of the country's most populous states. In many places, groundwater levels are falling by more than a meter a year, bringing major land-subsidence problems in their wake. In Mexico, 106 aquifers are currently being overdrawn, with 65% of overall demand from cities and industries being met through groundwater withdrawals. Land subsidence in Mexico City has reached 30 centimeters a year, with cumulative effects over the past 60 years totaling more than 20 meters.

4   **Severe storms and flooding:** Unprecedented volumes of rain coupled with severe winds have wreaked havoc across Central America, the Himalayas and the Philippines (Mindanao), displacing hundreds of thousands of inhabitants and destroying key infrastructure in the process, including bridges, roads, housing and hydroelectric dams. Floods in Pakistan inflicted billions of dollars in damage, affecting 10% of the country's population. In 2015, 1.8 million people were flooded out of the homes in southern states Tamil Nadu and Andhra Pradesh, with the city of Chennai particularly hard hit. The following year, more than 1.1 million people were displaced in India's northeastern state of Assam, causing billions of dollars in lost productivity and damage. Flooding in the Greater Mekong displaced hundreds of thousands of families in low-lying areas. Increased snowmelt and glacial lake outbreak floods washed away entire villages and transport infrastructure in the Himalayas and South and Central Asia.

5   **Salinization:** Drawing down groundwater in Pakistan and Mesopotamia and depleting the Aral Sea in Central Asia have rendered once-fertile land unusable due to increased salt and mineral deposits. Saltwater intrusion in low-lying areas of Mekong, Nigeria and Pakistan has changed the ecology and productivity of broad expanses of land. Declining recharge and

evaporation have rendered the Aral Sea and Lake Urmia highly salinized lakes surrounded by vast reaches of salt and mineral beds.

6   **Land degradation:** Erosion caused by intense rainfall and deforestation has severely reduced land productivity in many countries including Pakistan, the countries of Central Asia, Afghanistan, El Salvador, Panama and Guatemala.

7   **Desertification:** The southward desertification push across the Sahel has turned grasslands and savannas into desert. The increased frequency and strength of windstorms in Mesopotamia, Central Asia and Afghanistan continue to weaken the productivity and inhabitability of vast swaths of land. Windstorms and drought have reduced agricultural land in Iran, surrendering once-productive territory to the encroaching desert.

In addition to enumerating the ecological changes experienced in the various geographies, analysts highlighted the causes or drivers that led to changing ecological conditions, with particular emphasis on the following three issues.

Governance failures: Foremost among those drivers was a wide range of government failures in managing water resources. The governance failures that analysts underscored in their country research included shortcomings in legal frameworks, inadequate regulatory systems, inappropriate incentive structures, biases and corruption and protracted failure to enforce established laws and codes. Those shortcomings frequently held sway for decades. Not infrequently, those deficiencies arose from legal systems designed for an earlier epoch of the human experience when safe water was more abundant.

Some of the more egregious examples of governance failures include the grandiose debacle that occurred when the Central Asian countries under the former Soviet Union collectively drew down surface waters feeding the Aral Sea to support cotton production across the region. In equal measure, the depletion of Lake Chad for agriculture was supported by national governments and international development agencies and justified by poverty-reduction strategies and economic growth strategies. Incentives from the Indian, Pakistani, Syrian and Nigerian governments encouraging unregulated sinking of tube wells to support agricultural production accelerated the depletion of aquifers, eventually locking rural communities into nonsustainable livelihoods.

In Guatemala, governmental failure to properly assess and regulate the impacts of mining companies on water resources greatly prejudiced the access, health and livelihood opportunities for vulnerable indigenous communities. Moreover, the total absence of water laws, regulations and adjudication mechanisms have spawned multiple conflicts in which the powerful groups in Guatemalan society override with impunity the interests and needs of Mayan descendants.

Acquiescence by governments across the Greater Mekong to Chinese financial incentives to build water and energy infrastructure prejudiced the food security of thousands of communities across the Greater Mekong, reflecting the unequal distribution of costs and benefits to different sectors of society. Iran's food security policy included subsidies for food, water and energy for extracting groundwater, with the combined result of increased reliance on groundwater extraction and steadily falling water tables.

The experience in India provides a troubling tale of how the lack of a clear, national water management policy has given rise to rapidly growing institutional failures, leaving all parties unable to resolve even the most basic water rights issues. In that policy vacuum, India has experienced a crescendo of conflicts among its 29 states and between the states and the central government. In fact, the failed institutional arrangement actually encourages, if not obliges, states to subdivide into additional states, as witnessed by the creation of 29 states today from the original 14 in 1956, in order to resolve growing conflicts over resource allocation. The Indian analysts underscore how these unresolved conflicts are weakening the constitutional foundations of the country at the very time that drought, groundwater depletion and climate change are posing ever-sharper challenges to the economy and body politic.

These and myriad other institutional and regulatory failures continue to intersect with other forces active in the studies the analysts examined. While these shortcomings are, at times, recognized by government offices, efforts to change prevailing governance arrangements are frequently postponed to protect the careers of political elites and technical planners.

Rising demand and the supply response: Rising demographic pressure is a constant societal strain that contributes to longer-term ecological change. In fact, some countries, such as Nigeria and countries across Mesopotamia, host some of the highest birthrates in the world, resulting in major youth bulges that those struggling economies must try to absorb. Not only do those growing populations simply demand more water, but also the resulting rural migration and accompanying urbanization significantly change the composition and geography of demand as middle classes grow and lifestyles change. The increasing difficulty experienced by local governments to provide safe water to surging populations in Karachi, Lagos, Panama City, Tehran and Mexico City are endemic features of those rapidly expanding metropolitan areas.

Increased water demand arising from demographic growth goes hand in hand with long-term economic diversification and expansion. The studies are replete with examples that reflect efforts to expand mining, manufacturing and agribusiness, each generating new water demands. An initial response of many governments has been to increase the number of large-scale water storage infrastructure systems, build more extensive irrigation systems and expand generation of hydroelectric power through construction of dams and new coal-fired power plants. Iran's construction of almost 800 large dams since 1979, Turkey's Grand Anatolian Project (GAP) and India's plans covering some 20 years to tap the hydroelectric potential of rivers churning down from the Himalayas exemplify those efforts. Perhaps no region hosts as many plans for new water infrastructure as the Greater Mekong, which is rapidly being transformed by new east-west and north-south transport networks, new energy demand from within the region, trade routes that are deepening China's economic influence across South and Southeast Asia with funding made available through multiple Chinese financial windows.

In past decades, government action to address the rising demand for water was to build more infrastructure that, in turn, fostered increased economic activity to raise living standards. Now, experts writing about Pakistan, India and the Greater Mekong point out that combined economic and population growth, while yielding prized indicators of aggregate economic growth, are obliging governments and communities to make increasingly difficult and costly trade-off decisions about how to best use increasingly constrained water resources. The issue of trade-offs among multiple development and water allocation pathways—prioritizing energy generation, food production, or domestic consumption—is highlighted as a central challenge facing current and future generations.

Climate change: Initially, we attempted to separate long-standing ecological trends such as desertification, falling water tables and salinization from the impacts of climate change. That is not possible. The country and regional analyses affirm that the rapidly changing climate is expanding, accelerating and deepening the ongoing ecological shifts. Prolonged drought, accelerating glacial snowmelt, shifting patterns of monsoon rains, rising diurnal and nocturnal temperatures and devastating storms of greater strength and frequency are among the most frequent expressions of the changing climate. Based on the analyses in this volume, we can affirm that climate change is catalyzing ecological change that leads, through multiple paths of influence, to various forms of social disruption.

II. Impacts on Human Systems

Ecological change carries with it costs for society and potentially new opportunities. The experts' analyses, however, focused on the social dislocations that are being experienced in rural and urban areas because the changing ecological conditions quickly exposed vulnerabilities in those societies, with potential benefits not having yet materialized. As we trace the impact of ecological change on these societies, we use the term "human systems" to signify the organized institutions and processes that the societies employ to provide for their material well-being, including the production systems, governance arrangements and social structures, and ultimately to deliver prosperity and collective security. Our underlying interest is to understand how the forces driving ecological change converge to shape the nexus of social systems and how, in response to changes in social systems, communities and individuals respond. Only through those cumulative impacts and social responses can we better understand how U.S. interests are being put into play.

1 **Weakened rural livelihoods:** Given the intimate dependence of food production and rural livelihoods on stable water availability, it is not surprising that the agricultural sector is among the most immediately affected by ecological change. The central concern of virtually every analyst is the impact of changing ecological conditions—including deepening drought, declining water table levels, soil erosion and land degradation—on the most vulnerable social groups, particularly those in rural areas whose livelihoods depend directly on water availability and soil productivity. The breadth of weakened livelihoods translates into declining living standards, increased food insecurity and a general rise in vulnerability to any additional shocks. In India, 330 million farmers have been severely affected by drought, leaving hundreds of abandoned farming communities and putting farmer suicides on a dramatic upward tick. In the eastern part of El Salvador, drought, accompanied by extreme weather events, resulted in the collapse of rural livelihoods when the entire grain harvest was lost in 2014. El Niño's effects impacted 2 million Guatemalans, with rural families losing between 50% and 100% of their food reserves.

   The 2016 drought in Mindanao, affecting thousands of hungry farmers, has been exacerbated by the increased water use by agribusiness plantations, coal-fired power plants and mining operations. Millions of agriculturalists in Nigeria's north and northeast regions have been affected by encroachment of the Sahel and loss of agricultural land, leading to the abandonment of some 200 villages. In the country's Middle Belt, decreasingly fertile land, deforestation and encroachment of the Sahel have increased the precariousness of both Christian Yoruba farmers and Hausa-Fulani herders, as scarcity intensifies conflicts between agriculturalists and nomadic tribes. Crop and livestock failures across Mesopotamia reached beyond the crisis in Syria, punishing small-hold farmers in both Turkey and Iraq. Water mismanagement, drought and windstorms buried hundreds of Iranian rural communities and continue to threaten 110 million hectares with desertification.

2 **Structural inequalities:** Without exception, the studies find the greatest impact of ecological changes falls on the most vulnerable, often rural, populations, which rely directly on the productivity and integrity of ecosystems for their survival. In some cases, the poor and vulnerable have simply been the victims of decisions and natural processes far beyond their control. For example, Ishaq Khan, once a small-hold farmer in Baltistan, lost his land in the 2010 floods and is now pushed into increased vulnerability and uncertainty in the Pakistan countryside. But, far more

frequently, studies explored how communities have waged cycles of struggle to protect and reclaim their lands and water access, whether in the foothills of the Himalayas, the mountains of Mindanao, or the western highlands of Guatemala. Those communities faced daunting odds when confronting the international oil companies in the Niger Delta, the consortium of dam builders along the main stem of the Mekong and mining companies in Panama. Water infrastructure projects have forced thousands of farmers off their lands in Iran, reinforcing the access of elite sectors relative to the marginalized rural and ethnic communities.

The story of resource marginalization of the poor and resource concentration of elites has been told endless times through the lens of dependency theorists, political ecologists and community activists. However, the key message from the chapters in this volume is that the process of marginalization from natural resource wealth is not complete, but rather is continuing, if not accelerating, as ecological change ripples through society. The competition for natural resources is intensifying tensions between more vulnerable communities and the privileged sectors of those developing societies as the elites pursue economic activities designed, managed and financed from the urban centers. As water and the land that holds the water become increasingly scarce, there is little doubt that these long-standing struggles to control natural resources will intensify, with continued dispossession of the poor.

3 **Destroyed and disrupted infrastructure:** The studies in this volume highlight the vulnerability of the energy sector to drought and extreme weather events. Analysts underscored multiple events by which drought depleted the reservoirs designed to store water to drive the turbines in hydroelectric plants while also depriving coal-fired plants of the water needed to cool the fossil fuel systems. Reduced rainfall regularly affected provision of electricity to urban centers and mining operations in Guatemala, Afghanistan, Panama, Iran and Pakistan. India's energy-generating capacity was crippled both by falling reservoir levels and by extremely intense monsoon rains that washed away entire hydroelectric plants and damaged dozens of others. Those extreme flooding events eventually led to the abandonment of existing and planned hydroelectric plants across the Himalayan foothills.

Expert analysis in El Salvador underscored the destructive impact of hurricanes in 2009, 2010 and 2011 that washed away transport infrastructure, including bridges, roads and irrigation systems. The raging rivers in the Himalayas carried off major road, telecommunication and energy distribution networks. Floods in Pakistan destroyed irrigation systems and washed away bridges, roads, public buildings and water filtration systems.

Provision of water to important urban centers was a repeated challenged signaled by the analysts. Delivery of safe water to Panama City was interrupted frequently during the 2014–16 El Niño drought. Tehran and other Iranian cities experienced frequent challenges to provide citizens with a

376 David Reed

steady supply of water. Declining water quality across Pakistan is typified by the contamination of 90% of Karachi's water by bacteria, iron, nitrates and turbidity, while none of Hyderabad's water samples were considered safe.

In 2010, the Panama Canal was closed temporarily due to record-high water levels in Gatun Lake, the main water source in the Canal Watershed. During the 2014–16 El Niño drought, passage of large ships was curtailed and the hydroelectric plant on Gatun Lake was suspended for lack of water. Some 2,800 of the 14,000 ships that use the Canal annually were rerouted on longer journeys around South America.

An additional impact of changing ecological conditions is reflected in the rising burdens on health systems. For example, the rising frequency and intensity of windstorms in Iran's city of Zabol cost the city $100 million in lost economic activity between 2000 and 2005, with a rise in the rate of cardiac and respiratory illness and death up to 3%.

4   **Humanitarian crises of displaced peoples:** The severe tropical storms and excessive rainfall that punished communities across Mindanao created humanitarian crises for tens of thousands of homeless victims. Hundreds of thousands of Pakistanis were uprooted from their homes and forced into camps until floodwaters receded and rebuilding could begin. Entire rural communities were washed from hillsides in El Salvador and Guatemala, leaving thousands at the mercy of emergency governmental and international aid.

III. Human Responses

The transmission of ecological change through the many human systems of a given society eventually reaches down to community and household levels. When faced with reduced water supply and declining water quality, communities and individuals move to protect their interests—and they often move in forceful, even combative, ways in trying to regain control over this vital resource for their livelihoods. Their actions include seeking redress from governments, companies and other actors considered responsible for their increasingly precarious situation. It is, ultimately, the actions of major groups in society that can destabilize a country's government and thereafter jeopardize U.S. security interests. In this section, we highlight many of the significant reactions of important social groups in the countries examined to understand how their behavior opens paths to social disruption at subnational, national and regional levels.

1   **Rural-to-urban migration—and beyond:** The declining viability of rural livelihoods across Central America's Northern Triangle has spawned a sharp exodus from rural areas into the region's urban areas. In addition to the rapid growth of Guatemala's urban centers, population pressures and lack of economic opportunities are pushing poor residents in the country's

eastern provinces, the Dry Corridor, to migrate to the northern highlands, where conflicts with Mayan ethnic communities are rising sharply. Violence and urban lawlessness have spawned migration across the region and toward the United States, which, following the migratory influx in 2014, is now providing security support to Mexico to stop the northward flight before fleeing Central Americans reach the U.S. border. Mexico, as well, is coping with the lingering impacts of drought and the decline of rural livelihoods, which has spawned frequent rural-to-urban migration and, along with it, hopes of crossing the U.S. border.

Mesopotamia has come to symbolize how the convergence of intense water stress in an arid land intersects with long-term policy failures and corruption to generate great human tragedy. Those factors unleashed the flight of hundreds of thousands of Syrian farmers to urban centers, where they found equally difficult economic, social and environmental problems that erupted into civil war. Drought, diminished flows on the Tigris and Euphrates and sandstorms, coupled with pervasive security threats, forced widespread dislocation of Iraqi farmers to urban areas that offered little refuge from insurgency and terrorism. The tide of migrants emanating from the region pushed outward, carrying millions of refugees from the region into the Caucuses and eventually the heart of northern Europe.

The decades-long conflicts in Afghanistan pushed tens of thousands of citizens westward, forming part of the exodus to southern and northern Europe. In addition, several million Afghanis migrated eastward into Pakistan, fueling social tensions and, at times, supporting the insurgencies in both countries. Today, no longer able to absorb the growing pressures on its soils, Pakistan is forcing the more than 1 million Afghanis back into their country of origin.

Unpredictable violence, floods and drought have forced a growing movement of Pakistanis in the Northwest Territories, Balochistan and other provinces to move to the sprawling urban centers of Karachi, Hyderabad and Islamabad, among others. That influx of rural migrations has severely strained the water, electricity and social services of all of Pakistan's cities. Rural-to-urban migration is fueling rapid expansion of Iran's cities, with the rural population expected to fall some 24% in coming decades. Falling groundwater levels, drought and the increasing precariousness of agricultural livelihoods, all of which cause water stress, are the "push" and "pull" factors fueling the rural out-migration.

2 **Intensified domestic conflicts:** Analysts did not detail the efforts that aggrieved individuals and communities undertook to address their water-related problems prior to escalating their responses to the level of direct confrontation with government offices, mining companies, dam construction consortia, or neighboring communities. Rather, their reports highlight the direct encounters between competing parties, affirming activists' perceived need to use force to settle unresolved conflicts. Future analyses of pre-violence efforts to seek resolution would be very important in order

to identify reasons for derailed mediation efforts, blockages encountered by various parties and opportunities that were missed in seeking peaceful outcomes.

The overarching cause of increased water-related conflicts is sharpening competition between water users, including between those seeking to maintain rural livelihoods, urban consumers, energy producers and extractive industries. Although many of those competing interests are resolved through normal governance procedures, the studies seem to indicate that conflicts over water are becoming more frequent and intense. As in Iran, rapid urbanization has increasingly strained municipal infrastructure, resulting in urban water rationing, while massive water transfer systems further channel water away from rural communities. In response, violent opposition broke out in the 1980s as farmers protested the construction of the Khamirian Dam and, in 2013, farmers destroyed pumps carrying water to the city of Yazd some 200 miles away, resulting in multiple deaths.

Conflicts in the Darien region of Panama have their origins in the large influx of refugees who, in the 1990s, fled the Colombian civil war and pushed into Panamanian territory, where they collided with Wounaan villagers and other indigenous groups. Following that initial influx of guerrilla groups linked to the Revolutionary Armed Forces of Colombia (FARC), farmers and loggers, an ongoing conflict has continued wherein illegal loggers, paramilitary groups and drug traffickers have created a prevailing climate of lawlessness. Uncontrolled deforestation has degraded water resources, further intensifying competition for safe water. Moreover, mining, hydroelectric and water reservoir projects have provoked regular outbreaks of violence with indigenous peoples and farmers, as have plans to re-engineer the hydrology of the Panama Canal.

The Syrian civil war has galvanized understanding as to how drought and water mismanagement can intersect with other societal factors to create conditions of social disruption with an intensity seldom experienced by other countries. Although the spark of protest that ignited the civil war was centered in the Dara'a in southeastern Syria, the breadth and depth of the 2006–11 drought beset such a wide swath of communities that the entire region was quickly consumed by the uprising. Although the drought became a precipitating factor, underlying the social upheaval lay the years of mismanagement and corruption engrained in al-Assad's repressive regime, further fueling the conflict such that it acquired the dimension of a full-scale insurrection aimed at the government, not just a protest seeking redress of grievances on a local level.

Water governance failures in India have spawned ethnic/linguistic conflicts across the subcontinent. Social media has stoked the emotionally charged water rights issues, aggravating long-standing cultural animosities in a society without viable means of resolving most basic water conflicts.

A prominent feature of rising domestic turmoil is the pronounced ethnic dimension. With Iran's lakes and rivers under national government control, water allocation and infrastructure have triggered protests among ethnic communities in peripheral provinces. Arabs in southwestern Khuzestan have protested water diversion to distant cities and to the sugarcane industry, resulting in several riots. The collapse of the Lake Urmia ecosystem has sparked protests and Azeri nationalist fervor. The Baluch separatist insurgency has risen in Sistan and Baluchistan as desertification and lack of government support for development programs have deepened ethnic and religious grievances.

The roots of resource conflict in Mindanao (Philippines) reside in the decades-long biases that have marginalized the Muslim and Moro communities on the island. One of the more sinister efforts involved government resettlement schemes in the 1900s that sought to import Christian families into the region to increase the ratio of Christians to Muslims, thereby diluting ethnic Moro influence. A short-lived gold rush in the 1980s drew thousands of fortune seekers into the area, resulting over subsequent years in a multitude of illegal land claims, logging and mining activities, much to the prejudice of Moro communities. Structural biases have translated into efforts to erode the land claims of local communities and into pressure from logging companies to claim forested lands, from the local government to build a dam on the Pulangui River and from mining companies to extract oil and gas from the Liguasan Marsh, which would be opened up if the dam is constructed.

While the Boko Haram insurgency has captured considerable attention of the international press owing to its kidnappings, murders and shadowy presence in Nigeria, more deaths have been caused by conflicts in the country's Middle Belt where Muslim Hausa-Fulani herders have clashed with agriculturalist communities of Christian Yorubas. The annual southward migrations of the nomadic tribes trigger conflicts with the sedentary farmers, unleashing cycles of violence and retaliation involving assaults, and poisoning water and grass. Local government offices tend to favor the complaints of local residents, fueling the violence spiral leading to thousands of deaths.

The Mexican study asserts that water scarcity and dependence on trucked-in water are constant sources of social-political conflict that also fuels outward migration, particularly northward to the United States. The ethnic dimension to these conflicts is most sharply pronounced in Guatemala, where, for centuries, indigenous communities have been oppressed by the privileged Ladinos (descendants of Spanish colonizers). Today, the absence of any water regulation translates into a country-wide system of abuse and oppression of the Mayan descendants.

3 **Transboundary conflicts:** Changing ecological conditions and social systems were also reflected in dynamics between neighbors in shared watersheds. Transboundary water agreements have served to stabilize relations

among riverine states while, given their inadequacies and outdated terms, they also intensify tensions and conflicts. Water tensions between Iran and Iraq, though subject to a 1975 agreement between Tehran and Baghdad, have surfaced and calmed periodically over recent decades as downstream Iraqi farmers have protested water diversions along the Diyala River. Disputes and cross-border closings erupted in 2008 and again in 2011 and 2012. Potentially sharper conflicts may arise following the pending completion of a second Diyala Dam, which could further reduce flows to Iraq by as much as 22%. Further south, the dams on the Karkheh and Karun Rivers have created an even sharper diminution of waters flowing from Iran into Iraq that would, as a consequence, result in increased salinization of waters used for agricultural, industrial and domestic consumption.

In turn, conflicts between upstream Afghanistan and downstream Iran promise to intensify as Afghanistan moves to bring the Kamal Kahn and Kakhashabad dams online on the Helmand River and complete the restoration of the Kajaki Dam. Further, India's support for the completion of the Salma Dam will significantly reduce flows on the Harirud River into Iran. Paradoxically, despite sharing watercourses with its six neighbors, Afghanistan's only water treaty remains that with Iran, signed in 1973, governing use of the Helmand River on its southwest border. Water tensions between the two countries remained relatively quiescent after Afghanistan's 1973 coup, Iran's 1979 revolution, the Soviet invasion in 1979 and protracted civil war. In 2007, Taliban factions vying to topple the government used Iranian support to try to destroy the Kajaki, Kamal Khan and Kakhashahbad dams. During the past decade, Afghanistan's security problems and lack of water management capacity allowed its neighbors to increase their use of water originating inside the upstream country. Now, in the presence of American and coalition forces, the government is moving forward with various water storage and hydroelectric projects that would seem to indicate reduced flows to downstream Iran and other neighbors.

Conflicts in Mesopotamia date back 4,500 years, when the ancient cities of Umma and Lagash fought over irrigation canals diverted by the King of Lagash to support his kingdom's agriculture. More recent conflicts erupted between Syrian and Iraq in 1974 when Iraq threatened to bomb the Tabaqah Dam once Syrian dams reduced downstream flows. The completion of the Ataturk Dam, the linchpin of Turkey's GAP, in 1990 provoked protests from both Syria and Iraq as flows of the Euphrates were interrupted, causing major impacts on farming and energy generation in the downstream countries. While bilateral agreements have been reached between pairs of states, a comprehensive, multilateral agreement on water allocation and strategic planning has remained a distant project.

The India-Pakistan dispute over the waters of the Indus Basin figured prominently in the 2012 Intelligence Community Assessment, being one of seven major watersheds where interstate conflict was deemed most likely in the coming 10 years. The 1960 Indus Water Treaty has served as a moderating influence over past decades even in the face of strongly held claims by both parties, wherein Pakistan views the accord as posing security threats to its food, energy and overall economic well-being, and India sees the treaty as a challenge to its territorial sovereignty. As both sides engage in large-scale infrastructure development along the river, and as climate and demographic changes occur and recharge of aquifers diminishes, renegotiation of the treaty is urgently needed to adjust to rapidly changing conditions.

The central political issue facing the five Central Asian countries of Kyrgyzstan, Tajikistan, Uzbekistan, Kazakhstan and Turkmenistan is allocating scarce water resources to meet the very different energy and agricultural needs of those states. The downstream countries of Kazakhstan, Turkmenistan and Uzbekistan need water released in summer months to irrigate fields, while the upstream states of Kyrgyzstan and Tajikistan need to release water in the winter months to generate hydroelectric power. The long-standing resource-sharing agreement among the states brokered by Moscow that addressed those competing needs collapsed with the dissolution of the Soviet Union in 1991. The 1998 Long Term Framework Agreement brokered by USAID also fell apart. Today, fresh evidence indicates increasing instability in the region that threatens both the quality and quantity of safe water as climate change, infrastructure deterioration, failing agricultural yields, ethnic conflict on the borders between states, endemic corruption and personal animosity among presidents of the countries all converge. That immediate context is framed by the longer-term collapse of the Aral Sea and prevailing water governance challenges.

The six riparian countries of the Greater Mekong have a mixed record of cooperation over the past 30 years but now face greater tensions, as rapid economic growth and demographic changes unleash new domestic and regional pressures. Captured between the two large continental economies of China and India, the region is also buffeted by opportunities and potentially very high environmental costs as new infrastructure, opaque financial arrangements and political dynamics swirl in the context of long-standing mutual suspicion. The overall decline of water across the region is further intensifying tensions and forcing communities and governments to make increasingly difficult trade-offs among the choices of using water resources for energy, agriculture, or domestic consumption. One rising issue is the anti-dam protest movement that first emerged in 2003 with the construction of the Pak Mun Dam in Thailand. Similar protests have continued to the present, as Cambodian organizations protest work on the Xayaburi and Don Sahong projects, designed to send electricity to the Yunnan Province in China. In equal measure, protests in Myanmar against the Myitsone Dam on the Irrawaddy River have halted construction of that project. The absence of effective, region-wide water governance agreements has led to governments' regularly flouting protocols established by the Mekong River Commission.

4 **Rise of insurgencies:** The water-related crises unfolding in three geographies of Nigeria have each given rise to distinct yet now overlapping insurgencies. The water crisis in north and northeastern Nigeria, tied to the collapse of Lake Chad and expanding desertification, has contributed to spawning Boko Haram, a virulently anti-Western insurgency that recently pledged allegiance to the Islamic State. Conflicts between Muslim Hausa-Fulani tribes and Christian Yoruba farmers have intensified in recent years, as drought and desertification affect the country's Middle Belt. Penetration of Islamic militants associated with Boko Haram into the Middle Belt is feared to be rising. Active protests emerged in the Niger Delta in the early 1990s as communities protested the sharply rising contamination caused by repeated oil spills from wells owned by international petroleum companies. Today, three organized insurgent groups operate across the Delta, protesting contamination and the lack of development benefits provided by the extractive industries.

   Insurgencies in Mindanao predate by several decades the recent impacts of drought, typhoons and brazen efforts by local governments and companies to gain control over land and water resources claimed by Moro and Muslim communities. Creation of the Bangsamoro Autonomous Region in 2012 was a major concession to the resistance movement by the central Filipino government after decades of failed repression. Recently, however, rising natural resource competition has further fueled conflicts, as mining, logging and energy companies have tried to lay claim to land, water, forests and mineral rights. Recent affiliation of Abu Sayyaf, one of the main insurgent groups, with the Islamic State further raises the insurgency's salience in the eyes of both the Filipino government and the United States.

   Over centuries, water has been a central factor in shaping conflicts across the Mesopotamia, at times being the cause of conflicts, other times being an asset used by combatants to shape the outcomes of conflicts. The outbreak of the Syrian civil war and the subsequent emergence of the Islamic State have elevated the role of water as both a cause and a weapon of war. As water scarcity deepens its hold across the water-stressed region, this vital natural resource will continue to fuel conflict, possibly expanding into ongoing conflicts involving the Kurds and other separatist groups in the region.

5 **Expansion of the narcotics trade:** The growth of Afghanistan's poppy production, which supplies an estimated 80% to 90% of the world's heroin, was temporarily curtailed in 2014–15 due to disease affecting poppies in the southwestern part of the country. Production is expected to climb again as water scarcity further limits agricultural opportunities in the face of desertification that pushes across the country, given that poppy cultivation requires less water than other crops. The poppy trade is a fundamental source of financing for tribes affiliated with the Taliban, which, in turn,

provides financial and security incentives to maintain the allegiance of local farmers and communities.

Countries of Central America have been plagued by growing lawlessness and violence attributable to the effects of civil wars, corruption and human rights abuses for decades. Actively contributing to the violence and corruption is the steady expansion of international narcotics cartels that use the region as the major supply line to northern markets in the United States. As drought and extreme weather have diminished the viability of rural livelihoods, impoverished communities have embraced the economic benefits provided by affiliation with the narcotics trade. The grip of drug traffickers is so pervasive that experts calculate that 50% of Guatemalan territory is under the control of narcotics networks.

IV. Financing Water Infrastructure

The strategic importance to U.S. global interests of financing critical water infrastructure, including for water storage, disaster risk mitigation, urban water and wastewater projects, cannot be overstated. Failure to deliver safe water on a reliable basis to both rural and urban populations in developing countries will have serious social consequences. Without access to water sources for agricultural production, daily consumption and sanitation, vulnerable populations have demonstrated their willingness and ability to engage in social protest and actions that, if the causes aren't addressed, can threaten the rule of law, increase the incidence of violent engagements and open prospects for insurgencies and drug cartels. From those disruptive dynamics arise prospects of major migratory movements, strengthening non-state actors and contributing to the overall weakening of constructive U.S. influence.

The magnitude of financial resources required to address the growing investment needs is staggering. For water infrastructure of all classes, approximately $105 billion per annum is required; for water and sanitation, no less than $30 billion per annum—a doubling of current investment levels—is mandatory. Multiple studies have made clear that public funding alone cannot possibly meet the rising demand. Thus, the need to access private investment in the water sector in developing countries is uncontested.

The history of successfully engaging the private sector in water infrastructure is, at best, a highly checkered affair. As regards urban supply and sanitation, the financing challenges today are as overwhelming as in yesteryear, with no clear pathway for meeting the rapidly rising demand in growing metropolises. A paradigm change in the 1980s and '90s shifted prime responsibility from public ownership and management to privatized water utilities. That shift quickly crashed on numerous fronts, many of which continue to dog investors today. Foremost among the shortcomings was the failure to understand the "social dimension of water"—that is, the view that water is a public good, a necessity of life, which cannot be subject to the full costing of the value of water. Moreover, water's public-good dimension makes it highly susceptible to "politicization" of tariffs and often subject to widely vacillating political imperatives. The result is a vicious cycle of poor performance, insufficient funding and a deterioration of assets.

The public-good feature of water also creates false expectations that governments are responsible for and should finance water systems. In reality, only the private sector has the capital assets required to satisfy growing demand, albeit with public resources playing a supporting, targeted function. Today, two systemic challenges beset efforts to attract private financing into urban water infrastructure initiatives. First, many proposed water infrastructure projects do not meet necessary investment criteria. Weak administrative, planning and operational capacity beleaguers local governments, particularly as responsibility for water development has been steadily devolved downward to municipal governments. Operational problems, including the limited ability to set water tariffs, small scale of operations, lack of transparency and technical weaknesses, converge to deny municipal authorities creditworthiness capable of attracting private financing. Those endemic challenges further thwart development of demand for investment-grade projects that could ease a nation's water development burdens.

Second, the challenges in offering necessary financial returns to private investors are compounded by the nascent stage of debt capital markets in many developing countries. Specifically, water utilities are severely constrained in accessing local bond markets to finance their operations and, without the issuance of publicly held bonds, the risk of default remains prohibitively high for private investors. Moreover, although provision of water and sewer services has devolved to the local government level, there is no corresponding legislative authority to issue debt or raise revenue pledged as security against the debt obligations. Until these two problems are addressed, utilities will remain locked near the bottom of the ladder of financial sustainability, constantly subject to political pressures, with perennial difficulties in covering operating and maintenance costs and major constraints to extending service to growing populations.

Turning to the construction of river basin infrastructure, protests during the 1990s against social exclusion, distributional inequities and environmental costs catapulted large dam projects into a central place in debates about international development strategies. Eventually, a consensus on international norms for large infrastructure projects took hold among traditional multi- and bilateral lenders at the turn of the new millennium. By the same token, the public debate also revealed the long-standing tendency of planners to overstate the financial and social benefits of those projects while significantly understating the longer-term financial, social and environmental costs. Sharper recognition of the true costs of river basin infrastructure and pressure to comply with the higher standards pushed less-developed borrowing countries away from traditional, more exacting multilateral funding sources. Instead, they have moved toward public self-funding schemes and new multilateral banks such as the

Asian International Infrastructure Bank. Liberalized global financial markets and access to numerous funding streams have further locked in these less stringent arrangements as the preferred approach by a larger number of developing countries. Paying the true financial, social and environmental costs of those new projects is being postponed into the future for other generations to pay, as are potential conflicts with downstream riverine neighbors as new infrastructure comes online.

Today, urban and river basin infrastructure financing is enshrouded in uncertainty. Institutional development and reforms remain fundamental to increasing the capacity of municipal governments, which will increase their creditworthiness. Financial markets in most developing countries must undergo further maturation so that qualified utility borrowers can access domestic capital markets and comply with clear standards and lending requirements. Timely, targeted support from traditional international lenders can provide requisite guarantees for reducing risk, but no amount of public risk absorption by multilateral donors will guarantee the robust, sustained involvement of private investors that is required to meet burgeoning demand.

V. American Security Interests

Having followed this succession of impacts through which ecological change and water-related stress ripple through societies, we can now focus more directly on how the resulting social instability and disruption can pose immediate and longer-term threats to U.S. national security interests. The underlying premise of our analysis has been that U.S. security and prosperity depend directly on the prosperity and stability of our partners and competitors around the world. To the degree that water-driven stress destabilizes economies, undermines governance systems and weakens social cohesion, U.S. interests can come quickly into play. Increasingly, water scarcity has created a context of human and societal need wherein water has become a weapon of war, used as a way to deprive communities of a vital resource, as a threat to coerce political and economic concessions and as a source of revenue for emerging insurgencies and non-state actors. In this section we review threats that arise from migration pressures; the growth of insurgencies and terrorist networks; regional instability; supply chain disruption; and demand for U.S. development, humanitarian and military assistance.

1 **Regional instability:** We ascribe a rather general meaning to the term "regional instability," as a way of capturing the full range of social disruptions whose origins lie in ecological change but that translate into major challenges to national and regional economic prosperity, social well-being and stable governance arrangements. The Nigerian insurgencies, if allowed to fester and expand, can become a major challenge to American efforts to promote economic growth and democracy across West Africa through this pivotal nation. Should progress in Nigeria falter, the half dozen or

more insurgencies already active in West Africa can easily metastasize, producing a region crippled by rising lawlessness and direct attacks on U.S. interests.

The continued strength of the Moro Islamic Liberation Front and other local insurgencies coupled with the recent affiliation of Abu Sayyaf with the Islamic State represent direct challenges to the rule of law in southern Philippines and the opening of another front against U.S. and Western interests in Southeast Asia. Expansion of these insurgencies to other parts of the Philippines and other countries of the region will pose additional challenges to U.S. efforts to strengthen its presence in the face of aggressive Chinese policies in the South China Sea.

Instability in South Asia arises on several fronts. The continuing crisis in Indian agriculture brought on by drought and falling groundwater levels portends major challenges for the government to respond to rising internal migration, collapsing rural livelihoods and rising costs of importing foodstuffs, as well as challenges in delivering the required energy to support economic growth. However, the most salient feature of the instability on the subcontinent is internal governance dysfunction in calming rising water conflicts across the water-stressed country. A lack of effective resolution tribunals converges with rising ethnic/linguistic tensions, leading states to violate constitutional principles on which the nation was founded. Those internal tensions constitute a major challenge for the Indian government's ability to remain a steadfast, stabilizing ally in a highly contentious region. Pakistan's precarious agricultural sector, highly dependent on falling groundwater and susceptible to drought, amplifies current social uncertainty in South Asia. The volatility of the disputes over territorial claims, including waters of the Indus, underscores potential disruptive trends in the region.

Most conflicts in the Greater Mekong have developed largely at the subnational level, posing few direct challenges to regional stability for the time being. Protests against dam construction, major transport and energy infrastructure projects and resettlement programs have focused principally on local and national governments and supportive international agencies. Each construction project, however, represents one piece in a far larger infrastructure grid that will crisscross the subregion, east to west and north to south, which will be capable of moving raw materials and goods across all countries of the region, with China and India as the main economic poles. As the cumulative impacts, both environmental and social, of that integrated grid transform livelihoods, communities and ecosystems, the prospects of region-wide protest and instability may well figure in the future.

Ecological changes across Iran, Afghanistan and the five countries of Central Asia, combined with long-standing regional, water-based conflicts, represent a major challenge to stability in this water-stressed region. U.S. security interests in Afghanistan are under direct challenge to the degree that the Afghan government, with international assistance, is unable to expand water access to farmers and households whose livelihoods depend on it. A perfect storm of deteriorating ecological, social and political dynamics across Central Asia would seem to point to heightened transborder tensions and potential conflicts. While U.S. economic interests are modest, latent Islamic insurgencies across the region could benefit from increased instability, posing increased security threats to the United States, the Russian Federation and China.

The violent Middle East cauldron provides few prospects for enhanced U.S. interests across the region. Water scarcity will continue to intensify in coming years, combining with demographic pressures, governance failures and rapidly changing insurgencies such that the only hope seems to be to contain the contagion before it spreads to an even wider geography with even more profound consequences for U.S. interests. Continued, if not increased, instability will remain a challenge for U.S. engagements for years to come. Regardless of future political alignments, efforts must begin immediately to strengthen democratic governance of water resources and rebuild destroyed and degraded infrastructure.

2   **Migration pressures:** The most immediate migratory threat posed to the United States arises from the countries of Central America where water-related stress converges with lawlessness, urban violence and the narcotics trade. The surge of young immigrants during 2014 strained humanitarian and security facilities along the border and roiled domestic U.S. politics over immigration policy. While most of the young migrants fled the gang violence engulfing their cities, many sought economic opportunities that their countries could not provide, notably as rural livelihoods collapsed. That migratory influx prompted U.S. lawmakers to expand security and economic development programs, such as the Alliance for Prosperity, across the region while also increasing U.S. security assistance to the Mexican government to prevent refugees from Panama, El Salvador, Honduras and Guatemala from crossing Mexico's southern border on their northward flight to the United States. Such assistance has registered important gains in increasing judicial transparency and strengthening law enforcement, but nowhere has U.S. support been adequate to reverse either the broader environmental or social trends within those countries. Moreover, the surge in urban violence further undermines U.S. efforts to strengthen the rule of law and generally reduces U.S. influence while allowing the influence of narcotics networks to increase. This cycle of environmental degradation, weakened livelihood opportunities and continued urban violence may well erupt in subsequent tides of northward migration from our southern neighbors.

In addition to pressures mounting on the southern border, the United States is under increased international pressure to expand the number of refugees received from regions experiencing humanitarian crises, including the Middle East and Africa.

3 **Threats posed by strengthened insurgencies and criminal networks:** All of the insurgencies and criminal networks discussed earlier—including Abu Sayyaf, the Taliban, the Islamic State, Boko Haram and Central American drug cartels—existed prior to the onset of significant drought, extreme weather events and other water-driven conflicts. What the studies demonstrate is that the impacts of water stresses create social conditions that allow or encourage both insurgencies and criminal networks to flourish. These insurgencies directly affect U.S. interests by further weakening the oft-challenged governance systems in countries such as Afghanistan, the Philippines, Iraq, Nigeria and Guatemala. In weakening the rule of law, those networks create affinities and allegiances contrary to the national interests of partner countries and undermine formal economic activities, allowing corruption and intimidation to prevail. Moreover, the expansion of both criminal networks and insurgencies has gone hand in hand with those groups' identifying the United States as their sworn enemy, often targeting U.S. programs, bases and partner organizations. Were these movements self-contained and receding in geographic reach, specific attacks or threats to U.S. interests could be considered as increasingly isolated events. That is not the case, however, as both the insurgencies and water-related stress continue to grow, at times fueling each other and affecting wider territories and more countries. U.S. military interventions to suppress and destroy the networks have generated some gains in specific regions, for example in parts of Mesopotamia, while other networks, in Nigeria, the Philippines and the Northern Triangle, continue to flourish.

4 **Supply chain pressures:** The one direct impact for U.S. markets arose from the disruption of fresh fruit and vegetables grown in Central America (a prime sourcing geography for the United States), whose environmental changes translate into immediate supply and price dynamics in the United States. While supply was moderately affected during the recent years of El Niño, no major price shocks or longer-term supply disruptions were registered.

Water conditions in Mexico raise important questions about the future viability of U.S. manufacturing companies located in the country. U.S. multinational corporations are situated in two areas: the northern region along the U.S. border and the Valley of Mexico, where Mexico City is located. Both areas and in particular the northern border area, are experiencing severe water stress characterized by accelerating drawdown of aquifers. As groundwater levels fall, greater industrial contamination degrades water quality and drought continues, the Mexican study suggests that the companies will have to relocate to the southern regions of the country, where water is more abundant, or return to the United States. With almost $300 billion in Mexican goods exported northward each year, the impacts on U.S. companies and profits could be significant.

Of the countries covered by the studies, only India, Thailand and Vietnam—the three largest rice-exporting countries in the world—represent major exporters whose levels of production could seriously disrupt global supplies. Drought in all three countries significantly affected rice production in 2008. On one hand, reduced production obliged those countries, particularly India, to reverse outward commodity flows and import corn and other grains to meet domestic demand. On the other hand, their purchases increased prices across the weak global commodity markets, bringing benefits to major market suppliers, including U.S. farmers.

By the same token, almost all countries studied did experience important domestic commodity production disruptions as a result of drought, floods and severe storms. Iran, Afghanistan and the countries of Mesopotamia and Central Asia experienced difficulties in assuring ample supplies for local consumers. Again, as a major force in global commodity markets, U.S. growers reaped significant benefits from rising demand overseas.

Beyond short-term supply chain disruptions, the experience in Iran underscores a more fundamental challenge posed by the economic losses resulting from drought, groundwater depletion, unsafe water supplies and other causes. Total water-based economic loss to the economy totaled 2.82% of the country's gross domestic product (GDP). That is virtually the same annual GDP loss as that resulting from the imposition of U.S. and international sanctions over recent years. If comparable welfare loss impacts were aggregated across all of the countries covered in this study, the total loss of economic benefits to those countries and to the United States would be very far reaching.

Protests from the three insurgent groups in the Niger Delta created important impacts on global petroleum supply in earlier years of this decade. As the United States has steadily increased reliance on domestic natural gas supplies, those price signals had little impact on domestic fuel markets.

5   **Increased demand for U.S. humanitarian, development and military support:** With few exceptions, the analyses highlight ecological conditions that are not likely to contribute to stronger economies, build more stable rural livelihoods, provide an adequate water supply to burgeoning urban agglomerations, or increase resilience to the impacts of climate change. Quite to the contrary, the general trend lines sketched by the experts are for intensified competition for natural resources, increased vulnerability to environmental and social shocks, shorter time periods to recover from those shocks and governments weakened by social unrest, insurgency and lawlessness.

The fundamental link that these studies make is that when ecological conditions weaken the fabric of society; when development plans stall; when social cohesion frays as individuals and groups seek ways to protect their most basic survival needs, often in competition with others; and when ideologically inspired insurgencies or criminal enterprises capture the loyalties of honest citizens, U.S. efforts to promote prosperity and democratic forms of governance will falter. These studies carry us to the conclusion that when these conditions are allowed to fester unattended and unassisted by robust, sustained support by the U.S. government, our own stability and prosperity at home are under greater stress and threat.

The studies provide a clear rationale for a dramatic increase in U.S. assistance on all three dimensions of U.S. overseas engagements through humanitarian, development and military assistance. Demands from Mesoamerica are sharply rising to sustain the security and development assistance channeled to the region through the Central America Regional Security Initiative and the Alliance for Prosperity. Failure to maintain, if not dramatically expand, U.S. military and development support will invite greater lawlessness on our southern border and open the gates to renewed migration flows.

6   **Increased need for financial support:** Failure to respond to the diverse water infrastructure needs of developing countries results in an insidious set of outcomes. Water infrastructure is the vehicle for ensuring that safe water is available to growing urban populations, that water reserves can support energy generation systems and that water can be released in a timely manner to support agricultural production. In its absence, the full chain of social disruption presented in the case studies can unfold, as communities and companies are unable to satisfy their needs. The studies highlight the discomforting reality that unless there is a sustained, dedicated effort to create market and social conditions for private investment, urgently needed water infrastructure will fail to come online. While U.S. support has largely been channeled through multilateral institutions, the yawning need for far broader financial commitment has remained unaddressed, as the needs of hundreds of millions of vulnerable poor go unattended.

VI. Key Lessons

The analyses presented in the case studies illustrate the difficulties in establishing predictable sequences as to how water scarcity, extreme weather events and water-driven ecological change lead to social disruption. Despite the diversity of the national and regional experiences explored by the experts, there are a number of critical lessons that can be drawn from the studies that have enduring relevance in responding to these new water challenges.

1   **Governance failures are the key:** Long-standing, as well as more immediate, shortcomings in management of natural resources are the predominant drivers of ecological change and ensuing social disruption. The ways that failures in water governance influence society are multiple. An inherited yet prevailing policy orientation not adequately corrected by most governments is that water is or should be "free," not requiring that users pay for its full cost and that water is infinitely renewable. The failure to adequately price water goes hand in hand with lack of governmental

transparency about the factors that constitute water's costs and about the criteria and policies that actually determine how water prices, when applied, are set for different users. Other persistent governance shortcomings include failure to provide information on a society's baseline water balance, the provision of subsidies for water extraction and use, failure to monitor and regulate water extraction, the absence of adaptive regulatory systems and accompanying enforcement mechanisms, deficient water storage and management systems and ignoring the rising risks and costs of extreme water events, among many others.

Perhaps one of the most significant governance failures is the failure to ensure fair water access to all social groups. Resource-scarce societies understand how control of water translates into power and influence and, in response, governments distribute access to water resources in ways that often privilege the more powerful sectors, groups and ethnicities in the society. The links between wealth, political power and resource access are intimate and inseparable. As resource scarcity sharpens for a multiplicity of reasons, the exercise of power and privilege reinforces social dynamics that further concentrate resource access for the wealthy and powerful while marginalizing the least powerful, even from resources needed for survival. Water-related conflicts, as well as many humanitarian crises, often have their origins in the failure of governments to ensure fair access such that all citizens are guaranteed those natural assets on which stable livelihoods depend. As demographic pressures converge with the declining quantity and quality of natural resource wealth in a country, conflict often intensifies and the cycle of resource appropriation and marginalization continues.

While it is widely recognized that private sector investment is critical in allowing societies to develop and manage urgently needed water infrastructure, governments have seldom taken necessary steps to make private investment in the sector profitable. For example, as governments have devolved financial responsibility to subnational and municipal-level utilities, enabling legislation allowing those utilities to issue bonds and access local capital markets has seldom followed. While unburdening national governments through devolution to lower-level bodies, public agencies and authorities have not provided the technical, financial and managerial capacity that would make local utilities attractive to investors. In equal measure, opaque financial accountability, rampant inefficiencies and being subject to ever-changing political pressures—all responsibilities of public authorities—consistently erode the willingness of local investors to increase their presence in the sector. The societal cost is registered in stagnant investment levels, pent-up demand for water services and declining delivery and service records that can erupt in protest, black-market water operations and erosion of confidence in the authorities.

The governance failures experienced in Guatemala and India, albeit extreme, hold highly relevant lessons for other countries. The decades-long refusal by political and economic elites in Guatemala to establish a national water strategy with accompanying regulatory codes and enforcement mechanisms has pushed communities to resort to protest and violence to seek redress of their grievances against mining, agribusinesses and corrupt officials. As water stress rises, the government can expect protests to multiply and intensify. The water governance situation in India is significant because it reveals how failure to establish transparent adjudication systems for water conflicts can erode the credibility of local and national governments and enflame preexisting ethnic/linguistic tensions. The situation is such that, left unattended, the ongoing trend to subdivide existing states into multiple, albeit smaller, states will accelerate, while the constitutional foundations of Indian democracy are eroded.

Water governance failures are not emblematic of national governments alone. A number of countries lacked any recognized system of transboundary water management. Most, however, were party to some form of negotiated water agreement that provided a basic framework for water allocation among neighboring countries. Despite existing transboundary water agreements being in place for those countries, experts reported a clear increase in contentious relations with neighbors. For none did existing accords provide the foundation for longer-term strategic planning and monitoring that could allow for necessary adjustments in times of drought and extreme weather conditions. The absence of this foundation creates the conditions for unilateral action, misunderstandings and potential escalation of conflicts.

2 **There is never just one driver:** Ecological change and ensuing social disruption in a given society are caused by a convergence of multiple drivers. Changing water availability and extreme weather events take place in societies structured over decades, even centuries, by diverse social and economic incentives, opportunities, privilege and exclusion. Today, there is growing risk of assigning the cause of water scarcity and ecological change to one driver, notably to the impacts of climate change. Focusing on a solitary cause, particularly climate change, ignores the history of a country's social development and the underlying drivers of water scarcity. Moreover, focusing on one among many drivers distorts the way policymakers respond to crisis, often allowing underlying disparities and dysfunctional policies to remain firmly in place. Focusing on specific climate change impacts alone promises to divert attention away from governance failures and structural inequalities that have contributed to water stress over previous decades.

3 **Ensuing social disruptions fall disproportionately on the most vulnerable:** With good reason, every analyst focused on the impacts of water-driven stress on the most vulnerable sectors of the societies they examined. Because water and environmental stresses build slowly, at times imperceptibly, the increased precariousness of the most vulnerable groups may not be readily apparent or demonstrable. However, when

shocks—be they economic, environmental, or social—jolt a society, the increased vulnerability of these less-protected social groups is exposed in their diminished capacity to respond to and absorb the dislocations. Invariably, the resulting outcome is a decline in living standards and well-being. When experienced on a broad scale, such shocks can quickly translate into humanitarian crises of significant magnitude and social conflicts that spill well beyond national boundaries.

4   **Anticipate more frequent water conflicts:** The preceding analyses illustrate how social groups, when experiencing downward pressure on living standards, try to protect their livelihoods and eliminate the pressures that create uncertainty and dislocation. When the very foundations of communal livelihoods, in this case water, are denied, the studies showed that social groups and individuals will take direct action against other social actors, companies, or government offices they deem responsible for their hardships and deprivation. When serious environmental or economic shocks further weaken security and living standards, a wide range of responses, often unpredictable in focus and intensity, can convulse a community or society. Protest and direct action challenge the rule of law and prevailing leadership and seek solutions that will alleviate problems. As illustrated in several studies, community willingness to respond with violence is further exacerbated when public perceptions take hold that unanticipated shocks and deprivations are unevenly distributed across a society, falling unfairly on identifiable groups or sectors of society.

The water-based conflicts highlighted by analysts are characterized, in their various constellations, by the elements summarized earlier. In many cases, families responded through individualized reactions, including migration, engaging in narcotics production, or seeking alternative livelihoods. As in rural India, an increasing number of desperate farmers committed suicide, rather than face the tribulations of land dispossession and migration. In equal measure, however, collective actions of resistance and conflict entangled communities, spilling into direct confrontation with other rural communities, mining companies, loggers, dam-building consortia and government offices. Moreover, when communities perceived that actors on the other side of international boundaries were responsible, they launched forays across the borders to destroy irrigation systems or infrastructure.

Although the specific conditions and factors giving rise to violence differ, the experts' analyses underscore that the conditions giving rise to conflict are now affecting broader geographic areas and occurring with greater frequency than in the past. The analysts impart a clear message that, should these conditions continue to mount, increased conflict and violence are clearly etched into the future of these societies. They also point out that the growing impacts of climate change add new pressures that can provoke broader social upheaval and dislocation.

Analysts were careful to note that the great majority of water-driven conflict episodes they examined are mainly contained within national

394 *David Reed*

boundaries. That said, they also point out that cross-border water stress and tensions in Central Asia, Mesopotamia, the Greater Mekong, South Asia and West Africa are mounting, while governments have not taken necessary steps to update transboundary governance arrangements to manage more adverse water conditions and mitigate overt conflicts. In light of these considerations, there is good reason to anticipate that water-related violence will increase in the coming years. The intensity, scale and outcomes of these conflicts are far from certain, as are the potential future impacts on U.S. interests.

5 **There will be further ecosystem collapse:** Three geographies we studied have already experienced the collapse of important ecosystems—the Aral Sea, Lake Chad and Lake Urmia—on which the livelihoods of millions of people depended. Those collapses resulted from the confluence of multiple factors framed by multifaceted governance failures that resulted in major economic and social dislocations. Many other countries and regions, including Guatemala, India, Pakistan, Nigeria, Mesopotamia and El Salvador, experienced significant ecosystem disruptions that hold the potential for irreversible ecological change. At what point do additional weather events and governance failures guarantee that major regions of Iran, Pakistan, India and others draw down so much groundwater that tipping points are crossed beyond which ecological recovery is virtually impossible? At what point do continued desertification and drought in northern Nigeria, Afghanistan and Central Asia render ever-larger regions of those countries unproductive or ungovernable? We cannot make an accurate assessment as to where each country is along a pathway to experiencing ecological collapse within its borders. However, there are very troubling signs, particularly with the rising impacts of climate change, that indicate that further ecological collapse may be looming in the not-too-distant future. Those signs include two trends highlighted by the analysts:

- **Intensification of events:** One feature of the changing ecological experience in many countries is the intensification of weather events. El Salvador has experienced a continuum of severe weather events with an intensity that is without historical precedent. The tropical storms that hammered Mindanao were of unparalleled force and magnitude, bringing death and damage seldom known to the island. The monsoon rains that pummeled the Himalayas were of such force that multiple infrastructure systems, designed for earlier climatic conditions, were simply washed away under their ferocity. The droughts that beset Mesopotamia, Central America and Central Asia were of unprecedented duration and were accompanied by devastating windstorms and high temperatures.
- **Frequency of events:** Ecological changes and shocks occur across a broad temporal spectrum. Mesopotamia has experienced water-driven societal disruptions that date back millennia. In this sense, the recent

drought in Syria and across the region is but the most recent expression of a far-longer ecological challenge. However, in many cases that we studied, the increased frequency of extreme weather events has been registered only in recent years. During the past 15 years, El Salvador has recorded a frequency of extreme weather events unparalleled in its history. The succession of floods in Pakistan and India took both countries by surprise, as each was still recovering from the first events when the second floods struck. The number of windstorms that blanket parts of Iran has become more frequent and intense in recent years. The point to be made is that each event, whether temporally distant or near, can have impacts that are cumulative and reinforce effects that have been building over years, decades, or even centuries. With increased frequency, be it of droughts, floods, or storms, recovery becomes more difficult.

The frequency and intensity of extreme weather events transmit a clear message that without major water policy and management regime change, ecological collapse will become more frequent.

6   **The dimensions of needed response:** These key lessons point to important factors that must be included in successful long-term responses. First, dramatic changes in government and transboundary water management policies are the requisite for forestalling both ecosystem collapse and humanitarian crises. Those policy changes must be comprehensive and must be tailored to the conditions in each geography. Second, changes must address the multiple drivers that converge to unleash the unpredictable causes of ecologically driven social disruption. Addressing one dimension of the problem alone will simply transfer greater social and ecological costs to the future. Third, in conflict areas, constructive responses to water scarcity and mismanagement must begin even before the cessation of all hostilities and must lay the foundation for proactive citizen participation and the steady supply of water. Sustained water provision is, in itself, a precondition for mitigating lingering conflicts and transitioning to a period of reconstruction and stability. Fourth, responses, to be effective and enduring, must unfold over an extended period of time. Just as ecological change is cumulative, so, too, must be the corrective measures. Fifth, only if those measures directly target the most vulnerable members of society can broad-scale social disruptions be forestalled and the return to human development become possible.

21 Recommendations for Water, Security and U.S. Foreign Policy

David Reed

The basic premise of this volume is that U.S. prosperity and security depend directly on the prosperity and stability of both partner and competitor countries around the world. Moreover, we have affirmed that ecological changes associated with water scarcity and increasingly frequent and intense extreme weather events pose significant threats to U.S. security interests. As we have shown in the preceding chapters, those threats arise from regional instability, social disruptions taking the form of internal and transboundary migration, more frequent domestic conflicts (often with a discernible ethnic dimension), intensifying transboundary water conflicts, the rise of insurgencies and the expansion of the narcotics trade. The studies in this volume also highlight the unresolved challenges of financing water infrastructure, notably in engaging private investors, whose financing is central to responding to the rising demand for water.

This chapter presents recommendations as to ways the U.S. government can respond to the challenges, preferably mitigating further social disruption, as documented in earlier chapters. While it is challenging for any government to influence the behavior of another sovereign state, there are key areas where the U.S. government can and should expand current engagements, strengthen ongoing partnerships, provide additional technical and financial resources and create new responses to growing water-related threats that will affect the United States and partner countries alike. Strengthening such government-to-government collaboration is fundamental because governance failure is the common denominator in creating and, thereafter, amplifying the many social disruptions documented in the preceding chapters. Long-standing governance shortcomings invariably intersect with the new pressures on water resources generated by demographic growth and steady economic expansion at national and global levels. To a greater degree, the added impacts of climate change inject new dimensions of uncertainty and insecurity. While increasing water supply and managing demand must ultimately be grounded in the governance mechanisms and economic dynamics of each country, the United States can play an influential and highly constructive role. In that perspective of reinforcing the U.S. government's proactive engagements that support shared interests in supporting water security, in this chapter I offer recommendations organized into four categories.

1. Strengthen water governance to support partner country development goals and enhance U.S. interests.

    a   The United States should focus development assistance on building inclusive water governance mechanisms as a means of correcting distributional inequalities, building climate resilience and strengthening the rule of law. Through strengthened democratic water management, rural livelihoods can be improved and greater ownership and public accountability can be nurtured.
    b   Through bi- and multilateral assistance, the United States should promote clear natural resource ownership rights and responsibilities. Legally sanctioned public and private rights are imperative to nurture development of markets and national welfare. Official regulatory standards, enforcement mechanisms and penalties are mandatory for cultivating responsible public and private management of water resources and for reducing risks for enterprise and society.
    c   The United States should work through bi- and multilateral assistance to establish clear price signals for all social groups and individual users so that transparent water management becomes customary. Clear price signals should be made in line with the United Nations' recognition of access to safe and clean drinking water as a human right.
    d   U.S. development assistance must expand programs to ensure active engagement of civil society organizations and strengthen citizen dialogue with policy-makers at all levels of government. Past U.S. efforts have demonstrated their effectiveness in curtailing the rise of lawlessness, diminishing the appeal of migration and rendering illegal drug and criminal networks less attractive to disenfranchised populations, all of which are consistent with broader U.S. policy interests.
    e   The U.S. government should establish a National Water Commission to develop a national water strategy and to enhance coordination among the multilayered water management mechanisms in our own country. The increased frequency and intensity of ecological shocks demand more robust planning and prevention initiatives. Moreover, protracted drought conditions in the continental United States, with their multiple economic costs, raise planning and coordination imperatives to the level of a national crisis. This effort will not only strengthen domestic water and extreme weather event management but also increase U.S. credibility abroad.

2. Make water, sanitation and hygiene (WASH), a secure food supply and disaster risk reduction possible through sustainable, climate-resilient water management at the river basin level.

a   The U.S. government must lead an international effort to strengthen transboundary water agreements. Many existing transboundary agreements are in urgent need of revision to respond to rapidly changing ecological and climate conditions. Moreover, many regions experiencing water insecurity lack both bilateral and multilateral water accords that can ensure development goals and diminish prospects of interstate conflict.

b   To ensure delivery on its WASH and global food and water security objectives, the U.S. government should increase its focus on promoting long-term water management within partner countries from which focused project assistance for access and sanitation can be secured. These efforts should include the following:

   i   Establish new indicators for water resource management in development assistance programs.
   ii Establish long-term monitoring and evaluation systems to constantly sharpen project objectives and interventions.
   iii Establish an ongoing environmental flows monitoring system for distribution and use across government agencies.

c   The broadening impacts of water scarcity and weather extremes are increasing pressures to manage water resources at the river basin level to ensure equitable access to all riverine nations and users. Moreover, transboundary water management is needed to increase resilience to the growing number of ecological shocks. Preventing those shocks, and adapting to them, should include the following:

   i   Explicit evaluation of impacts and future projections of a changing climate.
   ii Adaptive management plans based on information and scenario planning for multiple possible future outcomes.
   iii Reevaluation of environmental flows based on likely future changes in precipitation and revision of e-flow regulations to ensure optimal seasonal access to flows for all users, including the environment.

d   Via the U.S. Agency for International Development (USAID) and the U.S. Department of State, significantly expand U.S. efforts to anticipate and mitigate local and regional water conflicts. Those programs—often in association with other development agencies and civil society organizations—provide technical, management and political expertise to defuse explosive conflicts between social groups before the dynamics of broader social disruption set in. U.S. development agencies should use these programs, proven effective under trying circumstances, to put in place cooperative water management systems prior to the cessation of all conflicts as a means of creating institutions for the post-conflict reconstruction process.

3   Address the underlying challenges that prevent private sector investment in water infrastructure and utilities, noting that public resources will fall far short of meeting growing water investment demand.

   a   U.S. efforts to engender private sector investment in the water sector require focused attention on the following:

   i   Strengthening the managerial, administrative and technical capacity of municipal water utilities to increase efficiency gains (reduce water loss), reinforce cost recovery practices and encourage long-term investment and expansion.
   ii Working with multilateral development institutions to mature local capital markets so that national and local financial institutions can lend to creditworthy utilities.

   b   Support the basic principle that access to safe and clean drinking water and sanitation is a human right while promoting the accompanying principle that the cost of water, in all of its physical expressions (sourcing, purifying, distributing) must be paid for on an affordable and fair basis.
   c   Strengthen international norms/standards for planning and constructing large water infrastructure. Those standards should include ensuring implementation of environmental impact assessments (EIAs) early in the planning phase, including upstream assessments and extending the geographic boundaries of EIAs to capture the full breadth of impacts.
   d   Conduct inclusive cost/benefit analyses that consider the full financial, social and ecological impacts over the usable life of a project. Total costs and benefits for the entire river basin should guarantee fair allocation of benefits to all users and a publicly agreed-upon distribution of full costs.
   e   Ensure that the siting, design and operation of all large infrastructure projects are based on scientific understanding of the potential impacts of climate change. An appropriate mix of gray and green "natural infrastructure" is necessary in order to maximize the ability of river basins to address storage, filtration and disaster mitigation needs.

4   Promote institutional reforms and structural changes in U.S. government offices and agencies.

   a   The Executive Office of the President:

   i   Establish and resource an entity reporting to the assistant to the president for national security affairs and the assistant to the president for science and technology to oversee all international activities addressing or supporting programs designed to stabilize and strengthen sustainable water management.

b   The U.S. Department of State:

   i   Satisfy the mandate of the Water for the World Act by adequately resourcing the special advisor for water resources to meet the specific duties required by the act.
   ii The special advisor for water resources should be involved in relevant operations of other State Department offices and should co-chair the Inter-Agency Water Working Group.
   iii Create regional water desk officers organized around priority water basins to monitor issues and identify opportunities for insertion of U.S. assets.
   iv In furtherance of the Water for the World Act, actively work with other donor governments to coordinate interventions and increase funding streams to address water security.

c. USAID:

   i   Based on the best available global water resources data, USAID must conduct a strategic assessment of water security risks as relates to USAID development assistance.
   ii Identify appropriate mechanisms by which USAID can effectively and efficiently invest in projects supporting all SDG 6 targets, including WASH and water resource management.

d. The U.S. Department of Defense:

   i   Require combatant commanders to include water threat assessments and appropriate responses in theater engagement plans.
   ii Provide expertise and logistical support to U.S. overseas development assistance workers addressing water security threats in collaborative programs with partner governments.
   iii Provide technical and security support when requested by partner governments to attenuate threats/actions that employ water as a weapon of war.

e. Intelligence community

   i   Generate annual reports on water and U.S. security interests, and make available to other federal agencies.

Contributors

Aldo Baietti recently retired as program manager and lead infrastructure specialist of the World Bank and has more than 37 years of experience in international development. His work has focused primarily on innovative financing mechanisms, utility reform, and private-sector participation in the infrastructure sectors, particularly in water and renewable energy. He has direct work experience in more than 40 developing countries across most continents. Mr. Baietti graduated from Georgetown University with a BA in Economics and obtained an MBA degree in Finance and International Business from George Washington University.

Asit K. Biswas is a distinguished visiting professor at Lee Kuan Yew School of Public Policy, NUS, Singapore, and co-founder of the Third World Centre for Water Management, Mexico.

E. Patrick Coady has had a lifelong career in investment banking and is currently Senior Director at Seale & Associates, Washington DC, as well as a senior fellow at Conservation International. Between 1989 and 1993, Pat was U.S. executive director of the World Bank. In 1994, he co-founded and served as chairman of the Northern Virginia Conservation Trust. He is a graduate of Massachusetts Institute of Technology and the Harvard Business School.

Ariel Cuschnir is an international environmental and social sustainability and risk management expert who has provided senior advisory support to international financial institutions, governments, and the private sector, across Latin America, Asia, and Africa. In the last 25 years he has focused his efforts in forecasting impacts and developing mitigation and risk reduction actions that ensure the objectives, opportunities, sustainability, and success of a program.

Roger-Mark De Souza is a recognized analyst, author, and speaker on resilience and sustainability. Currently, he is the director of Population, Environmental Change and Security at the Woodrow Wilson Center in Washington DC, where he leads programs on climate change resilience, environmental security, livelihoods, and reproductive and maternal health. Previously he worked at Population Action International, the Sierra Club, Population Reference Bureau, and the World Resources Institute. He holds graduate degrees in international relations and development policy from the George Washington University and the University of the West Indies in Trinidad.

Peter Gleick is co-founder, president emeritus, and chief scientist of the Pacific Institute, Oakland, California. He is a member of the U.S. National Academy of Sciences and a MacArthur Fellow. Gleick's work has focused on the risks of climate change for water resources, the links between environmental problems and international security, and strategies for moving to a sustainable water future. He is the author of many scientific papers and books.

Román Gómez González Cosío is currently Mexico country lead for the 2030WRG and consultant for the World Bank. Previously, he worked as deputy director for the Environmental Commission of the Megalopolis, Chief Technical Adviser for UN-Habitat's Water and Sanitation Branch, technical adviser at the National Water Commission of Mexico (CONAGUA), and member of the 4th World Water Forum Secretariat. He holds a PhD in Urban and Regional Planning, an MSc in Urban Planning, and a BSc in Architecture and Planning. His fields of expertise are public policy, governance, water resources planning, and urban and regional development.

Glen Hearns has worked in more than 23 countries throughout the globe consulting on a variety of assignments related to water, energy, mining, and health. Most recently, he leads the World Bank capacity building initiative for the government of Afghanistan on its transboundary water issues.

Farah F. Hegazi is a PhD student in environmental politics at Duke University. She seeks to understand, explain, and address the challenges that governments in the Middle East face in delivering water and sanitation services to unserved and underserved areas.

Marc Jeuland is assistant professor of Public Policy and Global Health Institute at Duke University. His primary research interests include water resources economics, nonmarket valuation, environment and development, environmental health, energy economics, and climate change.

General James L. Jones served his country with distinction for more than 30 years. After having served as the 32nd commandant of the U.S. Marine Corp from 1999 to 2003, he led all military operations of the North Atlantic Treaty Organization from 2003 to 2006. Upon retirement in 2007, he became president and chief executive officer of the U.S. Chamber of Commerce's Institute for 21st Energy, after which he served as President Barack Obama's National Security Advisor from 2009 to 2010.

Marcus King is John O. Rankin Associate Professor and director of the Master of Arts Program at George Washington University's Elliott School of International Affairs. Marcus King is also senior fellow at the Center for Climate and Security. He has held research and presidential appointments in the energy and environmental security field.

Lilian Marquez is a Guatemalan forester with a PhD in Public Policy from Indiana University, where she studied with Elinor Ostrom, the first woman to receive the Nobel Prize in Economics. She has been associated with the World Wildlife Fund (WWF) since 2008, and is currently the Design and Impact Officer at WWF Guatemala/Mesoamerica.

David Michel is an executive-in-residence with the Global Fellowship Initiative at the Geneva Centre for Security Policy in Switzerland, and Non-resident Fellow with the Environmental Security Program at the Stimson Center in Washington, DC.

Richard Kyle Paisley's academic research, teaching, graduate supervision, and advisory interests include international water and energy law, hydro-diplomacy, and conflict resolution. He has worked throughout Africa, Asia, and the Americas and published extensively in the scholarly academic literature.

David Reed is senior policy advisor for WWF-US. He holds a PhD and graduate degrees from the University of Geneva, Switzerland. He has worked for 35 years in social and economic development programs in Latin America, Africa, and Asia at both the grassroots and managerial levels. Reed is a widely published author and global expert on the complex relationships between macroeconomic policies, social structures, and the environment in the developing world. His most recent book, *In Pursuit of Prosperity: U.S. Foreign Policy in an Era of Natural Resource Scarcity* (Routledge, 2014), explores the impacts of resource scarcity and climate change on social stability and U.S. security interests around the world.

Herman Rosa has served as minister of the environment and natural resources of El Salvador (2009–2014). Previously he was senior researcher and director of PRISMA, a think tank on environment and development issues based in San Salvador.

Udisha Saklani is an independent policy researcher working with the Institute of Water Policy, Lee Kuan Yew School of Public Policy, NUS, Singapore.

Ali Hasnain Sayed is an engineer with an MBA and MA in Environment and Sustainability. He has more than 15 years of experience as a development practitioner with particular focus on policy, practice, and management issues of sustainable development. Currently he heads the Water Department of

WWF-Pakistan and is an adjunct faculty member at Lahore University of Management Sciences.

Keith Schneider is senior editor and chief correspondent at Circle of Blue, a U.S. news organization that reports globally on energy, water, and food. He has reported from the frontlines on six continents. A winner of numerous honors for his work, Keith also is a writer for *The New York Times,* where he served for ten years as a national correspondent, and has been a regular contributor since 1981.

Chelsea N. Spangler is a water and security specialist at World Wildlife Fund. She holds an MA in Global Environmental Policy from American University and an MA in ethnography from the University of California, Berkeley. Her previous work has focused on both public and private sector solutions to environmental problems, particularly surrounding agriculture.

Eduardo Stein is the former vice president (2004–2008) and Foreign Minister (1996–2000) of the Republic of Guatemala. He participated actively in the Esquipulas Peace Process in Central America and has worked in development agendas with rural communities and indigenous peoples.

William Streeter is an international infrastructure and municipal finance specialist with more than 30 years of experience, including with ratings agencies and an international infrastructure asset manager. Bill is a private-sector member of the infrastructure and capital markets development advisory groups for the APEC Finance Ministers, and is a visiting scholar at the Global Projects Center, Stanford University.

Arjun Thapan works on water and development, and he is the current chair of WaterLinks. He is a former director general of the Asian Development Bank, Manila, and a former Chair of the World Economic Forum's Global Council on Water Security.

Cecilia Tortajada is a senior research fellow at the Institute of Water Policy, Lee Kuan Yew School of Public Policy, NUS, Singapore, and Co-Founder of the Third World Centre for Water Management, Mexico.

Muhammad Faizan Usman received his BSc degree in Environmental Sciences from Forman Christian College, Lahore, Pakistan, in 2016. His research interests include water quality and quantity assessment, and water economics. He currently works at WWF-Pakistan as Community Mobilizer in the Freshwater department.

Erika Weinthal is Lee Hill Snowdon Professor of Environmental Policy and Associate Dean for International Programs at the Nicholas School of the Environment at Duke University. She received her PhD in Political Science from Columbia University. Her most recent book is the co-edited volume, *Water and Post-Conflict Peacebuilding: Shoring Up Peace* (Earthscan Press, 2014).

Lesha B. M. Witmer has worked as an independent senior advisor and facilitator on sustainable development focusing on issues of water governance, transboundary cooperation, and international water conventions. She studied Human Resources and Organization Development (Sittard 1976), Western Sociology (Erasmus Rotterdam), Labor and International Law (University of Amsterdam), and General Management (Nijenrode, the Netherlands).

Index

Note: Page numbers in italic indicate a figure or table on the corresponding page.

Abu Sayyaf 293, 295–6, 382
ACP *see* Panama Canal Authority
Adani Group 265
ADB *see* Asian Development Bank
administrative framework for water service debt 333–4
AES solar plant 266
Afghanistan 37, 189–205; agriculture 189–90, 192, 195; climate change 192–3, 195–6; conflict and meeting targets of SDGs 46; dams 178, 192, 193–4, 197; development statistics 190–2; Global Adaptation Index 195; Kabul River and Indus River system 233; Kandahar Food Zone 197, *198*; map *191*; National Development Strategy 192; need for water storage 189–90, 192–3, 197–8; paths of influence 200; poppy cultivation for opium 181, 189–90, 194–8, 382–3; population *192*, 196; transboundary arrangements 190, 193, 194, 196, 207; transboundary tensions 177–8, 181; U.S. interests 196–7, 199; Water Law 198; water resources 190
Agenda 21, 15
Agenda for Sustainable Development (2030) 39, 41
agroforestry system in El Salvador, Quesungual 71
Ahmad, N. 189
Ahmadinejad, Mahmoud 176
Ahmed, Shahid 226
AIIB *see* Asian Infrastructure Investment Bank
airports, as monopoly utility 328
Akbar the Great 237
Aleppo water treatment facility 158

Alliance for Prosperity Program 121, 122, 390
Almaty Agreement (1992) 194
al-Qaeda 160, 296
Al Qaeda in the Islamic Maghreb (AQIM) 140
American Near East Refugee Aid (ANERA) 28
Amu Darya River 49, 194, 209, 210, 211, 214
ANERA *see* American Near East Refugee Aid
AQIM *see* Al Qaeda in the Islamic Maghreb
AquaFed 41
Arab League 27, 153
Aral Sea 209, 210, 211, 369, 370
Argentina 22, 351
ARMM *see* Autonomous Region in Muslim Mindanao
Arroyo, Gloria Macapagal 296
ASEAN *see* Association of Southeast Asian Nations
Asia *see* Central Asia; Greater Mekong subregion
Asian Development Bank (ADB) 280, 282–5
Asian Infrastructure Investment Bank (AIIB) 346, 350–1, 385
Asian Water Development Outlook for Asia-Pacific 224
Asia-Pacific region: challenges faced 37
Association of Southeast Asian Nations (ASEAN) 272, *273*
Aung San Suu Kyi 279
Autonomous Region in Muslim Mindanao (ARMM) 288, 291, 292

Baietti, Aldo 311, 401
Bandhopadhyay, Gautam 260
Bangladesh 238, 254
Ban Ki-moon 42
Barrientos, Irene 117
Bechtel corporation 25
Bhutan 238, 259, 261, 351
Biden, Joseph 110, 113, 123
"bigger is better" economic principle 257, 266–7
birthrates 372
Biswas, Asit K. 237, 401
blended finance option 319–21, 324
bluewater withdrawals: in Iran 169
Boko Haram, Nigeria 128, 130, 135–6, 137, 142; summary 379, 382
Bolivia 22, 351
Bonga oil spill, Nigeria 140
Bosshard, Peter 257
BOT *see* build, operate, and transfer
Botswana 351
Boundary Waters Treaty (1909) 9
Brahmaputra River 49
Brazil 23
Bremer, Paul 25
build, operate, and transfer (BOT) 93, 316, 328
Burundi 10
Bush, George H. W. 14
Bush, George W. 16, 18

CA *see* Central Asia
CAB *see* Comprehensive Agreement on the Bangsamoro
Cabrera, Jorge 120
California 5–6, 11
Cambodia *see* Greater Mekong subregion
Cameroon 136
Campbell, John 135
Canada 9, 181
capital expenditures (CAPEX) investments 323
capital markets, for financing water utilities 316–17, 330–3, 384–5, 391
Caribbean Catastrophe Risk Insurance Facility (CCRIF) 322–3
CARSI *see* Central American Regional Security Initiative
Carter, Jimmy 3
Catastrophic Drawdown Option 322
catastrophic risk (CAT) insurance 322–3
Cauvery River 48, 241–2, 244, 255, 258, 352
CCRIF *see* Caribbean Catastrophe Risk Insurance Facility

Center for Naval Analyses (CNA) 160
Central American Regional Security Initiative (CARSI) 72, 122–3, 390
Central Asia (CA) 206–21; area covered 207; climate change 210, 211; GDP 207; legally binding international agreements 213–15; Long-Term Framework Agreement (1998) 206, 209; map *208*; mediation on Uzbekistan objections 215–18; paths of influence 217; population 207; problem evolution and analysis of impact 209–12; problem statement 207–9; resource-sharing system from Soviet Union 206, 209, 371, 381; Rogun and Kambarata hydroelectric power projects 206, 211–12, 213, 218; solutions 213–18; U.S. interests in and leadership recommended 206, 207, 212–13, 218; Uzbekistan's objections to upstream hydro projects 207, 211, 215–18; water-energy-agriculture nexus conundrum 206, 207, 209, 213
Central Asia Regional Economic Cooperation Program 199
Chad 136
Chardin, Jean 168–9
China 37; 21st Century Maritime Silk Road 282; challenges 238–9; as Dialogue Partner status in MRC 280–2; Harmon Doctrine principles 10; Lancang-Mekong River Community of Common Destiny 282; Nicaraguan canal 76; soft engineering for water storage 231; Three Gorges Dam 231; *see also* Greater Mekong subregion; South China Sea dispute
Chorabari Lake flood 262–3
civil unrest *see* protests and demonstrations
Clapper, James 24, 210
Clean Water Act (1987) 336
climate change 73, 89, 106, 122, 135, 137, 139, 143, 162, 182, 200, 217, 234, 250, 267, 284, 294; as cause of social upheaval and/or insurgent movements 4; and conflict, framing note on social dimensions of water 35–7; consideration in planning processes 347; as convergent trend 8; financial adaptation strategies for climate change risks 321–2; scientific evidence 155–7; summary as ecological change driver 373; summary chapter and perspectives

368–9; as threat multiplier 159–60; as threat to implementation of SDGs 47–8; U.S. State Department focus on 19; *see also individual countries*
climate finance and risk instruments 322–3
climate justified strategy, financial adaptation strategy 321–2
Clinton, Bill 15
Clinton, Hillary 18–19
CNA *see* Center for Naval Analyses
Coady, E. Patrick 307, 401
Coal *see* India, water-energy nexus in Himalayas
Coalition Provisional Authority (CPA) 24–5
Cold War: shift of threats 149–50, 160
Colombia 76, 78, 82, 197, 378
Colorado River 9, 13
COMEST *see* World Commission on the Ethics of Scientific Knowledge and Technology
commercial bank lending to water utilities 317
Comprehensive Agreement on the Bangsamoro (CAB) 292, 299
CONAGUA *see* National Water Commission, Mexico; National Water Council, Panama
concessional lending to water utilities 317, 320, 339
conditional grants 334
conflict: anticipation of more frequent 393–4; climate change, and social dimensions of water 35–7; domestic, as human response 377–9; in fragile states, and SDGs 45–7; Six-Day War (1967) 27; transboundary cooperation and global conflict 48–50; *see also* insurgencies; protests and demonstrations
Consultative Group on International Agricultural Research 162
Contract with America 15
convergent trends of water use and management 7–9
cooperation, transboundary, and global conflict 48–50
Cosío, Román Gómez Gonsález 91, 402
cotton production 223, 228
CPA *see* Coalition Provisional Authority
criminal networks as U.S. security interest 388
Curitiba Declaration (1997) 345

Curlier, Maaria Solin 218
Cuschnir, Ariel 76, 401

Dahla Dam 178, 192–3
dams: building period in U.S. (1930–1970) 10–12; development controversy 23, 308; International Commission on Large Dams 173; World Commission on Dams (WCD) 23, 308, 345, 348, 352; *see also individual countries*
data, digital elevation model and soil map in El Salvador 70
data collection and sharing 46, 354, 356
debt capital markets, for financing water utilities 316–17, 330–3, 384–5, 391
debt crisis (late 1980s) 307
debt service coverage ratio (DSCR) 320
debt service reserve fund 337
demand and supply: as ecological change driver 372–3
Democratic Republic of the Congo 49
desertification 132, 134, 168–9, 174, 374; summary of ecological change 371
De Souza, Roger-Mark 288, 401–2
development and diplomacy 39–55; Agenda 2030 39; climate change as challenge 47–8; cross-sectoral linkages across goals 43–5; fragile states, conflict, and SDGs 45–7; global conflict and cooperation 48–50; importance of U.S. support of SDGs 51–2; moving from MDGs to SDGs for water 40–3; partnerships and financing 50–1; SDGs 39–40
development of water infrastructure 344–64; background history of large infrastructure 344–7; consequences for political, socioeconomic, and environmental security 351–3; finance shift from official sources to alternate funding 345–6, 348–51; gap between financial concerns and global norms 353–5; planning process 347–8; in U.S. dam building period 10–11
direct grants for water utilities 317, 320
Director of National Intelligence (DNI) 39, 47
disaster risk instruments 322–3
displaced peoples 376; *see also* migrations; *individual countries*
DNI *see* Director of National Intelligence
domestic conflicts: as human response 377–9

domestic foundations of U.S. water policy 9–14
drivers of water scarcity, never just one 392, 395
drought, as ecological change 369–70; *see also individual countries*
DSCR *see* debt service coverage ratio
Dublin Conference, Ireland 14
Dublin Principles 14–15, 42, 44
Duterte, Rodrigo 296–7

Earth Day, April 22: Guatemalan demonstration 110, 117
ecological change 369–71; desertification 371; drought 369–70; ecosystem collapse 369, 394–5; flooding and storms 370; groundwater depletion 370; land degradation 371; salinization 370–1; and social disruption 369
ecological change drivers 371–3; climate change 373; governance failures 371–2; rising demand and supply response 372–3
Economic Cooperation Organization 199
economic development from water infrastructure 344–5
economic principle of 20th century, "bigger is better," *vs.* 21st century principle 257, 266–7
economic reforms (early 1990s) 307–8
ecoregions, defined, in Nigeria 134
ecosystem collapse 369, 394–5
education of girls, in SDGs 42
Egypt 218
EIA *see* environmental impact assessment
Eisenhower, Dwight 27
Eliasson, Jan 49
El Niño 56, 79, 116, 293–4, 370, 376
El Salvador 36, 56–75; bridges 61, 64; civil war (1980s) 65, 67, 71; coffee and coffee rust epidemics 59, *61*, 65; crop losses 56, 57; drought 56–7, *59*; ecological landscape restoration 56, 68–71, 72; economic growth 67–8; El Niño 56; extreme weather trends 56–65, 72; forest cover 68; homicide rates and violence 65, *66*, 67, 72, 112; maize yields 64; migrations 57, 59, 65–8, 71, 113; National Landscape and Ecosystem Restoration Program 69–70; Partnership for Growth Initiative in 71–2; paths of Influence in *73*; population 67; precipitation maps *58*, *62*, *63*; Quesungual agroforestry system 71; REDD+ program 70; reforestation hotspots 59, *60*; remittances 57, *60*, 65, 68, 71; San Vicente volcano 59, 61; transboundary problems with Guatemala 112, 113, 120; Tropical Storm Agatha 61; urbanization 59; U.S. involvement 71–3
El Tambor gold mine, Guatemala 115
ENCCP *see* National Strategy for Climate Change
Energy Regulatory Commission, Philippines (ERC) 328
environmental costs of water policies 21–2, 29
environmental impact assessment (EIA) 348, 399
environmentally sensitive planning 349–51
environmental security 158–9, 351–3
EPIRA *see* legislation, Electric Power Industry Reform Act
Equator Principles 88
ERC *see* Energy Regulatory Commission
Ethiopia 47, 211, 218
Euphrates River 6, 26, 49, 158; *see also* Mesopotamia
European Commission: on refugee crisis 160
Exxon multinational corporation 138

FACTS *see* Foreign Assistance Coordination and Tracking System
faith-based organizations 16
Falkenmark indicator: defined, in Iran 169
FAO *see* UN Food and Agriculture Organization
FARC *see* Revolutionary Armed Forces of Colombia
Farinotti, Daniel 211
FCCC *see* UN Framework Convention on Climate Change
FDI *see* foreign direct investment
Federal Water Resources Council, U.S. 12
Feed the Future, U.S. 19, 141
finance of water infrastructure: framing note 307–9; paths of influence, summary 383–5
financial help for weak water utilities 311–24; blended finance option 319–21; challenges and contradiction 311–12, 319; climate change

adaptation: climate justified strategy 321–2; climate change adaptation: no regrets strategy 321–2; climate change risks 321–3; climate finance and risk instruments 322–3; financing instruments and lending by donors and IFIs 317–19; ladder of financial sustainability 312–14; problem: limited financially viable instruments 315–16; problem: underdeveloped financial sectors and capital markets 316–17; problem: water is different 314–15; recommendations 323–4; U.S. foreign engagement 323–4; viability gap financing 319–21
financing water and sewer infrastructure in developing world 326–41; compared to other infrastructure assets 329–30, *331*; debt markets in nascent stage 330, 332–3; focus on quality service and adaptation 338–40; how water projects differ 328–9; India Water and Sanitation Pooled Fund 336–8; Kenya Pooled Water Fund 338; Manila Water Company 339; Mexican structured certificates 335–6; pooled project risk 336–8; Pune Municipal Corporation 339–40; recommendations 340–1; regulatory and administrative frameworks 333–4; securing project cash flows 334–6; underfunding of water projects 327–33
First Solar 266
Fitch Ratings 337
flooding and storms, as ecological change 370; *see also individual countries*
FMSO *see* Foreign Military Studies Office
food security initiative, Feed the Future, U.S. 19
Foreign Assistance Coordination and Tracking System (FACTS) 17
foreign direct investment (FDI): in Mexico 101; in Pakistan 223
foreign exchange (FX): risk in financing water utilities 316–17
Foreign Military Studies Office (FMSO) 212
foreign policy, recommendations for U.S. 396–400
Forest Carbon Partnership Fund 70, 81
forest cover in Central American Isthmus *68*

F-process 17
fragile states: conflict, and SDGs 45–7; role of water stress in creating xi–xii; *see also* governance failures
Frist, Bill 17
FX *see* foreign exchange

GAP *see* Great Anatolia Project, Turkey
gap financing *see* viability gap financing
Gates Foundation 50
Gaza desalination plant 47–8, 217
gender equality: included in SDGs 42
GERD *see* Grand Ethiopian Renaissance Dam
Germanwatch 254
Ghana 42, 72
Gingrich, Newt 15
Gini coefficient: in Guatemala 111
Gleick, Peter 149, 402
Global Adaptation Index: for Central Asia *195*
Global Climate Risk Index: for Guatemala 116; for Mindanao, Philippines 295; for Pakistan 229
global economy, as convergent trend 7
Global Food Security Index: in Pakistan 223
Global Terrorism Index: on Fulani militants in Nigeria 137
Global Water Partnership (GWP): in Panama 83
GMS *see* Greater Mekong subregion
GON *see* Nigeria, Government of
governance failures 318, 371–2, 390–2; recommendations for U.S. foreign policy 396–7; *see also* fragile states; *individual countries*
Grand Ethiopian Renaissance Dam (GERD) 211, 218, 349, 351
grants for water utilities 317, 320
Gravity Recovery and Climate Experiment 232
Great Anatolia Project, Turkey (GAP) 349, 351, 373, 380
Greater Mekong subregion (GMS) 37, 272–87; agriculture 272, 274, 275–6, 277; area covered 272; Basin Development Strategy 283; China as Dialogue Partner in MRC 280–2; climate change 275; coal-fired power plants 274, 278; conflict potential 278–9; dam development and biodiversity loss 276–8; dolphins and tourists 280; droughts 275; Economic

Cooperation Program 283–4; energy-food nexus 272–3, 276–8; fisheries 272, 276–8, 279; floods 275; GDP 272; giant catfish 279; hydropower 272, 274, 276; Lower Mekong Initiative (LMI) 282, 285; Mekong River Commission (MRC) 275, 278–9, 280–2; paths of influence 284; politics of Mekong River management 280–2; pollution 278; rice cultivation 272, 275–6, 279; strategic significance 272–3; transboundary water governance 280–2, 283; Upper Mekong hydropower projects 349, 350; U.S. roles 282–5; water resources and food insecurity 273–6; *see also* Mekong River
Great Migration 10
Green Revolution 232, 239, 246
green technologies: by U.S. in Panama 87–8
groundwater depletion: summary of ecological change 370
Guatemala 36, 57, *59*, 110–27; access to water 113–14; agriculture 111, 114; climate change 116; coffee and coffee rust 116; Congressional Commission on Water Resources 120; demographics 110–12; drug trafficking 112–13; El Niño 116; governance failure 112–13, 117–19, 120; homicide rates and violence 112–13; hydropower 114–15, 116; La Puya protests at El Tambor gold mine 115; local control of natural resources 116–17; malnutrition 111–12; migrations 113; mining 115; paths of influence 122; population 111; poverty in 111, 113, 122; remittances 121–2; river diversions 114, 116, 118; U.S., proposed agenda 121–3; Walk for the Water, Mother Earth, Land, and Life 110, 114, 117; water security 117–21
GWP *see* Global Water Partnership

Halliburton 25
Harirud River 170, 177–9, 181, 190, *191*, 193, 380
Harmon Doctrine of absolute sovereignty 9–10
Hausa-Fulani herders, Nigeria 128, 130, 136–7, 141; summary 379, 382
HDI *see* Human Development Index
Hearns, Glen 189, 402

Hegazi, Farah F. 39, 402
Helmand Commission 178
Helmand River 170, 174, 177–9, 190, *191*, 194, 197, 380
Helmand Treaty (1973) 194
Helms, Jesse 15
Helsinki Rules (1966) 214, 247
heroin, from Afghanistan 197; *see also* poppy cultivation for opium in Afghanistan
Himalayas *see* India, water-energy nexus in Himalayas
Hirakud Dam 260
holistic planning 348, 349
Honduras 57, *59*, 71, 112, 113, 120
Hoover, Herbert 10
Hoover Dam 11
Human Development Index (HDI): in Afghanistan 190; in Guatemala 111
human responses 376–83; domestic conflicts 377–9; migrations 376–7; narcotics trade 382–3; rise of insurgencies 382; transboundary conflicts 379–81
human rights: as goal in SDGs 42, 44
human systems, impacts on 373–6; displaced peoples 376; infrastructure destroyed and disrupted 375–6; rural livelihoods 374; structural inequalities 374–5
Hussein, Saddam 26
hydraulic mission paradigm, in Iran 173, 176
hydroelectric plants: in dam development controversy 23; Uzbekistan's objections to upstream hydro projects 207, 211, 215–18
hydropower: in Guatemala 114–15, 116; in India 256, 261–3, 265; in Iran 171, 175; in Panama 78, 79; Rogun project in Central Asia 206, 211–12, 213, 218

IADB *see* Inter-American Development Bank
IBNET *see* International Benchmarking Network for Water and Sanitation Utilities
IBWC *see* International Boundary and Water Commission
ICG *see* International Crisis Group
ICWC *see* Interstate Commission for Water Coordination
IDAAN *see* Panama, IDAAN

412 *Index*

IFAS *see* International Fund for Saving the Aral Sea
IFC *see* International Finance Corporation
IFI *see* international financial institutions
IITA *see* International Institute of Tropical Agriculture
IJC *see* International Joint Commission
indemnity insurance for climate risk 322–3
India: challenges faced 37; dam construction with Afghanistan 193; dam development controversy 23; Indus Basin Water Treaty, with Pakistan 49, 225, 229–30, 232; Ministry of Urban Development (MoUD) 333, 337; Pune Municipal Corporation 339–40; State Bank of, and infrastructure investment 330, 332; urbanization and climate change threat 48; Water and Sanitation Pooled Fund 336–8
India, regional security and water scarcity 237–52; ancient history 237; Cauvery Tribunal 241–2, 244, 352; Central Government involvement in water disputes 239–40, 241–3, 251; competitive federalism 243–4, 248; demand management 250–1; demands for statehood 245–6; economic growth 238–9; groundwater extraction 239; Inter-State Council (ISC) 243; interstate river conflicts 240–8; Inter-State Water Disputes (ISWD) Act 240; Krishna Water Disputes Tribunal (KWDT)-II 246–7; language and ethnicity divisions 244–5; media and violence over water 244–5; middle class use of water 238–9; Narmada River Tribunal 241; National Water Policy draft 242; paths of influence 250; pollution sources 239; population 237–8; Project Jalayagnam 247; Punjab SYL Canal Bill 247–8; Punjab Termination of Waters Agreement 247; Rashtriya Krishi Vikas Yojana (RKVY) program 243; Ravi-Beas interstate conflict 241–2, 247; Sardar Sarovar Dam 241, 349; state politics and water as political weapon 246–8, 251; Sutlej-Yamuna Link (SYL) canal 241, 247–8; tribunals 240–3, 244, 246–7, 352, 353; U.S. foreign policy 248–9; wastewater issues 238, 239

India, water-energy nexus in Himalayas 253–71; climate change 254, 255, 259, 268; coal, as energy path 259–61, 266; coal-fired power plants 254, 256, 257, 258, 265; distress at base of Himalayas 253–5; droughts 255–6; economic principles of 20th century *vs*. 21st century 257, 266–7; energy consequences 258; energy paths 258–66; floods 255, 262–3; government response 256–7; hydropower, as energy path 261–3; hydropower plants 256, 265; Paris climate agreement 263–4; paths of influence 267; problems anticipated 255–6; recommendations 266–8; renewable-energy industry 266; risk thesis 257–8; solar and wind, as energy path 259, 263–6; U.S. interest in India 254, 268
Indian Wind Power Association 265
Indonesia 23, 332
Indus River Basin 49, 222, 225, 227–8, 229–34, 254, 351
Indus Water Treaty 49, 284, 380–1
infrastructure: destroyed and disrupted as impact on human systems 375–6; power and transport compared to water 314, 326, 327–8; *see also* development of water infrastructure; financial help for weak water utilities; financing water and sewer infrastructure in developing world
insurance for climate risk 322–3
insurgencies: fueled by water stress xi–xii, 4, 36–7; as human response 382; as U.S. security interest 388, 389; *see also* Abu Sayyaf; al-Qaeda; Boko Haram; conflict; Taliban; *individual countries*
integrated water resources management (IWRM) 48, 345, 348; in Mexico 93, 104–5
Intelligence Community Assessment 210, 380
Inter-American Development Bank (IADB): in Panama 78–9, 82
Intergovernmental Panel on Climate Change (IPCC): on anticipated water crisis in India 255; on climate change in Greater Mekong subregion 275; on climate change in Iran 171; financially dealing with climate change risks

Index 413

321–2; on water stress in Africa 132, 156–7
International Benchmarking Network for Water and Sanitation Utilities (IBNET) 315
International Boundary and Water Commission (IBWC) 9
International Commission on Large Dams 173
International Committee of the Red Cross 299–300
International Court of Justice 353
International Crisis Group (ICG) 207, 209, 213, 215, 218
International Finance Corporation (IFC) 88, 351
international financial institutions (IFIs) 312, 313, 317, 318, 324
International Fund for Saving the Aral Sea (IFAS) 199, 213
International Institute of Tropical Agriculture (IITA) 141
International Joint Commission (IJC) 9
International Monetary Fund 173
International Organization for Migration 300
International Police Science Association 112
International Rivers 257
International Solar Alliance 264
International Union for Conservation of Nature and Natural Resources (IUCN) 345
International Water Management Institute 162
Interstate Commission for Water Coordination (ICWC) 199
IPCC *see* Intergovernmental Panel on Climate Change
Iran 168–88; agriculture 171–2, 173, 179–80; Allahverdi Khan Bridge 170; climate change 171; dams 173, 176–9; desertification and dust storms 174; development statistics *192*; drought 173, 178; GDP 169; Global Adaptation Index *195*; government policies 171–4; groundwater, aquifers, and tube wells 170, 172, 174–5; hydraulic mission paradigm 173, 176; hydropower 171, 175; Joint Comprehensive Plan of Action 168, 174, 179; Kajaki Dam 178, 192, 197, 380; Lake Urmia 170, 173–4, 176, 180; mountains and deserts 168–9; paths of influence 182; policy reforms 179–80; protests and demonstrations 176–8; Revolutionary Guards Corps, Khatam al-Anbia 173; rivers 170, 175, 177–9; salt marshes and salinization 174–5, 177; Shatt al-Arab River *151*, 177; socioeconomic impacts of water challenges 174–6; solar desalination 180, 181; subsidy programs 172–3, 179–80; Taliban 178, 181; Targeted Subsidies Reform Act 172–3; transboundary tensions 177–9, 181; urbanization and migration 175; U.S. collaborative opportunities 180–1; water resources 169–71; water stress 169; Zayendeh-Rud River 170, 175–6
Iraq: conflict and meeting targets of SDGs 46; conflicts over water 25–6, 153, 351; dams 26; Ministry of Water Resources 5; oil industry 25; population 155; Shatt al-Arab Watershed 150, *151*, 177; transboundary river basin management 49, 177; U.S. development assistance to 24–6; U.S. invasion of, and water policy 5–6, 30; water negotiations with neighbors 5–6, 26; water use 151–3
ISC *see* India, Inter-State Council
Islamic State (IS): dams in Iraq 25–6, 30; in Mindanao, Philippines 295, 382; transboundary river basin management 49; water control and terrorism 158, 160
Israel 27–8, 30, 47, 161
ISWD *see* India, Inter-State Water Disputes Act
IUCN *see* International Union for Conservation of Nature and Natural Resources
IWRM *see* integrated water resources management

Jawaharlal Nehru National Solar Mission 265
Jawaharlal Nehru National Urban Renewal Mission (JNNURM) 333–4
JCPOA *see* Iran, Joint Comprehensive Plan of Action
Jenča, Miroslav 211
Jeuland, Marc 344, 402
Johnston Mission, U.S. (1955) 27
Joint Monitoring Programme (JMP) 41, 45

414 *Index*

Jones, James L. xi–xii, 402
Jordan 27, *151*, 161
Jordan River Basin 27–8, 30, 49

Kahl, Colin 110
Kajaki Dam 178, 192, 197, 380
Kalantari, Issa 179
Kamal Khan Dam 178, *193*, 380
Kambarata hydroelectric power plant 206, 211–12, 213, 218
Kandahar Food Zone 197, 198
Karimov, Islam 206, 211, 218
Kazakhstan 37, *192*, 194, *195*, 196; *see also* Central Asia
KBR corporation 25
Kenya 338
Kerry, John 15, 19
key lessons, summary in paths of influence 390–5; dimensions of needed response 395; drivers, never just one 392; governance failures 390–2; more ecosystem collapse 394–5; more frequent water conflicts 393–4; social disruptions fall on most vulnerable 392–3
Khatam al-Anbia 173
Kim, Jim Yong 42
King, Marcus 128, 403
King Abdullah Canal, Jordan (1966) 27
KPWF *see* Kenya
KRG *see* Kurdish Regional Government
Krishna Water Disputes Tribunal, India (KWDT) 246–7
KU Projects Ltd. 260
Kurdish Regional Government (KRG) 26, 177
Kyrgyzstan 37, *192*, 194, *195*, 196, 209; *see also* Central Asia

Ladder of Financial Sustainability 312–14, 318
Lake Chad crisis, Nigeria 136, 369, 371
Lake Urmia, Iran 170, 173–4, 176, 180, 369, 379
Lancang-Mekong River Community of Common Destiny (LMCCD) 282–4
land degradation: summary of ecological change 371
Laos *see* Greater Mekong subregion
Lasswell, Harold 176
LEAD *see* Leadership for Environment and Development
Leadership for Environment and Development (LEAD): in Pakistan 226

legal foundations of U.S. transboundary relations 9–10
legislation: Electric Power Industry Reform Act, Philippines (EPIRA, 2001) 328; General Water Law, Mexico 100; Inter-State Water Disputes (ISWD) Act 240; Mexico Securities Market Law (2001) 335; National Water Policy draft, India (2012) 242; Punjab SYL Canal Bill, India (2016) 247–8; Punjab Termination of Waters Agreement, India (2004) 247; riparian rights for water division, in English common law 241; Targeted Subsidies Reform Act, Iran (TSR, 2010) 172–3; Water Law, Afghanistan (2009) 198
legislation, U.S.: Clean Water Act (1987) 336; Senator Paul Simon Water for the Poor Act (WPA, 2005) 18, 51; Senator Paul Simon Water for the World Act (WfWA, 2014) 19–20, 51, 400; Water Resources Planning Act (1965) 12
León, Gerson de 117
LGU *see* local government unit
Living River Siam Association 279
LMCCD *see* Lancang-Mekong River Community of Common Destiny
LMI *see* Lower Mekong Initiative
local government bond 334
local government units (LGUs) 328–9
loss financing instruments 322
Lower Colorado River Basin Development Fund, U.S. 13
Lower Mekong Initiative (LMI) 282, 285

Maldonado, Alejandro 115
Mali 136, 140
Maliki, Nouri al 26
Manila Water Company 339
Marcos, Ferdinand 293, 296
market economy principles 21–2
Marquez, Lilian 110, 403
Mauritania 140, 345
MDB *see* multilateral development bank
MDGs *see* Millennium Development Goals
mediation, in resolving conflicts 215–17
Mekong Agreement (1995) 278
Mekong Delta, saline ingress 275, 278
Mekong River 49, 272, 277; *see also* Greater Mekong subregion
Mekong River Commission (MRC) 275, 278–9, 280–2, 381

Index   415

MEND *see* Movement for the Emancipation of the Niger Delta
Mesopotamia 37, 149–67; ancient water conflict 149, 150, 380; changing landscape 154–7; climate change 155–7; demographic change 154–5; drought 156; evapotranspiration 156; interbasin water management 153–4; international impacts 158–60; links among climate, water, and security 157–60; paths of influence 162; regional impacts 157–8; security risk reduction 160–2; transboundary river management of Euphrates and Tigris Rivers 6, 26, 49, 150, *151*, 153, 154, 156, 158, 177; water use by Turkey, Syria, and Iraq 151–3; *see also* Iraq; Syria; Turkey
Mexican Institute for Competitiveness 329
Mexican Stock Exchange 335–6
Mexico 91–109; agriculture 98, 99, 100–1; budgetary cuts 99–100, 101; challenges faced 36, 91–2; climate change 98–9, 101; CONAGUA 93, 94, 97, 99, 101, 104; critical juncture, first 92; critical juncture, second 92–3; critical juncture, third 94–100; critical junctures, defined 91–2; drug trafficking offensive, effect on Guatemala 112, 121; General Water Law 100; groundwater and aquifers 95–7, 100, 102; Hydrological Administrative Regions 93; integrated water resources management 93, 104–5; macroeconomic conditions 99; maps *95*, *96*, 102, *103*; migrations 97, 113; multi-stakeholder platforms 93, 99; paths of influence 106; population 92, 98; poverty in 106, 122; public-private partnerships 101, 105; responses needed to pursue water security 100–1; River Basin Organizations 93, 99; Securities Market Law 335; structured certificates 335–6; trade with U.S. 101–3; urban and rural water insecurity 97–8, 100; U.S. border cooperation 9, 181; U.S. foreign policy in 101–3; U.S. involvement in water polity 103–6; Water Advisory Council 93; water polity, defined 91–2; Water Property Rights Registry 93; water resources 94–7

Mexico City: aquifer 97; budget cuts to water system 99–100; and climate change 101; population 92
Michel, David 168, 403
Middle East: challenges faced 36; conflict and meeting targets of SDGs 45; water scarcity and U.S. interests in 24
migrations: from El Salvador 57, 59, 65–8, 71, 113; Great Migration 10; as human response 376–7; International Organization for Migration 300; from Mexico 97, 113; in Syria 157; as U.S. security interest 387; *see also individual countries*
MILF *see* Moro Islamic Liberation Front
Millennium Challenge Corporation 72
Millennium Declaration 40
Millennium Development Goals (MDGs) 16, 29, 39–43, 45, 312
Mindanao, Philippines 288–305; agriculture 294–5, 299–300; area covered 288; Autonomous Region in Muslim Mindanao 288, 291, 292; Bangsamoro Autonomous Region 292, 299, 382; building trust 298–9; climate change 292, 293–4, 299; dams and National Power Cooperation 290; demand management plan 290–1; displacement of people 292, 295, 299–300; drought 290, 294, 296; ethnic and religious conflicts 292–3, 298–9; extremist and terrorist groups 295–6, 300; future directions 297–300; historical roots of instability 292–5; land tenure and ancestral domains 291, 299; Liguasan Marsh 290–1; map *289*; mining and logging 291; paths of influence 294; peacebuilding 297–9; rule of law absence 293; urban water security 290–1; U.S. strategic interests 295–7, 300; violence and conflict 288, 292–3; voter turnout and electoral process 296; water conflict dynamics 290–2; water linkages 299–300; watersheds 290, 294
Mississippi River Flood (1927) 10
Modi, Narendra 240, 248, 254, 255, 258, 261, 264
Moily, M. Veerappa 242
monitoring, reporting, and verification (MRV) 70
Moro Islamic Liberation Front (MILF) 291–2, 298

MoUD *see* India, Ministry of Urban Development
Movement for the Emancipation of the Niger Delta (MEND) 128, 139
MRC *see* Mekong River Commission
MRV *see* monitoring, reporting, and verification
MSP *see* multi-stakeholder platform
multilateral development bank (MDB) 20–4, 29, 307–8
multinational corporations (MNCs) 25; in Greater Mekong subregion 283–4; in Mexico 102–3; in Pakistan 226
multi-stakeholder platforms (MSP): in Mexico 105
municipal development funds for water utilities 317
municipal reforms for water service debt 333–4
municipal revenue bond 334
Myanmar 254; *see also* Greater Mekong subregion

Nadi Ghati Morcha 260
NAFTA *see* North American Free Trade Agreement
Namibia 351
narcotics trade 181, 189–90, 194–8, 382–3, 388
National Hydropower Corporation (NHPC) 262
National Meteorological Department 256
National Resources Planning Board, U.S. 12
national security: water declared as, by Secretary of State 18–19
National Strategy for Climate Change, Panama (ENCCP) 81–2
National Thermal Power Corporation 260
National Water Carrier, Israel (1964) 27
National Water Commission, Mexico (CONAGUA) 93, 94, 97, 99, 101, 104
National Water Commission, U.S. 13
National Water Council, Panama (CONAGUA) 80
National Water Security Plan, Panama (PNSH) 77, 80, 81
NATO *see* North Atlantic Treaty Organization
NDA *see* Niger Delta Avengers
ND-GAIN *see* Notre Dame Global Adaptation Index
NDN *see* Northern Distribution Network
NDPVF *see* Niger Delta People's Volunteer Force
NDV *see* Niger Delta Vigilante
Nepal 238, 254, 259, 261–2, 349
Netherlands Water Partnership 338
NHPC *see* National Hydropower Corporation
Niagara River 9
NIC *see* U.S. National Intelligence Council
Nicaragua: Chinese canal 76
Niger 136, 140
Niger Delta Avengers (NDA) 139
Niger Delta People's Volunteer Force (NDPVF) 139
Niger Delta Vigilante (NDV) 128
Nigeria 36, 128–48; Agricultural Transformation Agenda (ATA) 130; Biafran War 131; Boko Haram 128, 130, 135–6, 137, 142; Bonga oil spill 140; climate change 130, 131–2, 133–4; conflict flashpoints 130–1, 134–40; conflict linkages 133–4; demographics 132–3; disease in 131; droughts 134, 136; environmental protests 138–9; Federal Ministry of Water Resources 133; as geostrategic power 129–31; global implications of instability 140; governance 133, 143–4; Government of (GON) 130, 133, 138, 141–2; Hausa-Fulani herders 128, 130, 136–7, 141; indigenes 137; Lake Chad crisis 136, 369, 371; Middle Belt 130, 136–7, 141; Niger Delta 130–1, 138–9; north and northeast region 130, 134–6; oil production and oil spills 129–30, 138–40, 143; paths of influence 137, 139, 143; recommendations for GON 141–2; religious conflict 130, 132, 136; U.S. recommendations 142–3; water resources 131
Niger River Basin/Delta 128, 139, 142
Nile Basin Initiative (2010) 284
Nile River 49, 153, 351
"no regrets," financial adaptation strategy 321–2
North American Free Trade Agreement (NAFTA): with Mexico 101
North Atlantic Treaty Organization (NATO): in Central Asia 212
Northern Distribution Network (NDN): in Central Asia 212–13

Notre Dame Global Adaptation Index (ND-GAIN) 132
nuclear weapons 212, 222

OBA *see* output-based aid
Obama, Barack 19, 113, 259, 264
OBI for water utilities 318
OCHA *see* UN Office for the Coordination of Humanitarian Affairs
official development aid (ODA) 317–19
oil 25, 129–30, 138–40, 160, 212, 291
Omar, Mohammed Amin 189, 199
opium (poppy) production in Afghanistan 181, 189–90, 194–8, 382–3
Orontes River 153
Oslo II Agreement (1995) 27
output-based aid (OBA) 312
Oxfam 116, 299
Özal, Turgut 153

Paisley, Richard Kyle 206, 403
Pakistan 222–36; agriculture 222, 223, 226, 233; challenges faced 37, 222–5; climate change 228–9, 231; climate risk index 254; conflict with India 222, 225, 229; Council for Research in Water Resources (PCRWR) 224, 227; development statistics *192*; droughts 227, 229; floods 228–9, 231; GDP and Economic Survey 223; Global Adaptation Index *195*; groundwater 225–7, 230–1, 233; Indus Basin Water Treaty, with India 49, 225, 229–30, 232; map *224*; paths of influence 234; population 238; salinization and pollution 226–7; sanitation 222, 223; terrorism 222, 223; transboundary issues 229–30, 231–2; U.S. relations and recommendations 222, 232–3; waterborn disease 222, 223; weak governance 222–4, 225
Palestine 27–8, 30, 48
Panama 76–90; Arco Seco 77, 78; best practices 88; challenges faced 36; climate change 79, 81–3, 85–6; Darien Province 76, 77, 78, 89; deforestation 77, 78; drug trafficking 76, 78; El Niño 79; emergency response 88; green technologies 87–8; hydroelectric energy 78, 79; IDAAN 77–8, 82; illegal logging 77, 78, 89; MiAmbiente 78, 80, 81; National Strategy for Climate Change (ENCCP) 81–2; National Water Security Plan (PNSH) 77, 80–2, 84; paths of influence 89; private sector 87; recommendations 84–6; REDD+ office 81; rivers and watersheds 78–83; specific tribes 78–9, 82; U.S. involvement 79, 83–4, 86–9; water resource management 77, 79, 80–3, 84; water security 77–9
Panama Canal Authority (ACP) 76, 86–7, 376
Panama Canal Green Route 86
Panama Canal Treaty (1977) 83–4
Panama Canal Watershed 79, 82, 86
Pandit, Raj 257
parametric insurance for climate risk 322–3
Paris climate agreement 263–4, 268
Parsons corporation 25
partial credit and partial risk guarantees for water utilities 317
Partnership for Growth Initiative 71–2
partnerships and financing for SDGs 50–1
Partnership to Advance Clean Energy 264
paths of influence, in selected countries 37; in Afghanistan 200; in Central Asia 217; in El Salvador *73*; in Greater Mekong subregion 284; in Guatemala 122; in Himalayas 267; in India 250; in Iran 182; in Mesopotamia 162; in Mexico 106; in Mindanao, Philippines 294; in Nigeria 137, 139, 143; in Pakistan 234; in Panama 89
paths of influence, summary 367–95; ecological change 369–73; human responses 376–83; human systems impacts 373–6; key lessons 390–5; methodological approach 367–9; U.S. security interests 385–90; water infrastructure financing 383–5
Paz River 61
PCRWR *see* Pakistan, Council for Research in Water Resources
pension funds 330, 332
People's Republic of China *see* China
Philippines: challenges faced 37; Partnership for Growth Initiative in 72; privatization of water utilities 22; water commissions and regulation 328, 339; *see also* Mindanao, Philippines
planning processes 347–51
PNSH *see* National Water Security Plan
political risk insurance for water utilities 318

418 *Index*

political security: consequences of changing dynamics 351–3
politics of water management, U.S.: development assistance under siege 15–20; geostrategic interests and global threats 24, 30; global leadership (1978–1994) 14–15; influence on multilateral development banks 20–4; in Iraq 24–6; in Jordan River Basin 27–8
pooled finance investments for water utilities 318
pooled project risk 336–8, 341
poppy cultivation for opium in Afghanistan 181, 189–90, 194–8, 382–3
population of world 7, 48
Post-2015 Development Agenda 312
PPPs *see* public-private partnerships
President's Water Resources Policy Commission, U.S. 12
private financing of water utilities 309, 312–13, 315–16, 323; shift from official sources to alternative 348–51, *350*
private sector investment: recommendations for U.S. foreign policy 399
privatization of water utilities 22–3, 25, 29, 308
project cash flows 334–6
protests and demonstrations: anti-dam movement 278–9, 308, 381; civil unrest at base of Himalayas 254, 255; Curitiba Declaration (1997) 345; large projects with alternative funding 349, 351–2; *see also* conflict; *individual countries*
public-private partnerships (PPPs) 308–9, 384; financial help for weak water utilities 312, 314–16, 319–20; financing infrastructure in developing world 326, 333–4; in Mexico 101, 105
Pune Municipal Corporation, India 339–40
Purdue University 70
Putz, Catherine 211

QDR *see* U.S. Quadrennial Defense Review
Quesungual agroforestry system in El Salvador 71

Ramadi Dam, on Euphrates River 158
RBO *see* river basin organization

Reagan, Ronald 13, 21, 23
recommendations for U.S. foreign policy 396–400
Reed, David x, xi, xii, 3, 35, 88, 367, 396, 403
reforms in exchange for support 327, 332, 341
regional instability 385–7
regulation of environmental laws during construction 21–3, 350–1
regulatory framework for water service debt 333–4, 341
Reilly, Bill 8
Revolutionary Armed Forces of Colombia (FARC) 378
revolving funds for water utilities 317
Rice, Condoleezza 17
Rio Grande River 9
river basin agreements 49, 353–4, 384–5
river basin organization (RBO) 354
Rockefeller Foundation 11
Rogun hydroelectric power plant 206, 211–12, 213, 218
Roosevelt, Franklin D. 11, 12
Rosa, Herman 56, 403
Russian Federation 37, 158, 160

Sahel desertification 132, 134
Saklani, Udisha 237, 403
salinization: in Iran 174–5, 177; in Mekong Delta 275, 278; in Pakistan 226–7; summary of ecological change 370–1
Salma Dam 178, 192–3, 380
Samantara, Prafulla 260
sanitation: as target under MDGs 41–2, 312
Saro-Wiwa, Ken 138
Saudi Arabia *151*, 153
Sayed, Ali Hasnain 222, 403
Schneider, Keith 253, 404
SDGs *see* Sustainable Development Goals
SEA *see* strategic environmental assessment
security interests *see* United States security interests; *individual countries*
Senator Paul Simon Water Acts 18, 19–20, 51, 400
Senegal 136
Senegal River 352
Shack/Sum Dwellers International 338
Shah, Syed Mansoor Ali 231
Sharma, Shankar 256

Shatt al-Arab River and Watershed 150, *151*, 177
Shell multinational corporation 138, 139
Silk Road 207
Simon, Paul 19, 51
Singh, Manmohan 248
Singh, Rajendra 256
Sistan Region 46, 174, 176, 178, 194, 379
slash-and-burn system 71
Social Conflict Analysis Database: on African conflicts 134
social dimensions of water 35–7, 383–4
social disruption 367–8, 373, 392–3
socioeconomic security: consequences of changing dynamics 351–3
soil map of El Salvador 70
solar energy 180, 181, 259, 263–6
South Asia Network on Dams, Rivers, and People 253
South China Sea dispute 273, 279, 282, 295, 296–7
Soviet Union: resource-sharing system in Central Asia 206, 209, 371, 381
Spangler, Chelsea N. 222, 404
special-purpose funds for water utilities 317
special purpose vehicles (SPVs) 326, 327
Stein, Eduardo 110, 404
Stephenson, James 25
Stockholm Water Prize 256
storms, severe: summary of ecological change 370
strategic environmental assessment (SEA): in Pakistan 231
Streeter, William 326, 404
structural adjustment lending 21
structured certificates, Mexico 335–6
Subansiri dams, India 262
supply chains 372–3, 388–9
Sustainable Development Goals (SDGs) 29, 36, *40*; about universality and interdependence 39–40; climate change as threat 47–8; cross-sectoral linkages across goals 43–5; financial challenges 312, 313, 315; fragile states and conflict 45–7; global conflict and cooperation 48–50; importance of U.S. support 51–2; moving from MDGs to SDGs for water 40–3; partnerships and financing 50–1; SDG 6 36, 40, 42–3, 48, 400
Syr Darya River 49, 209, 211, 214
Syria: civil war 4, 6, 151, 161, 378, 382; conflict and meeting targets of SDGs 45–6; conflicts over water 153, 351; links among climate, water, and security 157–8, 160; population 154; Shatt al-Arab Watershed area 150, *151*; water allocation and Iraq 5–6, 26; water use 151–3

TA *see* technical assistance
Tajikistan 37, 179, *192*, 194, *195*; *see also* Central Asia
Taliban 178, 181, 380
Tanzania: Partnership for Growth Initiative in 72
TARWR *see* total actual renewable water resources
technical assistance (TA) 312, 318–19, 355
Tennessee Valley Authority (TVA) 10–11, 23, 348
terrorism *see* insurgencies
Thailand *see* Greater Mekong subregion
Thakkar, Himanshu 253, 254–5, 256–7, 263
Thapan, Arjun 272, 404
Thatcher, Margaret 21, 23
ThirdPole.net 257
Three Gorges Dam 231, 349
Tibet 272
Tigris River 6, 26, 49, 177; *see also* Mesopotamia
Tortajada, Cecilia 237, 404
total actual renewable water resources (TARWR): in Iran 169
Track Two diplomacy 216
transboundary agreements: recommendation for U.S. foreign policy 398
transboundary conflicts: as human response 379–81; *see also individual countries*
treaties: Almaty Agreement (1992) 194; Boundary Waters Convention (1906) 9; Boundary Waters Treaty (1909) 9; coordination and SDGs 50; Helmand Treaty (1973) 194; Indus Water Treaty 49, 284, 380–1; Nile Basin Initiative (2010) 284; Panama Canal Treaty (1977) 83–4; transboundary watershed management in Mesopotamia 161; U.S.-Canada Columbia River Treaty 181; U.S.-Mexico Border Environment Cooperation Commission 181
tribunals, in India 240–3, 244, 246–7, 352, 353

Truman, Harry 12
Trump, Donald: effect of wall on Guatemala 121
trust estate and escrow agreement 334
TSR *see* Iran, Targeted Subsidies Reform Act
Turkey: Ataturk Dam 150, 153; conflicts over water 153–4, 351; Great Anatolia Project (GAP) 349, 351, 373, 380; and Harmon Doctrine principles 10; population 155; Shatt al-Arab Watershed area 150, *151*; structural adjustment loan 21; transboundary relations 179, 181; water allocation and Iraq 5–6, 26; water use 151–3
Turkmenistan 37, 179, 181, *192*, 194, *195*; *see also* Central Asia
TVA *see* Tennessee Valley Authority
Twain, Mark 251

ULBs *see* urban local bodies
Ultra Mega Power Plants, India (UMPP) 261
UMPP *see* Ultra Mega Power Plants
UN Conference on Trade and Development 223
UN Convention on the Law of Non-Navigational Uses of International Watercourses 154, 161, 353–4
UN Department of Political Affairs (UN DPA): Mediation Support Unit in Central Asia 207, 213
UN Development Program (UNDP): in Greater Mekong subregion 280; in Panama 80, 82
UN Economic and Social Commission for Asia and the Pacific 280
UN Economic Commission for Europe (UNECE): and Panama 84; Water Convention in Central Asia 214
UN Environment Programme (UNEP) 46, 138, 140, 179, 210
UNESCO World Heritage site in India 237
UN Food and Agriculture Organization (FAO): expertise on water resources 162; on value of fisheries in Greater Mekong subregion 276; on water status in Nigeria 131; water use in Tigris-Euphrates 151
UN Framework Convention on Climate Change (FCCC): and Panama 84–5
UNICEF: on Aleppo water treatment facility 158

Unified Plan 27
United Nations: bombing of, in Abuja, Nigeria 135; Conference on the Environment and Development, Rio de Janeiro (1992) 14; International Decade for Drinking Water Supply and Sanitation 14; Water Conference, Argentina (1977) 14
United States: aircraft carrier *Kitty Hawk* to Pakistan and India 248; border treaties 181; competing governmental jurisdictional interests 12–14; Contract with America 15; counterinsurgency strategy 46–7; dam building period (1930–1970) 10–12; foreign water policy 4–5, 28–30; foreign water policy considerations 323–4; legal foundations of transboundary relations 9–10; politics of water management *see* politics of water management (U.S.); SDG, importance of support 51–2; trade with Mexico 101–3; water wealth of 3–4; *see also* legislation, U.S.
United States security interests 385–90; demand for humanitarian, development, and military support 389–90; financial support needs 390; migration pressures 387; regional instability 385–7; supply chain pressures 388–9; threats by insurgencies and criminal networks 388
UN Millennium Summit (2000) 40
UN Office for the Coordination of Humanitarian Affairs (OCHA) 57
UN Regional Centre for Preventive Diplomacy for Central Asia (UNRCCA) 207, 213
UN University 42
UN Water 43, 162
UN Watercourses Convention 48, 214, 215
Upper Mekong hydropower projects 349, 350
urban areas, and focus on quality service and adaptation 338–40
urbanization and climate change threat 48
urban local bodies (ULBs) 336–7
U.S. Africa Command 142
U.S. Agency for International Development (USAID): in Afghanistan 199; development assistance under siege 15–20; Development Innovation Ventures 50; in Guatemala 122;

Long-Term Framework Agreement, Central Asia (1998) 206, 209, 381; Mandatory Reference, climate change risk management 86; Mekong Adaptation and Resilience to Climate Change Project 285; in Mindanao, Philippines 290, 297–8; in Nigeria 141; in Pakistan 232; in Palestine 28; in Panama 78, 83–4, 85–6; recommendations for U.S. foreign policy 398, 400; strategic priority countries 51–2; Water and Sanitation for Health Program 14, 47; WSPF bonds in India 337
U.S. Arms Control and Disarmament Agency 15
U.S. Army Corps of Engineers (USACE): in dam-building period 10–11; in Iraq 5, 25; in Nigeria 142
U.S. Bureau of Reclamation 10–11
U.S. Bureau of the Budget 12–13
U.S. Defense Intelligence Agency 158
U.S. Department of Defense (DoD) 142, 159; recommendations for U.S. foreign policy 400
U.S. Department of State: recommendations for U.S. foreign policy 398, 400
U.S. Geological Survey 142, 199
U.S. Government (USG): on climate change as threat multiplier 159; recommendations for U.S. foreign policy 399–400
U.S. Information Agency 15
U.S. Joint Special Operations Task Force-Philippines 296
U.S. National Intelligence Council (NIC) 255
U.S. National War College 158
U.S. Office of Management and Budget 13
U.S. President's Water Resources Policy Commission 12
U.S. Quadrennial Defense Review (QDR, 2014) 24, 159
U.S. State Department 15–19
U.S. Water Partnership (USWP) 19, 50, 104
USACE *see* U.S. Army Corps of Engineers
USAID *see* U.S. Agency for International Development
U.S.-India Joint Strategic Vision 254, 264
Usman, Muhammad Faizan 222, 404

USWP *see* U.S. Water Partnership
utilities: water projects compared to other utilities 328–9; *see also* financial help for weak water utilities
Uzbekistan 37, *192*, 194, *195*, 196; objections to upstream hydro projects 207, 211, 215–18; *see also* Central Asia

Vajpayee, Atal Bihari 261
Venezuela: energy crisis 116
Verisk Maplecroft 254
Verma, Jai Shanker 266
viability gap financing 319–21, 324
Vietnam *see* Greater Mekong subregion

Wadia Institute of Himalayan Geology 263
WASH *see* water, sanitation, and hygiene
WASH for Life 50
Washington Consensus, U.S. 21–2, 308
WASREB *see* Kenya
water: assumption of limitless fresh water 7–8; average demand per day in four countries 172; perception as entitlement 328, 383–4; safe drinking, as target under MDGs 41–2; used as political weapon in India 246; used as weapon of war 26, 28, 159, 295
water, sanitation, and hygiene (WASH): recommendations for U.S. foreign policy 397–9, 400; USAID program 14, 47, 52
Water and Sanitation Pooled Fund, India (WSPF) 336–8
Water for the Poor Act, U.S. (WPA) 18, 51
Water for the World Act, U.S. (WfWA) 19–20, 51, 400
Water Quality Index: in Iran 170
water resource management (WRM): in Panama 77, 79, 80–3
Water Resources Planning Act, U.S. (1965) 12
water stress: defined, in Iran 169; role in creating fragile states xi–xii
water supply and sanitation (WSS): deterioration of services 315, 316
water user fees 335
Watt, James 13
WCD *see* World Commission on Dams
weather events: drought, flooding, and storms, as ecological change 369–70; intensification and frequency of 394–5; *see also individual countries*

Weinthal, Erika 39, 404
WfWA *see* Water for the World Act
Winpenny, James 329
Witmer, Lesha B. M. 39, 405
Wolfowitz, Paul 25
women and water: Dublin Principles 14–15, 42, 44
Women for Water 42
World Bank 18–19; Catastrophic Drawdown Option 322; in Central Asia 207, 211, 213; database of finance and operations of water utilities 315; database on infrastructure projects 329–30, *331*; ease of doing business ranking for India 244; Forest Carbon Partnership Facility program in Panama 81; GDP in Iran 171, 174; in Greater Mekong subregion 283–4; in Mexico 95; in Mindanao, Philippines 293, 299; in Pakistan 229, 232; river basin treaties 353, 354; urbanization and climate change threat 48; on water conditions in Mesopotamia 156; and WCD 345
World Bank Group: development lending 20–3, 29

World Commission on Dams (WCD) 23, 308, 345, 348, 352
World Commission on the Ethics of Scientific Knowledge and Technology (COMEST) 328
World Economic Forum 162
World Food Programme 113
World Health Organization 131, 227
World Internal Security and Police Index 112
World Resources Institute 170
World Water Week (2015) 50
Worldwide Threat Assessment, U.S. (2015) 24
World Wildlife Fund (WWF) viii, xi, 134
WPA *see* Water for the Poor Act
WRM *see* water resource management
WSPF *see* Water and Sanitation Pooled Fund
WSS *see* water supply and sanitation
WWF *see* World Wildlife Fund

Xi Jinping 259, 264

Zinni, Anthony 25